Function Spaces, Interpolation Theory and Related Topics

Function Spaces, Interpolation Theory and Related Topics

Proceedings of the International Conference
in honour of Jaak Peetre
on his 65th birthday

Lund, Sweden
August 17–22, 2000

Editors

Michael Cwikel
Miroslav Engliš
Alois Kufner
Lars-Erik Persson
Gunnar Sparr

Walter de Gruyter · Berlin · New York 2002

Editors

Michael Cwikel
Department of Mathematics
Technion – Israel Institute of Technology
Haifa 32000
Israel

Miroslav Engliš
Mathematical Institute
Czech Academy of Sciences
11567 Prague 1
Czech Republic

Alois Kufner
Mathematical Institute
Czech Academy of Sciences
11567 Prague 1
Czech Republic

Lars-Erik Persson
Department of Mathematics
Lulea University of Technology
97187 Lulea
Sweden

Gunnar Sparr
Department of Mathematics
Lund Institute of Technology/Lund University
22100 Lund
Sweden

Mathematics Subject Classification 2000:
00B30; 46-06; 47-06; 46B70; 46EXX, 47BXX

Keywords:
Interpolation theory, function spaces, operator theory

Ⓢ Printed on acid-free paper which falls within the guidelines of the ANSI to ensure permanence and durability.

Library of Congress – Cataloging-in-Publication Data

Function spaces, interpolation theory and related topics : proceedings of the international conference in honour of Jaak Peetre on his 65th birthday : Lund, Sweden, August 17–22, 2000 / editors, Michael Cwikel ... [et al.].
p. cm.
ISBN 3110171171 (acid-free paper)
1. Functions spaces – Congresses. 2. Interpolation spaces – Congresses. I. Peetre, Jaak, 1935– II. Cwikel, M. (Michael), 1948–
QA323 .F8652 2002
515′.73–dc21

2002067478

Die Deutsche Bibliothek – Cataloging-in-Publication Data

Function spaces, interpolation theory and related topics : proceedings of the International Conference in Honour of Jaak Peetre on His 65th Birthday, Lund, Sweden, August, 17–22, 2000 / ed. Michael Cwikel ... . – Berlin ; New York : de Gruyter, 2002
ISBN 3-11-017117-1

Printed in Germany.
Cover design: Thomas Bonnie, Hamburg.
Typeset using the authors' $\TeX$ files: I. Zimmermann, Freiburg.
Printing and binding: Hubert & Co. GmbH & Co. KG, Göttingen.

Preface

The 65th birthday of Jaak Peetre is a natural time for a very special celebration. As July 29, 2000 drew closer, this was quite obvious to all of us who work in the theory of interpolation spaces, or apply this theory to other fields, and also to many working in other fields where Jaak has also left an indelible mark.

So, indeed, on August 17–22, 2000, the Centre for Mathematical Sciences at Lund University held a conference in honour of Jaak's birthday. This auspicious event was attended by 130 mathematicians from 19 countries and these Proceedings contain (mostly) contributions delivered at the conference.

The conference was organized by Michael Cwikel, Björn Jawerth, Jacques-Louis Lions, Lars-Erik Persson, and Gunnar Sparr, with invaluable assistance from Genkai Zhang and also from Ann-Kristin Ottosson, Sven Spanne, Lars Vretare, and other members of the staff of the Centre for Mathematical Sciences at Lund University.

To enable these Proceedings to better serve as a record of the conference, we have recalled and presented various details of the speakers and the program in some of the pages to follow. The reader wishing to further participate in the special spirit of the conference is invited to visit the website `http://www.maths.lth.se/conferences/peetre65.html` including the pictures on display there, and to remember or imagine the wonderful banquet in the Trollenäs castle, complete with medieval music and memorable speeches by Lars Gårding, Jaak himself, and others.

Given the great depth and breadth of Jaak Peetre's mathematical work, a detailed survey and discussion of it could fill an entire volume, and it would be larger than this one. We have confined ourselves to giving a brief summary of this work, including a list of Jaak's publications. We are particularly grateful that Jaak himself agreed to augment our efforts here with an article giving some of his perspectives on the history of the theory to which he has contributed so substantially.

It seems particularly fitting that the article appearing immediately after Jaak's is a contribution from Jacques-Louis Lions, Jaak's co-author for the creation of the "espaces de moyennes" in their illustrious seminal paper about the "real" interpolation method. As already mentioned above, Jacques-Louis Lions was also actively involved in the organization of the conference, even though he was not able to actually attend it. Very sadly, he passed away soon after we received this paper from him.

There was also another tragic event which occurred just a short time before the conference. Yet another outstanding mathematician, Thomas Wolff, who was to have been one of the main speakers, perished in a road accident. On the day and at the time when Tom had been scheduled to speak, we honoured his memory by sharing recollections of him and recalling highlights of his work in a memorial session.

The main speakers at the conference were Jonathan Arazy, Yuri Brudnyi, Mischa Cotlar, Ciprian Foias, Svante Janson, Nigel Kalton, Sergey Kislyakov, Peter Lindqvist,

Vladimir Maz'ya, Vladimir Ovchinnikov, Vladimir Peller, Richard Rochberg, Evgueni Semenov, Hans Triebel, Hans Wallin, and Nahum Zobin. In all, 73 oral presentations were delivered, a full list of which can be found at the end of this volume.

We hope that this volume will give the reader valuable insights about new results and trends in some important fields of analysis which have all been strongly influenced by the work of Jaak Peetre.

Finally we would like to express our gratitude to the Publishing House Walter de Gruyter, Berlin, for fruitful collaboration.

Haifa, Prague, Luleå, Lund
April 2002

Michael Cwikel, Miroslav Engliš, Alois Kufner, Lars-Erik Persson, and Gunnar Sparr

Contents

Opening address

by Hans Wallin, Umeå University

Dear Jaak,
Dear fellow mathematicians,

We have come to this conference to celebrate Jaak as a mathematician and as our friend. Jaak is well-known and respected all over the mathematical world and it is a great pleasure for me to say a few words to him and about him at the opening of this conference.

I would like to start by quoting the author G. B. Shaw:

> *The reasonable man adapts himself to the world;*
> *the unreasonable one persists in trying to adapt the world to himself.*
>
> *Therefore all progress depends on the unreasonable.*

I think that a successful mathematician has to be a little unreasonable, and I think that Jaak has understood this.

Jaak is a great mathematician with a strong curiosity and a big appetite for mathematics. He is, together with Jacques-Louis Lions, the creator of the real method in interpolation theory. Jaak talked about that method as an invited speaker at the International Congress of Mathematicians in 1970. Jaak's fundamental contribution to interpolation theory and related fields such as function spaces and approximation theory has been very influential. It has inspired many mathematicians, including many attending this conference. In particular, I want to mention his charming monograph "New thoughts on Besov spaces". This book is written in such a good way that when you read it you can even forgive Besov spaces for having three indices.

Jaak's mathematical interests are very broad. He has made contributions on partial differential equations, spectral theory, singular integrals, differential geometry, Hankel operators, and other areas.

Let me give you some figures about Jaak:

- Jaak was born in Estonia and came to Sweden at the age of 9. He has kept his contact with Estonia and is an honorary member of the Estonian Mathematical Society. He is, by the way, also one of the three honorary members of the Swedish Mathematical Society.
- He became professor in Lund at the age of 27 – he was then the youngest professor in Sweden. Recently he became the youngest retired professor in Sweden; the age of retirement in Sweden is 65.
- He has been the thesis advisor of approximately 15 PhD students.

• He has written more than 200 mathematical papers.

• He has joint publications with 44 different mathematicians from 13 countries.

• He got 65 and retired in July this year. Somebody told me that before his retirement he had some kind of device in his office showing how many seconds there remained to his retirement.

• He has successfully run 19 marathon races. He is a tough long distance runner. Once at a conference in Budapest, on an extremely hot day, I was running with him. The only thing that kept me running with him the whole distance was his patience with me and the heat – when I tried to stop I felt the heat burning like a fire inside me.

In a newspaper interview recently Jaak said that the best part of summer is June with its changes in nature and the worst part is July which he claims is in some sense a dead month. He did not say anything about August. However, I feel that this year August, with this conference, will be the best part of summer.

Dear Jaak,
I am proud that I know you.
I am happy that you are my friend.
I look forward to this week.

I declare the conference opened.

In the rose garden (photo by Eila)

Jaak Peetre

Jaak Peetre, the man and his work

Michael Cwikel, Lars-Erik Persson, Richard Rochberg and Gunnar Sparr

1. A brief biography – from 1935 to today

Jaak Peetre was born on July 29, 1935 in Tallinn, Estonia. He grew up in the ancient town of Pärnu, some 120 kilometers south of Tallinn.

The turmoil of World War II did not spare Estonia. On September 15, 1944 Jaak and his family left Pärnu for Tallinn and soon afterwards they sailed from there to Sweden while Tallinn harbour was attacked by aircraft. Two days later most of the mediæval quarter of their former home town Pärnu was destroyed in another air raid. On January 13, 1945, the family settled in Lund, which has been Jaak's home and place of work for most of the time ever since.

By the time Jaak had completed high school, his passion and talent for mathematics were completely apparent, and he enthusiastically and unhesitatingly set out along the path towards making mathematics his life's work.

By 1959 he had already completed his undergraduate and graduate studies at Lund University with a Ph.D. in the area of partial differential equations. In that same year the university appointed him as a docent in mathematics.

A number of mathematicians greatly influenced and inspired Jaak's early research. These included Åke Pleijel, who was his supervisor for licentiate studies, and also Lars Gårding, Lars Hörmander, Jan Odhnoff, and Bernard Malgrange.

In 1963 Jaak Peetre became the youngest professor of mathematics in Sweden. His appointment was at the newly created Lund Institute of Technology (Lunds Tekniska Högskola), which is a part of Lund University. Apart from the period of his professorial appointment at Stockholm University (1988–1992), Jaak has worked at Lund University ever since.

In 1962 Jaak married Irene Kunnos. Their three children, Mikaela, Jakob (Oppi) and Benjamin were born in 1963, 1964 and 1968.

Irene's tragic and untimely death in 1972 meant that Jaak, for many years, would combine his mathematical research and other duties with the no less demanding and essential rôle of being *both* parents for his young children. Among mathematicians, Jaak likes to refer fondly to his children as "BMO".

In August 2000, Hans Wallin gave an eloquent and heartfelt welcoming address which opened the conference in Lund celebrating Jaak's 65th birthday, a celebration which continues in these pages. We are very happy to reproduce Hans' text elsewhere in this volume. It includes a list of impressive "facts and figures" about Jaak. We would now like to expand upon some of the items in that list, and add a few more:

- Yes, as may surprise those who know him only by his publications, Jaak is indeed a long distance runner and he has run 19 marathon races. His best time for a marathon is 2 hours 59 minutes!
- The list of languages that he knows and uses obviously includes Estonian, Swedish and English, but there are also Finnish, French, German, Latin, Russian and Spanish and even a few words of Hebrew[1] and quite possibly some other languages that he has not yet told us about.
- Among his other diverse skills we can mention, for example, baking. His repertoire here includes a wonderful flax seed bread, and "glace au four", a remarkable cake which, despite spending considerable time in the oven, emerges with frozen ice cream in its interior.
- Jaak has maintained his links with his native Estonia in various ways:

  He is an honorary member of the Estonian Mathematical Society.

  Last year, on February 24, Estonia's national day, Jaak was among those who received the III Class of the Order of the White Star (Valgetähe orden) from President Lennart Meri.

  Many of Jaak's colleagues recall that the letters we received from him in past times would often bear stickers with pictures of Estonian dissidents, calling for their freedom. This cause may have sometimes seemed quite hopeless to many of us then, but Jaak's persistence would be well justified and bear its fruits in due course.
- Apart from being one of the few (now five in all) honorary members of the Swedish Mathematical Society, Jaak was also president of the society during the years 1984–1987.
- Jaak was elected to membership of the Royal Swedish Academy of Sciences in 1983.
- Jaak Peetre has supervised 16 graduate students. We give some details about them below (see Section 3). To this day, each of Jaak's former students is working and/or teaching enthusiastically and significantly in some branch of mathematics or some other related field. Several of them have apparently been "infected" by Jaak's taste for a wide and imaginative range and interplay of interests, since they have headed in widely divergent directions, some of them reaching fascinating new interfaces of mathematics with other disciplines.

[1] Many years ago when Jaak attended a lecture, in Hebrew, given in Jerusalem by one of the authors of this article, another member of the audience, namely Joram Lindenstrauss, interrupted the lecturer and said, in Hebrew, that he should be lecturing in English for the benefit of our guest from abroad. But that guest immediately called out "Lo!" ("no" in Hebrew).

- We will later have more to say about the fundamental paper [21] by Jacques-Louis Lions and Jaak Peetre which founded the real interpolation method and developed a substantial part of its theory. This paper apparently has the added distinction of containing the only research in pure mathematics to have ever been supported by Interpol! The reader is invited to read the footnote on the first page of [21] and draw his or her own conclusions!

On September 22, 1985, Jaak met Eila, and they have shared life together ever since. Nowadays they live in the former fishermen's village of Kåseberga, 80 kilometers from Lund. All those who wish to become acquainted with Jaak and Eila's rose garden, and many other things, are warmly invited to visit Jaak's homepage!

`http://www.maths.lth.se/matematiklu/personal/jaak/engJP.html`

2. Research

We hope that this section will help you share our great enthusiasm about Jaak's mathematical creations and Jaak's own great enthusiasm about mathematics. We should stress that we have not sought to engage in any rigorous processes of allocating credits or priorities. We of course connect Jaak's work with the work of others, but inevitably we can only indicate some small part of the big picture. We certainly do not intend to make any judgements, even implicitly or by omission, concerning the works of other mathematicians.

Let us begin by attempting to divide Jaak's scientific output into a number of main subjects or subject groups. Of course, in the nature of things, there are overlaps between these groups and some papers may be listed under more than one heading. The numbers here refer to the list of Jaak's publications which appears as a separate item immediately after this article[2].

1. *Partial differential equations.* ([1–8], [10–13], [15], [18], [19], [38], [43], [44], [46], [87], [98]).

2. *Interpolation spaces and interpolation of operators.* ([9], [14], [16], [17], [20], [21], [23–25], [28], [31], [33–35], [37], [39], [41], [42], [45–47], [50–53], [56–58], [60–63], [65], [67], [70], [71], [78], [79], [81], [85], [86], [88], [90], [92], [94], [95], [97], [98], [100], [105], [107], [126], [140], [149], [164], [169], [173], [174]).

3. *Spectral theory and distribution semigroups.* ([22], [26], [27], [29], [36], [37], [54], [59]).

4. *Besov spaces and related function spaces.* ([9], [32], [40], [48], [59], [68], [73–78], [80], [84], [90], [115]).

[2] We have not systematically included the research reports and other less formal bibliographical items ([201] to [289]) in this classification. Many of them also fit into one or another of the listed categories. Some deal with entirely different topics, e.g. automata theory.

5. *Approximation theory.* ([49], [55], [63], [64], [75], [93], [117]).

6. *Fock spaces and Clifford analysis.* ([116], [120], [123], [124], [142], [146], [147], [151], [157],[167]).

7. *Inequalities, means and iteration of means.* ([66], [95], [102], [106], [114], [132], [133], [144]).

8. *Hankel operators and invariant function spaces.* ([93], [95], [96], [99], [101], [103], [104], [108], [109], [110], [111], [112], [116], [118], [121], [122], [124], [125], [127], [128], [130], [136], [138], [139], [141], [143], [145], [148], [152], [155], [156], [159], [160], [162], [165], [168], [178], [182]).

9. *Green's functions and the Berezin transform.* ([134], [136], [166], [170], [171], [172], [178], [179], [187]).

10. *Multilinear forms, trilinear forms in Hilbert spaces.* ([93], [104], [136], [150], [175], [180], [181], [184], [210],[217], [220], [221], [222], [227–230]).

11. *Special functions.* ([69], [89], [91], [96], [135], [145]).

12. *Fourier analysis and more general harmonic analysis.* ([51], [141], [98], [155], [165]).

13. *History of mathematics and related questions.* ([113], [158], [161], [176], [177], [186], [206], [207], [208]).

14. *Other miscellaneous topics.* (Differential geometry: [82], [83]. Generalized translations: [49], [72]. Ordinary differential equations: [158]. Singular integrals: [30]. Stieltjes algorithm: [137]. None of the above: [119], [129], [131], [153], [154], [157], [163], [183], [185].)

15. *Editorial work and translations.* ([188–200]).

Our original hope was to provide a survey of Jaak's work in all or most of the above categories. But the constraints of space and time have obliged us to very reluctantly settle for a partial description. In the following four subsections we have concentrated on items 1, 2, 8 and 10. The reader should be well aware that many of Jaak's other fascinating and significant results have not been mentioned here. Thus, alas, for example, we have said nothing about his important book "New thoughts on Besov spaces", and we have hardly discussed his more recent works with Miroslav Engliš and with Peter Lindqvist, and so on and so on . . . .

Partial differential equations

Jaak's career began in the realm of p.d.e. His very first papers are the short notes [1] and [2] which appeared in 1957. They extend a result of his supervisor Åke Pleijel to the case of Riemannian manifolds and give asymptotic estimates for the number of nodal domains for eigenfunctions of the Laplace–Beltrami operator. One important ingredient is an extension of the Rayleigh–Faber–Krahn estimate for the lowest eigenvalue of the Dirichlet problem to this same context. (Coincidentally,

Krahn had been Jaak's mother's teacher in Tartu in 1925–1929.) Jaak himself has tended to dismiss these youthful beginnings as rather insignificant, but others have taken them rather more seriously. They found their place among a distinguished family of results discussed by Robert Osserman in his address to the ICM in Helsinki in 1978.

A result about elliptic operators by Felix Browder [Br] was the starting point for Jaak's Ph.D. thesis [6]. He extended Browder's result to the class of (formally) hypoelliptic operators. It was in this connection that he introduced the class of "operators of type (P)", which have since come to be more commonly referred to as "operators of constant strength" and have been extensively studied. (See, for example, the chapters about them in both of Hörmander's books [Hö1] and [Hö2] (part II, Chap. XIII).) The thesis also dealt with elliptic boundary value problems, to which its author would later return in greater detail.

There are also a number of other more "technical" innovations in the thesis which should be particularly mentioned:

1° a useful definition of the Sobolev spaces $H^s(\Omega)$, where Ω is a domain in $\mathbb{R}^n$, as a quotient of $H^s(\mathbb{R}^n)$.

2° the introduction of the operator $\partial/\partial t + i\sqrt{1-\Delta'}$ (where $\Delta' = \partial^2/\partial x_1^2 + \cdots + \partial^2/\partial x_{n-1}^2$ and $t = x_n$), which preserves the class of distributions in $H^s(\mathbb{R}^n)$ which are supported in the halfspace $\{t \geq 0\}$.

3° the definition of distributional boundary values for solutions of a general p.d.e. in a domain with noncharacteristic boundary, along with an accompanying Green's formula. (A microlocal version of this result was subsequently obtained by Melrose [Me].)

We have already alluded in Section 1 to the important influence of several mathematicians on Jaak's early research. More specifically, in connection with his thesis, we should particularly mention Lars Gårding and Lars Hörmander and also Jan Odhnoff, a fellow student whom Jaak especially appreciated. Furthermore, some discussions with Bernard Malgrange during a brief visit – at the initiative of Laurent Schwartz – to Strasbourg in April 1959, played an especially crucial rôle. Malgrange suggested investigating quasi-linear hypoelliptic operators also. In fact that same month saw another event which would be pivotal for Jaak's later development, a meeting with Jacques-Louis Lions and Emilio Gagliardo in Nancy.

During the preparation of his thesis Jaak also obtained a "local" characterization of linear partial differential operators. He published it separately as the paper [7]. As with [1] and [2], Jaak has tended to consider this as a minor result, but again he seems to have underestimated it. Actually it turns out that [7] is one of his most cited papers. A gap in the proof was overcome in [8] using a different approach suggested by Lennart Carleson. But Jaak's preferred way of fixing this gap, an easy modification of the original proof, would be given later in [254]. Another later sequel to [7] was the joint work [237] with Jan Persson, where the result was extended to a non-linear situation. But this paper has remained a preprint since some of its content turned out to overlap with a paper of Marchaud [Ma] (cf. also [Per]).

Jaak spent most of the academic years 1960/61 and 1961/62 in the United States, first at New York University, in fact at the Courant Institute before it became the Courant Institute, and then at the University of Maryland. During this time he continued his work on p.d.e., further exploiting techniques developed in his preceding papers. The papers [10–13] and [19] were written during this period. The paper [12], in particular, has had some interesting sequels. Although the situation it treats may at first seem rather "artificial" (constant coefficients and a halfspace), [12] has nevertheless triggered a body of further research. For a survey, see the book of Gelman and Mazya [GM], as well as the work of T. Donaldson (e.g. [Do]). (Note also the connection with [43] discussed below.) In [13] and [19] Jaak studied mixed elliptic problems in two dimensions. He dealt only with the L^2-theory. The corresponding L^p-estimates were obtained shortly afterwards by Eli Shamir. (See [Sham].)

It also seems that a chapter on elliptic boundary value problems in Dieudonné's treatise [Di] was probably influenced by this part of Jaak's research.

The extensive lecture notes [18], written soon after these papers (spring 1962), gave Jaak an opportunity to summarize and express the perspectives and insights which he had acquired by this stage in his study of elliptic p.d.e.s.

One more p.d.e. paper from this epoch is the brief note [43], written during a visit to Brazil. It suggests an alternative approach, based on the formulæ in [12], to the Agmon–Douglis–Nirenberg theory of L^p-estimates for elliptic boundary value problems. The details of this were later developed by Jaak's student Leif Arkeryd [Ark1]. Cf. also [Ark2], [Ark3].

By this time Jaak's research had already begun to branch out into interpolation and several other fields, which will be described, at least in part, in the remaining three subsections. However since then he has also returned from time to time to study p.d.e., for example in the papers [44], [46], [87], [98] which consider quite a variety of questions, and also the lecture notes [38]. These notes had been intended to be of an elementary nature. But one idea in them turned out to be useful in the research of Winzell [Wi].

A number of Jaak's papers in spectral theory which are listed above under item 3 also have connections with p.d.e.

Interpolation spaces and interpolation of operators

For many of us the words "interpolation spaces" and "Jaak Peetre" are almost synonymous. Indeed, a very substantial part of Jaak's research deals with aspects, many aspects, of the theory and applications of interpolation spaces. Even a mere glance at any of the monographs [BSh], [BL], [BKr], [BB], [KPS], [Ov2] or [Tri], or at the survey article [BKS], suffices to reveal the depth and breadth of Jaak's essential contributions to this subject.

The foundations of interpolation space theory were laid in the early 1960's. It arose from the classical interpolation theorems of Riesz–Thorin and Marcinkiewicz originally developed for use in harmonic analysis, and also from questions in p.d.e. and traces of Sobolev spaces. Among those who contributed to its "birth" we can mention, apart from Jaak, Nachman Aronszajn, Alberto Calderón, Mischa Cotlar, Emilio Gagliardo, Selim Grigorievich Kreĭn and Jacques-Louis Lions. Other early pioneers of interpolation whom Jaak also particularly appreciated included Martin Schechter and Martin Zerner.

At about this time a general framework for the theory was being defined by Aronszajn and Gagliardo. Their fundamental paper [AG] about this would appear in 1965. Calderón, Kreĭn and Lions were each developing their respective versions of complex interpolation spaces, distilled from Thorin's beautiful proof[3] of the Riesz–Thorin theorem.

As mentioned above, Jaak had already met Gagliardo and Lions in 1959. They had both made a deep and lasting impression upon him, and he subsequently maintained regular contacts and correspondence with Lions.

In the paper [9] Jaak took his first steps from the topics of his thesis towards the realm of interpolation. He dealt with inclusions between various function spaces which he had introduced in his thesis. This work, in parallel with Lions' work on trace spaces, and the correspondence between the two of them, apparently helped to create the fertile environment in which both of them simultaneously discovered the "espaces de moyennes". They announced these parallel discoveries jointly in [14] and went on to give a definitive and far reaching account of the theory of these spaces. This was done in the now classical paper [21], truly a landmark in interpolation theory. Their "espaces de moyennes" are now referred to more often as "real method interpolation spaces" and denoted by $(A_0, A_1)_{\theta,p}$.

Many very important properties of these spaces, including reiteration formulæ, description of their duals, and initial versions of multilinear interpolation theorems and interpolation of compactness, can be found already in [21].

If the Calderón (and also Lions and Kreĭn) complex method spaces $[A_0, A_1]_\theta$, are clearly "descendants" of the Riesz–Thorin theorem, it could also be claimed that the Lions–Peetre real method spaces $(A_0, A_1)_{\theta,p}$ can be considered as "descendants" of the Marcinkiewicz interpolation theorem. But in this case the "genealogy" is rather less transparent. Certainly, one of the several equivalent ways of constructing $(A_0, A_1)_{\theta,p}$ given in [21] is quite reminiscent of Marcinkiewicz' technique of decomposing a function into a sum of two parts and applying different norm estimates to the action of an operator on each of those parts. Furthermore the Marcinkiewicz theorem can be deduced from the reiteration formula for real method spaces. But the initial definition of $(A_0, A_1)_{\theta,p}$ in [21] also looks very much like a "Fourier transform" of the definition of the spaces $[A_0, A_1]_\theta$ and indeed this fact is exploited in [21] to establish inclusions between these two kinds of interpolation spaces. It is intriguing to wonder how and to what extent these two different facts, and/or other facts, such as a number of related

[3] This proof has aroused much enthusiasm and been warmly praised, e.g. by Littlewood. See Jaak's article in this volume for some details about that. Thorin's proof, understandably, has come to overshadow Riesz' original proof of 1927, but later we will say more about Jaak's interaction with that original proof.

ideas introduced by Gagliardo (cf. [28]), or trace spaces, or other problems related to elliptic p.d.e.s and the Sobolev spaces $H^s(\Omega)$ actually motivated the authors of [21] at the time when these spaces were first defined. At various times in the past we have asked both Jaak and Jacques-Louis Lions about this. Neither could confirm our "pet" theory, that the real method spaces emerged naturally as "Fourier transforms" of the complex method spaces. In Jaak's case the main inspiration was apparently from Lions' trace spaces, and it was only afterwards that he saw the connection with the complex method[4].

Jaak recalls deciding in about 1962 to devote his research entirely to interpolation space theory[5]. One avenue of this research was the refinement and generalization of the real method introduced in [21]. Thus in [17] and [20] he showed how to obtain all Lions–Peetre spaces using only two parameters, instead of the three or four used in their original definition. (It is clear from standard examples that it is impossible in general to manage with fewer than two parameters, but a stronger version of this fact, answering a long standing question of Lions, would only be obtained much later by Per Nilsson in [94].) Another innovation in [17] was the reformulation of the construction of the spaces $(A_0, A_1)_{\theta,p}$ in terms of certain equivalent norms on $A_0 + A_1$ and on $A_0 \cap A_1$. Jaak called these the "K-functional", and the "J-functional". He is able to recall [P1] the train of thought that led him to them. He surely did not guess then how very important and popular they would become. Although at first their introduction may seem like little more than a rewriting of definitions, these two functionals, especially the K-functional, have turned out to be extremely helpful conceptually, exactly the right vehicle for understanding how and why the real interpolation method works, and expressing it concisely. The K-functional has also had a major impact beyond the realm of interpolation spaces, notably in approximation theory[6].

The papers [23], [25] and [41], are surveys presenting various versions of the real interpolation method via K- and J-functionals. In particular [41] is based on a series of lectures which Jaak gave in Brasilía in 1963. Here the theory is presented in the rather general context of what have since been come to be known as "function parameters". There is also some implicit anticipation of the "Boyd indices" [Bo].

In the next stage of evolution of the real interpolation method, beginning in the late 1960's, various seemingly natural conditions, which had hitherto been imposed on the spaces A_0 and A_1, were successively discarded, until they were no longer even spaces[7]. At the same time the hypotheses on the (originally linear) operators acting on A_0 and A_1 in interpolation theorems could also be weakened. Such developments

---

[4]Jaak has recorded and may some day [P1] publish some recollections of his meetings and correspondence with Lions at this exciting time.

[5]But of course he would also continue working in parallel on other topics. Jaak's creativity by its very nature cannot be confined to any one "box". Conversely, many years later he told one of us that he had finished working on interpolation and given the subject over to his students. But of course in his "spare time" he continued to contribute significantly and substantially to interpolation, so no one but himself noticed that he had "left" the field.

[6]The K-functional is implicit in Gagliardo's work (cf. also the much used "Gagliardo diagram"). It was also rediscovered later by Oklander [Ok].

may at first sight seem quite pointless – exotic or even crazy generalizations for their own sake[8]. However they turn out, surprisingly, to be very fruitful for a variety of natural applications, e.g. in approximation theory and harmonic analysis.

This process apparently began with a paper by Paul Krée [Krée], who pointed out that almost all of the machinery of real interpolation still works when the spaces being interpolated are neither complete nor even normed. The norm is replaced by a "quasi-norm". This approach was continued by Jaak's student Tord Holmstedt [Ho] with some very useful estimates for certain K-functionals. The next step was Jaak's paper [53] where he replaced the (quasi-)norms of the spaces A_0 and A_1 by those same expressions raised to some powers. This transforms the K-functional into what he termed the "L-functional". Subsequently, in collaboration with one of us, in [67] (cf. also [79]) the need for multiplication by scalars was dispensed with, so that A_0 and A_1 were now merely groups, normed abelian groups. (Here, "normed" and "quasi-normed" can be seen to be essentially the same thing.) The theory is fully developed with many applications. One intriguing feature in this new context is that there is now a natural way to define and interpolate L^p spaces, not only for all p in the expanded range $p \in (0, \infty]$, but also even for $p = 0$. A parallel development in similar spirit was made by Yoram Sagher in [Sa1] and [Sa2]. In his case A_0 and A_1 become cones rather than groups, and there are applications to Fourier series.

As far as we know, the farthest steps taken so far in this program, even beyond normed groups, are in [61] and [Kru]. In [61] Jaak began constructing a theory of interpolation of metric spaces. One application which he had in mind was to obtain the Nash–Moser implicit function theorem as a (nonlinear) interpolation theorem in this context. It is precisely this topic which is considered in rather more detail by Natan Krugljak in [Kru].

In parallel with the introduction and development of real interpolation, Jaak was also studying and sometimes also creating other interpolation methods. In [51] he refined the inclusions between real and complex interpolation spaces which he and Lions had established in [21]. This further clarified the "Fourier transform" connections between these two kinds of spaces: If A_0 and A_1 have a property which Jaak called "Fourier type p" then the inclusions of [21] can be sharpened, to an extent depending on the value of p. The strongest result (equality between the two methods) occurs when $p = 2$, for example when both the spaces are Hilbert spaces. (Jaak conjectured that in fact the only Banach spaces of Fourier type 2 are Hilbert spaces and, at his suggestion, Kwapień [Kw] proved this.) The paper [65] is a continuation of [51] and is inspired in part by a late paper [Ga] of Gagliardo. Here Jaak introduces the new interpolation spaces $\langle A_0, A_1 \rangle_\theta$ which are related to but different [Jan] from both the real and complex method spaces. He hints at a common framework for defining

[7] There had also been some earlier partial results extending interpolation to locally convex topological vector spaces (see [21] and some references there).

[8] At that time there also was a sort of belief, promoted at least implicitly by *Bourbaki* that the only "good" spaces are locally convex ones. So some degree of audacity was required to pursue these kinds of generalizations.

these three different methods, an idea which would later be developed in [CKMR]. The spaces $\langle A_0, A_1\rangle_\theta$ still seem quite mysterious. Jaak and Jan Gustavsson introduced slight variants of them in [81] and applied them to interpolation of Orlicz spaces. Very soon after this Vladimir Ovchinnikov [Ov1] introduced some other new interpolation methods which were apparently even better for interpolating Orlicz spaces. Ovchinnikov used a form of the orbit construction of [AG] and also Grothendieck's inequality in impressive and surprising ways. He has also introduced many other new ideas into interpolation. See also, for example, his survey [Ov2].

This is a good place to mention the fundamental paper of Svante Janson[9] [Jan] which would later reveal that *all* the different interpolation methods for pairs of Banach spaces mentioned so far in this subsection can be obtained by suitable concrete versions of the "minimal" or "maximal" (also called "orbit" or "co-orbit") constructions of Aronszajn and Gagliardo [AG].

But Jaak has also developed at least one kind of interpolation [86] which, as far as we know so far, does not seem to fit naturally into the framework of [Jan]. This is his remarkable celebration of the jubilee of interpolation. Fifty years after Marcel Riesz gave the original proof of what is now known as the Riesz–Thorin theorem, Jaak turned his attention to that classical but rather forgotten proof and generalized it to create two new "semi-interpolation" methods which lead to interpolation theorems.

Having been so active in extending the real interpolation method beyond the original confines of couples of Banach spaces it was natural for Jaak to consider the possibility of analogous developments for the complex method. The work of [90] indicates the kinds of obstacles which can be encountered in such an enterprise. A much more recent contribution to this topic is [173]. The paper [98] offers some intriguing insights about the complex method in its more usual Banach space setting. The standard definition of this method uses Banach space valued analytic functions. But Jaak and Svante Janson show that there are alternative definitions using either harmonic functions or caloric functions (i.e. solutions of the heat equation).

In a different vein, Jaak considered a problem originally raised by Ciprian Foiaş and Lions [FL], of characterizing all weighted L^p spaces which are *exact* interpolation spaces with respect to a given couple of weighted L^p spaces. The papers [24], [34], [42], [52] (and their later sequels [106] and [114]) are all connected to [FL] and this problem. But apparently the complete solution of all cases of the original problem has yet to be found.

The interpolation methods mentioned so far are all defined in the context of couples (A_0, A_1) of (Banach or more general) spaces compatibly embedded in some larger space. But another possible generalization of interpolation theory occurs when these couples are replaced by n-tuples $(A_0, A_1, \ldots, A_{n-1})$ or even infinite families $\{A_\gamma\}_{\gamma\in\Gamma}$ of spaces which are, again, all embedded in some larger space. At first sight one would expect the definition and theory of interpolation spaces with respect to such finite or infinite families to be a straightforward generalization of the case of couples.

[9]This paper was unexpectedly catalysed by Svante's visit to Lund to work on very different things, and his resulting meeting by chance with Jaak. More details may someday be told in [P1].

But it turns out that there are many surprises and difficulties. Indeed, after working with n-tuples a researcher may have some difficulty believing that the case of couples works so well. Despite these frustrations, quite a number of mathematicians, including Jaak, Selim Kreĭn, Lions, Svante Janson, and also all the authors of this article, have been enticed to work at various times with this exotic variant of interpolation theory. In many cases we have to "blame" Jaak for inspiring us to get caught up in this beautiful but complicated topic. More specifically:

- G.S. wrote his doctoral thesis under Jaak's supervision in exactly this subject.
- L.-E.P. worked on it together with Maria Carro, Ludmila Nikolova and Jaak in [169] benefiting greatly from Jaak's great enthusiasm and deep involvement.
- M.C. and Svante Janson read a "casual" remark of Jaak's which was in fact a beautiful proof of an inclusion between the different Lions–Favini and Kreĭn–Nikolova/ Cwikel–Coifman–Rochberg–Sagher–Weiss versions of the complex method for n-tuples. The elegance of this proof somehow "obliged" them to extensively investigate further connections between the various real and complex methods for n-tuples and for infinite families.
- R.R.'s additional investigations of (mainly complex) interpolation of infinite families, have taken him far afield, e.g. to connections with function theory, differential geometry and vector bundles.

There are very intriguing applications of all this, but they are far fewer and much harder to find, so far, than in the case of couples.

Jaak has coupled his enthusiasm about interpolation space theory *per se* with a keen awareness of its applications and potential applications. Thus, for example, his early note [16] already indicated that interpolation spaces could be used to prove theorems about approximation of functions. This message was also taken up in the book [BB] by Paul Butzer and Hubert Berens, which appeared four years later. Ultimately, interpolation spaces, and the K-functional (as we indicated above), would be widely recognized as important tools in approximation theory. (See the papers listed above under item 5.) Later Jaak wrote a series of papers [31], [33], [35], [37] and [39] giving additional applications of interpolation to a variety of other topics in analysis.

After mentioning so many important things which Jaak did in the realm of interpolation, we should also mention one important thing which he did not do, although in [88] he came within an "ε" of it, as did Svante Janson in [Jan]. This is the remarkable theorem about the "K-divisibility" of the K-functional, proved in 1981 by Yuri Brudnyi and Natan Krugljak. This result is something of a revolution in the real interpolation method. Characteristically, Jaak took vigorous action to promote the publication in the West of a book by Brudnyi and Krugljak to make their work better known [BKr]. He had earlier acted similarly to encourage the appearance of Ovchinnikov's book [Ov2].

What is K divisibility? For a long time, Jaak, and also quite a number of other mathematicians, had been considering a large class of real method interpolation spaces which generalize the spaces $(A_0, A_1)_{\theta,p}$. These are the spaces $(A_0, A_1)_{\Phi,K}$ whose norms are given by composing the K-functional with a lattice norm $\|\cdot\|_\Phi$ on the measurable functions on $(0,\infty)$, i.e. $\|a\|_{(A_0,A_1)_{\Phi,K}} = \|K(\cdot, a; A_0, A_1)\|_\Phi$. The K-divisibility theorem makes it possible to establish all sorts of results for these spaces, reiteration, duality, equivalence of J-functional and K-functional constructions, which had earlier been known only for the spaces $(A_0, A_1)_{\theta,p}$ and various rather specialized variants of them. It also leads to a much neater formulation of a whole family of previously known results about "Calderón couples" or "K-monotone couples", i.e. couples (A_0, A_1) for which *all* interpolation spaces can be described in terms of a certain monotonicity condition with respect to the K-functional. In fact all the interpolation spaces of such couples are of the relatively simple form $(A_0, A_1)_{\Phi,K}$. K-divisibility enables great flexibility in the handling of K-functionals, and related important quantities like moduli of continuity, and thus is potentially very useful also beyond the realm of interpolation.

Many years earlier, Jaak had himself planned to write a book about interpolation. For a while he and Lions considered doing so jointly. The "applications" papers [31], [33], [35], [37] mentioned above were also conceived as forerunners for some of the contents of such a book. Ultimately, for a combination of diverse reasons, Jaak decided not to pursue this project, but his former students Jöran Bergh and Jörgen Löfström were able to take it over and produce the important and useful book [BL] that has served many of us well for many years.

In parallel with Jaak's published papers there is also a large collection of less formal documents which he has written. These are mainly technical reports or files posted on his website. We felt strongly that these too should be included in the list of Jaak's publications in this volume because some of them have been almost as influential and important as his formal publications, and others contain material which may possibly yet have an important impact[10]. A notable example is the technical report [240] where Jaak suggests formulating interpolation theory in the language of category theory. Among other things, with the help of this viewpoint, he elegantly and easily obtains descriptions of all interpolation spaces for couples of weak L^p spaces, and shows that *relative* interpolation spaces for operators mapping an arbitrary Banach couple to a couple of weighted L^∞ spaces have an analogous description. Though it is informal and unpolished, this note has catalysed a lot of research by a number of mathematicians, on Calderón couples and various other topics. (See also [62] for a more concise discussion of these matters.)

[10] The reader of these less formal documents can also enjoy, from time to time, a more light-hearted sometimes almost cheeky style of writing, the sorts of things which sometimes do and sometimes do not cross into journals through the filter of serious and formal editing and "self-editing". One example: "I thank myself for my excellent typing." Another example: It is customary to use the notation $A \stackrel{\text{def}}{=} B$ to indicate that A is defined by taking it to be equal to B. But Jaak sometimes also chooses to use his alternative notation $B \stackrel{\text{fed}}{=} A$.

Suppose that the two "scales" of complex interpolation spaces $\{[A_0, A_1]_\theta\}_{0\le\theta\le1}$ and $\{[B_0, B_1]_\theta\}_{0\le\theta\le1}$ satisfy $A_0 \cap B_1 \subset A_1 \cap B_0$ and $[A_0, A_1]_\beta = B_0$ and $[B_0, B_1]_\alpha = A_1$ for some α and β in $(0, 1)$. Then the two scales can be "glued" together, i.e. all members of both scales are of the form $[A_0, B_1]_\theta$ for suitable values of θ. This fact, perhaps analogously to K-divisibility, probably seemed too good to be true, but it was conjectured by Yoram Sagher, and proved by Tom Wolff [Wo] in 1981. The result is very convenient e.g. for understanding interpolation between L^p spaces and Hardy spaces H^p. There is in fact a second version of the theorem for real method interpolation spaces.

Jaak, together with Svante Janson, Per Nilsson and Misha Zafran, has given us deeper insights into this phenomenon and widely generalized Wolff's theorem in their amusingly titled paper [94]. Then later in [126], with Fernando Cobos, Jaak even extended the theorem to a "multidimensional" context (couples of Banach spaces replaced by n-tuples). Among Jaak's papers with Cobos we also particularly mention [140] which deals with the problem of interpolating compact operators in a very imaginative and successful way.

Hankel operators and invariant function spaces

Hankel operators have always been an important topic in the function theoretic operator theory (= operator theoretic function theory) of the unit disk. In the late '70's Vladimir Peller [Pel] showed that the properties of Hankel operators, in particular their Schatten–von Neumann properties, and the Besov spaces of holomorphic functions on the disk, in particular the diagonal Besov spaces $B^{p,p}_{1/p}$, were intimately connected. At about that time, influenced in part by Peller's work, (and also apparently because of some casual remarks by Jöran Bergh and one of us) Jaak began to work actively on Hankel operators. A substantial fraction of his research since then has been related to Hankel operators. His earliest work in the area centered on classical Hankel operators and he continued to make contributions in that area ([93], [101], [115]). However he soon began to study broader and deeper issues of which Hankel operators were merely a single manifestation. Rather than try to enumerate all of his contributions in this area, or even prepare an annotated list of our favorites, we will try to point to some of the basic themes introduced, developed, and championed by Jaak over the last 20^+ years. These themes have done a great deal to shape the current landscape in this area.

One theme Jaak returned to repeatedly was that there was a systematic relation between the theory of Hankel operators and certain commutators of multiplication operators and singular integral operators which arise in real variable harmonic analysis. An early contribution in this area was [99]. He and Svante Janson soon took this theme much much further. In [104], [118], and [136] there is presented a broad theory of "paracommutators"; a theory of a class of operators defined by Fourier analytic means which includes the paraproducts and the commutators (from whence the name) of real variable harmonic analysis as well as versions of both the Toeplitz and the Hankel operators of function theoretic operator theory. (Jaak long championed calling the operators in these last two classes, along with various generalizations, "Ha-plitz" operators [134], [141]. The fate of that verbal coinage remains in doubt.) For a connection between paracommutators and compensated compactness see [PW].

A second major theme was that what was known about Hankel operators on the Hardy space of the disk could be viewed as the simplest and most well understood instance of a broad range of questions in function theoretic operator theory. Some of his first work in this direction involved operators on function spaces other than the circle [103] as well as studying Hankel type operators on Bergman spaces [112], [121], [125], [128], [138]. His research on Hankel operators broadened to include work on Hankel operators in venues quite different from those previously considered, Hankel forms on function spaces on the entire plane (the Fock space (a.k.a. the Bargmann-Segal space)) as well as Bergman type spaces associated with bounded symmetric domains. We will return to these in a moment.

The spaces $B_{1/p}^{p,p}$ which play a fundamental role in Peller's work on Hankel operators are Möbius invariant. This fact, along with conversations with Jonathan Arazy, helped focus Jaak's work in the early 80's on the role of symmetry groups in the study of function spaces and in function theoretic operator theory. That theme is fundamental in much of his work since then.

In the classical theory of Hankel operators on the Bergman space the Möbius group acts transitively on the underlying space (the disk), irreducibly on the Bergman space, and nicely on the Besov spaces and other natural symbol classes in the theory. This theme shows up in [101], [108], and [109]. It is then taken much further in [116] and later papers. In fact, one way to view Hankel operators on the classical Bergman space is as a theory about a domain, the disk, with a transitive automorphism group, the Möbius group, a class of spaces that transform nicely under the group action, the weighted Bergman spaces, and a class of operators, the Hankel operators, that transform nicely under the group action. One then finds that the symbol classes of holomorphic functions that correspond to bounded, compact, Hilbert–Schmidt, etc. operators can be described quite conveniently in terms of the group action and differential operators associated to the group action. Of course it is natural to wonder if there are other instances of this type of situation. The surprising conclusion in [116] is that their are quite a number. A theory of Hankel forms can be developed for any bounded symmetric domain along with its automorphism as well as for the Fock space of entire functions on $\mathbb{C}^n$ with the group action of the Heisenberg group. The presence of the transitive automorphism induces most of the analytical tools that are needed. The group action can be used to prove that the square of the Bergman reproducing kernel can be realized as the Bergman kernel for an associated weighted Bergman space. That fact is the analytical engine that drives the theory. Although the theory of [116] is quite abstract and general it instantiates quite easily to give much of the classical theory on the Bergman space of the disk. Jaak continued to explore extensions of these themes in [120], [123], [124], [143] and [148], often investigating contexts where some but not all of the structure used in [116] was available. These themes also continue to be studied by others ([HR2], [HR1], [DG]).

In his work on these topics, as with his work in other areas, Jaak was joined by many collaborators. In some cases the collaborators later took the ideas much further and into very interesting new directions. However one can often see the shaping influence of their interaction with Jaak. In particular in the work of Jonathan Arazy, Miroslav Engliš, and Genkai Zhang one sees the development of a very rich and interesting function theoretic operator theory for domains with large symmetry groups. Although much of this work was not in collaboration with Jaak, the body of work seems to be in "the Peetre tradition", richly informed by Jaak's views about mathematics. Jaak always prefers to think (and talk, and write) of Hankel bilinear forms (rather than linear or conjugate linear Hankel operators). Although these formulations are equivalent, the viewpoint of forms led to a profound extension. A Hankel form is a bilinear form B acting on a function space such that $B(f, g)$ is a linear function of the product fg. A bilinear form on a Hilbert space H is defined on the product $H \times H$ and extends to (at least) the algebraic tensor product. If the form is Hilbert–Schmidt then it extends to the Hilbert space tensor product $H \otimes H$; let us assume that that is the case. Let us now marry this viewpoint with the group action mentioned in the previous paragraph. If H is, for instance, the Bergman space of the disk then the Möbius group acts irreducibly on H. This action induces an action on $H \otimes H$ which is no longer irreducible. However representation theory tells us how to decompose this action into irreducible components, and, in fact, how to order them from simplest to more complicated. It turns out that the simplest irreducible component is (naturally identified with) the space of (Hilbert–Schmidt) Hankel forms. This fact raises several very fundamental new questions: How can one describe and analyze the bilinear forms corresponding to the more complicated irreducible components (the Hankel forms of "higher order" or "higher weight")? What happens when you decompose a general bilinear form by projecting on these irreducible summands (i.e. a call for Harmonic Analysis on the space of bilinear forms)? What happens for domains more complicated than the disk but with a full automorphism group? What happens if there is no automorphism group?

These higher order forms were introduced in [110] and there the basic theory is worked out for the Bergman space of the disk (using, among other things, the authors' theory of paracommutators [118]). Much of Jaak's work since that time has involved questions and ideas that flow from that work on higher order Hankel forms on the Bergman space, [130], [141], [152], [159], [160], [156], [168], [182]. This is a very broad program with surprising (to us) connections to a huge range of other areas of analysis including representation theory, special function theory (one and several variables), Hilbert module theory, and classical invariant theory. (It was [110] which sent one of us (R.R.) to the library to find out what a "transvectant" was.) We suspect that there are some really interesting discoveries to be made in this area. Also, this expansion into the study of a broad range of generalized Hankel forms makes the distinction between Hankel and Toeplitz operators becomes less sharp; it moves us closer to seeing both operators as part of a single larger family – see [182].

Of course once one is thinking of Hankel's as bilinear forms it is quite natural to wonder what a corresponding theory of trilinear or multilinear forms might be. It is clear that those are hard questions and it is not clear where they will lead. With his usual intellectual curiosity and courage Jaak is continuing to look at those issues [150], [175], [180]. (We will have more to say about Jaak's work on trilinear and multilinear forms in our next and final subsection.)

Two final comments should be made about this body of work. First, interpolation theory has always been one of the great intellectual loves of Jaak's life. The work described here is not about interpolation but is pervaded by ideas from interpolation. Sometimes interpolation provides the viewpoint; for instance the focus on the minimal space in [103]. Other times it provides the analytic tool to get the desired result, see for instance [118]; or to see a very delicate interpolation proof done by a true master, look at [112].

Finally, Jaak is an analyst and most of analysis deals with estimates, comparisons, and inequalities. It is quite striking to find in analysis papers elegant, suggestive, and rather unexpected equalities. In [138] an elegant equality is presented for the Hilbert–Schmidt (i.e. $\mathfrak{S}_2$) norm of certain Hankel forms in planar domains. A similarly elegant and similarly surprising formula for an $\mathfrak{S}_4$ norm of a Hankel operator on the Fock space is in [148]. Finally, in [JUW], a set of similar equalities for $\mathfrak{S}_4$, and $\mathfrak{S}_6$ are described. As was noted at the conference which we celebrate in this volume; these striking results appear as ad hoc isolates. One would certainly like to see them tied to each other, to know if there are more such results or why not, and in general to know what is going on. It has been suggested informally that there may even be an index theorem lurking here.

Multilinear forms, trilinear forms in Hilbert space

Jaak's interest in trilinear forms began almost twenty years ago, via his work on Hankel theory [93], [104].

Since the early 1990's it has been one of his main mathematical preoccupations; indeed, Jaak sees this as perhaps his most important piece of research. It involves a bringing together of various areas of his previous interests, such as interpolation and Hankel theory, with a surprisingly wide and varied collection of other topics and structures, including Jordan triple systems, algebraic geometry, invariant theory, spherical trigonometry and more. It thus provides a striking illustration of the *unity of mathematics*.

A considerable part of Jaak's work here has been in collaboration with others, notably Fernando Cobos and Thomas Kühn, and at a later stage Bo Bernhardsson. He has also had some correspondence with I. M. Gel'fand.

The matrix of a Hankel operator or a bilinear Hankel form with respect to orthonormal bases is of the form $a_{jk} = a_{j+k}$. It is natural to ask what happens if we replace a_{j+k} by a tensor of rank 3 (or, as Gel'fand calls it, a 3-dimensional matrix) $a_{jk\ell}$ of the form $a_{jk\ell} = a_{j+k+\ell}$. This leads to the notion of a trilinear Hankel form. The study of such forms was begun in [104]. The notions introduced there included Schatten–von Neumann classes $\mathfrak{S}_p$ of (general) trilinear forms. (See also [136] and [150].) The classes $\mathfrak{S}_1$ and $\mathfrak{S}_\infty$ (compact forms) have natural definitions. Then $\mathfrak{S}_p$ for $1 < p < \infty$ can be defined by interpolation. But it is still not known if real and complex interpolation give the same result.

To take this theory further it is first necessary to understand the properties of *bounded* trilinear forms. This is surprisingly difficult to do, even when the three underlying Hilbert spaces are finite dimensional, say of dimension n. Rather extensive results have been obtained for $n = 2$ ([180], [181], [210]), though several intricate problems still remain. But almost nothing is known yet for $n = 3$. This case is already enormously complicated, and it is apparently no longer possible to do explicit calculations, not even symbolically. Probably entirely new ideas are needed here. Furthermore, no corresponding results are known for quadrilinear forms.

It is possible to define singular values of trilinear forms. This is done in [184]. These turn out to be roots of an algebraic equation analogous to the secular equation, which is called the *millenial equation*. Some numerical computations suggest that the singular values can be obtained via a min-max principle. At this stage an interesting open problem would be to determine the singular values and the millenial equation for specific trilinear Hankel forms in low dimensions, for example the Hilbert form $a_{jk\ell} = \frac{1}{j+k+\ell}$.

Jaak's work on trilinear forms has given him new insights concerning invariant theory in general. In particular he has determined some (linear) invariants of trilinear forms in low dimensions. (See [220], [221], [222]. The fact that such invariants do arise in a trilinear context was clear from [180].)

3. Graduate students

Jaak has had sixteen graduate students. We list them here, with the titles of their theses, also specifying the degrees which they obtained, and when they obtained them.

We should perhaps explain, for those not familiar with the Swedish system, that fil.dr. and tekn.dr. are doctoral degrees (of philosophy or technology respectively) and fil.lic. and tekn.lic. are corresponding licentiate degrees. In the framework of the older system of degrees, which was in force till about 1975, the licentiate degree was approximately equivalent to the current doctoral degree.

Leif Arkeryd, fil.lic. 1966.
On the L^p estimates for elliptic boundary problems.
Published in Math. Scand. **19** (1966), 59–76.

Tord Holmstedt, fil.lic. 1969.
Interpolation of quasi-normed spaces.
Published in Math. Scand. **26** (1970), 177–199.

Barbro Grevholm, fil.lic. 1970.
On the structure of the spaces $\mathcal{L}_k^{p,\lambda}$.
Published in Math. Scand. **26** (1970), 241–254.

Sigrid Sjöstrand, fil.lic. 1970.
On the Riesz means of the solutions of the Schrödinger equation.
Published in Ann. Scuola Norm. Sup. Pisa (3) **24** (1970), 331–348.

Jöran Bergh, fil.lic. 1971.
On the interpolation of normed linear spaces. Thesis, Lund, 1971.

Jörgen Löfström, fil.dr. 1971.
Besov spaces in the theory of approximation.
Published in Ann. Mat. Pura Appl. **85** (1970), 93-184.

Annika Haaker-Sparr, tekn.lic. 1971.
On the conjugate space of Lorentz space[11]. Thesis, Lund, 1971.

Gunnar Sparr, tekn.dr. 1972.
Interpolation of several Banach spaces.
Published in Ann. Mat. Pura Appl. **99** (1974), 247-316.

Lars Vretare, fil.lic. 1972.
Multiplier theorems connected with generalized translations.
Thesis, Lund, 1972.

Björn Jawerth, tekn.dr. 1977.
On Besov spaces. Thesis, Lund, 1977.

Jan Gustavsson, fil.dr. 1980.
On some interpolation methods. Thesis, Lund, 1979.

Wilhelm Kremer, fil.dr. 1980.
Topics in the calculus of variations. Thesis, Lund, 1980.

[11]One of us, (M.C.) would like to add a comment about this so far unpublished thesis: It contains a number of fine results, which even today, thirty years later, have still not all been discovered by other people. In the early 1970's Yoram Sagher and I were excited to discover that Weak L^1 has non trivial dual. Soon afterwards I was happy to show that Weak L^p has trivial dual for $p \in (0, 1)$ on a nonatomic measure space. The first of these results intrigued and surprised at least one very famous mathematician. But both of these facts turn out to be special cases of a more general theorem in Annika's thesis which she possibly obtained before us. Several people, including Jaak and myself, very much hope that Annika will make this work more widely available before too long.

Per Nilsson, tekn.dr. 1982.
A study of real interpolation spaces.
Published in Ann. Mat. Pura Appl. **132** (1982), 291–330 and **134** (1983), 201–232.

Märta Ahlman, fil.lic. 1984.
The trace ideal criterion for Hankel operators on the weighted Bergman space $A^{\alpha 2}$ in the unit ball of C^n. Thesis, Lund, 1984.

Genkai Zhang, fil.dr. 1991.
Hankel operators and weighted Plancherel formula. Thesis, Stockholm, 1991.

Hjalmar Rosengren, fil.dr. 1999.
Multivariable Orthogonal Polynomials as Coupling Coefficients for Lie and Quantum Algebra Representations. Thesis, Lund, 1999.

Acknowledgements

It was with considerable trepidation that we approached the task of trying, in this short article, to adequately summarize forty-four years of Jaak's intense mathematical activities in a very broad range of fields, many of them quite distant from the collectivity of all our own research interests.

Fortunately, at various stages, we have been helped by insights and details kindly offered by a number of our colleagues. These include Jonathan Arazy, Fernando Cobos, Miroslav Engliš, Thomas Kühn, Doron Lubinsky, Per Nilsson, Mario Milman, Genkai Zhang, and, last but not least, Jaak himself. Given the limited space available for this article, we have not been able to incorporate all the interesting details that they provided. Nor, as already explained, have we been able to cover all fields of Jaak's endeavour. We hope to make more of this material available at some time in the future, possibly as a web document.

We are also very grateful for additional help from Miroslav Engliš and Lech Maligranda.

Finally the authors of this article beg our readers' indulgence, as we digress for a moment to address a few words to our two friends and colleagues from Prague: We express our very warmest thanks and appreciation to Alois Kufner for agreeing to head the team that created this volume, and to Alois and, yet again, to Miroslav Engliš for their great expertise, devotion and hard work. Without their essential contribution this book would not have happened!

References

(References [1] to [289] are from the list of all Jaak's publications to date, which follows this bibliography.)

[Ark1] L. Arkeryd, On L^p estimates for elliptic boundary problems, Math. Scand. 19 (1966), 59–76.

[Ark2] L. Arkeryd, On L^p estimates for quasi-elliptic boundary problems, Math. Scand. 24 (1969), 141–144.

[Ark3] L. Arkeryd, A priori estimates for hypoelliptic differential equations in a half-space, Ann. Scuola Norm. Sup. Pisa, Sci. Fis. Mat., III. Ser., 22 (1968), 409–424.

[AG] N. Aronszajn, E. Gagliardo, Interpolation spaces and interpolation methods, Ann. Mat. Pura Appl. 68 (1965), 51–118.

[BSh] C. Bennett, R. Sharpley, Interpolation of operators, Academic Press, New York, 1988.

[BL] J. Bergh, J. Löfström, Interpolation spaces. An introduction, Grundlehren Math. Wiss. 223, Springer-Verlag, Berlin–Heidelberg–New York, 1976.

[Bo] D. W. Boyd, Indices of function spaces and their relationship to interpolation, Canad. J. Math. 21 (1969), 1245–1254.

[Br] F. E. Browder, La théorie spectrale des opèrateurs aux dérivées partielles du type elliptique, C.R. Acad. Sci. Paris 246 (1958), 526–528 [French].

[BKS] Yu. Brudnyĭ, S. G. Kreĭn, E. M. Semenov, Interpolation of linear operators, in: Itogi Nauki i Tekhniki Ser. Mat. Anal. 24, Moscow, VINITI, 1986, 3–163 [Russian]; English translation: J. Soviet Math. 42 (1988), 2009–2112.

[BKr] Yu. Brudnyĭ, N. Ja. Krugljak, Interpolation functors and interpolation spaces, Vol. I, North-Holland, Amsterdam, 1991.

[BB] P. L. Butzer, H. Berens, Semi-groups of operators and approximation, Grundlehren Math. Wiss. 145, Springer-Verlag, Berlin–Heidelberg–New York, 1967.

[CKMR] M. Cwikel, N. J. Kalton, M. Milman and R. Rochberg, A unified theory of commutator estimates for a class of interpolation methods, Adv. in Math., to appear.

[DG] T. Deck and L. Gross, Hankel Operators over Complex Manifolds, preprint 2001.

[Di] J. Dieudonné, Élements d'analyse, t. 8. Gauthiers-Villars, Paris, 1978.

[Do] T. Donaldson, A Laplace transform calculus for partial differential operators, Mem. Amer. Math. Soc. 143 (1974).

[FL] C. Foiaş, J. L. Lions, Sur certains théorèmes d'interpolation, Acta Sci. Math. Szeged 22 (1961), 269–282.

[Ga] E. Gagliardo, Caratterizzazione construttiva di tutti gli spazi di interpolazione tra spazi di Banach, Sympos. Math. 2 (1968), 95–106.

[GM] I. W. Gelman, W. G. Mazja, Abschätzungen für Differentialoperatoren im Halbraum, Akademie-Verlag, Berlin, 1981.

[HR1] F. Holland and R. Rochberg, Bergman kernels and Hankel forms on generalized Fock spaces, in: Function spaces (K. Jarosz, ed.), Contemp. Math. 232, Amer. Math. Soc., Providence, RI, 1999, 189–200.

[HR2] F. Holland and R. Rochberg, Bergman kernel asymptotics for generalized Fock spaces, J. Anal. Math. 83 (2001), 207–242.

[Ho] T. Holmstedt, Interpolation of quasi-normed spaces, Math. Scand. 26 (1970), 177–199.

[Hö1] L. Hörmander, Linear partial differential operators, Grundlehren Math. Wiss. 116, Springer, Berlin–Göttingen–Heidelberg, 1963.

[Hö2] L. Hörmander, The analysis of linear partial differential operators I–IV, Grundlehren Math. Wiss. 256, 257, 274, 275, Springer-Verlag, Berlin–Heidelberg–New York–Tokyo, 1983, 1985.

[Jan] S. Janson, Minimal and maximal methods of interpolation, J. Funct. Anal. 44 (1981), 50–73.

[JUW] S. Janson, H. Upmeier and R. Wallstén, Schatten-norm identities for Hankel operators, J. Funct. Anal. 119 (1) (1994), 210–216.

[Jaw] B. Jawerth, On Besov spaces, Technical report, Lund, 1977.

[Krée] P. Krée, Interpolation d'espaces vectoriels qui ne sont ni normés, ni complets. Applications, Ann. Inst. Fourier (Grenoble) 17 (2) (1967), 137–174 [French].

[KPS] S. G. Kreĭn, Yu. I. Petunin and E. M. Semenov, Interpolation of linear operators, Nauka, Moscow, 1978 [Russian]. English translation: Transl. Math. Monogr. 54, American Mathematical Society, Providence, R.I., 1982.

[Kre] W. Kremer, A variational proof of the Gauss-Bonnet formula, Technical report, Lund, 1979.

[Kru] N. Ya. Krugljak, Imbedding thorems, interpolation of operators and the Nash-Moser implicit function theorem, Dokl. Akad. Nauk SSSR 226 (1976), 771–773 [Russian].

[Kw] S. Kwapień, Isomorphic characterization of inner product spaces by orthogonal series with vector valued coefficients, in: Sém. Maurey-Schwartz Année 1972–1973: Espaces L^p et applications radonifiantes. Exp. No. 7, 7 pp. Centre de Math., École Polytechn., Paris, 1973.

[Ma] A. Marchaud, Sur les champs de demi-droites et les équations différentielles du premier ordre, Bull. Soc. Math. France 62 (1934), 1–38.

[Me] R. B. Melrose, Singularities of solutions to boundary value problems, in: Proc. Internat. Congress of Mathematicians, Helsinki, 1978, Vol. 2, Helsinki, 1980, 785–790.

[Ok] E. T. Oklander, Interpolacion, espacios de Lorentz y teorem de Marcinkiewicz, Cursos y seminarios de matematica 20. Univ. de Buenos Aires, 1965.

[Os] R. Osserman, Isoperimetric inequalities and eigenvalues of the Laplacian, in: Proc. Internat. Congress of Mathematicians, Helsinki, 1978, Vol. 1, Helsinki, 1980, 435–442.

[Ov1] V. I. Ovchinnikov, Interpolation theorems that arise from Grothendieck's inequality, Funkcional. Anal. Prilozhen. 10 (1976), 45–54 [Russian].

[Ov2] V. I. Ovchinnikov, The method of orbits in interpolation theory, Math. Reports 1 (1984), 349–515.

[Pl] J. Peetre, Early encounters with mathematics, in preparation.

[Pel] V. V. Peller, Hankel operators of class $\mathfrak{S}_p$ and their applications (rational approximation, Gaussian processes, the problem of majorization of operators), Mat. Sb. (N.S.) 113 (4) (155 (12)) (1980), 538–581 (637) [Russian].

[PW] L. Peng and M. W. Wong, Compensated compactness and paracommutators, J. London Math. Soc. (2) 62 (2) (2000), 505–520.

[Per] J. Persson, A generalization of Carathéodory's existence theorem for ordinary differential equations, J. Math. Anal. Appl. 49 (1975), 496–503.

[Sa1] Y. Sagher, Some remarks on interpolation of operators and Fourier coefficients, Studia Math. 44 (1972), 239–252.

[Sa2] Y. Sagher, An application of interpolation theory to Fourier series, Studia Math. 41 (1972), 169–181.

[Sham] E. Shamir, Mixed boundary value problems for elliptic equations in the plane. The L^p theory, Ann. Scuola Norm. Sup. Pisa 17 (1963), 117–139.

[Sj] S. Sjöstrand, On the Riesz means of the solutions of the Schrödinger equation, Ann. Scuola Norm. Sup. Pisa 24 (1970), 331–348.

[Span] S. Spanne, Sur l'interpolation entre les espaces $\mathcal{L}_k^{p\Phi}$, Ann. Scuola Norm. Sup. Pisa 20 (1966), 625–648.

[Spar] G. Sparr, Interpolation of several Banach spaces, Ann. Mat. Pura Appl. 99 (1974), 247–316.

[Tri] H. Triebel, Interpolation Theory. Function spaces. Differential operators. Deutscher Verlag der Wissenschaften, Berlin, 1978.

[Wi] B. Winzell, Finite element Galerkin methods for multiphase Stefan problems, Appl. Math. Modelling 7 (1983), 329–344.

[Wo] T. H. Wolff, A note on interpolation spaces, in: Harmonic analysis (G. Weiss and F. Ricci, eds.), Lecture Notes in Math. 908, Springer-Verlag, Berlin–New York, 1982, 199–204.

M. Cwikel, Department of Mathematics, Technion - Israel Institute of Technology, Haifa 32000, Israel

E-mail: `mcwikel@math.technion.ac.il`

L.-E. Persson, Department of Mathematics, Luleå University of Technology, SE-971 87 Luleå, Sweden

E-mail: `larserik@sm.luth.se`

R. Rochberg, Department of Mathematics, Washington University in St. Louis, Campus Box 1146, 1 Brookings Drive, St.Louis, MO 63130, U.S.A.

E-mail: `rr@math.wustl.edu`

G. Sparr, Centre for Mathematical Sciences, Lund University, Box 118, SE-221 00 Lund, Sweden

E-mail: `gunnar@maths.lth.se`

List of publications of Jaak Peetre

The following list contains, to the best of our knowledge, all the papers, books and other items that Jaak has published so far. Its first part, with items numbered [1]–[187], includes his more formal publications. The second part, with items numbered [188]–[200], consists of Jaak's most important translations and editoral works. The third part, with items numbered [201]–[289], consists of various preprints, technical reports and also documents, marked by ⋆, which are currently available from Jaak's website. The second and third parts contain a small number of items, a translation and some earlier technical reports etc., which do not appear in Jaak's own list, currently posted on his website. To avoid confusion, in the first part we use numbering which is consistent with the numbering in Jaak's own list. This means in particular that the order is not always strictly chronological.

We invite the reader to visit Jaak's website

`http://www.maths.lth.se/matematiklu/personal/jaak`

for further updates in the future.

I. Scientific publications

[1] A generalization of Courant's nodal domain theorem, Math. Scand. 5 (1957), 15–20.

[2] Estimates for the number of nodal domains, Comptes Rendus XIII Congr. Math. Scand., Helsinki, 1957, 198–201.

[3] Une classe d'opérateurs différentiels, C.R. Acad. Sci. Paris 248 (1959), 1102–1103.

[4] Opérateurs différentiels quasi linéaires hypoelliptiques, C. R. Acad. Sci. Paris 248 (1959), 3401–3403.

[5] Comparaison d'opérateurs différentiels, Faculté des Sciences de Paris, Séminaire d'Analyse (P. Lelong), Année 1958/59, exp. 16, 7 pp.

[6] Théorèmes de regularité pour quelques classes d'opérateurs différentiels, Ph. D. thesis, Department of Mathematics, Lund University, Sweden, 1959, 122 pp.

[7] Une caractérisation abstraite des opérateurs différentiels, Math. Scand. 7 (1959), 211–218.

[8] Rectification à l'article "Une caractérisation abstraite des opérateurs différentiels", Math. Scand. 8 (1960), 116–120.

[9] Relations d'inclusion entre quelques espaces fonctionnels, Kungl. Fysiogr. Sällsk. i Lund Förh. 30 (1960), 47–50.

[10] A proof of the hypoellipticity of formally hypoelliptic differential operators, Comm. Pure Appl. Math. 14 (1961), 737–744.

[11] Another approach to elliptic boundary problems, Comm. Pure Appl. Math. 14 (1961), 711–731; Russian transl. in: Matematika 7 (1963), 43–65.

[12] On estimating the solutions of hypoelliptic differential equations near the plane boundary, Math. Scand. 9 (1961), 337–351.

[13] Mixed problems for higher order elliptic equations in two variables, I, Ann. Scuola Norm. Sup. Pisa 15 (1961), 337–353.

[14] Propriétés d'espaces d'interpolation, C. R. Acad. Sci. Paris 253 (1961), 1747–1749 (coauthor: J.-L. Lions).

[15] On the differentiability of the solutions of quasilinear partial differential equations, Trans. Amer. Math. Soc. 104 (1962), 476–482.

[16] Espaces intermédiaires et la théorie constructive des fonctions, C. R. Acad. Sci. Paris 256 (1963), 54–55.

[17] Nouvelles proprietés d'espaces d'interpolation, C. R. Acad. Sci. Paris 256 (1963), 1424–1426.

[18] Elliptic partial differential equations of higher order, Lecture notes, University of Maryland, 1962, 1–122.

[19] Mixed problems for higher order elliptic equations in two variables, II, Ann. Scuola Norm. Sup. Pisa 17 (1963), 1–12.

[20] Sur le nombre de paramètres dans la définition de certains espaces d'interpolation, Ricerche Mat. 12 (1963), 248–261.

[21] Sur une classe d'espaces d'interpolation, Inst. Hautes Etudes Sci. Publ. Math. 19 (1964), 5–68 (coauthor: J.-L. Lions).

[22] Sur la théorie des semi-groupes distributions, Séminaire sur les Équations aux dérivées partielles (1963–1964), II, Collège de France, Paris, 1964, 76–99.

[23] Espaces d'interpolation, généralisations, applications, Rend. Sem. Mat. Fis. Milano 34 (1964), 133–164.

[24] On an interpolation theorem of Foias and Lions, Acta Sci. Math. (Szeged) 25 (1964), 255–261.

[25] On the theory of interpolation spaces, Rev. Un. Mat. Argentina 23 (1966/1967), 49–66.

[26] Remark on eigenfunction expansions for elliptic operators with constant coefficients, Math. Scand. 15 (1964), 83–92.

[27] Some remarks on continuous orthogonal expansions and eigenfunction expansions for positive self-adjoint elliptic operators with variable coefficients, Math. Scand. 17 (1965), 56–64.

[28] Relations entre deux méthodes d'interpolation, Inst. Hautes Etudes Sci. Publ. Math. 29 (1966), 49–53.

[29] Some estimates for spectral functions, Math. Z. 92 (1966), 146–153.

[30] Pointwise convergence of singular convolution integrals, Ann. Scuola Norm. Sup. Pisa 20 (1966), 45–61.

[31] Espaces d'interpolation et théoreme de Soboleff, Ann. Inst. Fourier 16 (1966), 279–317.

[32] On convolution operators leaving $L^{p,\lambda}$ spaces invariant, Ann. Mat. Pura Appl. 72 (1966), 295–304.

[33] Applications de la théorie des espaces d'interpolation dans l'Analyse Harmonique, Ricerche Mat. 15 (1966), 3–36.

[34] On interpolation functions, Acta Sci. Math. (Szeged) 27 (1966), 167–171.

[35] Applications de la théorie des espaces d'interpolation aux développements orthogonaux, Rend. Sem. Mat. Univ. Padova 37 (1967), 133–145.

[36] Estimates for eigenfunctions, Studia Math. 28 (1967), 169–182.

[37] Absolute convergence of eigenfunction expansions, Math. Ann. 169 (1967), 307–314.

[38] Introduction to Hilbert space methods in partial differential equations, Monografia No. 1, Instituto Central de Matematica, Universidade de Brasilia, 1965, 100 pp.

[39] On interpolation of L^p spaces with weight functions, Acta Sci. Math. (Szeged) 28 (1967), 61–69.

[40] Sur les espaces de Besov, C. R. Acad. Sci. Paris 264 (1967), 281–283.

[41] A theory of interpolation of normed spaces, Notas de Matematica No. 39, Rio de Janeiro, 1968, 88 pp.

[42] On interpolation functions II, Acta Sci. Math. (Szeged) 29 (1968), 91–92.

[43] A formula for the solution of Dirichlet's problem, in: Proc. Third Brazilian Math. Colloq. (1961), Summa Brasil. Math. 5 (1962), 195–201.

[44] Sur la positivité de la fonction de Green, Math. Scand. 21 (1967), 80–89 (coauthor: I. A. Rus).

[45] ε-entropie, ε-capacité et espaces d'interpolation, Ricerche Math. 17 (1968), 216–220.

[46] On the rate of convergence for discrete initial-value problems, Math. Scand. 21 (1967), 159–176 (coauthor: V. Thomée).

[47] On certain functionals arising in the theory of interpolation spaces, J. Funct. Anal. 4 (1969), 88–94 (coauthor: T. Holmstedt).

[48] On the theory of $\mathcal{L}_{p\,\lambda}$-spaces, J. Funct. Anal. 4 (1969), 71–87.

[49] Approximation theorems connected with generalized translations, Math. Ann. 181 (1969), 255–268 (coauthor: J. Löfström).

[50] Exact interpolation theorems for Lipschitz continuous functions, Ricerche Mat. 18 (1969), 239–259.

[51] Sur la transformation de Fourier des fonctions à valeurs vectorielles, Rend. Sem. Mat. Univ. Padova 42 (1969), 15–26.

[52] On interpolation functions III, Acta Sci. Math. (Szeged) 30 (1969), 235–239.

[53] A new approach in interpolation spaces, Studia Math. 34 (1970), 23–42.

[54] On the value of a distribution at a point, Portugal. Math. 27 (1968), 149–159.

[55] Approximation of norms, J. Approx. Theory 3 (1970), 243–260.

[56] Concave majorants of positive functions, Acta Math. Acad. Sci. Hung. 21 (1970), 327–333.

[57] Zur Interpolation von Operatorenräumen, Arch. Math. (Basel) 21 (1970), 601–608.

[58] Interpolationsräume, in: Elliptische Differentialgleichungen (G. Anger, ed.), Band I, Kolloq. Berlin 1969, Schriftenreihe Inst. Math. Reihe A: Reine Mathematik. Heft 7. Akademie-Verlag, Berlin, 1970, 93–101.

[59] Some classes of operators in Banach space, in: Elliptische Differentialgleichungen (G. Anger, ed.), Band I, Kolloq. Berlin 1969, Schriftenreihe Inst. Math. Reihe A: Reine Mathematik. Heft 7. Akademie-Verlag, Berlin, 1970, 103–109.

[60] Non-commutative interpolation, Matematiche 25 (1970), 159–173.

[61] Interpolation of Lipschitz operators and metric spaces, Mathematica (Cluj) 12 (1970), 325–334.

[62] Interpolation functors and Banach couples, in: Actes du congrès International des Mathématiciens (Nice, 1970), Tome 2, Gauthiers-Villars, Paris, 1971, 373–378.

[63] On the connection between the theory of interpolation spaces and approximation theory, in: Constructive Theory of Functions, Proc. Int. Conf. (Budapest 1969), 1972, 351–363.

[64] Approximation of linear operators, in: Constructive Theory of Functions, Proc. Int. Conf. (Varna 1970), Publishing House of the Bulgarian Academy of Sciences, Sofia, 1972, 245–263.

[65] Sur l'utilisation des suites inconditionellement sommables dans la théorie des espaces d'interpolation, Rend. Sem. Mat. Padova 46 (1971), 173–190.

[66] The best constant in some inequalities involving L_q norms of derivatives, Ricerche Mat. 21 (1972), 176–183.

[67] Interpolation of normed Abelian groups, Ann. Mat. Pura Appl. 92 (1972), 217–262 (coauthor: G. Sparr).

[68] Remarques sur les espaces de Besov. Le cas $0 < p < 1$, C. R. Acad. Sci. Paris 277 (1973), 947–949.

[69] The Weyl transform and Laguerre polynomials, Matematiche 27 (1972), 301–323.

[70] Remark on the dual of an interpolation space, Math. Scand. 34 (1974), 124–128.

[71] Über den Durchschnitt von Interpolationsräumen, Arch. Math. (Basel) 25 (1974), 511–513.

[72] A remark on generalized translation operators (Delsarte-Levitan operators), Mathematica (Cluj) 15 (1973), 241–255.

[73] A remark on Sobolev spaces. The case $0 < p < 1$, J. Approx. Theory 13 (1975), 218–228.

[74] Classes de Hardy sur les variétés, C. R. Acad. Sci. Paris 280 (1975), 439–441.

[75] On the spaces V_p $(0 < p \leq \infty)$, Boll. Un. Mat. Ital. 10 (1974), 632–648 (coauthor: J. Bergh).

[76] On spaces of Triebel-Lizorkin type, Ark. Mat. 13 (1975), 123–130.

[77] On the trace of potentials, Ann. Scuola Norm. Sup. Pisa 2 (1975), 33–43.

[78] F-modular spaces, Funct. Approx. Comment. Math. 1 (1974), 67–73 (coauthor: J. Musielak).

[79] Interpolation and non-commutative integration, Ann. Mat. Pura Appl. 104 (1975), 187–207 (coauthor: G. Sparr).

[80] New Thoughts on Besov spaces, Duke University Mathematics Series 1, Mathematics Department, Duke University, Durham, 1976, 304 pp.

[81] Interpolation of Orlicz spaces, Studia Math. 60 (1977), 33–59 (coauthor: J. Gustavsson).

[82] The Euler derivative. An intrinsic approach to the Calculus of Variations, Math. Scand. 42 (1978), 313–333.

[83] La derivée d'Euler, le complexe de Lagrange et les points proches d'André Weil, C. R. Acad. Sci. Paris 286 (1978), 1241–1242.

[84] Two observations on a theorem by Coifman, Studia Math. 64 (1979), 191–194.

[85] A counter-example connected with Gagliardo's trace theorem, Special issue dedicted to Professor W. Orlicz on the occasion of his seventy-fifth birthday, Comment. Math. Special Issue 2 (1979), 277–282.

[86] Two new interpolation methods based on the duality map, Acta Math. 143 (1979), 73–91.

[87] On Hadamard's variational formula, J. Differ. Equations 36 (1980), 335–346.

[88] Abstract K and J spaces, J. Math. Pures Appl. 60 (1981), 1–49 (coauthor: M. Cwikel).

[89] A class of kernels related to the inequalities of Beckner and Nelson, in: A tribute to Åke Pleijel, Uppsala, 1980, 171–210.

[90] Locally analytically pseudo-convex topological vector spaces, Studia Math. 73 (1982), 253–262.

[91] Properties of the L function, Studia Math. 74 (1982), 105–121 (coauthor: J. Gustavsson).

[92] Recent progress in real interpolation, in: Methods of Functional Analysis and Theory of Elliptic Equations (D. G. Liguori, ed.), Proc. Int. Meeting (Naples, 1982), dedicated to the memory of Carlo Miranda, 1983, 231–263.

[93] Hankel operators, rational approximation and allied questions of analysis, in: Second Edmonton Conference on Approximation Theory (Z. Ditzian et al., eds.), CMS Conf. Proc. 3, Amer. Math. Soc., Providence, RI, 1983, 287–332.

[94] Notes on Wolff's note on interpolation spaces, Proc. London Math. Soc. 48 (1984), 283–299 (coauthors: S. Janson, P. Nilsson and M. Zafran).

[95] On the generalized Hardy's inequality of McGehee, Pigno and Smith and the problem of interpolation between BMO and a Besov space, Math. Scand. 54 (1984), 221–241 (coauthor: E. Svensson).

[96] A reproducing kernel, Boll. Un. Mat. Ital. 3A (1984), 373–382.

[97] The theory of interpolation spaces – its origin, prospects for the future, in: Interpolation Spaces and Allied Topics in Analysis (M. Cwikel et al., eds.), Proc. Int. Conf. (Lund 1983), Lecture Notes in Math. 1070, Springer-Verlag, Berlin–Heidelberg–New York 1984, 1–9.

[98] Harmonic interpolation, in: Interpolation Spaces and Allied Topics in Analysis (M. Cwikel et al., eds.), Proc. Int. Conf. (Lund 1983), Lecture Notes in Math. 1070, Springer-Verlag, Berlin–Heidelberg–New York 1984, 92–124 (coauthor: S. Janson).

[99] Higher order commutators of singular integral operators, in: Interpolation Spaces and Allied Topics in Analysis (M. Cwikel et al., eds.), Proc. Int. Conf. (Lund 1983), Lecture Notes in Math. 1070, Springer-Verlag, Berlin–Heidelberg–New York 1984, 125–142 (coauthor: S. Janson).

[100] Duality for Fernandez type spaces, Math. Nachr. 119 (1984), 231–238.

[101] Invariant function spaces connected with the holomorphic discrete series, in: Anniversary volume on approximation and functional analysis (P. L. Butzer et al., eds.), Oberwolfach, 1983, Birkhäuser, Basel–Boston–Stuttgart, 1984, 119–134.

[102] On Asplund's averaging method – the interpolation (function) way, in: Constructive Theory of Functions, Proc. Int. Conf. (Varna, 1984), Publishing House of the Bulgarian Academy of Sciences, 1984, 664–671.

[103] On the action of Hankel and Toeplitz operators on some function spaces, Duke Math. J. 51 (1984), 937–958 (coauthors: S. Janson and S. Semmes).

[104] Paracommutators and minimal spaces, in: Operators and function theory (S. C. Power, ed.), Proc. Int. Conf. (Lancaster, 1984), NATO Adv. Sci. Inst. Ser. C Math. Phys. Sci. 153, Reidel, Dordrecht, 1985, 163–224.

[105] On the K functional between L^1 and L^2 and some other K functionals, J. Approx. Theory 85 (1986), 322–327 (coauthor: P. Nilsson).

[106] Means and their iterations, in: Proceedings of the Nineteenth Nordic Congress of Mathematicians (Jon R. Stefanson, ed.), Reykjavík, 1984, Icelandic Math. Soc., Science Inst., University of Iceland, Reykjavík, 1985, 191–212 (coauthors: J. Arazy, T. Claesson and S. Janson).

[107] Some unsolved problems, in: Alfred Haar Memorial Conference, Vol. I, II, Proc. Int. Conf. (Budapest,1985), Colloq. Math. Soc. Janos Bolyai 49, North-Holland, Amsterdam, 1987, 711–735.

[108] Möbius invariant function spaces, J. Reine Angew. Math. 363 (1985), 110–145 (coauthors: J. Arazy and S. Fisher).

[109] Invariant function spaces and Hankel operators – a rapid survey, Exposition. Math. 5 (1987), 3–16.

[110] A new generalization of Hankel operators (the case of higher weights), Math. Nachr. 132 (1987), 313–328 (coauthor: S. Janson).

[111] Generalizations of Hankel operators, in: Nonlinear analysis, function spaces and applications, Vol. 3, Proc. Int. Spring School (Litomyšl, 1986), Teubner-Texte zur Mathematik 93, Teubner, Leipzig, 1986, 85–108.

[112] Hankel operators in weighted Bergman spaces, Amer. J. Math. 110 (1988), 989–1053 (coauthors: J. Arazy and S. Fisher).

[113] Marcel Riesz in Lund, in: Function Spaces and Applications (M. Cwikel et al., eds.), Proc. Int. Conf. (Lund, 1986), Lecture Notes in Math. 1302, Springer-Verlag, Berlin–Heidelberg–New York–London–Paris–Tokyo, 1988, 1–10.

[114] The classes V_a are monotone, in: Function Spaces and Applications (M. Cwikel et al., eds.), Proc. Int. Conf. (Lund, 1986), Lecture Notes in Math. 1302, Springer-Verlag, Berlin–Heidelberg–New York–London–Paris–Tokyo, 1988, 153–157 (coauthor: L. Bondesson).

[115] Besov norms of rational functions, in: Function Spaces and Applications (M. Cwikel et al., eds.),, Proc. Int. Conf. (Lund, 1986), Lecture Notes in Math. 1302, Springer-Verlag, Berlin–Heidelberg–New York–London–Paris–Tokyo, 1988, 125–129 (coauthors: J. Arazy and S. Fisher).

[116] Hankel forms and the Fock space, Rev. Mat. Iberoamericana 3 (1987), 61–138 (coauthors: S. Janson and R. Rochberg).

[117] Rational approximation – analysis of the work of Pekarskiĭ, Rocky Mountain J. Math. 19 (1989), 313–333 (coauthor: J. Karlsson).

[118] Paracommutators – boundedness and Schatten-von Neumann properties, Trans. Amer. Math. Soc. 305 (1988), 467–504 (coauthor: S. Janson).

[119] Reproducing formulae for holomorphic tensor fields, Boll. Un. Mat. Ital. (7) 2-B (1988), 345–359.

[120] Some calculations related to Fock space and the Shale-Weil representation, Integral Equations Operator Theory 12 (1989), 67–81.

[121] Hankel forms on multiply connected domains I. The case of connectivity two, Complex Variables Theory Appl. 10 (1988), 123–139.

[122] Möbius invariant spaces of analytic functions, in: Complex analysis I (C. A. Berenstein, ed.), Proc. Spec. Year (College Park, Md., 1985–86), Lecture Notes in Math. 1275, Springer-Verlag, Berlin–Heidelberg–New York–London–Paris–Tokyo, 1987, 10–22 (coauthors: J. Arazy and S. Fisher).

[123] Weak factorization in periodic Fock space, Math. Nachr. 146 (1990), 159–165 (coauthor: S. Janson).

[124] Hankel forms on Fock space modulo C_N, Resultate Math. 14 (1988), 333–339.

[125] Hankel operators on planar domains, Constr. Approx. 6 (1990), 113–138 (coauthors: J. Arazy and S. Fisher).

[126] A multidimensional Wolff's theorem, Studia Math. 94 (1989), 273–290 (coauthor: F. Cobos).

[127] A new look on Hankel forms over Fock space, Studia Math. 95 (1989), 33–41 (coauthors: S. Janson and R. Wallstén).

[128] Hankel forms on multiply connected domains II. The case of higher connectivity, Complex Variables Theory Appl. 13 (1990), 239–250 (coauthor: B. Gustafsson).

[129] Notes on projective structures on complex manifolds, Nagoya Math. J. 116 (1989), 63–88 (coauthor: B. Gustafsson).

[130] Big Hankel operators of higher weight, Rend. Circ. Mat. Palermo 3 (1989), 65–78 (coauthors: J. Boman and S. Janson).

[131] Möbius invariant differential operators on Riemann surfaces, in: Function spaces, differential operators and non-linear analysis (L. Päivärinta, ed.), Sodankylä, 1988, Pitman Res. Notes Math. Ser. 211, Longman Sci. Tech. Harlow, 1989, 14–75 (coauthor: B. Gustafsson).

[132] A general Beckenbach's inequality with applications, in: Function spaces, differential operators and non-linear analysis (L. Päivärinta, ed.), Sodankylä, 1988, Pitman Res. Notes Math. Ser. 211, Longman Sci. Tech. Harlow, 1989, 125–139 (coauthor: L. E. Persson).

[133] Generalizing the arithmetic geometric mean – a hapless computer experiment, Internat. J. Math. Math. Sci. 12 (1989), 235–245.

[134] The Berezin transform and Ha-plitz operators, J. Operator Theory 24 (1990), 165–186.

[135] A submultiplicative function, Nederl. Akad. Wetensch. Indag. Math. 51 (1989), 435–442 (coauthors: J. Gustavsson and L. Maligranda).

[136] Paracommutators – brief introduction, open problems, Rev. Mat. Univ. Complut. Madrid 2 (1989), No. Suppl. 201–211.

[137] On an algorithm considered by Stieltjes, J. Math. Anal. Appl. 150 (1990), 481–493 (coauthor: T. Claesson).

[138] An identity for reproducing kernels in planar domains and Hilbert-Schmidt Hankel operators, J. Reine Angew. Math. 406 (1990), 179–199 (coauthors: J. Arazy, S. Fisher and S. Janson).

[139] Membership of Hankel operators in the ball in unitary ideals, J. London Math. Soc. 43 (1991), 485–508 (coauthors: J. Arazy, S. Fisher and S. Janson).

[140] Interpolation of compactness using Aronszajn-Gagliardo functors, Israel J. Math. 68 (1989), 220–240 (coauthor: F. Cobos).

[141] Fourier analysis of a space of Hilbert-Schmidt operators – new Ha-plitz type operators, Publ. Mat. 34 (1990), 181–197.

[142] The Fock bundle, in: Analysis and partial differential equations (Cora Sadosky, ed.), A collection of papers dedicated to Mischa Cotlar, Lecture Notes in Pure and Appl. Math. 122, Marcel Dekker, New York and Basel, 1990, 301–326.

[143] Projective structures on an annulus and Hankel forms, Glasgow Math. J. 33 (1991), 247–266 (coauthor: G. Zhang).

[144] Some observations on algorithms of the Gauss-Borchardt type, Proc. Edinburgh Math. Soc. 34 (1991), 415–431.

[145] Orthogonal polynomials arising in connection with Hankel forms of higher weight, Bull. Sci. Math. (2) 116 (1992), 265–284.

[146] A quasi-linearizable pair connected with 3-line theorems, Complex Variables Theory Appl. 19 (1992), 143–149 (coauthor: P. Sjölin).

[147] Three-line theorems and Clifford analysis, Complex Variables Theory Appl. 19 (1992), 151–163 (coauthor: P. Sjölin).

[148] On the S_4-norm of a Hankel form, Rev. Mat. Iberoamericana 8 (1992), 121–130.

[149] Interpolation of compact operators: the multidimensional case, Proc. London Math. Soc. 63 (1991), 371–400 (coauthor: F. Cobos).

[150] Schatten-von Neumann classes of multilinear forms, Duke Math. J. 65 (1992), 121–156 (coauthors: F. Cobos and T. Kühn).

[151] The periodic Fock bundle, Algebra i Analiz 3(1991), 134–154; English transl. in: St. Petersburg Math. J. 3 (1992), 1069–1088.

[152] Möbius invariant function spaces – the case of hyperbolic space, Proc. Roy. Irish Acad. Sect. 92A (1992), 243–265.

[153] Reproducing formulae for monogenic functions, Bull. Soc. Math. Belg. Ser. B 44 (1992), 171–192.

[154] Möbius covariance of iterated Dirac operators, J. Austral. Math. Soc. Ser. A 56 (1994), 403–414 (coauthor: T. Qian).

[155] Harmonic analysis on the quantized Riemann sphere, Internat. J. Math. Math. Sci. 16 (1993), 225–243 (coauthor: G. Zhang).

[156] Comparison of two kinds of Hankel operators, Rend. Circ. Mat. Palermo 42 (1993), 181–194.

[157] The Bellaterra connection, Publ. Mat. 37 (1993), 165–176.

[158] The differential equation $y'^p - y^p = \pm 1$ $(p > 0)$, Ricerche Mat. 43 (1994), 91–128.

[159] Hankel forms of arbitrary weight over a symmetric domain via the transvectant, Rocky Mountain J. Math. 24 (1994), 1065–1085.

[160] Hankel kernels of higher weight for the ball, Nagoya Math. J. 130 (1993), 183–192.

[161] Outline of a scientific bibliography of E. Meissel (1826–1895), Historia Math. 22 (1995), 154–178.

[162] A reproducing formula for vector valued functions over a Cartan domain, Eesti Tead. Akad. Toimetised füüs. Mat. 42 (1993), 213–221.

[163] Correspondence principle for the quantized annulus, Romanovski polynomials and Morse potential, J. Funct. Anal. 117 (1993), 377–400.

[164] ϵ-entropy, ϵ-rate and interpolation spaces revisited with an application to linear communication channels, J. Math. Anal. Appl. 186 (1994), 265–276 (coauthors: T. Koski and L. E. Persson).

[165] A weighted Plancherel formula III. The case of the hyperbolic matrix ball, Collect. Math. 43 (1992), 273–301 (coauthor: G. Zhang).

[166] On the correspondence principle for the quantized annulus, Math. Scand. 78 (1996), 183–206 (coauthor: M. Engliš).

[167] Generalized Fock spaces, interpolation, multipliers, circle geometry, Rev. Mat. Iberoamericano 12 (1996), 63–110 (coauthors: S. Thangavelu and N.-O. Wallin).

[168] Higher order Hankel forms, in: Multivariable operator theory (R. Curto et al., eds.), Proc. Int. Conf. (Seattle, WA, 1993), Contemp. Math. 185, Amer. Math. Soc., Providence, RI, 1995, 283–306 (coauthor: R. Rochberg).

[169] Some real interpolation methods for families of Banach spaces. A comparison, J. Approx. Theory 89 (1997), 26–57 (coauthors: M. Carro, L. I. Nikolova and L. E. Persson).

[170] A Green's function for the annulus, Ann. Mat. Pura Appl. 171 (1996), 313–377 (coauthor: M. Engliš).

[171] Covariant differential operators and Green's functions, Ann. Polon. Math. 66 (1997), 77–103 (coauthor: M. Engliš).

[172] Covariant Cauchy-Riemann operators and higher Laplacians on Kähler manifolds, J. Reine Angew. Math. 478 (1996), 17–56 (coauthor: M. Engliš).

[173] On the connection between real and complex interpolation of quasi-Banach spaces, Bull. Sci. Math. 122 (1998), 17–37 (coauthos: F. Cobos and L. E. Persson).

[174] Some compactness results in real interpolation for families of Banach spaces, J. London Math. Soc. 58 (1998), 451–466 (coauthor: M. Carro).

[175] On $\mathcal{S}_p$-classes of trilinear forms, J. London Math. Soc. 59 (1999), 1003–1022 (coauthors: F. Cobos and T. Kühn).

[176] On a number theoretic sum considered by Meissel – a historical observation, Nieuw Arch. Wiskd. 15 (1997), 175–179 (coauthor: P. Lindqvist).

[177] On the remainder in a series of Mertens, Exposition. Math. 15 (1997), 467–478 (coauthor: P. Lindqvist).

[178] Invariant Cauchy-Riemann operators and relative discrete series of line bundles over the unit ball of C^d, Michigan Math. J. 45 (1998), 387–397 (coauthor: G. Zhang).

[179] Green's functions for powers of the invariant Laplacian, Canad. J. Math. 50 (1998), 40–73 (coauthor: M. Engliš).

[180] Extreme points of the complex binary trilinear ball, Studia Math. 138 (2000), 81–92 (coauthors: F. Cobos and T. Kühn).

[181] Remarks on symmetries of trilinear forms, Revista R. Acad. Cienc. Fisicas Nat. Madr. 94 (2000), 441–449 (coauthors: F. Cobos and T. Kühn) .

[182] A new kind of Hankel-Toeplitz type operator connected with the complementary series, Arab. J. Math. Sci. 6 (2000), 49–80 (coauthors: M. Engliš, S. C. Hille, H. Rosengren and G. Zhang).

[183] Two remarkable identities, called twos, for inverses to some Abelian integrals, Am. Math. Monthly 108 (2001), 403–410 (coauthor: P. Lindqvist).

[184] Singular values of trilinear forms, Experimental Math. 10 (2001), 509–517 (coauthor: B. Bernhardsson) .

[185] Généralisation de la méthode d'Héron pour toute racine simple d'une équation non linéaire, Ann. Sci. Math. Que. 25 (2001), 91–99.

[186] On the development of interpolation – instead of a history three letters, in: Function Spaces, Interpolation Theory and Related Topics (A. Kufner, M. Cwikel, M. Engliš, L.-E. Persson, G. Sparr, eds.), Proc. Conf. in honour of Jaak Peetre on the occasion of his 65th birthday (Lund, 2000), Walter de Gruyter, Berlin, 2002, 41–50.

[187] Green functions and eigenfunction expansions for the square of the Laplace-Beltrami operator on plane domains, Ann. Mat. Pura Appl., to appear (coauthor: M. Engliš).

II. Translations and editorial work

[188] Translation from the Russian original: B. S. Mityagin, An interpolation theorem for modular spaces, in: Interpolation Spaces and Allied Topics in Analysis (M. Cwikel et al., eds.), Proc. Int. Conf. (Lund 1983), Lecture Notes in Math. 1070, Springer-Verlag, Berlin–Heidelberg–New York 1984, 10–23 (the Russian original appeared in Mat. Sb. 66 (1965), 472–482).

[189] Translation from the Russian original: N. K. Nikol'skiĭ, Treatise on the shift operator, Spectral function theory, Grundlehren Math. Wiss. 273, Springer-Verlag, Berlin, 1986, 491 pp.

[190] Translation from the Russian original: Several complex variables III. Geometric function theory (G. M. Khenkin, ed.), Encyclopaedia Math. Sci. 9, Springer-Verlag, Berlin, 1988, 261 pp.

[191] Translation from the Russian original: Analysis III. Spaces of differentiable functions (S. M. Nikolskiĭ, ed.), Encyclopaedia Math. Sci. 26, Springer-Verlag, Berlin, 1991, 221 pp.

[192] Translation from the Russian original: Commutative harmonic analysis IV. Harmonic analysis in R^n (N. K. Nikolskiĭ, V. P. Khavin, eds.),, Encyclopaedia Math. Sci. 42, Springer-Verlag, Berlin, 1992, 228 pp.

[193] Translation to Swedish of the paper: Arnold, V.I., A mathematical trivium, Normat 47 (1999), 111–121 (the Russian original appeared in Uspekhi Mat. Nauk. 46 (1991), 225–232).

Editorial work etc.

[194] Editor of the book: V. I. Ovchinnikov, The Method of Orbits in Interpolation Theory, Harwood Academic Publ., 1984.

[195] Editor of the book: Interpolation Spaces and Allied Topics in Analysis, Proceedings of the international conference held in Lund August 1983, Lecture Notes in Math. 1070, Springer-Verlag, Berlin–Heidelberg –New York–Tokyo, 1984 (coeditor: M. Cwikel).

[196] Editor of the book: Function Spaces and Applications, Proceedings of the U.S.-Swedish Seminar held in Lund June 1986, Lecture Notes in Math. 1302, Springer-Verlag, Berlin–Heidelberg–New York–London–Paris–Tokyo, 1988 (coeditors: M. Cwikel, Y. Sagher and H. Wallin).

[197] Preface of the book: Yu. A. Brudnyi and N. Ya. Krugljak, Interpolation Functors and Interpolation spaces, Vol. I, North-Holland Math. Library, 47, North-Holland Publ. Co., Amsterdam, 1991.

[198] Editor of the book: Edgar Krahn 1894–1961. A centenary volume. IOS Press, Amsterdam; a copublication with the Estonian Mathematical Society, Tartu, 1994 (coeditor: Ü. Lumiste).

[199] * Editor of: Index of the Gösta Mittag-Leffler separate collection. Part I, Small boxes, University of Lund, 1996 (coeditor: T. Claeson).

[200] * Editor of: Index of the Gösta Mittag-Leffler separate collection. Big boxes. Part IIA (subjects), University of Lund, 2000.

III. Miscellanea

[201] Fermat and Mordell conjectures and a sensation: Falting's theorem, Elementa 68 (1985), 24–26 (in Swedish).

[202] Rational and spline-approximation – a breakthrough, Normat 34 (1986), 80–82 (in Swedish).

[203] Fast tunnels II, Normat 42 (1994), 13–18 (in Swedish).

[204] Faber Krahni võrratuse ja selle ühe autori lugu: Edgar Krahni 100. sünniaasta-päevaks (The story of the Faber-Krahn inequality and of its author, marking Edgar Krahn's 100th birthday), in: Koolimatemaatika XXI. Tartu Ülikool, Tartu, 1994, 10–13 (coauthors: Ü. Lumiste and E. Tamme).

[205] A remarkable differential equation, Normat 43 (1995), 182–183 (in Swedish).

[206] The Estonian mathematician Edgar Krahn, The Baltic Eye (1996), 10–14.

[207] Erik Lundberg and the hypergeometric functions, Normat 45 (1997), 1–24 (in Swedish) (coauthor: P. Lindqvist).

[208] On the Eve of a new millenium – was Julius Caesar a mathematician, Normat 47 (1999), 145–151 (in Swedish).

[209] * Ernst Meissel and the Pythagorean problem – the Drei-Körper-Problem in the Nachlass Meissel, Typoscript, 1997.

[210] * On the structure of bounded trilinear forms, Typoscript, 1997 (coauthors: F. Cobos and T. Kühn).

[211] p-arclength of the q-circle, Technical Report, Lund, 2000, submitted (coauthor: P. Lindqvist).

[212] * Erik Lundberg, On hypergoniometric functions of complex variables, Translation by Jaak Peetre with the assistance of Julia and Peter Lindqvist.

[213] * On Fourier's discovery of Fourier series and Fourier integrals.

[214] * On the equation $x^3 - 3x + 1 = 0$ and other similar equations.

[215] * In defence of Rabbi Levi ben Gerson – reply to Opinion 36 by Doran Zeilberger.

[216] * Vad är matematisk sanning? – ett kåseri, [Swedish] (What is mathermatical thruth? – A causerie).

[217] * On binary trilinear forms – the case of quaternions.

[218] * J. Plonck, A remark on parallel addition.

[219] * Göran Dillner, Min levnadsbeskrivning, tillğnad min familj (My autobiography, devoted to my family), [in Swedish] (Revised and commented by Jaak Peetre).

[220] * A brief survey of invariants of trilinear forms.

[221] * On a certain invariant of degree 6.

[222] * More on invariants of ternary trilinear forms.

[223] * On matrices with isotropic eigenvectors.

[224] * What is known about square free quadratic forms.

[225] * Remarques sur les espaces de traces.

[226] * E. Meissel, Fragment über die Riemannsche Xi-Funktion (Übersetz und herausgegeben von J. Peetre).

[227] * Milleniar equation and curves of finite species – outline of a program.

[228] * A new approach to bounded trilinear forms – forms that peak at several places.

[229] * Milleniar equation revisited.

[230] * More general nonsense – a general formulation of singular value problem for trilinear forms.

Research reports etc. (up to 1988)

[231] A duality theorem in the theory of differential operators, Manuscript, ca. 1960.

[232] Interpolation av Banach-rum, Mimeographed lecture notes, Lund, 1962 (coauthor: A. Persson).

[233] Intermediate spaces and approximation, (Preliminary version), Mimeographed, Lund, ca. 1964.

[234] On an equivalence theorem of Taibleson, Mimeographed, Lund, ca. 1965.

[235] On a fundamental lemma in the theory of interpolation spaces, Mimeographed, Lund, ca. 1969.

[236] Integration in quasi-Banach space, Technical report, Lund, ca. 1970.

[237] The Peano existence theorem under weaker assumptions, Mimeographed, Lund, 1971 (coauthor: J. Persson).

[238] Multiplier theorems connected with generalized translations, Technical report, Lund, 1971 (coauthor: L. Vretare) .

[239] Estimates for the modulus of continuity, Technical report, Lund, 1971.

[240] Banach couples, I. Elementary theory, Technical report, Lund, 1971.

[241] The Bolzman equation (the physical background), Mimeographed, Lund, ca. 1971.

[242] Analysis in quasi-Banach space and approximation theory (preliminary version), Technical report, Lund, 1972.

[243] Kombinatorisk teori, Mimeographed lecture notes, Lund, 1973.

[244] H_p spaces, Mimeographed lectures notes, Lund, 1974.

[245] Bounded operators in $L_p, 0 < p < 1$, Technical report, Lund, 1975.

[246] The trace of a Besov space – a limiting case, Technical report, Lund, 1975.

[247] Algebra, Mimeographed lecture notes, 2nd edition, Lund, ca. 1975.

[248] Inledning till värdefördelningsläran (Nevanlinna teori), Mimeographed, Lund, ca. 1975.

[249] The Fermi connection, Technical report, Lund, 1976.

[250] Hilberts 23-e problem, Mimeographed, Lund, ca. 1976.

[251] Some open problems and questions realted to interpolation spaces, Mimeographed, Lund, ca. 1976.

[252] On É. Cartan's approach to Finsler spaces, Technical report, Lund, 1977.

[253] Dissections, connections, and sprays, Technical report, Lund, 1977.

[254] On abstract differential operators, Technical report, Lund, 1977.

[255] On the ring of germs of differentiable functions, Technical report, Lund, 1977.

[256] Om divergens, Mimeographed, Lund, ca. 1977.

[257] Om Killing-vektorer, Mimeographed, Lund, ca. 1977.

[258] Om vektorfält i $\mathcal{TM}$ och om konnektioner, Mimeographed, Lund, ca. 1977.

[259] On the general theory of lifts, Manuscript, ca. 1977.

[260] On the infinitesimal automorphims of a G-structure, Technical report, Lund, 1978.

[261] Pre-symplectic bundles as models for classical particles, Technical report, Lund, 1978.

[262] On imbedding a Kähler manifold in a Hilbert space, Technical report, Lund, 1978.

[263] Kategorier och algebraisk systemteori, Lecture notes, Lund, 1976.

[264] Sannolikhetslärans gränsvärdessatser med rest, behandlade med K-funktionalen, Mimeographed, Lund, ca. 1979.

[265] An intrinsic volume element on complex manifolds, Mimeographed, Lund, ca. 1979.

[266] Further comments on the Euler derivative, Technical report, Lund, 1980.

[267] Investigation of the power series $\sum_{n=1}^{\infty} z^{p^n}$ and some allied series, Technical report, Lund, 1980 (coauthor: A. Grgic) .

[268] On the action of the Mehler transform on some special functions, Technical report, Lund, 1980.

[269] Remarks on Ovcinnikov's theorem, Abstracts, 19th Scand. Congr. Math., Aarhus, Aug. 18–22, 1980. Various Publication Series No. 33, Matematisk institut, Aarhus universitet, 1980.

[270] Problems of hypoellipticity in differential geometry, Abstracts, 19th Scand. Congr. Math., Aarhus, Aug. 18–22, 1980. Various Publication Series No. 33, Matematisk institut, Aarhus universitet, 1980.

[271] Generalizing Ovcinnikov's theorem, Technical report, Lund, 1981.

[272] H^∞ and complex interpolation, Technical report, Lund, 1981.

[273] Van der Waerden's conjecture and hyperbolicity, Technical report, Lund, 1981.

[274] Hankel operators, weak factorizations and Hardy's inequality in Bergman classes, Technical report, Lund, 1982.

[275] Complex section theory, a generalization of complex function theory, Technical report, Lund, 1982.

[276] Investigation of some special Fourier integrals, Manuscript (fragment), Lund, 1982 (coauthor: J. Gustavsson).

[277] Paracommutators and invariant function spaces, Abstracts, XIXth Nordic Congress of Mathematicians, Reykjavik, Aug. 13–17, 1984. Raunivísindastofun Háskólans, Reykjavik, 1984.

[278] On generalized majorization, Technical report, Lund, 1985.

[279] Minikurs i Hankeloperatorer, Lund, 1986.

[280] Complex extrapolation, Manuscript (fragment), 1986 (coauthors: M. Cwikel and M. Milman).

[281] Hankel forms on Fock space modulo C_N, Technical report, Lund, 1987.

[282] Hankel forms over line bundles and vector bundles, Technical report, Lund, 1987.

[283] Recent progress in Hankel forms, Technical report, Lund, 1987.

[284] Hankel forms in weaker assumptions, Technical report, Lund, 1987.

[285] Weak factorization in periodic Fock space, Technical report, Lund, 1987 (coauthor: S. Janson).

[286] A quasi linearizable pair, Technical report, Uppsala, 1987 (coauthor: P. Sjölin) .

[287] A submultiplicative function, Technical report, Stockholm, 1987 (coauthor: L. Maligranda) .

[288] Notes on projective structures on complex manifolds, Technical report, Stockholm, 1988 (coauthor: B. Gustafsson).

[289] Big Hankel operators of higher weights, Technical report, Stockholm, 1988 (coauthors: J. Boman and S. Janson).

On the development of interpolation – instead of a history three letters

Edited and/or translated by Jaak Peetre

Abstract. This compilation should be viewed as a kind of prolegomena to a history of interpolation. Three letters from Mischa Cotlar, Antoni Zygmund and Olof Thorin from c. 1980 are reproduced. Thorin's letter is given in translation. These letters were written as a reply to an inquiry by the editor/translator. It is remarkable that while Thorin spent his whole professional life as an actuary, neither of his two publications in interpolation were covered by *Mathematical Reviews* nor by the *Zentralblatt für Mathematik.* On the other hand, Cotlar, in his letter, gives a disclaimer to a rumor that he had rediscovered Marcinkiewicz's interpolation theorem while studying under Zygmund in Chicago in the mid 1950s.

The editor/translator makes an Appeal to all Readers who might have further information about the founding years of the theory of interpolation spaces, *i.e.* the years just before or after 1960, to send him similar letters.

2000 Mathematics Subject Classification: 33A99; 53A99.

Introduction

By interpolation we shall mean here *modern* interpolation, *i.e.* the theory of interpolation spaces. Most authorities agree that this theory came into existence around 1960 and, among the "founding fathers", the names Alberto Calderón, Emilio Gagliardo, Jacques-Louis Lions and Selim Gregorovič Kreĭn are often mentioned.

Classical interpolation was highlighted by two fundamental theorems, the Riesz–Thorin theorem [Ri27], [Th39], [Th48] (with later complements by Stein and Weiss [S56], [SW57], [SW58]) and Marcinkiewicz's theorem [Ma39][1]. The motivation for them came from problems in harmonic analysis (Hausdorff–Young theorem etc.). Moreover, *non-linear* interpolation was considered already by Banach (see [Scot], p. 161, along with the comments by Maligranda there).

On the other hand, in the theory of interpolation spaces, at least in the early stage, the prime sources of inspiration were a number of questions related to (elliptic) PDE, especially the trace problem.

[1] Marcinkiewicz was, probably, the first to use the word "interpolation" in its present sense; Riesz and Thorin spoke of "convexity theorems", not interpolation theorems.

It is not clear why one should study the history of mathematics at all. The history of mathematics developed into a vigorous discipline in its own right in the mid 19th century but most research was devoted to antiquity. In the 20th century this changed and nowadays much work is devoted to foundational questions, how this or that concept developed historically (the notion of function, uniform convergence etc.). However, it is equally interesting to study the history of an individual branch. Let us mention such a magnificent work as Dieudonné's history of algebraic topology [D], or Lützen's book on distributions [Lü].

Generally speaking, there are several ways to approach the study of the history of mathematics. Either one can base one's investigation on written sources only. Or, in the case of living persons, one can make interviews. For instance, [Lü] is to a considerable extent based on interviews with the one main figure, Laurent Schwartz. But there are *a priori* no reasons that one method should be better than the other.

In the case of interpolation there are special reasons for studying its history. One reason is that there are some misconceptions about how this discipline came into being. Moreover it is a comparatively narrow field from a "sociological" point of view (*e.g.* how many people are involved[2]; etc.). In particular, there seems to be numerous misunderstandings regarding the early steps of development. For instance, there has been a sort of rumor that Mischa Cotlar rediscovered the theorem of Marcinkiewicz. A denial of this will be issued below (Section 3). Regarding this theorem itself, there are also some uncertainties. Marcinkiewicz announced his theorem without proof in a note [Ma39] immediately before the outbreak of the war in September, 1939. He died soon after (Katyn, spring 1940) and his result did not arouse any attention. Its importance was realized only much later by Zygmund, who gave it a sort of revival in [Z56].[3] Also Thorin's rôle in the Riesz–Thorin theorem does not seem to be clear to the general public.[4]

As for me, I have been interested in the history of mathematics all my life; however now that I am older this interest has gradually become more and more pronounced – indeed, it is almost an obsession. Already around 1980 I had the idea of writing a history of interpolation, or at least its history up to c. 1960. I decided to follow the second method just indicated and wrote letters to each of the three main characters involved, to wit Olof Thorin, Antoni Zygmund, and Mischa Cotlar. I got prompt answers from all of them. I do not recall what were the reasons or circumstances which prevented me from continuing with this project at that time. In particular, it is unforgivable that I did not follow Zygmund's suggestions to write to Calderón; today it is, regretfully, too late.

---

[2]According to one authority "an area of math[ematics] isn't important until at least a hundred people are working at it" [H], p. 204.

[3]Concerning the life of Marcinkiewicz, see [Z64].

[4]Thorin's two interpolation papers [Th39] and [Th48] have not been reviewed in *Mathematical Reviews* nor in *Zentralblatt für Mathematik*. The fact that there are only reviews of his 12 papers in probability from the period 1968–1982, gives the false impression that he was a probabilist only.

Regarding Thorin's proof of the convexity theorem, Littlewood [L], p. 20 speaks of it as *the most impudent idea in mathematics*. *Cf.* also a laudatory mention in the preface of [K].

Now, twenty years later, when I return to these matters, it is with a much more modest objective. I want only to make these three letters public, hoping in this way to encourage others to continue with this investigation from where I stopped.

Appeal. I hereby invite all Readers who might have further information about the founding years of the theory of interpolation spaces, i. e. the years just before or after 1960, to send me similar letters.

The present compilation should therefore be viewed at most as a prolegomena to a genuine history of interpolation. As for myself, at some stage I might write something about my early contacts with Gagliardo and Lions.

Acknowledgment. I would like to thank Mischa Cotlar and Olof Thorin for their kind permission to publish their letters; the latter also pointed out several obscurities in my manuscript.

1. Letter from O. Thorin dated Dec. 20, 1979 (translation by J. P.)

Dear Jaak,

Many thanks for your letter from Sept. 28 along with two attached papers[5] "On two interpolation methods related to Marcel Riesz's proof" and "On Barry Simon's new interpolation method and Marcel Riesz's proof". In your letter you ask if I have some information to give on my own contribution to the "Riesz–Thorin theorem" and if I have heard anything from Marcel Riesz (MR) about what could have influenced him etc.

Please, excuse that I have delayed my answer but so much time has passed after the events in question so I have needed some time to go over the material that I have stored away in my hiding places. I do not possess any first hand information regarding what did influence MR in his work on the convexity theorem. I do not recall that he ever should have given me any confidences in this direction (and I probably never tried to press him in this respect, for sure I was too shy to do so). Therefore I shall, in what follows, communicate my own involvement to the extent that I understand it. In order not to make the presentation too chaotic I feel that I have to give, in parallel, as a frame of reference, a quick sketch of my own educational development. Of course, I realize that what I say cannot be of any greater interest for you, but in this way it will be easier for me to present things which are perhaps more interesting.

I was born in 1912, and by and by I found my way to the secondary school in Halmstad where, only in the gymnasium, I became excited about mathematics (teacher: O. I. Holmqvist). I matriculated in 1929 and began my studies at Lund University in the fall of the same year with a career as a teacher in mind. As the combination mathematics, physics and chemistry seemed to be the most immediate in view of my expected future career as a teacher, following what was customary

[5] *Editor's Note.*Preliminary versions of my paper [Pe86], originally planned as a joint work with Simon.

I ought to have chosen these topics in the inverse order. Eager as I was to begin with the study of mathematics, and with a certain aversion towards the laboratory subjects, I first threw myself into mathematics. After some time I learnt that the insurance companies (in the first place the life insurance companies) hired people with mathematical training. Therefore I thought that I could skip physics and chemistry with a clear conscience. So after four years I had graduated with the degree of Candidate of philosophy (filosofie kandidat; abbrev.: fil.-kand.) in mathematics (N. O. Zeilon); mechanics (V. W. Ekman); and mathematical statistics (an exemption subject with S. D. Wicksell as examiner).

In the fall of 1933 I began my studies in earnest for the degree of Licentiate ((filosofie) licentiat; abbrev.: fil.-lic.) in mathematics under MR. Already since the fall of 1931 (during which term I had been examined by Zeilon in the course for the mark three in the fil.-kand.-examination) I had participated in the higher mathematical seminar (directed by MR) and already in April 1932 I had delivered a talk on "the Hilbert–Hellinger theory of quadratic forms in infinitely many variables". I continued the direction indicated there in two talks about various transformations in Hilbert space in November–December 1933. I believe that two then rather recent papers by J. von Neumann and M. H. Stone had attracted my interest. Quite naively I thought that I would spontaneously discover something in this area as a foundation for my thesis. However, it turned out that I was not capable of such spontaneity in reasonable time.

Therefore I called upon MR in the latter half of the spring term of 1934 in order to obtain a suitable subject. After a short introductory dialogue MR showed me his – then quite unknown [to me] – paper [Ri27] and pointed out several points that ought to be further penetrated, in particular the following two. In the first place MR asked for a multilinear counterpart of his bilinear theorem. In the second place he demanded a counterexample showing that the convexity theorem could <u>not</u> hold in the whole square even if the variables were allowed to be complex. Indeed, in the last paragraph of the paper he had designated such an extension of the convexity theorem as "peu probable" (hardly probable). This was also the opinion that he expressed during our conversation. He mentioned also that Zygmund had told him in Zürich about a paper with a new proof of the convexity theorem. However, he did not recall the details so he advised me to write to Zygmund in order to get the precise information which I also did in due time. Zygmund, who worked in Vilnius in the then Polish part of Lithuania, sent a most polite reply on June 18, 1934 where he told that he knew only about R. E. A. C. Paley's two papers [Pa31] and [Pa33]. I quote the following from Zygmund's postcard: "Curiously enough the author himself thought that the proof given in the first of the papers was not complete, whereas it was 'overcomplete' and should have stopped at formula 4.1 ".[6]

During the summer holidays of 1934 I devoted myself to the multilinear case and in the fall I could show MR a result to the effect that convexity holds in the region $\alpha_1 \leq 1, \ldots, \alpha_n \leq 1, \alpha_1 + \cdots + \alpha_n \geq n - 1$. The method of proof closely

[6]A facsimile of this postcard is on page 43.

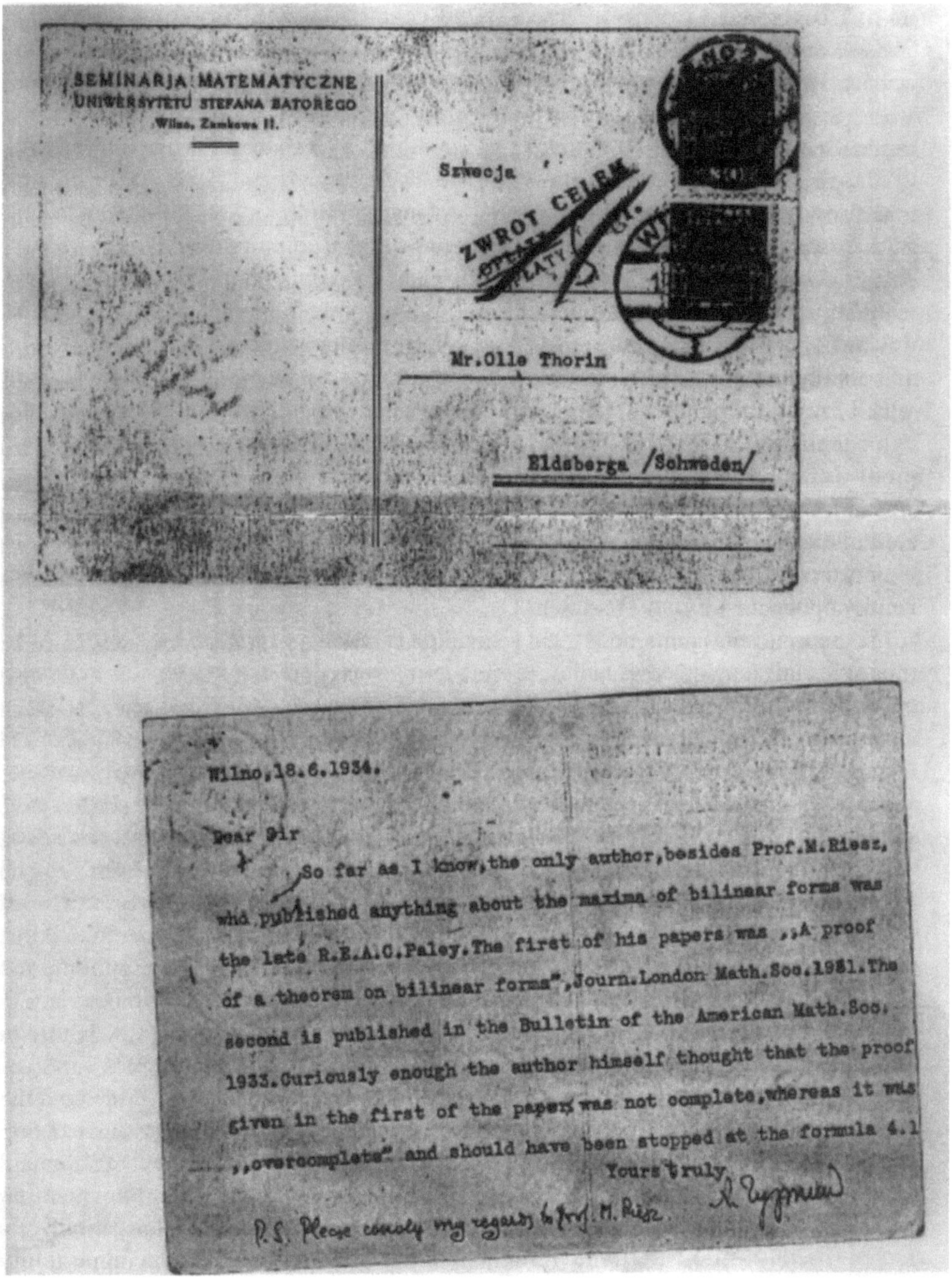

SEMINARJA MATEMATYCZNE
UNIWERSYTETU STEFANA BATOREGO
Wilno, Zamkowa 11.

Szwecja

ZWROT CELNEJ OPŁATY

Mr.Olle Thorin

Eldsberga /Schweden/

Wilno,18.6.1934.

Dear Sir

So far as I know,the only author,besides Prof.M.Riesz,
who published anything about the maxima of bilinear forms was
the late R.E.A.C.Paley.The first of his papers was ,,A proof
of a theorem on bilinear forms",Journ.London Math.Soc.1931.The
second is published in the Bulletin of the American Math.Soc.
1933.Curiously enough the author himself thought that the proof
given in the first of the paper was not complete,whereas it was
,,overcomplete" and should have been stopped at the formula 4.1
Yours truly,
A. Zygmund

P.S. Please convey my regards to Prof. M. Riesz.

Facsimile of a postcard from Zygmund to Thorin

followed MR's own proof in the case $n = 2$. MR was naturally moderately impressed and so I continued working. The expected counterexample turned out to have a ghostlike ability to slip out of the hands of the researcher. In April 1935 I had the opportunity of presenting my multilinear result in a seminar talk. In connection with an application for a position in an insurance company, MR even wrote in my examination book in June 1935 that I had submitted a Licentiate thesis with the title "On the maxima of multilinear forms". However, he emphasized to me that he did not make any commitments thereby regarding the mark. While looking for the expected counterexample I told myself that a direct approach based on traditional differential calculus ought to be tried (in contrast to a sophisticated manipulation with Hölder's inequality). This not very profound idea eventually turned out to lead to the result. In this way I obtained, in the course of 1935, a partial result to which I wish to refer a little and possibly remove it from the dust. Namely I found a new proof of MR's theorem which I myself thought was superior to Paley's and which I also thought could stand as an alternative to MR's own proof. It likewise worked in the multilinear case. This proof was published only in my Doctoral dissertation [Th48] in 1948, namely in the Appendix dealing with the real case (Theorem A1). There it appears in the multilinear case in the context of a somewhat complicated apparatus of notation. I presume that the only person ever to read it was the second opponent Carl Hyltén-Cavallius (the faculty opponent was Otto Frostman).

In the spring and summer of 1936 I was called to military service in Stockholm as an office assistant (statistics assistant). As this activity was not overly burdensome I rented a room in town together with two comrades and so I was able to continue to some extent my differential calculus approach. Upon my return to Lund in the fall of 1936 I managed to find a proof by the method indicated of the convexity theorem in the whole square when the variables were allowed to vary in the complex domain. I delivered a seminar talk about this in Oct. 1936. In this talk I concentrated on the bilinear case. While walking to the post seminar one of the listeners (Otto Frostman himself[7]) made "A casual remark" to the effect that the proof ought to work for analytic functions in general. Otto never returned later to this remark. MR, to whom I mentioned this incident afterwards, wanted to minimize its importance and was of the opinion that anybody could have said such a thing. As a consequence no "acknowledgement"[8] was made to Otto. However, precisely this remark made my tardy mind slowly turn in such a direction that I began to treat maxima of a general function of complex variables under such subsidiary conditions as MR had in his theorem. However, I devoted the spring term of 1937 to studying the literature course in the Licentiate exam and to being examined, which led to the degree of fil. lic. (3 marks) in May, 1937. After that I applied for a job in the life insurance business and on Sep. 1 1937 I got an employment on the recommendation of MR (I remained in this trade, however with various employers and in various positions, until my retirement in 1977). Therefore I had no opportunity to devote myself to preparing a publication until some time in the first half of 1938

[7] *Translator's Note.* In English in the original.

[8] *Translator's Note.* In English in the original.

when I handed MR a manuscript which, after slight modifications, became my paper [Th39], printed on Oct. 29, 1938. This paper contains a very general result. However, I think that, from the point of view of proof, it is not at par with what I obtained later and which I published in [Th48] where already in the introduction there is a proof of MR's theorem in the complex case which later won considerable popularity. I did not discover this last proof until 1942. It was not published then and I concentrated on writing a more extensive memoir with interesting applications. In the meantime Tamarkin and Zygmund published their note [TZ44] containing similar simplifications which I had found already in 1942. Because of the special conditions prevailing then in connection with the World War, I only learned later about their contribution. In any case I received a reprint from Zygmund only in Feb. 1947, which induced me to write to Zygmund on June 20, 1947 with thanks for the reprint, indicating my related results. As far as I know there was never any dispute concerning priority. Some time in 1945 I discovered that G. Valiron [V], in 1923, had generalized the Hadamard three circle theorem to several dimensions using a similar method. Had MR known this paper and realized the connection with the complex case, he could have been able to settle the complex case also and so made my intervention superfluous! It was thus my great luck that this was not the case.

Concerning the now so popular proof of the complex version of MR's theorem I believe that, although one can say that the embryo of it can be found in [Th39], the mature plant has to be located only in [Th48]. Therefore I think that your description in the Introduction to "On two interpolation methods related to Marcel Riesz's proof" – even if I realize that the presentation must be brief – gives a Reader, who has not studied the original papers, a far too flattering view of my achievement.

If you should want clarifications on any other points or some amplifications to the above already overly voluminous account, I am of course willing to try to give you, to the extent I am capable, the information desired.

Please, receive my best wishes for Christmas and the New Year.

Sincerely yours

Olof Thorin
Falkvägen 19
183 52 TÄBY

2. Letter from A. Zygmund dated Chicago, Mar. 22, 1980

Dear Prof. Peetre,

Thank you for your letter dated, March 7, 80. I am happy that you are planning a historical article about the origins of the Riesz–Thorin theorem and I am looking forward to its appearance. Unfortunately, I myself could not contribute much more. I was always full of admiration for that theorem and consider it one of the most beautiful result[s] of Analysis, but I myself never obtained anything interesting in that direction. Perhaps my main contribution here was educating Marcinkiewicz and, as you know, [I] found a different method of proof. He was my student (prior to 1940) at the University of Wilno, then in Poland, and he found a novel (purely 'real variable') approach to the theory, an approach which can supplement (but of course never replace) the Riesz–Thorin ideas.

I showed your letter to Prof. Calderón, and I have the feeling that if you wrote him he might supply you with some facts here. Possibly he can even suggest new problems which could be at least as interesting. He is a very friendly person, and it might be useful to develop correspondence with you (and perhaps some other younger people) on the topic.

I wish you (and your work) every success here.

Sincerely A. Zygmund

P.S. Prof. Calderón just (this morning) left for Argentina and will not be back before Oct. 1, but he suggested that I might supply you with his address there. It is Prof. A. P. Calderón, Malabia 2791, Buenos Aires 1425, Argentina.

3. Letter from M. Cotlar dated Caracas, Apr. 4, 1982

Dear Prof. Peetre,

I was very happy to receive a letter from you (only that our postman delivered it as usual with much delay). It is good to know that you are planning to write a survey on interpolation theory: a survey by a most competent person, and one of the creators of the field will be an important event indeed.

I feel ashamed that an uncareful expression in a footnote, on page 65 of my 1955 paper [C2], could cause some misunderstanding. I always wanted to clear up that vague expression but didn't find an occasion for it, and am glad that now you grant me a chance to do it. A few weeks before starting my thesis under Zygmund, and without knowing Marcinkiewicz's theorem, I was trying to get some generalized forms of Riesz's interpolation theorem (with the aim to generalize the ergodic theorems), *but without succeeding [in getting] concrete statements nor proofs.*[9] A few weeks later Zygmund started a short Seminar where I learned about Marcinkiewicz's theorem and Zygmund's proof, and then I realized that what I was trying to get was already done. However my first attempts, before attending Zygmund's Seminar, contained some other ideas of a different type, and on returning to Argentina I decided to detail them in the parts 2 and 4 [C2], [C4] my 1955 papers in the *Revista Cuyana* [C1-4], as well as an article with M. Bruschi [BC], quoted by Zygmund in his book and in his biography of Marcinkiewicz [10]. But, again before knowing about Marcinkiewicz–Zygmund's theorem I was unable to give a concrete form nor proofs to those first

[9] *Editor's Note.* Author's emphasis.

[10] *Editor's Note.* Probably [Z64] is intended.

vague intentions. And of course it was a mistake to mention these things in my paper. Since I am talking of my less than modest work, I want to express my gratitude, and how greatly I am touched, by the generosity and benevolence shown to me by such an eminent and distinguished mathematician as you. I only regret that health problems didn't allow me to be more worthy of such kindness.

My wife Yanny asks me to give her kindest regards to you and your children Mikaela, Jakob and Benjamin. With all best wishes for your work and your personal happiness I remain as ever

Mischa Cotlar

P.S. Under separate cover I am sending you a reprint of a survey article of some results concerning generalized Toeplitz kernels, obtained in collaboration with Cora Sadosky and Rodrigo Arocena. I will be grateful to receive any preprints of your works or lecture notes.

References

[BC] M. Bruschi, M. Cotlar, On the convexity theorems of Riesz-Thorin and Marcinkiewicz, Rev. Union Mat Argent. 18 (1957), 162–172.

[C1] M. Cotlar, A combinatorial inequality and its applications to L^2-spaces, Rev. Mat. Cuyana 2 (1955), 41–55.

[C2] M. Cotlar, A general interpolation theorem for linear equations, Rev. Mat. Cuyana 2 (1955), 57–84.

[C3] M. Cotlar, Some generalizations of the Hardy-Littlewood maximal theorem, Rev. Mat. Cuyana 2 (1955), 85–104.

[C4] M. Cotlar, A unified theory of Hilbert transforms and ergodic theorems, Rev. Mat. Cuyana 2 (1955), 105–167.

[D] Jean Dieudonné, A history of algebraic and differential topology 1900-1960, Birkhäuser, Boston–Basel, 1989.

[H] P. Hoffman, The man who loved only numbers. The story of Paul Erdös and the search for mathematical truth, Fourth Estate, London, 1998.

[K] P. Koosis, Introduction to H_p spaces, London Math. Soc. Lecture Notes Ser. 40, Cambridge University Press, Cambridge, 1980.

[L] J. E. Littlewood, A mathematician's miscellany, Methuen, London, 1953.

[Lü] J. Lützen, The prehistory of the theory of distributions, Stud. Hist. Math. Phys. Sci. 7, Springer-Verlag, New York–Berlin, 1982.

[Ma39] Józef Marcinkiewicz, Sur l'interpolation d'opérations, C. R. Acad. Sci. Paris 208 (1939), 1272-1273, in: [Ma64], 539–540.

[Ma64] Józef Marcinkiewicz, Collected papers, Pańatowe Wydawnictwo Naukowe, Warszawa, 1964.

[Pa31] R. E. A. C. Paley, A proof of a theorem on bilinear forms, J. London Math. Soc. 6 (1931), 226–30.

[Pa33] R. E. A. C. Paley, A note on bilinear forms, Bull. Am. Math. Soc. 39 (1933), 259–260.

[Pe86] J. Peetre, Two new interpolation methods based on the duality map, Acta Math. 143 (1979), 73–91.

[Ri27] M. Riesz, Sur les maxima des formes bilinéaires et sur les fonctionnelles linéaires, Acta Math. 49 (1927), 465–497.

[Scot] The Scottish book. Mathematics from the Scottish café (R. D. Mauldin, ed.), Birkhäuser, Boston–Basel–Stuttgart, 1981.

[S56] E. Stein, Interpolation of linear operators, Trans. Amer. Math. Soc. 83 (1956), 482–492.

[SW57] E. Stein, G. Weiss, On the interpolation of analytic families of operators acting on H^p-spaces, Tohoku Math. J. (2) Ser. 9 (1957), 318–339.

[SW58] E. Stein, G. Weiss, Interpolation of operators with change of measure, Trans. Amer. Math. Soc. 87 (1958), 159-172.

[TZ44] J.-D. Tamarkin, A. Zygmund, Proof of a theorem of Thorin, Bull. Amer. Math. Soc. 50, (1944). 27–282.

[Th39] O. Thorin, An extension of the convexity theorem of M. Riesz, Kungl. Fysiogr. Sällskp. Lund Förh. 8 (1939), 166–170.

[Th48] O. Thorin, Convexity theorems, Thesis. Med. Lunds Univ. Mat. Sem. 9 (1948), 1–58.

[V] G. Valiron, Sur un théorème de M. Hadamard, Bull. Sci. Math. 47 (1923), 177–192.

[Z35] A. Zygmund, Trigonometric series. 1935.

[Z56] A. Zygmund, On a theorem of Marcinkiewicz concerning interpolation of operations, J. Math. Pures Appl. 35 (1956), 223–248.

[Z59] A. Zygmund, Trigonometric series II, Cambridge University Press, Cambridge, 1959.

[Z64] A. Zygmund, Józef Marcinkiewicz, in: [Ma64], 1–30.

Jaak Peetre, Centre for Mathematical Sciences, Lund University, Box 118, S-221 00 Lund, Sweden

e-mail: `jaak@maths.lth.se`

Remarks on reproducing kernels of some function spaces

J.-L. Lions

Dedicated to Jaak Peetre with all my friendship

Abstract. Reproducing kernels of Hilbert spaces of continuous functions (cf. N. Aronszajn [1] and the bibliography therein) can be introduced as the solution of problems of the Calculus of Variations. From the optimality system of these problems (i.e. Euler equations and Lagrange multipliers) one can deduce a strategy for computing these kernels. We obtain in this way explicit formulae for reproducing kernels of families of Hilbert spaces.

2000 Mathematics Subject Classification: 46E22.

Contents

1. Introduction

Let $\mathcal{H}$ be a Hilbert space of functions defined on a set $X \subset \mathbb{R}^n$ (to fix ideas), such that

$$\mathcal{H} \subset C(X) = \text{space of continuous functions on } X. \tag{1.1}$$

The elements of $\mathcal{H}$ can be real valued or complex valued. We assume here that they are real valued.

Because of (1.1), if b is given in X, then

$$u \to u(b)$$

is a continuous linear form on $\mathcal{H}$, so that there exists a unique element $K(b) \in \mathcal{H}$ such that

$$u(b) = (K(b), u)_{\mathcal{H}} \quad \forall u \in \mathcal{H}, \tag{1.2}$$

where $(\varphi, \psi)_{\mathcal{H}}$ denotes the scalar product on $\mathcal{H}$.

Let $x \to K(x, b)$ be the function $K(b) \in \mathcal{H}$. The *kernel* $K(x, b)$ *is the reproducing kernel* of $\mathcal{H}$.

This definition is due to N. Aronszajn [1] who studied general properties of reproducing kernels.

In particular cases, i.e., for *specific spaces* $\mathcal{H}$, such notions have been introduced by S. Bergman [2], G. Szegö [11], M. Schiffer [9], S. Zaremba [12], where the corresponding reproducing kernels are computed and estimated; cf. N. Aronszajn, loc. cit., and P. Garabedian [3].

There is a systematic way to (try to) compute explicitly the kernel $K(b)$, in the following fashion.

Let us consider the problem

$$\inf_u \|u\|^2_{\mathcal{H}}, \quad (K(b), u)_{\mathcal{H}} = 1, \quad (\text{or } u(b) = 1). \tag{1.3}$$

Let w_0 *be the solution of* (1.3), i.e. the projection of the origin on the linear variety $(K(b), u)_{\mathcal{H}} = 1$. Then

$$w_0 = \frac{K(b)}{\|K(b)\|^2_{\mathcal{H}}} \tag{1.4}$$

which (essentially) gives $K(b)$ if w_0 is known.

One uses next *methods of Calculus of Variations for solving* (1.3).

This is the strategy which has been followed in the Author [5] for a specific function space of harmonic functions.

In the present paper we give a formula for a whole family of Hilbert spaces of harmonic functions (cf. Section 2), whose proof is given in Section 3 following the above strategy. Further remarks and comparison with previous results are given in Section 4.

2. Reproducing kernels of the spaces $\mathcal{H}^s(\Omega)$

Let Ω be a bounded open set of $\mathbb{R}^n$, with smooth boundary Γ. Let $\mathcal{D}'_{\mathcal{H}}(\Omega)$ be the space of distributions u on Ω, such that

$$\Delta u = 0 \quad \text{in } \Omega. \tag{2.1}$$

If Γ is real analytic, there exists a continuous linear mapping

$$\left|\begin{array}{l} u \to \gamma u \\ \text{from } \mathcal{D}'_{\mathcal{H}}(\Omega) \to \mathcal{H}'(\Gamma) \end{array}\right. \tag{2.2}$$

where $\mathcal{H}'(\Gamma)$ denotes the space of analytic functionals on Γ (cf. J.-L. Lions and E. Magenes [7], Chapter 7, Section 3.2.), such that

$$\gamma u = \text{trace of } u \text{ on } \Gamma \text{ if } u \text{ is continuous in } \overline{\Omega}. \tag{2.3}$$

One then introduces the spaces $H^S(\Gamma)$. They are the spaces of functions on Γ whose all tangential derivatives (taken on Γ) of order $\leq S$ are in $L^2(\Gamma)$, a definition which makes sense only for S = a positive integer. But let Δ_Γ be the Laplace–Beltrami operator on Γ. We set

$$\Lambda = 1 - \Delta_\Gamma \quad \text{defined as an unbounded operator in } L^2(\Gamma) \tag{2.4}$$

and then Λ^S makes sense $\forall S \in \mathbb{R}$. Then we define

$$H^S(\Gamma) = \text{Domain of } \Lambda^S \tag{2.5}$$

cf. J.-L. Lions and E. Magenes [6], Chapter 1, Section 7.3. These spaces are simple *interpolation spaces*, a very particular case of spaces introduced in J. Peetre and the Author [8].

We then introduce

$$\mathcal{H}^S(\Omega) = \{u \mid u \in \mathcal{D}'(\Omega),\ \Delta u = 0 \text{ in } \Omega,\ \gamma u \in H^S(\Gamma)\}. \tag{2.6}$$

For $u, v \in \mathcal{H}^S(\Omega)$, we introduce the scalar product

$$(u, v)_{\mathcal{H}^S(\Omega)} = (\gamma u, \gamma v)_{\mathcal{H}^S(\Gamma)} = (\Lambda^S \gamma u, \gamma v)_\Gamma = (\gamma u, \Lambda^S \gamma v)_\Gamma. \tag{2.7}$$

Provided with (2.7), $\mathcal{H}^S(\Omega)$ *is a Hilbert space.*

Remark 2.1. When Γ is real analytic – or even only C^∞ – the definitions (2.6) make sense $\forall s \in \mathbb{R}$. If Γ is, say, m times continuously diferentiable then $\mathcal{H}^S(\Omega)$ makes sense for $|S| \leq m$. □

Remark 2.2. In (2.7), $(\Lambda^S \gamma u, \gamma v)_\Gamma$ denotes the scalar product between $\Lambda^S \gamma u \in H^{-S}(\Gamma)$ and $\gamma v \in H^S(\Gamma)$. □

Remark 2.3. One has (with the notations of J.-L. Lions, E. Magenes [6])

$$\mathcal{H}^S(\Omega) = \{u \mid u \in \mathcal{D}'(\Omega),\ \Delta u = 0,\ u \in H^{S+1/2}(\Omega)\} \tag{2.8}$$

and one obtains the *same* Hilbert space (up *to an equivalence of scalar product*) by taking

$$[u, v]_{\mathcal{H}^S(\Omega)} = (u, v)_{H^{S+1/2}}(\Omega). \tag{2.9}$$

But different scalar products lead *to different reproducing kernels.* □

Of course one has

$$\mathcal{H}^S(\Omega) \subset C(\Omega) = \text{space of continuous functions on } \Omega. \tag{2.10}$$

Therefore $\mathcal{H}^S(\Omega)$ *has a reproducing kernel.* Given $b \in \Omega$, the form

$$u \to u(b)$$

is continuous on $\mathcal{H}^S(\Omega)$, so that *there exists a unique element* $K(b) \in \mathcal{H}^S(\Omega)$ *such that*

$$u(b) = (K_S(b), u)_{\mathcal{H}^S(\Omega)} \quad \forall\, u \in \mathcal{H}^S(\Omega). \tag{2.11}$$

For $b \in \Omega$, $K_S(b)$ is a function, or a distribution that we still write as a function,

$$K_S(b) : x \to K_S(x, b). \tag{2.12}$$

For every $S \in \mathbb{R}$, $K_S(x, b)$ *is the reproducing kernel of* $\mathcal{H}^S(\Omega)$. We are going to prove in the next section *the formula*

$$K_S(x, b) = \int_\Gamma \Lambda_y^{-S/2} \frac{\partial G}{\partial n_y}(y, x) \Lambda_y^{-S/2} \frac{\partial G}{\partial n_y}(y, b)\, d\Gamma_y \tag{2.13}$$

where $G(x, b)$ denotes the Green's function of $-\Delta$ in Ω, i.e.

$$\left| \begin{array}{l} -\Delta_x G(x, b) = \delta(x - b) \\ G(x, b) = 0 \quad \text{for } x \in \Gamma. \end{array} \right. \tag{2.14}$$

Remark 2.4. Since the function $y \to \dfrac{\partial G}{\partial n_y}$ (or $\dfrac{\partial G}{\partial n_y}(y, b)$) is C^∞ on Γ, formula (2.13) makes sense. □

We now proceed with the Proof of (2.13).

3. Proof of formula (2.13)

Step 1. Following the strategy indicated in the Introduction we consider the problem

$$\inf_u \|u\|^2_{\mathcal{H}^S(\Omega)}, \quad u(b) = 1. \tag{3.1}$$

Since $u(b) = 1$ is equivalent to $(K_S(b), u)_{\mathcal{H}^S(\Omega)} = 1$ the unique solution w of (3.1) is given by the projection of the origin on the variety $(K_S(b), u)_{\mathcal{H}^S(\Omega)} = 1$, i.e.

$$w = \frac{1}{\|K_S(b)\|^2_{\mathcal{H}^S(\Omega)}} K_S(b) \tag{3.2}$$

or

$$w(x) = \frac{1}{\|K_S(b)\|^2_{\mathcal{H}^S(\Omega)}} K_S(x,b). \tag{3.3}$$

Since $w(b) = 1$ it follows that $K_S(b,b) = \|K_S(b)\|^2_{\mathcal{H}^S(\Omega)}$, so that one can write as well

$$w(x) = \frac{K_S(x,b)}{K_S(b,b)}. \tag{3.4}$$

All these formulas are valid for every value of $S \in \mathbb{R}$.

Step 2. We want now to obtain *an optimality system* for problem (3.1). We penalize the condition $\Delta u = 0$, i.e. we introduce, for $\varepsilon > 0$ "small",

$$J_\varepsilon(u) = \|u\|^2_{\mathcal{H}^S(\Omega)} + \frac{1}{\varepsilon}\,\|\Delta u\|^2_{L^2(\Omega)} \tag{3.5}$$

on the space of functions u such that

$$\Delta u \in L^2(\Omega), \quad \gamma u \in H^S(\Gamma). \tag{3.6}$$

The writing (3.5) is slightly ambiguous since $u \notin \mathcal{H}^S(\Omega)$ in $J_\varepsilon(u)$ (since $\Delta u \neq 0$).

It follows from (3.6) that

$$u \text{ is locally in } H^2(\Omega) \cap H^{S+1/2}(\Omega). \tag{3.7}$$

It follows (by Sobolev's imbedding theorem) that $u \in C(\Omega)$

$$\text{either if } n \leq 3 \quad \text{or if} \quad s > \frac{n-1}{2}. \tag{3.8}$$

Then we can consider the problem

$$\inf J_\varepsilon(u),\ u \text{ subject to (3.6) and to } u(b) = 1. \tag{3.9}$$

Remark 3.1. If (3.8) does not hold true, one approximates the condition $u(b) = 1$ by the fact that averages of u in smaller and smaller balls around b converge to 1 as $\varepsilon \to 0$. Cf. J.-L. Lions [5] (for a different space, but the same technique applies). □

Let w_ε be the unique solution of (3.9). Since

$$\inf J_\varepsilon(u) \leq J_\varepsilon(1) = \|1\|^2_{\mathcal{H}^S(\Omega)}$$

it follows that

$$\|w_\varepsilon\|_{\mathcal{H}^S(\Omega)} \leq C = \text{constant independent of } \varepsilon \tag{3.10}$$

$$\|\Delta w_\varepsilon\|_{L^2(\Omega)} \leq C\sqrt{\varepsilon}. \tag{3.11}$$

Let us define

$$p_\varepsilon = \frac{1}{\varepsilon}\Delta w_\varepsilon. \tag{3.12}$$

The Euler equation is then written

$$(w_\varepsilon, v)_{\mathcal{H}^S(\Omega)} + (p_\varepsilon, \Delta v) = 0 \quad \forall\, v \text{ such that (3.6) is satisfied for } v \text{ and } v(b) = 0. \tag{3.13}$$

If v is arbitrary, then $v - v(b)$ is acceptable in (3.13), and it gives

$$(\Lambda^S w_\varepsilon, v - v(b))_\Gamma + (p_\varepsilon, \Delta v) = 0 \quad \forall\, v. \tag{2.14}$$

Using integrations by parts, it follows that

$$(\Lambda^S w_\varepsilon, v)_\Gamma - v(b)\int_\Gamma \Lambda^S W_\varepsilon \, d\Gamma + \int_\Gamma \left(p_\varepsilon \frac{\partial v}{\partial n} - \frac{\partial p_\varepsilon}{\partial n} v\right) d\Gamma + (\Delta p_\varepsilon, v) = 0$$

i.e.

$$\left| \begin{array}{l} \Delta p_\varepsilon = \left(\int_\Gamma \Lambda^S w_\varepsilon \, d\Gamma\right)\delta(x-b) \text{ in } \Omega \\ p_\varepsilon = 0 \text{ on } \Gamma \end{array} \right. \tag{3.15}$$

and

$$\frac{\partial p_\varepsilon}{\partial n} = \Lambda^S w_\varepsilon \quad \text{on } \Gamma. \tag{3.16}$$

Let us set

$$\int_\Gamma \Lambda^S w_\varepsilon \, d\Gamma = d_\varepsilon. \tag{3.17}$$

We can extract a subsequence, still denoted by w_ε, such that

$$w_\varepsilon \to w \quad \text{in } \mathcal{H}^S(\Omega) \text{ weakly}. \tag{3.18}$$

Then

$$(w_\varepsilon, 1)_{\mathcal{H}^S} = d_\varepsilon \to d = (w, 1)_{\mathcal{H}^S}. \tag{3.19}$$

Using all this information, we can let $\varepsilon \to 0$, and we obtain (the *optimality system*)

$$\left| \begin{array}{l} \Delta w = 0, \;\; w = \Lambda^{-S} \frac{\partial p}{\partial n} \text{ on } \Gamma \\ -\Delta p = -d\delta(x-b), \;\; d = (w, 1)_{\mathcal{H}^S}, \\ p = 0 \;\; \text{on } \Gamma. \end{array} \right. \tag{3.20}$$

This optimality system *characterizes* (3.4). Therefore

$$(w, 1)_{\mathcal{H}^S} = (K_S(b), 1)_{\mathcal{H}^S} \frac{1}{K_S(b, b)} = \frac{1}{K_S(b, b)}. \tag{3.21}$$

Step 3. Proof of formula (2.13). We first obtain from (3.20) that

$$p(x) = -d\, G(x, b). \tag{3.22}$$

By Green's formula we have

$$w(y) = -\int_\Gamma \frac{\partial G}{\partial n_x}(x, y) w(x)\, d\Gamma_x. \tag{3.23}$$

But $w(x) = \Lambda_x^{-S} \dfrac{\partial p}{\partial n_x} = -d\Lambda_x^{-S} \dfrac{\partial G}{\partial n_x}(y, b)$, so that (3.23) gives

$$w(y) = d \int_\Gamma \frac{\partial G}{\partial n_x}(x, y) \Lambda_y^{-S} \frac{\partial G}{\partial n_x}(x, b)\, d\Gamma_x, \tag{3.24}$$

or exchanging x and y

$$w(x) = d \int_\Gamma \frac{\partial G}{\partial n_y}(y, x) \Lambda_y^{-S} \frac{\partial G}{\partial n_y}(y, b)\, d\Gamma_y. \tag{3.25}$$

We now use (3.4) and observe by (3.21) that $d = \dfrac{1}{K_S(b, b)}$. It follows that

$$K_S(x, b) = \int_\Gamma \frac{\partial G}{\partial n_y}(y, x) \Lambda_y^{-S} \frac{\partial G}{\partial n_y}(y, b)\, d\Gamma_y. \tag{3.26}$$

which can be written in the form (2.13) as well. □

4. Further remarks

Remark 4.1. Let us take $s = -\frac{1}{2}$. Then (cf. Remark 2.3)

$$\mathcal{H}^{-1/2}(\Omega) = \{u \mid u \in L^2(\Omega),\ \Delta u = 0\}. \tag{4.1}$$

The Hilbertian norm

$$[u]_{\mathcal{H}^{-1/2}(\Omega)} = \|u\|_{L^2(\Omega)} \tag{4.2}$$

is equivalent to the one given by (2.7) (with $S = -\frac{1}{2}$). But the corresponding reproducing kernel, which is given by

$$u(b) = \int_\Omega R(x, b) u(x)\, dx \tag{4.3}$$

is different from $K_{-1/2}(x, b)$.

One can follow the same strategy as before to compute $R(x, b)$. One finds in this way that $R(x, b)$ can be expressed (cf. J.-L. Lions [5]) in terms of the Green functions for $-\Delta$ and for Δ^2, for the Dirichlet Boundary Conditions. One recovers in this way a formula of S. Zaremba [12]. □

Remark 4.2. One can also define $\mathcal{H}^S(\Omega)$ using the Neumann conditions. It gives

$$\mathcal{H}^S(\Omega) = \left\{u \mid u \in \mathcal{D}'(\Omega),\ \Delta u = 0,\ \frac{\partial u}{\partial n} \in H^{S-1}(\Gamma)\right\}. \tag{4.4}$$

Provided with the scalar product

$$\{u, v\}_{\mathcal{H}^S} = \left(\frac{\partial u}{\partial n}, \frac{\partial v}{\partial n}\right)_{H^{S-1}(\Gamma)}. \tag{4.5}$$

which is still equivalent to (2.7) after taking the quotient by the constants, the Hilbert space $\mathcal{H}^S(\Omega)$ admits *another* family of reproducing kernels

$$u(b) = \{N_S(b), u\}_{\mathcal{H}^S}\ . \tag{4.6}$$

The kernel $N_S(x, b)$ can be expressed in terms of the Green's function for $-\Delta$ this time with the Neumann boundary conditions. □

Remark 4.3. In a domain Ω of $\mathbb{R}^2$ one can consider the complex valued distributions u such that

$$\frac{\partial u}{\partial \bar{z}} = 0 \quad \text{in } \Omega \tag{4.7}$$

(notations of L. Schwartz [10], Vol. 1) and introduce the space

$$\mathcal{S}(\Omega) = \{u \mid \frac{\partial u}{\partial \bar{z}} = 0 \quad \text{in } \Omega, \quad \gamma u \in L^2(\Gamma)\}, \tag{4.8}$$

provided with the scalar product $(\gamma u, \gamma v)_{L^2(\Gamma)}$.

It admits a reproducing kernel

$$u(b) = (S(b), u)_{L^2(\Gamma)} \tag{4.9}$$

cf. N. Aronszajn [1], G. Szegö [11]. The function $S(x, b)$ is the Szegö kernel for Ω.

One can also introduce the Bergman kernel (cf. S. Bergman [2]) as follows. One defines

$$\mathcal{B}(\Omega) = \left\{u \mid \frac{\partial u}{\partial \bar{z}} = 0 \text{ in } \Omega,\ u \in L^2(\Omega)\right\}, \tag{4.10}$$

provided with the scalar product

$$(u, v)_{L^2(\Omega)}.$$

It admits a reproducing kernel, the so-called Bergman kernel:

$$u(b) = (B(b), u) = \int_\Omega B(x, b) u(x)\, dx. \tag{4.11}$$

All these kernels can be computed by the same strategy as above. But we have not been able to recover by this method the results of P. Garabedian [3], which give the connections between $S(x, b)$ and $B(x, b)$. We hope to return on this point. □

Remark 4.4. In all what has been said one can replace the condition

$$\Delta u = 0$$

by

$$Au = 0 \tag{4.12}$$

where A is any *elliptic* operator (with C^∞ coefficients), of *any order*.

For instance one can consider the space

$$\mathcal{K}^S(\Omega) = \left\{u \mid u \in \mathcal{D}'(\Omega),\ \Delta^2 u = 0,\ \gamma u,\ \gamma\frac{\partial u}{\partial n} \in H^S(\Gamma) \times H^{S-1}(\Gamma)\right\}. \tag{4.13}$$

Provided with the scalar product

$$(u, v)_{\mathcal{K}^S(\Omega)} = (\gamma u, \gamma v)_{H^S(\Gamma)} + \left(\gamma\frac{\partial u}{\partial n}, \gamma\frac{\partial v}{\partial n}\right)_{H^{S-1}(\Gamma)} \tag{4.14}$$

it becomes a Hilbert space which admits a reproducing kernel. It can be computed by the same strategy.

A similar remark applies if one replaces (4.12) by a hypoelliptic equation, such as

$$\frac{\partial u}{\partial t} - \Delta u = 0 \quad \text{in} \quad \Omega \times (0, T). \tag{4.15}$$

(Cf. an example in J.-L. Lions, loc. cit.) □

Remark 4.5. Let us give in conclusion an example where there is *no* reproducing kernel. Let us consider smooth functions in $\overline{\Omega}$, such that

$$\Delta u = 0 \quad \text{in } \Omega. \tag{4.16}$$

Let us assume that Γ_0 is a (smooth) subset of $\Gamma = \partial\Omega$. Let us provide the set of functions u with the scalar product

$$C_{\Gamma_0}(u, v) = \int_{\Gamma_0} \left(uv + \frac{\partial u}{\partial n}\frac{\partial v}{\partial n}\right) d\Gamma_0. \tag{4.17}$$

It defines a pre-hilbertian norm $c_{\Gamma_0}(u, u)^{1/2}$, since $c_{\Gamma_0}(u, u) = 0$ implies $u = \frac{\partial u}{\partial n} = 0$ on Γ_0, which implies that $u = 0$ in Ω by a uniqueness theorem.

We can then consider the space $c_{\Gamma_0}(\Omega) =$ the completion of the smooth functions u for the structure (4.17). If $\Gamma_0 \subset \Gamma$ strictly, $c_{\Gamma_0}(\Omega)$ is a space of ultra distributions, whose elements are not necessarily continuous in Ω, so that *there is no reproducing kernel here*. For $u \in c_{\Gamma_0}(\Omega)$ one cannot define $u(b)$, but one can only consider functionals (averages) of u. □

Remark 4.6. The notion of reproducing kernel makes sense in the non-hilbertian case. For instance, if $\mathcal{D}(\Omega)$ denotes the L. Schwartz[10] space of C^∞ functions with compact support in Ω, and if b is given in Ω, $u \to u(b)$ is a continuous linear form on Ω, so that there exists a kernel $R(b)$ such that

$$u(b) = < R(b), u > .$$

with the usual notations

$$u(b) = \int_{\Omega} \delta(x - b) u(x)\, dx,$$

so that $R(b) =$ the Dirac measure $\delta(x - b)$.

If, in other function spaces, every function u admits a Fourier type expansion

$$u = \sum (u, w_j) w_j$$

then the corresponding reproducing kernel is given by $\int_j w_j \otimes w_j$.

Suitable approximations of $\delta(x - b)$ lead to approximation methods for the solutions of partial differential equations. Let us mention in particular the Reproducing Kernel Particle Method; cf. F. Günther et al. [4] and the bibliography therein. □

References

[1] N. Aronszajn, Theory of reproducing kernels, Trans. Amer. Math. Soc. 68 (1) (1950), 337–404.

[2] S. Bergman, Über die Entwicklung der harmonischen Funktionen der Ebene und des Raumes nach Orthogonalfunktionen, Math. Ann. 86 (1922), 238–271.

[3] P. Garabedian, Schwarz's lemma and the Szegö kernel function, Trans. Amer. Math. Soc. 67 (1949), 1–35.

[4] F. Günther, W. K. Liu, D. Diachin and M. A. Christon, Multi scale mesh free parallel computations for viscous, compressible flows, Comp. Math. Appl. Mech. Eng. 190 (2000), 279–303.

[5] J.-L. Lions, Noyau reproduisants et système d'optimalité, in: Aspects of mathematics and its applications (J. A. Barroso, ed.), North-Holland Math. Library 34, Elsevier, Amsterdam, 1986, 573–582.

[6] J.-L. Lions and E. Magenes, Problèmes aux limites non homogènes et applications, Vol. 1, Dunod, Paris, 1968.

[7] J.-L. Lions and E. Magenes, Problèmes aux limites non homogènes et applications, Vol. 3, Dunod, Paris, 1970.

[8] J.-L. Lions and J. Peetre, Sur une classe d'espaces d'interpolation, Inst. Hautes Études Sci. Publ. Math. 19 (1964), 5–68.

[9] M. Schiffer, The kernel function of an orthogonal system, Duke Math. J. 13 (1945), 529–540.

[10] L. Schwartz, Théorie des distributions, t. 1, Hermann, Paris, 1950.

[11] G. Szegö, Über orthogonale Polynome, die zu einer gegebenen Kurve der komplexen Ebene gehören, Math. Z. 9 (1921), 218–270.

[12] S. Zaremba, Sur le calcul numérique des fonctions demandées dans le problème de Dirichlet et le problème hydrodynamique, Bull. Inst. Acad. Sci. Cracovie (1908), 125–195.

An interesting class of operators with unusual Schatten–von Neumann behavior

A. B. Aleksandrov, S. Janson, V. V. Peller and R. Rochberg *

We dedicate this paper to Jaak Peetre on the occasion of his 65th birthday and to the memory of Tom Wolff. Both helped shape the mathematics of our time and profoundly influenced our mathematical thoughts. Each, through his singular humanity, helped our hearts grow.

Abstract. We consider the class of integral operators Q_φ on $L^2(\mathbb{R}_+)$ of the form $(Q_\varphi f)(x) = \int_0^\infty \varphi(\max\{x, y\}) f(y)\, dy$. We discuss necessary and sufficient conditions on φ to insure that Q_φ is bounded, compact, or in the Schatten–von Neumann class S_p, $1 < p < \infty$. We also give necessary and sufficient conditions for Q_φ to be a finite rank operator. However, there is a kind of cut-off at $p = 1$, and for membership in S_p, $0 < p \leq 1$, the situation is more complicated. Although we give various necessary conditions and sufficient conditions relating to $Q_\varphi \in S_p$ in that range, we do not have necessary and sufficient conditions. In the most important case $p = 1$, we have a necessary condition and a sufficient condition, using L^1 and L^2 modulus of continuity, respectively, with a rather small gap in between. A second cut-off occurs at $p = 1/2$: if φ is sufficiently smooth and decays reasonably fast, then Q_φ belongs to the weak Schatten–von Neumann class $S_{1/2,\infty}$, but never to $S_{1/2}$ unless $\varphi = 0$.

We also obtain results for related families of operators acting on $L^2(\mathbb{R})$ and $\ell^2(\mathbb{Z})$.

We further study operations acting on bounded linear operators on $L^2(\mathbb{R}_+)$ related to the class of operators Q_φ. In particular we study Schur multipliers given by functions of the form $\varphi(\max\{x, y\})$ and we study properties of the averaging projection (Hilbert–Schmidt projection) onto the operators of the form Q_φ.

2000 Mathematics Subject Classification: Primary 47G10; Secondary 47B34, 47B10, 34B24, 34L15, 30D55.

*The first author is partially supported by Grant 99-01-00103 of Russian Foundation of Fundamental Studies and by Grant 326.53 of Integration. The third author is partially supported by NSF grant DMS 9970561. The fourth author is partially supported by NSF grant DMS 9970366

Contents

1. Introduction

For a function $\varphi \in L^1_{\text{loc}}(\mathbb{R}_+)$, which means that φ is a locally integrable function on $\mathbb{R}_+ = (0, \infty)$, we define the operator Q_φ on the set of bounded compactly supported functions f in $L^2(\mathbb{R}_+)$ by

$$(Q_\varphi f)(x) = \int_0^\infty \varphi(\max\{x, y\}) f(y)\, dy; \tag{1.1}$$

equivalently,

$$(Q_\varphi f)(x) = \varphi(x) \int_0^x f(y)\, dy + \int_x^\infty \varphi(y) f(y)\, dy. \tag{1.2}$$

We are going to study when Q_φ is (i.e., extends to) a bounded operator in $L^2(\mathbb{R}_+)$, and when this operator is compact, or belongs to Schatten–von Neumann classes $\boldsymbol{S}_p$.

We will also consider the corresponding Volterra operators Q_φ^+ and Q_φ^- defined by

$$(Q_\varphi^+ f)(x) = \varphi(x) \int_0^x f(y)\,dy,$$
$$(Q_\varphi^- f)(x) = \int_x^\infty \varphi(y) f(y)\,dy;$$

thus $Q_\varphi = Q_\varphi^+ + Q_\varphi^-$.

It is straightforward to see (and proved more generally in Theorem 2.4) that if any of these three operators is bounded on $L^2(\mathbb{R}_+)$, then $\int_a^\infty |\varphi|^2 < \infty$ for any $a > 0$, and thus the integrals above converge for any $f \in L^2(\mathbb{R}_+)$ and define all three operators on $L^2(\mathbb{R}_+)$.

We find in §3 simple necessary and sufficient conditions for Q_φ to be bounded or compact, and for $Q_\varphi \in \boldsymbol{S}_p$, $1 < p < \infty$. The conditions are $\varphi \in X_\infty$, $\varphi \in X_\infty^0$ and $\varphi \in X_p$, respectively, where the spaces X_p, X_∞, and X_∞^0 are defined as follows.

Definition. If $0 < p < \infty$, let X_p be the linear space of all measurable functions on $\mathbb{R}_+$ that satisfy the equivalent conditions

$$\sum_{n\in\mathbb{Z}} 2^{np/2}\left(\int_{2^n}^{2^{n+1}} |\varphi(x)|^2\,dx\right)^{p/2} < \infty; \tag{1.3}$$

$$\sum_{n\in\mathbb{Z}} 2^{np/2}\left(\int_{2^n}^{\infty} |\varphi(x)|^2\,dx\right)^{p/2} < \infty; \tag{1.4}$$

$$x^{1/2}\left(\int_x^\infty |\varphi(y)|^2\,dy\right)^{1/2} \in L^p(dx/x). \tag{1.5}$$

Similarly, let X_∞ be the linear space of all measurable functions on $\mathbb{R}_+$ that satisfy the equivalent conditions

$$\sup_{n\in\mathbb{Z}} 2^n\left(\int_{2^n}^{2^{n+1}} |\varphi(x)|^2\,dx\right) < \infty; \tag{1.6}$$

$$\sup_{x>0} x \int_x^\infty |\varphi(y)|^2\,dy < \infty. \tag{1.7}$$

Let X_∞^0 be the subspace of X_∞ consisting of the functions that satisfy the equivalent conditions

$$\lim_{n\to\pm\infty} 2^n\left(\int_{2^n}^{2^{n+1}} |\varphi(x)|^2\,dx\right) = 0; \tag{1.8}$$

$$\lim_{x\to 0} x \int_x^\infty |\varphi(y)|^2\,dy = \lim_{x\to\infty} x \int_x^\infty |\varphi(y)|^2\,dy = 0. \tag{1.9}$$

The equivalence of the different conditions is an exercise. For $1 \le p \le \infty$, X_p is a Banach space with the norm

$$\|\varphi\|_{X_p} = \left\| x^{1/2}\left(\int_x^\infty |\varphi(y)|^2\,dy\right)^{1/2} \right\|_{L^p(dx/x)};$$

for $0 < p < 1$, this is a quasi-norm and X_p is a quasi-Banach space. X_∞^0 is a closed subspace of X_∞, and thus a Banach space too. Note that $X_p \subset X_q$ if $0 < p \le q \le \infty$.

Remark. It is well known [Pee] that $\varphi \in X_p$ if and only if the Fourier transform $\mathcal{F}\varphi$ belongs to the homogeneous Besov space $B_{2\,p}^{1/2}$ (here we identify φ with the function extended to $\mathbb{R}$ by zero on $\mathbb{R}_-$).

Note that the operators Q_φ appear in a natural way in [MV] when studying the boundedness problem for the Sturm–Liouville operator $\mathcal{L}$ from $\overset{\circ}{L}{}_2^1(\mathbb{R}_+) \to L_2^{-1}(\mathbb{R}_+)$ defined by $\mathcal{L}u = -u'' + qu$. To be more precise, Maz'ya and Verbitsky studied in [MV] the problem of identifying potentials q for which the inequality

$$\left|\int_{\mathbb{R}_+} |u(t)|^2 q(t)\,dt\right| \le C \int_{\mathbb{R}_+} |u'(t)|^2\,dt$$

holds for any C^∞ compactly supported function u on $(0, \infty)$. This inequality is in turn equivalent to the boundedness of the quadratic form

$$\left|\int_{\mathbb{R}_+} u(t)\overline{v(t)}q(t)\,dt\right| \le C\|u'\|_{L^2(\mathbb{R}_+)}\|v'\|_{L^2(\mathbb{R}_+)}. \tag{1.10}$$

In [MV] under the assumption that the limit

$$\lim_{y\to\infty} \int_x^y q(t)\,dt = \int_x^\infty q(t)\,dt \overset{\text{def}}{=} \varphi(x)$$

exists for any $x > 0$ the problem of boundedness (compactness) of the quadratic form (1.10) was reduced to the problem of boundedness (compactness) of the operator Q_φ on $L^2(\mathbb{R}_+)$. Note that in [MV] the authors also obtained boundedness and compactness criteria for the operators Q_φ in terms of conditions (1.7) and (1.9).

We also mention here the papers [OP] and [AO] where the authors study the properties of the imbedding operators from $\overset{\circ}{L}{}_p^l(\Omega)$ to $L^p(\Omega)$, where Ω is a domain in $\mathbb{R}^n$. There results also lead to certain estimates of $s_n(Q_\varphi^+)$ for φ supported on $[0, a]$, $a \in \mathbb{R}_+$.

The conditions in the above definition are conditions on the size of φ only, and define Banach lattices of functions on $\mathbb{R}_+$. Thus, if $|\psi| \le |\varphi|$ and Q_φ is bounded, compact, or belongs to $\boldsymbol{S}_p$, $p > 1$, then Q_ψ has the same property and, for example, $\|Q_\psi\|_{\boldsymbol{S}_p} \le C_p\|Q_\varphi\|_{\boldsymbol{S}_p}$. Moreover, we will see that the same conditions are necessary and sufficient for these properties for the operators Q_φ^+ and Q_φ^- too; thus, if one of the three operators has one these properties, then all three have it. These results for

Q_φ^+ are not new, see for example [ES, EEH, No, NeS, St], and the results for Q_φ can easily be derived. Nevertheless we give complete proofs, by another method, as a background for the case $p \le 1$.

At $p = 1$, there is a kind of threshold, and for $\boldsymbol{S}_p$, $p \le 1$, the situation is much more complex. First, Q_φ^+ and Q_φ^- never belong to $\boldsymbol{S}_1$, except in the trivial case $\varphi = 0$ a.e., when the operators vanish (Theorem 6.6). Secondly, although $\varphi \in X_1$ is a necessary condition for $Q_\varphi \in \boldsymbol{S}_1$, it is not sufficient. Indeed, for $p \le 1$, we have not succeeded in finding both necessary and sufficient conditions for $Q_\varphi \in \boldsymbol{S}_p$, and any such conditions would have to be fairly complicated. For one thing, the property $Q_\varphi \in \boldsymbol{S}_p$ does not depend on the size of φ only; although $\varphi_0(x) = \chi_{[0,1]}$, the characteristic function of the unit interval, yields an operator Q_{φ_0} of rank 1, we show (see the example following Theorem 6.5) that there exists a function ψ with $|\psi| = |\varphi_0|$ such that $Q_\psi \notin \boldsymbol{S}_1$. In the positive direction we show (§5 and §10) that if φ is sufficiently smooth and decays sufficiently rapidly at infinity, then $Q_\varphi \in \boldsymbol{S}_p$, $1/2 < p \le 1$. Conversely, we give in §15 (Theorem 15.22) a necessary condition on the L^1 modulus of continuity for $Q_\varphi \in \boldsymbol{S}_1$.

At $p = 1/2$ there is a second threshold. We prove in §8 and §9, by two different methods, that if φ is smooth (locally absolutely continuous is enough), then Q_φ never belongs to $\boldsymbol{S}_{1/2}$ except when $\varphi = 0$ a.e. More precisely, if φ is sufficiently smooth and decays sufficiently rapidly at infinity, and does not vanish identically, then the singular numbers $s_n(Q_\varphi)$ decay asymptotically exactly like n^{-2} (Theorem 9.3).

On the other hand, Q_φ may belong to $\boldsymbol{S}_{1/2}$ for non-smooth functions: It is easily seen that if φ is a step function, then Q_φ has finite rank, and thus $Q_\varphi \in \boldsymbol{S}_p$ for every $p > 0$. Taking suitable infinite sums of step functions we find also other functions in $\boldsymbol{S}_p$, $p \le 1/2$.

The role of smoothness is thus complicated and not well understood. It seems to be a help towards $Q_\varphi \in \boldsymbol{S}_p$ for $1/2 < p \le 1$, but it is not necessary and it completely prevents $Q_\varphi \in \boldsymbol{S}_p$ for $p \le 1/2$. On the other hand, it is irrelevant for $p > 1$.

As said above, Q_φ has finite rank when φ is a step function. We show in Theorem 12.2 that this is the only case when Q_φ has finite rank.

The kernel in Definition (1.1) is symmetric, and thus Q_φ is self-adjoint if and only if φ is real. In §4 we show that Q_φ is a positive operator if and only if φ is a non-negative non-increasing function. In this special case, for each $p > 1/2$, $Q_\varphi \in \boldsymbol{S}_p$ if and only if $\varphi \in X_p$. In this case we also give an even simpler necessary and sufficient conditions for boundedness, compactness and $Q_\varphi \in \boldsymbol{S}_p$, $p > 1/2$ (Theorem 4.6). In particular, for positive operators we have a necessary and sufficient condition for $p = 1$ too.

When φ is real and thus Q_φ is self-adjoint, the singular values are the absolute values of the eigenvalues. In §9, we study the eigenvalues, which leads to a Sturm–Liouville problem that we study. We include one example (see §9), where the singular values can be calculated exactly by this method. We give also another example (Theorem 6.5) where the singular values are calculated within a constant factor by Fourier analysis.

In §13 and §14, we consider related families of operators acting on $L^2(\mathbb{R})$ and $\ell^2(\mathbb{Z})$; the latter operators include some given by weighted Hankel matrices.

We further study operations acting on bounded linear operators on $L^2(\mathbb{R}_+)$ related to the class of operators Q_φ. We study Schur multipliers given by functions of the form $\varphi(\max\{x, y\})$ in §7 and properties of the averaging projection onto the operators of the form Q_φ in §11.

We give in this paper several necessary conditions and sufficient conditions for properties such as $Q_\varphi \in \boldsymbol{S}_p$. In all cases there are corresponding norm estimates, which follow by inspection of the proofs or by the closed graph theorem, although we usually do not state these estimates explicitly.

We denote by $|I|$ the length of an interval I. We also use $|S|$ for the cardinality of a finite set S; there is no danger of confusion.

We use c and C, sometimes with subscripts or superscripts, to denote various unspecified constants, not necessarily the same on different occurrences. These constants are universal unless we indicate otherwise by subscripts.

2. Preliminaries

Definition (1.1) shows that the adjoint $Q_\varphi^* = Q_{\bar\varphi}$; in particular, Q_φ is self-adjoint if and only if φ is real. Similarly, $(Q_\varphi^+)^* = Q_{\bar\varphi}^-$; which has the same norm and singular numbers as Q_φ^-. Hence, we will mainly consider Q_φ^+; all results obtained in this paper for Q_φ^+ immediately holds for Q_φ^- too.

Schatten classes

We denote the singular numbers of a bounded operator T on a Hilbert space (or from one Hilbert space into another) by $s_n(T)$, $n = 0, 1, 2, \ldots$; thus $s_n(T) \stackrel{\text{def}}{=} \inf\{\|T - R\| : \operatorname{rank}(R) \le n\}$. We will frequently use the simple facts

$$s_{m+n}(T + R) \le s_m(T) + s_n(R), \quad m, n \ge 0, \tag{2.1}$$

and

$$s_{m+n}(TR) \le s_m(T)s_n(R), \quad m, n \ge 0, \tag{2.2}$$

Recall that the *Schatten–von Neumann classes* $\boldsymbol{S}_p$, $0 < p < \infty$ are defined by

$$\boldsymbol{S}_p = \Big\{T : \sum_{n\ge 0} s_n(T)^p < \infty\Big\},$$

and the *Schatten–Lorentz classes* $\boldsymbol{S}_{p,q}$ are defined by

$$\boldsymbol{S}_{p,q} = \Big\{T : \sum_{n\geq 0}(s_n(T))^q(1+n)^{q/p-1} < \infty\Big\}, \quad 0 < p < \infty,\ 0 < q < \infty,$$

$$\boldsymbol{S}_{p,\infty} = \Big\{T : s_n(T) \leq C(1+n)^{-1/p}\Big\}, \quad 0 < p < \infty.$$

See for example [GK1] and [BS].

Other intervals

We have defined our operators for the interval $\mathbb{R}_+ = (0, \infty)$. More generally, for any interval $I \subseteq \mathbb{R}$ and a function $\varphi \in L^1_{\text{loc}}(I)$, we define Q^I_φ to be the integral operator on $L^2(I)$ with kernel $\varphi\big(\max(x, y)\big)$.

It is easily seen that if $I = (-\infty, a)$ with $-\infty < a \leq \infty$, then Q^I_φ is bounded only for $\varphi = 0$ a.e. By translation invariance, it remains only to consider the cases $I = (0, \infty)$, as above, and $I = (0, a)$ for some finite a. The latter case will be used sometimes below, but it can always be reduced to the case $(0, \infty)$. Indeed, if we extend φ to $(0, \infty)$ letting $\varphi = 0$ on (a, ∞), then Q^I_φ and Q_φ may be identified. (Formally, they are defined on different spaces, and Q^I_φ is the restriction of Q_φ to $L^2(I)$, but the complementary restriction to $L^2(a, \infty)$ vanishes. In particular, Q^I_φ and Q_φ have the same singular numbers.)

The case of a finite interval can also be reduced to $[0, 1]$ by the following simple homogeneity result.

Lemma 2.1. *If $t > 0$, and $\varphi_t(x) = t\varphi(tx)$, then Q_φ and Q_{φ_t} are unitarily equivalent. Similarly, for a subinterval $I \subset (0, \infty)$, Q^I_φ and $Q^{t^{-1}I}_{\varphi_t}$ are unitarily equivalent.*

Proof. The mapping $T_t : f(x) \mapsto t^{1/2} f(tx)$ is a unitary operator in $L^2(\mathbb{R}_+)$, and $Q_{\varphi_t} = T_t Q_\varphi T_t^{-1}$. □

Note that the spaces X_p have the homogeneity property exhibited in this lemma: if $\varphi \in X_p$, then $\varphi_t \in X_p$ with the same norm. Of course, it is natural to have this property for any necessary or sufficient condition for $Q_\varphi \in \boldsymbol{S}_p$.

Distributions

We can also define the operators Q_φ, Q_φ^+, and Q_φ^- in the case when φ is a distribution.

For an open subset G of $\mathbb{R}^n$ we denote by $\mathcal{D}(G)$ the space of compactly supported C^∞ functions in G and denote by $\mathcal{D}'(G)$ the space of *distributions* on G, i.e., continuous linear functionals on $\mathcal{D}(G)$. We refer the reader to [Sch] for basic facts about distributions. We use the notation $\langle\varphi, f\rangle$ for $\varphi(f)$, where $\varphi \in \mathcal{D}'(G)$ and $f \in \mathcal{D}(G)$.

Suppose now that Σ and Ω are open subsets of $\mathbb{R}$. In this paper we usually consider the case when $\Sigma = \Omega = \mathbb{R}_+$ or $\Sigma = \Omega = \mathbb{R}$. Let $\Phi \in \mathcal{D}'(\Sigma \times \Omega)$. We say that Φ determines a bounded linear operator from $L^2(\Omega)$ into $L^2(\Sigma)$ if there exists a constant C such that $|\langle\Phi(x, y), f(y)\overline{g(x)}\rangle| \le C\|f\|_{L^2}\|g\|_{L^2}$ for any $f \in \mathcal{D}(\Omega)$ and any $g \in \mathcal{D}(\Sigma)$; the corresponding operator T then is given by $\langle Tf, g\rangle = \langle\Phi(x, y), f(y)\overline{g(x)}\rangle$ and Φ is called the *kernel* of T. We say that Φ determines an operator in $\boldsymbol{S}_p$ if this operator is an operator from $L^2(\Omega)$ into $L^2(\Sigma)$ of class $\boldsymbol{S}_p$. Note that for any bounded operator $T : L^2(\Omega) \to L^2(\Sigma)$ there exists a distribution $\Phi \in \mathcal{D}'(\Sigma \times \Omega)$ which determines the operator T (a special case of Schwartz's kernel theorem). Indeed, it is easy to see that any function in $\mathcal{D}(\Omega \times \Sigma)$ defines an operator from $L^2(\Sigma)$ into $L^2(\Omega)$ of class $\boldsymbol{S}_1$, and we have a continuous imbedding $j : \mathcal{D}(\Omega \times \Sigma) \to \boldsymbol{S}_1(L^2(\Sigma), L^2(\Omega))$. We may define the distribution $\Phi_T \in \mathcal{D}'(\Sigma\times\Omega)$ by the following formula $\langle\Phi_T(x, y), f(x, y)\rangle \stackrel{\text{def}}{=} \operatorname{trace}(TA)$, where $f \in \mathcal{D}(\Sigma\times\Omega)$ and A is the integral operator with kernel function $f(y, x)$. Clearly, Φ_T determines the operator T.

We also consider the space $\mathcal{S}(\mathbb{R}^n)$ of infinitely smooth functions whose derivatives of arbitrary orders decay at infinity faster than $(1 + |x|)^{-n}$ for any $n \in \mathbb{Z}_+$ and the dual space $\mathcal{S}'(\mathbb{R}^n)$ of *tempered distributions* (see [Sch] for basic facts). Recall that the Fourier transform

$$f \mapsto (\mathcal{F} f)(x) \stackrel{\text{def}}{=} \int_{\mathbb{R}^n} f(t)e^{-2\pi i(t,x)}\,dt \tag{2.3}$$

where (t, x) is the scalar product of x and t in $\mathbb{R}^n$, is an isomorphism of $\mathcal{S}(\mathbb{R}^n)$ onto itself, and that it can be extended to $\mathcal{S}'(\mathbb{R}^n)$ by duality.

We need the following elementary facts.

Lemma 2.2. *Suppose that a distribution $\Phi \in \mathcal{D}'(\mathbb{R}^2)$ determines a bounded operator on $L^2(\mathbb{R})$. Then Φ is a tempered distribution.*

Proof. It is easy to see that any function $\Phi \in \mathcal{S}(\mathbb{R}^2)$ determines an operator of class $\boldsymbol{S}_1$ and that the corresponding imbedding of $\mathcal{S}(\mathbb{R}^2)$ into $\boldsymbol{S}_1$ is continuous. The result follows now by duality. □

Lemma 2.3. *Let $\Phi \in \mathcal{S}'(\mathbb{R}^2)$ and consider the distribution Ψ on $\mathbb{R}^2$ defined by $\Psi(x, y) \stackrel{\text{def}}{=} (\mathcal{F}\Phi)(x, -y)$. Then Φ determines a bounded operator on $L^2(\mathbb{R})$ if and only if Ψ does. Moreover, these two operators are unitarily equivalent.*

Proof. It suffices to observe that

$$\begin{aligned}\langle\Psi(x,y), f(y)\overline{g(x)}\rangle &= \langle\Phi(x,y), \mathcal{F}(f(-y)\overline{g(x)})\rangle \\ &= \langle\Phi(x,y), (\mathcal{F}f)(-y)\overline{\mathcal{F}g(-x)}\rangle. \qquad \square\end{aligned}$$

Now we are ready to define the operators Q_φ, Q_φ^+ and Q_φ^- in the case where φ is distribution.

It is not hard to see that the operator $f \mapsto \int_0^x f(x,y)\,dy$ is a continuous operator from $\mathcal{D}(\mathbb{R}_+ \times \mathbb{R}_+)$ into $\mathcal{D}(\mathbb{R}_+)$. Hence, with any $\varphi \in \mathcal{D}'(\mathbb{R}_+)$ we can associate the distributions Λ_φ^+, Λ_φ^-,and Λ_φ in $\mathcal{D}'(\mathbb{R}_+ \times \mathbb{R}_+)$ defined by

$$\langle\Lambda_\varphi^+, f(x,y)\rangle \stackrel{\text{def}}{=} \langle\varphi, \int_0^x f(x,y)\,dy\rangle, \tag{2.4}$$

$$\langle\Lambda_\varphi^-, f(x,y)\rangle \stackrel{\text{def}}{=} \langle\varphi, \int_0^x f(y,x)\,dy\rangle \tag{2.5}$$

and

$$\Lambda_\varphi \stackrel{\text{def}}{=} \Lambda_\varphi^+ + \Lambda_\varphi^-. \tag{2.6}$$

For a distribution φ in $\mathcal{D}'(\mathbb{R}_+)$, we can consider now the operators Q_φ^+, Q_φ^-, and Q_φ determined by the distributions Λ_φ^+, Λ_φ^-,and Λ_φ respectively. It is easy to see that in case $\varphi \in L^1_{\text{loc}}(\mathbb{R}_+)$, the new definition coincides with the old one. The following theorem shows however that if one of those operators is bounded on $L^2(\mathbb{R}_+)$, then φ must be a locally integrable function on $\mathbb{R}_+$, and so we have not enlarged the class of bounded operators of the form Q_φ.

Theorem 2.4. *Let $\varphi \in \mathcal{D}'(\mathbb{R}_+)$. Suppose that at least one of the distributions Λ_φ^+, Λ_φ^-, or Λ_φ determines a bounded operator on $L^2(\mathbb{R}_+)$. Then $\varphi \in L^2_{\text{loc}}(\mathbb{R}_+)$.*

Proof. We consider the cases of the distributions Λ_φ^+ and Λ_φ. For Λ_φ^- the proof is the same as for Λ_φ^+. Let $a \in \mathbb{R}_+$. Fix a function $f_0 \in \mathcal{D}(\mathbb{R}_+)$ such that $\operatorname{supp} f_0 \subset [0,a]$ and $\int_{\mathbb{R}_+} f_0(x)\,dx = 1$. We have

$$\langle\Lambda_\varphi^+, f_0(y)\overline{g(x)}\rangle = \langle\Lambda_\varphi, f_0(y)\overline{g(x)}\rangle = \langle\varphi, \bar{g}\rangle$$

for any $g \in \mathcal{D}(\mathbb{R}_+)$ with $\operatorname{supp} g \subset [a,+\infty)$. Therefore

$$|\langle\varphi, \overline{g}\rangle| \le C\|f_0\|_{L^2(\mathbb{R}_+)}\|g\|_{L^2(\mathbb{R}_+)}$$

for any $g \in \mathcal{D}(a,\infty)$. Thus, $\varphi\big|(a,\infty) \in L^2(a,\infty)$. $\square$

Triangular projection

On the class of operators on $\boldsymbol{S}_p(L^2(\mathbb{R}_+))$, $p < \infty$, we define the operator of triangular projection $\mathcal{P}$ as follows. Consider first the case $p \le 2$. Let T be an operator on $L^2(\mathbb{R}_+)$ of class $\boldsymbol{S}_p$, $p \le 2$. Then T is an integral operator with kernel function k_T:

$$(Tf)(x) = \int_0^\infty k_T(x, y) f(y)\, dy, \quad f \in L^2(\mathbb{R}_+).$$

Then by definition

$$(\mathcal{P}Tf)(x) = \int_0^x k_T(x, y) f(y)\, dy, \quad f \in L^2(\mathbb{R}_+). \tag{2.7}$$

It is well known that

$$\|\mathcal{P}T\|_{\boldsymbol{S}_p} \le c_p \|T\|_{\boldsymbol{S}_p}, \quad 1 < p \le 2, \tag{2.8}$$

where c_p depends only on p. This allows one to extend by duality the definition of $\mathcal{P}$ and inequality (2.8) to the case $2 \le p < \infty$. Note also that $\mathcal{P}$ has weak type $(1, 1)$, i.e.,

$$s_n(\mathcal{P}T) \le C(1+n)^{-1} \|T\|_{\boldsymbol{S}_1}, \quad T \in \boldsymbol{S}_1. \tag{2.9}$$

We will need these results on the triangular projection $\mathcal{P}$ in a more general situation. Let μ and ν be regular Borel measures on $\mathbb{R}_+$. As above we can associate with any operator T from $L^2(\mu)$ to $L^2(\nu)$ of class $\boldsymbol{S}_2$ the operator $\mathcal{P}T$ by multiplying the kernel function of T by the characteristic function of the set $\{(x, y) \in \mathbb{R}_+^2 : 0 < y < x\}$.

Theorem 2.5. *$\mathcal{P}$ is a bounded linear projection on $\boldsymbol{S}_p(L^2(\mu), L^2(\nu))$ for $1 < p < \infty$ and $\mathcal{P}$ has weak type $(1, 1)$, i.e., $\mathcal{P}$ is a bounded linear operator from $\boldsymbol{S}_1(L^2(\mu), L^2(\nu))$ to $\boldsymbol{S}_{1,\infty}(L^2(\mu), L^2(\nu))$.*

Theorem 2.5 is well known at least when $\mu = \nu$. Let us explain how to reduce Theorem 2.5 to the case of the triangular projection onto the upper triangular matrices.

Let $\{\mathcal{K}_j\}_{j\ge 0}$ and $\{\mathcal{H}_k\}_{k\ge 0}$ be Hilbert spaces. We put $\mathcal{K} \stackrel{\text{def}}{=} \bigoplus_{j\ge 0} \mathcal{K}_j$ and $\mathcal{H} \stackrel{\text{def}}{=} \bigoplus_{k\ge 0} \mathcal{H}_k$ and identify operators $A \in \mathcal{B}(\mathcal{H}, \mathcal{K})$ with their block matrix representation $\{A_{jk}\}_{j,k\ge 0}$, where $A_{jk} \in \mathcal{B}(\mathcal{H}_k, \mathcal{K}_j)$. We define the triangular projection $\boldsymbol{\mathcal{P}}$ by $(\boldsymbol{\mathcal{P}}A)_{jk} \stackrel{\text{def}}{=} A_{jk}$ for $j > k$ and $(\boldsymbol{\mathcal{P}}A)_{jk} \stackrel{\text{def}}{=} 0$ for $j \le k$.

Lemma 2.6. *Let $1 < p < \infty$. Then $\boldsymbol{\mathcal{P}}$ is bounded on $\boldsymbol{S}_p(\mathcal{H}, \mathcal{K})$ and has weak type $(1, 1)$. Moreover, the norms of $\boldsymbol{\mathcal{P}}$ can be bounded independently of $\mathcal{H}$ and $\mathcal{K}$.*

In the case $\dim \mathcal{K}_j = \dim \mathcal{H}_k = 1$, this is the Krein–Matsaev theorem (it is equivalent to Theorems III.2.4 and III.6.2 of [GK2], see also Theorem IV.8.2 of [GK1]). In general the result can be reduced easily to this special case. Indeed, it is easy to reduce the general case to the case when $\dim \mathcal{H}_j = \dim \mathcal{K}_j < \infty$. Then it is easy to see that if $A \in \boldsymbol{S}_p$, $1 \le p < \infty$, then the diagonal part of A also belongs to $\boldsymbol{S}_p$, and so we may assume without loss of generality that $A_{jj} = 0$, $j \in \mathbb{Z}_+$. We can take an orthonormal basis in each $\mathcal{H}_j$ and consider the orthonormal basis of $\mathcal{H}$ that consists of those basis vectors of $\mathcal{H}_j$, $j \in \mathbb{Z}_+$. Then we can consider the matrix representation of A with respect to this orthonormal basis. We have now two triangular projections: with respect to the orthonormal basis and the projection $\mathcal{P}$, the triangular with respect to the decomposition $\mathcal{H} = \bigoplus_{k \geq 0} \mathcal{H}_k$. It is not hard to check that since the diagonal part of A is zero, both triangular projections applied to A give the same result. This reduces the general case to the Krein–Matsaev theorem mentioned above. □

Now it is easy to deduce Theorem 2.5 from Lemma 2.6.

Proof of Theorem 2.5. Let T be an integral operator with kernel function k. For $\varepsilon > 0$ we put $k_\varepsilon(x, y) \stackrel{\text{def}}{=} k(x, y)\chi_{\{(x,y)\in\mathbb{R}_+^2 : [\frac{x}{\varepsilon}]\varepsilon > y > 0\}}$, where $[a]$ denote the largest integer that is less than or equal to a. Suppose that $p > 1$. It is sufficient to consider the case $1 < p \leq 2$ and then use duality. Let T_ε be the integral operator with kernel function k_ε. It follows easily from Lemma 2.6 that

$$\|T_\varepsilon\|_{S_p(L^2(\mu),L^2(\nu))} \leq c_p \|T\|_{S_p(L^2(\mu),L^2(\nu))}$$

for any $\varepsilon > 0$. Clearly, $T_\varepsilon \to T$ in the weak operator topology as $\varepsilon \to 0$. It follows that $\|\mathcal{P}T\|_{S_p(L^2(\mu),L^2(\nu))} \leq c_p \|T\|_{S_p(L^2(\mu),L^2(\nu))}$. The case $p = 1$ may be considered in the same way. □

We have $Q_\varphi^+ = \mathcal{P} Q_\varphi$, which together with the equivalence

$$Q_\varphi^+ \in \boldsymbol{S}_p \iff Q_\varphi^- \in \boldsymbol{S}_p$$

yields

$$Q_\varphi^+ \in \boldsymbol{S}_p \iff Q_\varphi \in \boldsymbol{S}_p$$

for $1 < p < \infty$. We will give a direct proof of this in Theorem 3.3.

We introduce a more general operation. Let A be a measurable subset of $\mathbb{R}_+ \times \mathbb{R}_+$. For an operator T on $L^2(\mathbb{R}_+)$ of class $\boldsymbol{S}_2$ with kernel function k_T we consider the integral operator $\mathcal{P}_A T$ whose kernel function is $\chi_A k_T$, where χ_A is the characteristic function of A. In other words,

$$(\mathcal{P}_A T f)(x) = \int_0^\infty \chi_A(x, y) k_T(x, y) f(y)\, dy.$$

If $0 < p \leq 2$ and $\mathcal{P}_A$ maps $\boldsymbol{S}_p$ into itself, it follows from the closed graph theorem that the linear transformation $\mathcal{P}_A$ is a bounded linear operator on $\boldsymbol{S}_p$. If $1 < p \leq 2$ and $\mathcal{P}_A$ is a bounded linear operator on $\boldsymbol{S}_p$, then by duality we can define in a natural way the bounded linear operator $\mathcal{P}_A$ on $\boldsymbol{S}_{p'}$. If $\mathcal{P}_A$ is bounded on $\boldsymbol{S}_1$, we can define by duality $\mathcal{P}_A$ on the space $\mathcal{B}(L^2(\mathbb{R}_+))$ of bounded linear operators on $L^2(\mathbb{R}_+)$. Note that the projection $\mathcal{P}$ defined by (2.7) is equal to $\mathcal{P}_A$ with $A = \{(x, y) : x \in \mathbb{R}_+, y \leq x\}$.

3. Boundedness, compactness, and $p > 1$

Recall the spaces X_p defined in the Introduction.

Theorem 3.1. *Let $\varphi \in L^2_{\mathrm{loc}}(\mathbb{R}_+)$. The following are equivalent:*

(i) *Q_φ is bounded on $L^2(\mathbb{R}_+)$;*

(ii) *Q^+_φ is bounded on $L^2(\mathbb{R}_+)$;*

(iii) *$\varphi \in X_\infty$.*

Recall that the equivalence of (i) and (iii) was also established in [MV] by a different method.

Proof. Let us show that (ii) implies (i). If Q^+_φ is bounded, then the integral operator $Q^-_\varphi \stackrel{\mathrm{def}}{=} Q_\varphi - Q^+_\varphi$ is also bounded, since its kernel function is the reflection of the kernel function of Q^+_φ with respect to the line $\{x = y\}$. Hence, Q_φ is bounded.

Let us deduce now (iii) from (i). For $n \in \mathbb{Z}$ put

$$A_n = [2^n, 2^{n+1}] \times [2^{n-1}, 2^n].$$

Certainly, if Q_φ is bounded, then

$$\sup_{n\in\mathbb{Z}} \|\mathcal{P}_{A_n} Q_\varphi\| < \infty.$$

It is easy to see that $\mathcal{P}_{A_n} Q_\varphi$ is a rank one operator and

$$\|\mathcal{P}_{A_n} Q_\varphi\| = \left(2^{n-1} \int_{2^n}^{2^{n+1}} |\varphi(x)|^2 \, dx\right)^{1/2}$$

which implies (1.6).

It remains to prove that (1.6) implies (ii). Put

$$B = \bigcup_{n\in\mathbb{Z}} B_n, \tag{3.1}$$

where

$$B_n = \{(x, y) : \ 2^n \le x \le 2^{n+1}, \ 2^n < y < x\}. \tag{3.2}$$

We also define the sets

$$A_n^{(k)} = [2^n, 2^{n+1}] \times [2^{n-k}, 2^{n-k+1}] \tag{3.3}$$

and

$$A^{(k)} = \bigcup_{n\in\mathbb{Z}} A_n^{(k)}. \tag{3.4}$$

Clearly,

$$\{(x, y) : x > 0,\ 0 < y < x\} = B \cup \bigcup_{k \geq 1} A^{(k)},$$

and so

$$\|Q_\varphi^+\| \leq \|\mathcal{P}_B Q_\varphi\| + \sum_{k \geq 1} \|\mathcal{P}_{A^{(k)}} Q_\varphi\|. \tag{3.5}$$

Since the projections of the B_n onto the coordinate axes are pairwise disjoint, it is straightforward to see that

$$\|\mathcal{P}_B Q_\varphi^+\| = \sup_{n \in \mathbb{Z}} \|\mathcal{P}_{B_n} Q_\varphi^+\|.$$

Let R_n be the integral operator with kernel function

$$k_{R_n}(x, y) = \varphi(x) \chi_{[2^n, 2^{n+1}]}(x) \chi_{[2^n, 2^{n+1}]}(y). \tag{3.6}$$

Obviously, rank $R_n = 1$ and $\|R_n\|_{S_2} = \|R_n\| = \left(2^n \int_{2^n}^{2^{n+1}} |\varphi(x)|^2\,dx\right)^{1/2}$. It is also evident that $\mathcal{P} R_n = \mathcal{P}_{B_n} Q_\varphi^+$, and since $\mathcal{P}$ an orthogonal projection on S_2, we have

$$\begin{aligned} \|\mathcal{P}_{B_n} Q_\varphi^+\| &= \|\mathcal{P} R_n\| \leq \|\mathcal{P} R_n\|_{S_2} \leq \|R_n\|_{S_2} \\ &= \left(2^n \int_{2^n}^{2^{n+1}} |\varphi(x)|^2\,dx\right)^{1/2} \leq C\|\varphi\|_{X_\infty}. \end{aligned}$$

Next, it is also easy to see that

$$\|\mathcal{P}_{A^{(k)}} Q_\varphi\| = \sup_{n \in \mathbb{Z}} \|\mathcal{P}_{A_n^{(k)}} Q_\varphi\|.$$

Also, $\mathcal{P}_{A_n^{(k)}} Q_\varphi$ has rank one and norm $\left(2^{n-k} \int_{2^n}^{2^{n+1}} |\varphi(x)|^2\,dx\right)^{1/2}$, and so

$$\begin{aligned} \sum_{k \geq 1} \|\mathcal{P}_{A^{(k)}} Q_\varphi\| &= \sum_{k \geq 1} \sup_{n \in \mathbb{Z}} \left(2^{n-k} \int_{2^n}^{2^{n+1}} |\varphi(x)|^2\,dx\right)^{1/2} \\ &= \sum_{k \geq 1} 2^{-k/2} \sup_{n \in \mathbb{Z}} \left(2^n \int_{2^n}^{2^{n+1}} |\varphi(x)|^2\,dx\right)^{1/2} \\ &= C \sup_{n \in \mathbb{Z}} \left(2^n \int_{2^n}^{2^{n+1}} |\varphi(x)|^2\,dx\right)^{1/2}. \end{aligned}$$

The result follows now from (3.5). □

Theorem 3.2. *Let $\varphi \in L^1_{\text{loc}}(\mathbb{R}_+)$. The following are equivalent:*

(i) *Q_φ is compact on $L^2(\mathbb{R}_+)$;*

(ii) *Q_φ^+ is compact on $L^2(\mathbb{R}_+)$;*

(iii) *$\varphi \in X_\infty^0$.*

Recall that the equivalence of (i) and (iii) was also established in [MV] by a different method.

Proof. It is easy to see that the estimates given in the proof of Theorem 3.1 actually lead to the proof of Theorem 3.2; for the step (i) $\Rightarrow$ (iii) we observe that if Q_φ is compact, then $\lim_{n\to\pm\infty} \|\mathcal{P}_{A_n} Q_\varphi\| = 0$. □

Theorem 3.3. *Let* $1 < p < \infty$ *and let* $\varphi \in L^1_{\text{loc}}(\mathbb{R}_+)$. *Then the following conditions are equivalent:*

(i) $Q_\varphi \in \boldsymbol{S}_p$;

(ii) $Q_\varphi^+ \in \boldsymbol{S}_p$;

(iii) $\varphi \in X_p$.

Note that the fact that (ii) $\Leftrightarrow$ (iii) was proved in [No] by a different method, see also [NeS] and [St] for the case of more general Volterra operators.

Proof. The fact that (ii) $\Rightarrow$ (i) can be proved exactly as in the proof of Theorem 3.1. Let us show that (i) implies (iii). Consider the sets $A_n = A_n^{(1)}$ introduced in the proof of Theorem 3.1 (see (3.3)). Recall that $A^{(1)} = \bigcup_{n\in\mathbb{Z}} A_n^{(1)}$. It is easy to see that

$$\|\mathcal{P}_{A^{(1)}} Q_\varphi\|_{\boldsymbol{S}_p} \le \|Q_\varphi\|_{\boldsymbol{S}_p}.$$

Clearly,

$$\|\mathcal{P}_{A^{(1)}} Q_\varphi\|^p_{\boldsymbol{S}_p} = \sum_{n\in\mathbb{Z}} \|\mathcal{P}_{A_n} Q_\varphi\|^p_{\boldsymbol{S}_p},$$

the operator $\mathcal{P}_{A_n} Q_\varphi$ has rank one and

$$\|\mathcal{P}_{A_n} Q_\varphi\|_{\boldsymbol{S}_p} = \left(2^{n-1}\int_{2^n}^{2^{n+1}} |\varphi(x)|^2\,dx\right)^{1/2}.$$

This implies (1.3).

Let us show that (1.3) implies (ii). Consider the sets, B_n, B, $A_n^{(k)}$, $A^{(k)}$ defined in (3.2), (3.1), (3.3), and (3.4). Clearly,

$$\|Q_\varphi^+\|_{\boldsymbol{S}_p} \le \|\mathcal{P}_B Q_\varphi^+\|_{\boldsymbol{S}_p} + \sum_{k\ge 1} \|\mathcal{P}_{A^{(k)}} Q_\varphi^+\|_{\boldsymbol{S}_p}.$$

Let us first estimate $\|\mathcal{P}_B Q_\varphi^+\|_{\boldsymbol{S}_p}$. Clearly,

$$\|\mathcal{P}_B Q_\varphi^+\|^p_{\boldsymbol{S}_p} = \sum_{n\in\mathbb{Z}} \|\mathcal{P}_{B_n} Q_\varphi^+\|^p_{\boldsymbol{S}_p}.$$

Consider the rank one operators R_n defined in (3.6). As in the proof of Theorem 3.1 we have $\mathcal{P} R_n = \mathcal{P}_{B_n} Q_\varphi^+$ and since $\mathcal{P}$ is bounded on S_p, we obtain

$$\|\mathcal{P}_{B_n} Q_\varphi^+\|_{S_p} = \|\mathcal{P} R_n\|_{S_p} \le C_p \|R_n\|_{S_p} = C_p\Big(2^n \int_{2^n}^{2^{n+1}} |\varphi(x)|^2\,dx\Big)^{1/2},$$

and so

$$\|\mathcal{P}_B Q_\varphi^+\|_{S_p}^p \le C_p \sum_{n\in\mathbb{Z}} \Big(2^n \int_{2^n}^{2^{n+1}} |\varphi(x)|^2\,dx\Big)^{p/2}.$$

It is also easy to see that

$$\|\mathcal{P}_{A^{(k)}} Q_\varphi^+\|_{S_p}^p = \sum_{n\in\mathbb{Z}} \|\mathcal{P}_{A_n^{(k)}} Q_\varphi^+\|_{S_p}^p$$

and, since $\mathcal{P}_{A_n^{(k)}} Q_\varphi$ has rank 1,

$$\|\mathcal{P}_{A_n^{(k)}} Q_\varphi^+\|_{S_p} = \|\mathcal{P}_{A_n^{(k)}} Q_\varphi\|_{S_p} = \Big(2^{n-k} \int_{2^n}^{2^{n+1}} |\varphi(x)|^2\,dx\Big)^{1/2},$$

and so

$$\begin{aligned}\|\mathcal{P}_{A^{(k)}} Q_\varphi^+\|_{S_p} &= \Big(\sum_{n\in\mathbb{Z}} \Big(2^{n-k} \int_{2^n}^{2^{n+1}} |\varphi(x)|^2\,dx\Big)^{p/2}\Big)^{1/p} \\ &= 2^{-k/2}\Big(\sum_{n\in\mathbb{Z}} \Big(2^{n} \int_{2^n}^{2^{n+1}} |\varphi(x)|^2\,dx\Big)^{p/2}\Big)^{1/p}\end{aligned} \tag{3.7}$$

and

$$\sum_{k\ge 1} \|\mathcal{P}_{A^{(k)}} Q_\varphi^+\|_{S_p} \le C\Big(\sum_{n\in\mathbb{Z}} \Big(2^{n} \int_{2^n}^{2^{n+1}} |\varphi(x)|^2\,dx\Big)^{p/2}\Big)^{1/p} \tag{3.8}$$

which completes the proof. □

Remark. The same proof shows that for $p = 1$, (ii) ⇒ (i) ⇒ (iii), but the final part of it fails because the triangular projection is not bounded on S_1. We will see later that, indeed, none of the implications can be reversed for $p = 1$.

In the Hilbert–Schmidt case $p = 2$, the result simplifies further. Indeed, we have the equalities

$$\|Q_\varphi\|_{S_2} = \Big(\int_0^\infty \int_0^\infty |\varphi(\max(x,y))|^2\,dx\,dy\Big)^{1/2} = \Big(2\int_0^\infty x|\varphi(x)|^2\,dx\Big)^{1/2}$$

and

$$\|Q_\varphi^+\|_{S_2} = 2^{-1/2}\|Q_\varphi\|_{S_2} = \|x^{1/2}\varphi(x)\|_2 = \|\varphi\|_{X_2}.$$

4. Positive operators

We consider the special case when Q_φ is a positive operator, i.e., $\langle Q_\varphi f, f\rangle \geq 0$ for every $f \in L^2(\mathbb{R}_+)$. In this case we obtain rather complete results. We first characterize the corresponding symbols φ.

Theorem 4.1. *Suppose that $\varphi \in L^1_{\mathrm{loc}}(\mathbb{R}_+)$ is such that Q_φ is a bounded operator. Then Q_φ is a positive operator if and only if φ is a.e. equal to a non-increasing, non-negative function.*

Proof. Suppose that Q_φ is positive. Define, for $z, h > 0$, $f_{z,h} = h^{-1}\chi_{(z,z+h)}$ and let $\mathrm{Leb}(\varphi)$ be the set of Lebesgue points of φ. Then, if $z \in \mathrm{Leb}(\varphi)$,

$$
\begin{aligned}
|\langle Q_\varphi f_{z,h}, f_{z,h}\rangle - \varphi(z)| &= h^{-2}\left|\int_z^{z+h}\int_z^{z+h}\bigl(\varphi\{\max(x,y)\} - \varphi(z)\bigr)\,dx\,dy\right| \\
&\leq h^{-2}\int_z^{z+h}\int_z^{z+h}\bigl(|\varphi(x) - \varphi(z)| + |\varphi(y) - \varphi(z)|\bigr)\,dx\,dy \\
&= 2h^{-1}\int_z^{z+h}|\varphi(x) - \varphi(z)|\,dx \to 0
\end{aligned}
$$

as $h \to 0$. Since $\langle Q_\varphi f_{z,h}, f_{z,h}\rangle \geq 0$ for all $h > 0$, this implies $\varphi(z) \geq 0$.

Moreover, if $z_1, z_2 \in \mathrm{Leb}(\varphi)$ are two Lebesgue points with $z_1 < z_2$ and $0 < h < z_2 - z_1$, then, similarly,

$$
\langle Q_\varphi f_{z_1,h}, f_{z_2,h}\rangle = \langle Q_\varphi f_{z_2,h}, f_{z_1,h}\rangle = h^{-1}\int_{z_2}^{z_2+h}\varphi(y)\,dy \to \varphi(z_2)
$$

as $h \to 0$, and thus, with $g_h = f_{z_1,h} - f_{z_2,h}$,

$$
\langle Q_\varphi g_h, g_h\rangle \to \varphi(z_1) + \varphi(z_2) - 2\varphi(z_2) = \varphi(z_1) - \varphi(z_2).
$$

Hence, $\varphi(z_1) \geq \varphi(z_2)$.

It follows that the function $\tilde\varphi(x) = \sup\{\varphi(z) : z \geq x,\ z \in \mathrm{Leb}(\varphi)\}$ is non-negative and non-increasing, and that $\varphi = \tilde\varphi$ a.e.

Conversely, if φ is non-negative and non-increasing, then $\lim_{x\to\infty}\varphi(x) = 0$, since a positive lower bound is impossible by Theorem 3.1. Thus there exists a measure μ on $(0,\infty)$ such that $\varphi(x) = \mu(x,\infty)$ a.e. If, say, f is bounded with compact support in $(0,\infty)$, then by Fubini's theorem

$$
\begin{aligned}
\langle Q_\varphi f, f\rangle &= \iint \varphi\bigl(\max\{x,y\}\bigr) f(x)\bar f(y)\,dx\,dy = \iiint\limits_{\max\{x,y\}<z} f(x)\bar f(y)\,dx\,dy\,d\mu(z) \\
&= \int_0^\infty\left|\int_0^z f(x)\,dx\right|^2 d\mu \geq 0.
\end{aligned}
$$

Hence, Q_φ is a positive operator. □

We have used the fact that Q_φ is a sum of the Volterra operators Q_φ^+ and Q_φ^-. Operators of the type Q_φ also appear as the composition of Volterra operators.

Theorem 4.2. *Suppose that ψ_1 and ψ_2 are functions on $\mathbb{R}_+$ such that Q_{ψ_1} and Q_{ψ_2} are bounded linear operators. Let φ be the function defined by*

$$\varphi(x) = \int_x^\infty \psi_1(t)\psi_2(t)\,dt. \tag{4.1}$$

Then the operator Q_φ is bounded and admits a factorization

$$Q_\varphi = Q_{\psi_1}^- Q_{\psi_2}^+.$$

Proof. Let k_1 be the kernel function of $Q_{\psi_1}^-$ and k_2 the kernel function of $Q_{\psi_2}^+$. We have

$$k_1(x,t) = \begin{cases} \psi_1(t), & t \geq x \\ 0, & t < x \end{cases}$$

and

$$k_2(t,y) = \begin{cases} 0, & t < y \\ \psi_2(t), & t \geq y. \end{cases}$$

Then the kernel function k of the product $Q_{\psi_1}^- Q_{\psi_2}^+$ is given by

$$k(x,y) = \int_{\mathbb{R}_+} k_1(x,t)k_2(t,y)\,dt = \int_{\max\{x,y\}}^\infty \psi_1(t)\psi_2(t)\,dt = \varphi(\max\{x,y\})$$

by the hypotheses; the integrals converge by Theorem 3.1 and the Cauchy–Schwarz inequality. □

The function φ in (4.1) is always locally absolutely continuous. In order to treat more general non-increasing φ, we define, for a positive measure μ on $(0,\infty)$, the operator $\mathbf{Q}_\mu^+ : L^2(0,\infty) \to L^2(\mu)$ by $\mathbf{Q}_\mu^+ f(x) = \int_0^x f(y)\,dy$. (Thus, the operator itself does not depend on μ; only its range space does.) We have the following analogues of Theorems 3.1 and 3.3. (We leave the corresponding criterion for compactness to the reader.)

Theorem 4.3. *Let μ be a positive measure on $\mathbb{R}_+$. The following are equivalent:*

(i) $\mathbf{Q}_\mu^+$ *is bounded operator from $L^2(\mathbb{R}_+)$ to $L^2(\mu)$;*

(ii) $\sup_{n\in\mathbb{Z}} 2^n \mu[2^n, 2^{n+1}) < \infty$;

(iii) $\sup_{x>0} x\mu[x,\infty) < \infty$.

Theorem 4.4. *Let $1 < p < \infty$ and let μ be a positive measure on $\mathbb{R}_+$. The following conditions are equivalent:*

(i) $\mathbf{Q}_\mu^+ \in \boldsymbol{S}_p$;

(ii) $\sum_{n\in\mathbb{Z}} 2^{np/2}\big(\mu[2^n, 2^{n+1})\big)^{p/2} < \infty$;

(iii) $\sum_{n\in\mathbb{Z}} 2^{np/2}\big(\mu[2^n, \infty)\big)^{p/2} < \infty$;

(iv) $x^{1/2}\big(\mu(x,\infty)\big)^{1/2} \in L^p(dx/x)$.

The proofs of Theorems 4.3 and 4.4 are almost the same as the proofs of Theorems 3.1 and 3.3. The main difference is that we have to apply the theorem on the boundedness of the triangular projection on $\boldsymbol{S}_p$, $1 < p < \infty$, in the case of weighted L^2 spaces (see Theorem 2.5). □

Furthermore, the factorization in Theorem 4.2 extends.

Theorem 4.5. *Suppose that μ is a positive measure on $\mathbb{R}_+$ such that $\mathbf{Q}_\mu^+$ is a bounded linear operator. Let φ be the function defined by $\varphi(x) = \mu(x,\infty)$. Then the operator Q_φ is bounded and $Q_\varphi = (\mathbf{Q}_\mu^+)^*\mathbf{Q}_\mu^+$.*

Proof. By Theorem 4.3, $0 \le \varphi(x) \le C_\varphi/x$, and thus Q_φ is bounded by Theorem 3.1.
If, say, $f, g \in L^2(\mathbb{R}_+)$ are non-negative, then by Fubini's theorem

$$\begin{aligned}\langle(\mathbf{Q}_\mu^+)^*\mathbf{Q}_\mu^+ f, g\rangle = \langle\mathbf{Q}_\mu^+ f, \mathbf{Q}_\mu^+ g\rangle &= \int_0^\infty \int_0^z f(x)\,dx \int_0^z \bar{g}(y)\,dy\,d\mu(z)\\ &= \iiint\limits_{\max\{x,y\}<z} f(x)\bar{g}(y)\,dx\,dy\,d\mu(z) = \langle Q_\varphi f, g\rangle.\end{aligned}$$ □

For positive operators Q_φ, we have a simple result, Theorem 4.6 below. (For (i), cf. the discussion of the Hille condition in [MV].)

Theorem 4.6. *Suppose that φ is a non-negative, non-increasing function on $\mathbb{R}_+$.*

(i) *Q_φ is bounded if and only if $x\varphi(x)$ is bounded.*

(ii) *Q_φ is compact if and only if $x\varphi(x) \to 0$ as $x \to 0$ and as $x \to \infty$.*

(iii) *If $1/2 < p < \infty$, then the following are equivalent:*

(a) $Q_\varphi \in \boldsymbol{S}_p$;

(b) $\varphi \in X_p$;

(c) $x\varphi(x) \in L^p(dx/x)$.

Proof. The equivalence of $\varphi \in X_p$ and $x\varphi(x) \in L^p(dx/x)$ for non-increasing, non-negative φ is elementary, using $\varphi(2^n)^2 \geq 2^{-n}\int_{2^n}^{2^{n+1}} |\varphi|^2 \geq \varphi(2^{n+1})^2$. Hence, (i) follows from Theorem 3.1, and (iii) for $p > 1$ from Theorem 3.3; furthermore, (ii) follows similarly from Theorem 3.2.

For (iii) for a general $p > 1/2$, we first note that any of the three conditions (a), (b) and (c) implies that $x\varphi(x)$ is bounded. (This follows by (i) for (a), and by elementary estimates for (b) and (c).) We can assume without loss of generality that φ is right-continuous on $(0, \infty)$. If we let μ be the measure on $\mathbb{R}_+$ with $\mu(x, \infty) = \varphi(x)$, then by Theorems 4.3 and 4.5, $\mathbf{Q}_\mu^+$ is bounded and $Q_\varphi = (\mathbf{Q}_\mu^+)^*\mathbf{Q}_\mu^+$. Hence, $Q_\varphi \in \boldsymbol{S}_p \Leftrightarrow \mathbf{Q}_\mu^+ \in \boldsymbol{S}_{2p}$, and the result follows by Theorem 4.4. □

We will see in the example given at the beginning of §9 that Theorem 4.6 (iii) does not extend to $p \leq 1/2$.

5. A sufficient condition, $1/2 < p \leq 1$

By linearity, we immediately obtain from Theorem 4.6 a sufficient, but not necessary, condition for general symbols φ.

Definition. Y_p is the subspace of X_p spanned by non-increasing functions. I.e., $\varphi \in Y_p$ if and only if $\operatorname{Re}\varphi$ and $\operatorname{Im}\varphi$ both are differences of non-increasing functions in X_p.

Theorem 5.1. *Let $p > 1/2$. If $\varphi \in Y_p$, then $Q_\varphi \in \boldsymbol{S}_p$.* □

The condition $\varphi \in Y_p$ can be made more explicit and useful as follows. We denote by $\|\varphi\|'_{BV(I)}$ the total variation of a function φ over an interval I, and let $\|\varphi\|_{BV(I)} \stackrel{\text{def}}{=} \|\varphi\|'_{BV(I)} + \sup_I |\varphi|$. Moreover, we let $V_\varphi(x)$ denote the total variation of a function φ over the interval $[x, \infty)$. Note that if φ is locally absolutely continuous, then $V_\varphi(x) = \int_x^\infty |\varphi'(y)|\, dy$.

Lemma 5.2. *Let $0 < p < \infty$. If φ is non-increasing, then*

$$\varphi \in Y_p \Leftrightarrow \varphi \in X_p \Leftrightarrow \int_0^\infty |x\varphi(x)|^p \frac{dx}{x} < \infty.$$

Proof. For non-increasing φ, the first equivalence follows from the definition of Y_p, while the second equivalence was noted in the proof of Theorem 4.6. □

Theorem 5.3. *Let φ be a function on $\mathbb{R}_+$ and let $0 < p < \infty$. The following are equivalent:*

(i) $\varphi \in Y_p$;

(ii) $V_\varphi \in X_p$ *and* $\lim_{x\to\infty} \varphi(x) = 0$;

(iii) $xV_\varphi(x) \in L^p(dx/x)$ *and* $\lim_{x\to\infty} \varphi(x) = 0$;

(iv) φ *has locally bounded variation,* $\lim_{x\to\infty} \varphi(x) = 0$ *and*

$$\sum_{n\in\mathbb{Z}} 2^{np}\left(\int_{2^n}^{\infty} |d\varphi(x)|\right)^p < \infty$$

(v) φ *has locally bounded variation,* $\lim_{x\to\infty} \varphi(x) = 0$ *and*

$$\sum_{n\in\mathbb{Z}} 2^{np}\left(\int_{2^n}^{2^{n+1}} |d\varphi(x)|\right)^p < \infty;$$

(vi) $\sum_{n\in\mathbb{Z}} 2^{np}\|\varphi\|^p_{BV[2^n,2^{n+1}]} < \infty$;

(vii) $\sum_{n\in\mathbb{Z}} \|x\varphi(x)\|^p_{BV[2^n,2^{n+1}]} < \infty$.

Proof. To show that (i) implies (ii), it suffices to consider a non-increasing $\varphi \in X_p$; it is easily seen that then $\lim_{x\to\infty} \varphi(x) = 0$ and $V_\varphi = \varphi$, whence (ii) follows.

Conversely, suppose that (ii) holds. By considering real and imaginary parts, we may assume that φ is real. Then $\varphi = V_\varphi - (V_\varphi - \varphi)$, where V_φ and $V_\varphi - \varphi$ are non-increasing functions in X_p; note that $0 \le V_\varphi - \varphi \le 2V_\varphi$. Consequently (i) holds.

Since V_φ is non-increasing, (ii) $\Leftrightarrow$ (iii) follows by Lemma 5.2.

Next, (iii) $\Leftrightarrow$ (iv) follows easily because $V_\varphi(x) = \int_x^\infty |d\varphi(y)|$, and (iv) $\Leftrightarrow$ (v) is easily verified.

If (iv) holds, then $\|\varphi\|_{BV[2^n,2^{n+1}]} = \int_{2^n}^{2^{n+1}} |d\varphi(x)| + \sup_{[2^n,2^{n+1}]} |\varphi| \le 2\int_{2^n}^{\infty} |d\varphi(x)|$ and (vi) follows. Conversely, (vi) immediately implies (v).

Finally, for any functions φ and ψ on an interval I, we have $\|\psi\varphi\|_{BV(I)} \le \|\psi\|_{BV(I)}\|\varphi\|_{BV(I)}$, and the equivalence (vi) $\Leftrightarrow$ (vii) follows by taking $\psi(x) = x$ and $\psi(x) = 1/x$. □

We can define a norm in Y_p (a quasi-norm for $p < 1$) by

$$\|\varphi\|_{Y_p} \stackrel{\text{def}}{=} \left(\sum_{n\in\mathbb{Z}} 2^{np}\left(\int_{2^n}^{2^{n+1}} |d\varphi(x)|\right)^p\right)^{1/p}; \tag{5.1}$$

an alternative is

$$\left(\int_0^\infty |xV_\varphi(x)|^p \frac{dx}{x}\right)^{1/p},$$

which yields an equivalent (quasi-)norm.

We obtain as corollaries to Theorems 5.1 and 5.3 the following simple sufficient conditions for $Q_\varphi \in \boldsymbol{S}_p$.

Corollary 5.4. *Suppose φ is absolutely continuous on* $[0, \infty)$, $\lim_{x\to\infty} \varphi(x) = 0$ *and* $\sup_{x>0} x^\gamma |\varphi'(x)| < \infty$ *for some* $\gamma > 2$. *Then* $\varphi \in Y_p$ *for every* $p > 0$ *and thus* $Q_\varphi \in S_p$ *for every* $p > 1/2$.

Proof. $V_\varphi(x)$ is bounded and $|V_\varphi(x)| \le C \cdot x^{1-\gamma}$, and thus $x V_\varphi(x) \in L^p(dx/x)$ for every $p > 0$. □

Corollary 5.5. *If φ has bounded variation and support in a finite interval, then* $\varphi \in Y_p$ *for every* $p > 0$ *and thus* $Q_\varphi \in S_p$ *for every* $p > 1/2$. □

6. $p = 1$, first results

Let us now consider the case $p = 1$. We know already that $\varphi \in X_1$ is a necessary and $\varphi \in Y_1$ a sufficient condition for $Q_\varphi \in S_1$. We will later see that neither condition is both necessary and sufficient (see the example following Theorem 6.5). We restate these results as follows.

Theorem 6.1. *Suppose φ has locally bounded variation,* $\int_0^\infty x|d\varphi(x)| < \infty$ *and* $\lim_{x\to\infty} \varphi(x) = 0$. *Then* $Q_\varphi \in S_1$.

Proof. It follows from Theorem 5.3 and the calculation

$$\int_0^\infty V_\varphi(x)\,dx = \int_0^\infty \int_x^\infty |d\varphi(y)|\,dx = \int_0^\infty y|d\varphi(y)|$$

that the assumption is equivalent to $\varphi \in Y_1$, so the result follows from Theorem 5.1. □

Theorem 6.2. *If* $Q_\varphi \in S_1$, *then* $\varphi \in X_1$. *Furthermore,* $\varphi \in L^1(0, \infty)$ *and*

$$\operatorname{trace} Q_\varphi = \int_0^\infty \varphi(x)\,dx.$$

Proof. By the Remark at the end of §3, $Q_\varphi \in S_1 \Rightarrow \varphi \in X_1$. Next, $X_1 \subset L^1(0, \infty)$, since by the Cauchy–Schwarz inequality and (1.3)

$$\int_0^\infty |\varphi| \le \sum_{n\in\mathbb{Z}} 2^{n/2} \Big(\int_{2^n}^{2^{n+1}} |\varphi(x)|^2\,dx \Big)^{1/2} < \infty.$$

Finally, the trace formula follows from Theorem 6.3 below, since with $k(x, y) = \varphi(\max\{x, y\})$ for $x, y > 0$,

$$\int_{-\infty}^\infty k(x, x + a)\,dx = \int_{|a|}^\infty \varphi(x)\,dx. \qquad \square$$

Remark. This theorem gives a formula for the trace of Q_φ if that operator has a trace; i.e., if it is in the trace class, S_1. The theorem also shows that if Q_φ is in the trace class then we must have $\varphi \in X_1$. We will see later in this section that $\varphi \in X_1$ is not enough to insure that Q_φ is in the trace class. However, we will see later, Corollary 10.2, that $\varphi \in X_1$ is sufficient to insure that Q_φ and related operators do have a *Dixmier trace*.

In the previous theorem we used following fact from [A], improving an earlier result in [B]:

Theorem 6.3. *If T is an integral operator on $L^2(\mathbb{R})$ of class S_1 with kernel function k, then the function*

$$x \mapsto k(x, x+a), \quad x \in \mathbb{R},$$

is in $L^1(\mathbb{R})$ for almost all $a \in \mathbb{R}$ and the function

$$a \mapsto \int_{\mathbb{R}} k(x, x+a)\,dx, \quad a \in \mathbb{R},$$

is almost everywhere equal to the Fourier transform $\mathcal{F}h$ of a function $h \in L^1(\mathbb{R})$, in particular, it coincides a.e. with a continuous function on $\mathbb{R}$. Moreover,

$$\operatorname{trace} T = (\mathcal{F}h)(0).$$

Proof. It is sufficient to prove the result when $k(x, y) = f(x)g(y)$ with f and g in $L^2(\mathbb{R})$. It is then straightforward to verify, with $h(\xi) = (\mathcal{F}f)(\xi)(\mathcal{F}g)(-\xi)$. □

We can reduce the estimation of $\|Q_\varphi\|_{S_1}$ to the estimation of $\|Q_\varphi^I\|_{S_1}$ for dyadic intervals I.

Theorem 6.4. *Let $\varphi \in L^1_{\text{loc}}(\mathbb{R}_+)$ and let $I_n = [2^n, 2^{n+1}]$. Then $Q_\varphi \in S_1$ if and only if $\varphi \in X_1$ and*

$$\sum_{n\in\mathbb{Z}} \|Q_\varphi^{I_n}\|_{S_1} < \infty. \tag{6.1}$$

Proof. Consider the sets $A^{(k)}$ defined by (3.4) and consider their symmetric images $A^{(-k)}$ about the line $\{(x, y) : x = y\}$. As in (3.8) we have

$$\sum_{k\geq 1} \|\mathcal{P}_{A^{(k)}} Q_\varphi\|_{S_1} \leq C \sum_{n\in\mathbb{Z}} \left(2^n \int_{2^n}^{2^{n+1}} |\varphi(x)|^2\,dx\right)^{1/2}.$$

Similarly,

$$\sum_{k\geq 1} \|\mathcal{P}_{A^{(-k)}} Q_\varphi\|_{S_1} \leq C \sum_{n\in\mathbb{Z}} \left(2^n \int_{2^n}^{2^{n+1}} |\varphi(x)|^2\,dx\right)^{1/2}.$$

It thus follows from $\varphi \in X_1$ that

$$\mathcal{P}_A Q_\varphi \in S_1 \quad \text{and} \quad \mathcal{P}_{\check{A}} Q_\varphi \in S_1.$$

where

$$A = \bigcup_{k \geq 1} A^{(k)} \quad \text{and} \quad \check{A} = \bigcup_{k \leq -1} A^{(k)}.$$

Consequently, using Theorem 6.2, $Q_\varphi \in S_1$ if and only if $\varphi \in X_1$ and $\mathcal{P}_B Q_\varphi \in S_1$, where B is defined by (3.1). Since the projections of the sets B_n onto the coordinate axes are disjoint, and $\mathcal{P}_{B_n} Q_\varphi = Q_\varphi^{I_n}$, it follows that $\mathcal{P}_B Q_\varphi \in S_1$ if and only if (6.1) holds. □

Let $n \in \mathbb{Z}$. It is easy to see that $Q_\varphi^{I_n} \in S_1$ if and only if $Q_{\varphi_n} \in S_1$, where

$$\varphi_n(x) \stackrel{\text{def}}{=} \begin{cases} \varphi(x + 2^n), & x \in [0, 2^n] \\ 0, & \text{otherwise.} \end{cases}$$

Hence, the question of when Q_φ belongs to S_1 reduces to the question of estimating $\|Q_\varphi\|_{S_1}$ for functions φ supported on finite intervals.

Remark. For $0 < p < 1$, it can similarly be shown that $Q_\varphi \in S_p$ if $\varphi \in X_p$ and $\sum_{n \in \mathbb{Z}} \|Q_\varphi^{I_n}\|_{S_p}^p < \infty$. We do not know whether the converse holds.

We next show that $\varphi \in X_1$ is not sufficient for $Q_\varphi \in S_1$.

Theorem 6.5. *Let $\varphi_N(x) = e^{2\pi i N x} \chi_{[0,1]}(x)$ for $N = 1, 2, \ldots$. Then*

$$s_n(Q_{\varphi_N}) \asymp \min\left\{\frac{1}{n+1}, \frac{N}{(n+1)^2}\right\},$$

and so

$$\|Q_{\varphi_N}\|_{S_1} \asymp \log(N+1). \tag{6.2}$$

Note that $\asymp$ means that the ratio of the two sides are bounded from above and below by positive constants (not depending on n or N). Clearly,

$$\|\varphi_N\|_{X_1} = \|\chi_{[0,1]}\|_{X_1} = C$$

is independent of N. This shows, by the closed graph theorem, that

$$\varphi \in X_1 \not\Rightarrow Q_\varphi \in S_1;$$

a concrete counterexample is given in the example following the proof of Theorem 6.5. Moreover, $\varphi_N \in X_p$ for any $p > 0$, again with norm independent of N, so for every $p \leq 1$

$$\varphi \in X_p \not\Rightarrow Q_\varphi \in S_1 \quad \text{and} \quad \varphi \in X_p \not\Rightarrow Q_\varphi \in S_p.$$

It also follows from Theorem 6.5 that

$$\|Q_{\varphi_N}\|_{S_p} \asymp N^{(1-p)/p}, \quad \tfrac{1}{2} < p < 1. \tag{6.3}$$

Proof of Theorem 6.5. Since multiplication by a unimodular function is a unitary operator, the singular values of Q_{φ_N} are the same as the singular values $s_n(T_N)$, where T_N is the integral operator on $L^2[0,1]$ with kernel

$$\begin{aligned} e^{-\pi \mathrm{i} N x}\varphi_N\big(\max(x,y)\big)e^{-\pi \mathrm{i} N y} &= \exp\big(2\pi \mathrm{i} N\max(x,y) - \pi \mathrm{i} N x - \pi \mathrm{i} N y\big) \\ &= \exp\big(\pi \mathrm{i} N|x-y|\big). \end{aligned}$$

Let $g_N(x) = e^{\pi \mathrm{i} N|x|}$ for $|x| \le 1$, and extend g_N to a function on $\mathbb{R}$ with period 2. Let T_N' be the integral operator on $L^2[-1,1]$ with kernel $g_N(x-y)$. If $I_+ = [0,1]$, $I_- = [-1,0]$ and $A_{\alpha\beta} = I_\alpha \times I_\beta$, $\alpha, \beta \in \{+,-\}$, then $\mathcal{P}_{A_{++}}T_N' = T_N$ and thus

$$s_n(Q_{\varphi_N}) = s_n(T_N) \le s_n(T_N'). \tag{6.4}$$

Moreover, each $\mathcal{P}_{A_{\alpha\beta}}T_N'$ is by a translation unitarily equivalent to either T_N or the integral operator on $L^2[0,1]$ with kernel $g_N(x-y-1) = \exp\big(\pi \mathrm{i} N(1-|x-y|)\big) = (-1)^N\overline{g_N}(x-y)$, which has the same singular values. Hence, by (2.1),

$$s_{4n}(T_N') \le 4s_n(T_N) = 4s_n(Q_{\varphi_N}). \tag{6.5}$$

T_N' is a convolution operator on the circle $\mathbb{R}/2\mathbb{Z}$, so the elements of the orthonormal basis $\{2^{-1/2}\exp(\pi i k x)\}_{k=-\infty}^{\infty}$ in $L^2[-1,1]$ are eigenvectors with eigenvalues

$$\begin{aligned} \hat{g}_N(k) &\stackrel{\text{def}}{=} \int_{-1}^{1} g_N(x)e^{-\pi i k x}\,dx = \int_{-1}^{1} e^{\pi \mathrm{i} N|x| - \pi i k x}\,dx \\ &= \int_{-1}^{0} e^{-\pi \mathrm{i}(N+k)x}\,dx + \int_0^1 e^{\pi \mathrm{i}(N-k)x}\,dx \\ &= \begin{cases} 0, & k \equiv N \pmod 2,\ k \ne \pm N, \\ 1, & k = \pm N, \\ \frac{-2}{\pi \mathrm{i}(N+k)} + \frac{-2}{\pi \mathrm{i}(N-k)} = \frac{4\mathrm{i}N}{\pi(N^2-k^2)}, & k \equiv N+1 \pmod 2). \end{cases} \end{aligned}$$

Consequently, the singular values $s_n(T_N)$ are the absolute values $|\hat{g}_N(k)|$, $k \in \mathbb{Z}$, arranged in decreasing order. This easily yields

$$s_n(T_N') \asymp \min\left\{\frac{1}{n+1}, \frac{N}{(n+1)^2}\right\},$$

and the result follows by (6.4) and (6.5). □

Remark. A related method is used in a more general context in §14. Indeed, the estimates (6.2) and (6.3) follow easily from Theorem 14.10 (with the norm estimates implicit there).

Moreover, (6.2) and (6.3) also follow from the results in §16, obtained by a different method.

Example. For a concrete counterexample we let $N_k \geq 2$ be integers and define

$$\varphi = \sum_{k=1}^{\infty} \exp(2\pi i 2^k N_k x)\chi_{(2^{-k},2^{1-k})}(x).$$

Then $|\varphi| = \chi_{[0,1]}$; in particular, $\varphi \in X_p$ for every $p > 0$. On the other hand, for every k, by Lemma 2.1 and Theorem 6.5,

$$\|Q_\varphi\|_{S_1} \geq \left\|Q_\varphi^{[2^{-k},2^{1-k}]}\right\|_{S_1} = 2^{-k}\|Q_{\varphi_{N_k}}\|_{S_1} \geq c2^{-k}\log N_k.$$

Choosing $N_k = 2^{3^k}$, we find that $Q_\varphi \notin S_1$.

Other choices yield further interesting examples. Thus, $N_k = 2^k$ yields a symbol $\varphi \in X_1$ but $\varphi \notin Y_1$ such that, by Theorem 6.4, $Q_\varphi \in S_1$. In fact, using the Remark followed by Theorem 6.5, we can conclude that $Q_\varphi \in S_p$ for every $p > 1/2$.

The choice $N_k = 2^{k^2}$ yields a symbol $\varphi \in X_1 \setminus Y_1$ such that $Q_\varphi \in S_1$ but $Q_\varphi \notin S_p$ for $p < 1$.

Remark. Theorem 6.5 implies also that $\varphi \in X_1$ does not imply $Q_\varphi \in S_{1,q}$ for any Schatten–Lorentz space $S_{1,q}$ with $q < \infty$.

Let us prove now that the condition $Q_\varphi \in S_1$ does not imply that $Q_\varphi^+ \in S_1$. Moreover, as previously shown by Nowak [No], we show that there are no non-zero operators Q_φ^+ of class S_1. (A more refined result will be given in Theorem 10.1.)

Theorem 6.6. *Suppose that $\varphi \in L^1_{\text{loc}}(\mathbb{R}_+)$. If $Q_\varphi^+ \in S_1$, then φ is the zero function.*

Proof. Let $Q_\varphi^+ \in S_1$, and let k be the kernel function of Q_φ^+, extended by 0 to $\mathbb{R}^2$, i.e.

$$k(x,y) = \begin{cases} \varphi(x), & 0 < y < x \\ 0, & \text{otherwise.} \end{cases}$$

Let $\Delta \subset \mathbb{R}_+$ be a compact interval. Consider the operator $P_\Delta Q_\varphi^+ P_\Delta$, where P_Δ is multiplication by χ_Δ. Clearly, $P_\Delta Q_\varphi^+ P_\Delta$ is an integral operator with kernel function $k_\Delta \stackrel{\text{def}}{=} \chi_{\Delta\times\Delta}k$, (recall that χ_A is the characteristic function of a set A) and $P_\Delta Q_\varphi^+ P_\Delta \in S_1$. It follows from Theorem 6.3 that the function u

$$u(a) \stackrel{\text{def}}{=} \int_{\mathbb{R}} k_\Delta(x, x+a)\,dx$$

is a.e. equal to a continuous function $\mathbb{R}$. Clearly, $u(a) = 0$ if $a > 0$, and for $a \uparrow 0$

$$u(a) \to \int_\Delta \varphi(x)\,dx.$$

Hence,

$$\int_\Delta \varphi(x)\,dx = 0, \quad \text{for any interval } \Delta \subset \mathbb{R}_+.$$

Consequently, $\varphi = 0$. ∎

7. Schur multipliers of the form $\psi(\max\{x, y\})$, $x, y \in \mathbb{R}_+$

Let $0 < p \le 2$. Recall that a function ω on $\mathbb{R}^2$ is called a *Schur multiplier of* $\boldsymbol{S}_p$ if the integral operator on $L^2(\mathbb{R})$ with kernel function ωk belongs to $\boldsymbol{S}_p$ whenever the integral operator with kernel function k does. If $2 < p < \infty$, the class of Schur multipliers of $\boldsymbol{S}_p$ can be defined by duality: ω is a Schur multiplier of $\boldsymbol{S}_p$ if ω is a Schur multiplier of $\boldsymbol{S}_{p'}$, where $p' \stackrel{\text{def}}{=} p/(p-1)$. We say that ω is a *Schur multiplier of weak type* (p, p), $0 < p \le 2$, if the integral operator with kernel function ωk belongs to $\boldsymbol{S}_{p,\infty}$ whenever the integral operator with kernel function k belongs to $\boldsymbol{S}_p$. Note that in a similar way one can define Schur multipliers for an arbitrary measure space $(\mathcal{X}, \mu)$.

In this section for a function $\psi \in L^\infty(\mathbb{R})$ we find a sufficient condition for the function $(x, y) \mapsto \psi(\max\{x, y\})$, $(x, y) \in \mathbb{R}^2$, to be a Schur multiplier of $\boldsymbol{S}_p$. We also obtain a sufficient condition for this function to be a Schur multiplier of weak type $(1/2, 1/2)$.

Theorem 7.1. *Let* $1 < p < \infty$ *and let* $\psi \in L^\infty(\mathbb{R})$. *Then the function*

$$(x, y) \mapsto \psi(\max\{x, y\}), \quad (x, y) \in \mathbb{R}^2,$$

is a Schur multiplier of $\boldsymbol{S}_p$.

Proof. Since the triangular projection $\mathcal{P}$ is bounded on $\boldsymbol{S}_p$, the characteristic function of the set $\{(x, y) : x > y\}$ is a Schur multiplier of $\boldsymbol{S}_p$. It remains to observe that

$$\psi(\max\{x, y\}) = \psi(x)\chi_{\{(x,y):x>y\}} + \psi(y)\chi_{\{(x,y):x<y\}}. \tag{7.1}$$

□

Theorem 7.2. *Let* $\psi \in L^\infty(\mathbb{R})$. *Then the function*

$$(x, y) \mapsto \psi(\max\{x, y\}), \quad (x, y) \in \mathbb{R}^2,$$

is a Schur multiplier of weak type $(1, 1)$.

Proof. The result follows from (7.1) and the fact that the triangular projection $\mathcal{P}$ has weak type $(1, 1)$ (see (2.9)). □

Theorem 7.3. *Let* $1/2 < p < \infty$ *and let* ψ *be a function of bounded variation. Then the function* $(x, y) \mapsto \psi(\max\{x, y\})$ *on* $\mathbb{R}^2$ *is a Schur multiplier of* $\boldsymbol{S}_p$.

Proof. By Theorem 7.1 we may assume that $p \le 1$.

We consider first the case when ψ is absolutely continuous, i.e.,

$$\psi(x) = \int_x^\infty h(t)\,dt + C, \quad h \in L^1(\mathbb{R}). \tag{7.2}$$

We may assume that $C = 0$.

Let ξ and η be functions in L^2 and let T be the integral operator defined by

$$(Tf)(x) = \int_{\mathbb{R}} \xi(x)\eta(y)\psi(\max\{x,y\})f(y)\,dy, \quad f \in L^2(\mathbb{R}). \tag{7.3}$$

We have to prove that

$$\|T\|_{S_p} \le C(p,h)\|\xi\|_{L^2(\mathbb{R})}\|\eta\|_{L^2(\mathbb{R})},$$

where $C(p,h)$ may depend only on p and h.

We can factorize the function h in the form $h = uv$, where $u, v \in L^2(\mathbb{R})$. Put

$$k_1(x,y) \stackrel{\text{def}}{=} \begin{cases} 0, & y < x, \\ \xi(x)u(y), & y > x, \end{cases}$$

and

$$k_2(x,y) \stackrel{\text{def}}{=} \begin{cases} \eta(y)v(x), & y < x, \\ 0, & y > x. \end{cases}$$

Let T_1 and T_2 be the integral operators on $L^2(\mathbb{R})$ with kernel functions k_1 and k_2. It follow from the boundedness of the triangular projection that if $1 < q < \infty$, then $\|T_1\|_{S_q} \le C(q)\|\xi\|_{L^2}\|u\|_{L^2}$ and $\|T_2\|_{S_q} \le C(q)\|\eta\|_{L^2}\|v\|_{L^2}$, where $C(q)$ may depend only on q. It is also easy to verify that $T = T_1T_2$. It follows that

$$\|T\|_{S_p} \le \|T_1\|_{S_{2p}}\|T_2\|_{S_{2p}} \le (C(2p))^2\|\xi\|_{L^2}\|\eta\|_{L^2}\|u\|_{L^2}\|v\|_{L^2}.$$

To reduce the general case to the case of an absolutely continuous function ψ, we can consider a standard regularization process. □

We complete this section with the following result.

Theorem 7.4. *Suppose that ψ is a function of bounded variation. Then the function $(x,y) \mapsto \psi(\max\{x,y\})$ is a Schur multiplier of weak type $(1/2, 1/2)$.*

We need two lemmata.

Lemma 7.5. *Let $0 < p < 1$ and let $A \in S_{p,\infty}$. Set*

$$\|A\|^*_{S_{p,\infty}} \stackrel{\text{def}}{=} \sup_{t>0}\left(t^{p-1}\sum_{n\ge 0}\min\{t, s_n(A)\}\right)^{1/p}.$$

Then

$$\|A\|_{S_{p,\infty}} \le \|A\|^*_{S_{p,\infty}} \le (1-p)^{-1/p}\|A\|_{S_{p,\infty}}.$$

Proof. Taking $t = s_n(A)$ in the definition of $\|\cdot\|^*_{S_{p,\infty}}$, we obtain

$$(n+1)^{\frac{1}{p}}s_n(A) \le \|A\|^*_{S_{p,\infty}}.$$

Consequently, $\|A\|_{S_{p,\infty}} \le \|A\|^*_{S_{p,\infty}}$. Next, we have

$$t^{p-1}\sum_{n\ge 0}\min\{t, s_n(A)\} \le t^{p-1}\sum_{n\ge 0}\min\{t, \|A\|_{S_{p,\infty}}(n+1)^{-\frac{1}{p}}\}$$

$$\le t^{p-1}\int_0^\infty \min\{t, \|A\|_{S_{p,\infty}}x^{-1/p}\}\,dx = (1-p)^{-1}\|A\|^p_{S_{p,\infty}}.$$

Hence, $\|A\|^*_{S_{p,\infty}} \le (1-p)^{-1/p}\|A\|_{S_{p,\infty}}$. □

Lemma 7.6. *If* $0 < p < 1$, *then* $\|\cdot\|^*_{S_{p,\infty}}$ *is a p-norm, i.e.,*

$$\|A_1 + A_2\|^{*p}_{S_{p,\infty}} \le \|A_1\|^{*p}_{S_{p,\infty}} + \|A_2\|^{*p}_{S_{p,\infty}}, \quad A_1, A_2 \in S_{p,\infty}.$$

Proof. By Rotfeld's theorem [R], if Φ is a concave nondecreasing function on $\mathbb{R}_+$ such that $\Phi(0^+) = 0$, then

$$\sum_{j=0}^m \Phi(s_j(A_1 + A_2)) \le \sum_{j=0}^m \Phi(s_j(A_1)) + \sum_{j=0}^m \Phi(s_j(A_2)), \quad m \in \mathbb{Z}_+. \tag{7.4}$$

For $t > 0$ we define the function Φ_t on $\mathbb{R}_+$ by

$$\Phi_t(x) = t^{p-1}\min\{t, x\}.$$

Clearly,

$$\|A\|^{*p}_{S_{p,\infty}} = \sup_{t>0}\sum_{j\ge 0}\Phi_t(s_j(A)).$$

It remains to apply 7.4 for Φ_t and take the *supremum* over $t > 0$. □

Note that the fact that for $p < 1$ the space $L^{p,\infty}$ has a p-norm that is equivalent to the initial quasi-norm is well known (see [K]).

Proof of Theorem 7.4. The proof is similar to the proof of Theorem 7.3. Again, it is sufficient to assume that ψ has the form (7.2) with $C = 0$.

By Lemma 7.5 and Lemma 7.6, to prove that our function is a Schur multiplier of weak type $(1/2, 1/2)$, it is sufficient to prove that if T is defined by (7.3), then $T \in S_{1/2,\infty}$. Let u, v, k_1, k_2, T_1, and T_2 be as in the proof of Theorem 7.3. Let $n \ge 2$ and $n = m_1 + m_2$, where $|m_1 - n/2| \le 1/2$ and $|m_2 - n/2| \le 1/2$. Since the triangular projection $\mathcal{P}$ has weak type $(1, 1)$ (see (2.9)), we have

$$\|T_1\|_{S_{1,\infty}} \le C\|\xi\|_{L^2}\|u\|_{L^2} \quad \text{and} \quad \|T_2\|_{S_{1,\infty}} \le C\|\eta\|_{L^2}\|v\|_{L^2}.$$

Hence, by (2.2),

$$s_n(T) \le s_{m_1}(T_1)s_{m_2}(T_2) \le C\frac{1}{m_1m_2}\|\xi\|_{L^2}\|u\|_{L^2}\|\eta\|_{L^2}\|v\|_{L^2}$$

$$\le C\frac{1}{n^2}\|\xi\|_{L^2}\|u\|_{L^2}\|\eta\|_{L^2}\|v\|_{L^2}$$

which completes the proof. □

We let $\mathfrak{M}_p$ denote the space of Schur multipliers of $\boldsymbol{S}_p$, and put

$$\|\omega\|_{\mathfrak{M}_p} \stackrel{\text{def}}{=} \sup \|\omega k\|_{\boldsymbol{S}_p},$$

where the *supremum* is taken over all integral operators with kernel $k \in L^2$ such that $\|k\|_{\boldsymbol{S}_p} = 1$. Here by $\|k\|_{\boldsymbol{S}_p}$ we mean the $\boldsymbol{S}_p$ norm (quasi-norm if $p < 1$) of the integral operator with kernel function k. If ω is a Schur multiplier of weak type (p, p), we put

$$\|\omega\|_{\mathfrak{M}_{p,\mathrm{w}}} \stackrel{\text{def}}{=} \sup \|\omega k\|_{\boldsymbol{S}_{p,\infty}},$$

where the *supremum* is taken over all integral operators with kernel $k \in L^2$ such that $\|k\|_{\boldsymbol{S}_p} = 1$.

Remark. It is clear from the proofs of Theorems 7.1, 7.2, 7.3, and 7.4 that under the hypotheses of Theorem 7.1 we have

$$\|\psi(\max\{x, y\})\|_{\mathfrak{M}_p} \le C(p)\|\psi\|_{L^\infty}$$

and

$$\|\psi(\max\{x, y\})\|_{\mathfrak{M}_{1,\mathrm{w}}} \le C\|\psi\|_{L^\infty}$$

while under the hypotheses of Theorem 7.3 we have

$$\|\psi(\max\{x, y\})\|_{\mathfrak{M}_p} \le C(p)\|\psi\|_{BV},$$

and

$$\|\psi(\max\{x, y\})\|_{\mathfrak{M}_{1/2,\mathrm{w}}} \le C\|\psi\|_{BV}.$$

Here we use the notation

$$\|\varphi\|'_{BV} = \int_{\mathbb{R}} |d\varphi| \quad \text{and} \quad \|\varphi\|_{BV} = \|\varphi\|'_{BV} + \|\varphi\|_{L^\infty}.$$

The following result gives us a more accurate estimate for $\|\psi(\max\{x, y\})\|_{\mathfrak{M}_p}$ in the case $1/2 < p \le 1$.

Theorem 7.7. *Let ψ be a function of bounded variation on $\mathbb{R}$. Then*

$$\|\psi(\max\{x, y\})\|_{\mathfrak{M}_1} \le C\|\psi\|_{L^\infty} \log\left(2 + \frac{\|\psi\|'_{BV}}{\|\psi\|_{L^\infty}}\right)$$

and

$$\|\psi(\max\{x, y\})\|_{\mathfrak{M}_p} \le C\|\psi\|_{L^\infty}^{2-1/p}\|\psi\|_{BV}^{1/p-1}, \quad \frac{1}{2} < p < 1.$$

Proof. Let ξ and η be function in L^2 such that $\|\xi\|_{L^2} = \|\eta\|_{L^2} = 1$. We have to estimate the S_p-norm of the integral operator with kernel $\psi(\max\{x, y\})\xi(x)\eta(y)$. Let $\{s_n\}_{n\ge0}$ be the sequence of s-numbers of this integral operator. Theorem 7.2 implies that

$$s_n \le C\frac{\|\psi\|_{L^\infty}}{n+1}.$$

Theorem 7.4 implies that

$$s_n \le C\frac{\|\psi\|_{L^\infty} + \|\psi\|'_{BV}}{(n+1)^2}.$$

Consequently,

$$s_n \le C\min\left\{\frac{\|\psi\|_{L^\infty}}{n+1}, \frac{\|\psi\|_{L^\infty} + \|\psi\|'_{BV}}{(n+1)^2}\right\}.$$

The rest of the proof is an easy exercise. □

8. The case $p = 1/2$

Theorem 8.1. *Let φ be a function of bounded variation on $[0, 1]$. Then $Q_\varphi^{[0,1]} \in S_{1/2,\infty}$.*

Proof. We may extend the function φ by putting $\varphi(t) = 0$ for $t \in \mathbb{R}_+ \setminus [0, 1]$. Clearly, the integral operator with kernel function $\chi_{[0,1]^2}$ has rank one, and so it belongs to $S_{1/2}$. Consequently, by Theorem 7.4, the integral operator with kernel function $\chi_{[0,1]^2}\varphi(\max\{x, y\})$ belongs to $S_{1/2,\infty}$. □

To see that this result cannot be improved to $Q_\varphi^{[0,1]} \in S_{1/2}$, we begin with two extensions of Theorem 6.6.

Lemma 8.2. *Let $\varphi, \psi \in L^1_{\text{loc}}(\mathbb{R})$. Suppose that the integral operator with kernel function k,*

$$k(x, y) = \begin{cases} \varphi(x)\psi(y), & y \le x, \\ 0, & \text{otherwise}, \end{cases}$$

belongs to S_1. Then $\varphi\psi = 0$ almost everywhere.

Proof. First we assume that $\varphi, \psi \in L^2(\mathbb{R})$. By Theorem 6.3 we have

$$\lim_{a\to 0-} \int_{\mathbb{R}} k(x, x+a)\,dx = \lim_{a\to 0+} \int_{\mathbb{R}} k(x, x+a)\,dx = 0,$$

whence $\int_{\mathbb{R}} \varphi(x)\psi(x)\,dx = 0$. Now let φ and ψ be arbitrary functions in $L^1_{\mathrm{loc}}(\mathbb{R})$. Suppose that f and g are functions in $L^\infty(\mathbb{R})$ such that $f\varphi \in L^2(\mathbb{R})$ and $g\psi \in L^2(\mathbb{R})$. Consider the integral operator with kernel function

$$(x, y) \mapsto f(x)k(x, y)g(y).$$

Clearly, it belongs to $\boldsymbol{S}_1$. It follows from what we have just proved that

$$\int_{\mathbb{R}} \varphi(x)f(x)\psi(x)g(x)\,dx = 0.$$

Since f and g are arbitrary, this implies the result. □

Lemma 8.3. *Let $\varphi \in L^1_{\mathrm{loc}}(\mathbb{R}_+)$ and let $A = (0, \infty) \times \Delta$, where Δ is a measurable subset of $(0, \infty)$. Suppose that $\mathcal{P}_A Q^+_\varphi \in \boldsymbol{S}_1$. Then $\varphi = 0$ almost everywhere on Δ.*

Proof. The result follows easily from Lemma 8.2 with $\psi = \chi_\Delta$. □

Lemma 8.4. *Let φ be a nonincreasing locally absolutely continuous function on $\mathbb{R}_+$ and let Δ be a measurable subset of $\mathbb{R}_+$. Suppose that the integral operator with kernel function*

$$(x, y) \mapsto \varphi(\max\{x, y\})\chi_\Delta(x)\chi_\Delta(y)$$

belongs to $\boldsymbol{S}_{1/2}$. Then $\varphi' = 0$ almost everywhere on Δ.

Proof. By replacing Δ with $\Delta \cap (a, b)$, we may assume that $\Delta \subset [a, b]$ where $0 < a < b < \infty$. We may then subtract $\varphi(b)$ and modify φ outside $[a, b]$ so that φ becomes constant on $(0, a]$ and zero on $[b, \infty)$. Let $\psi = (-\varphi')^{1/2}$. Since $x\varphi(x)$ is bounded, we have $\psi \in X_\infty$. Thus Q^+_ψ is bounded and by Theorem 4.2, Q_φ admits a factorization $Q_\varphi = (Q^+_\psi)^* Q^+_\psi$. Let M be multiplication by χ_Δ. It follows that $MQ_\varphi M = (Q^+_\psi M)^*(Q^+_\psi M)$, and so $Q^+_\psi M \in \boldsymbol{S}_1$. The result follows now from Lemma 8.3. □

Corollary 8.5. *Let φ be as in Lemma 8.4 and let ψ be a function on $\mathbb{R}_+$ such that $Q_\psi \in \boldsymbol{S}_{1/2}$. Set $\Delta \stackrel{\text{def}}{=} \{x \in \mathbb{R}_+ : \psi(x) = \varphi(x)\}$. Suppose that ψ is differentiable almost everywhere on Δ. Then $\varphi' = \psi' = 0$ almost everywhere on Δ.*

Proof. It suffices to apply Lemma 8.4. □

Theorem 8.6. *Let ψ be an absolutely continuous function on $[0, 1]$. Suppose $Q^{[0,1]}_\psi \in \boldsymbol{S}_{1/2}$. Then ψ is constant.*

Proof. Clearly, we may assume that ψ is a real function. Suppose that ψ is not constant. Then $\max\psi > \psi(1)$ or $\min\psi < \psi(1)$. To be definite, suppose that $\max\psi > \psi(1)$. We use the "sun rising method". Let

$$\Delta \stackrel{\text{def}}{=} \{x \in [0,1] : \psi(x) \geq \psi(t) \quad \text{for all} \quad t \geq x\}.$$

Clearly, Δ is closed and $1 \in \Delta$. Moreover, the restriction $\psi\big|[\alpha,\beta]$ is constant for any interval (α,β) such that $\alpha,\beta \in \Delta$ and $\Delta \cap (\alpha,\beta) = \varnothing$. Set

$$\varphi(x) = \max_{[x,1]\cap\Delta} \psi.$$

Clearly, φ is non-increasing, $\varphi|_\Delta = \psi|_\Delta$ and $\varphi\big|[\alpha,\beta]$ is constant for any interval (α,β) such that $\alpha,\beta \in \Delta$ and $\Delta \cap (\alpha,\beta) = \varnothing$. Consequently, φ is absolutely continuous. Thus, $\varphi' = 0$ almost everywhere on Δ by Corollary 8.5. Moreover, $\varphi' = 0$ outside Δ because φ is locally constant outside Δ. Consequently, $\varphi' = 0$ almost everywhere on $[0,1]$, and so $\varphi(t) = \varphi(1) = \psi(1)$ for any $t \in [0,1]$ which contradicts the condition $\max\psi > \psi(1)$. □

Corollary 8.7. *Suppose that $Q_\psi \in \boldsymbol{S}_{1/2}$. Then ψ is constant on any interval $I \subset \mathbb{R}_+$ on which ψ is absolutely continuous.* □

Corollary 8.8. *Suppose that ψ is locally absolutely continuous and $Q_\psi \in \boldsymbol{S}_{1/2}$. Then $\psi = 0$ everywhere on $\mathbb{R}_+$.* □

Lemma 8.9. *Let φ be a nonincreasing function on $\mathbb{R}_+$ with $\lim_{t\to+\infty}\varphi(t) = 0$ and let $\varepsilon > 0$. Then there exists a nonincreasing absolutely continuous function ψ on $\mathbb{R}_+$ such that $\lim_{t\to+\infty}\psi(t) = 0$ and $\boldsymbol{m}\{\varphi \neq \psi\} < \varepsilon$.*

Note that here and in what follows $\boldsymbol{m}$ denotes Lebesgue measure on $\mathbb{R}$ or normalized Lebesgue measure on the unit circle $\mathbb{T}$.

Proof. We may assume that φ is right-continuous on $(0,\infty)$ and that φ is bounded. Consider the positive measure μ on $\mathbb{R}_+$ such that $\varphi(t) = \mu(t,\infty)$ for any $t > 0$. Denote by μ_s the singular part of μ. There exists a Borel set $E \subset \mathbb{R}_+$ such that $\boldsymbol{m}(E) = 0$ and $\mu_s((0,+\infty)\setminus E) = 0$. We may find an open set U such that $E \subset U \subset \mathbb{R}_+$ and $\mu(U) < \varepsilon$. Let $U = \bigcup_{n\geq1}(a_n,b_n)$, where (a_n,b_n) are mutually disjoint. Set

$$f(t) = \begin{cases} -\varphi'(t), & t \in \mathbb{R}_+ \setminus E, \\ \dfrac{\mu[a_n,b_n)}{b_n - a_n}, & t \in (a_n,b_n). \end{cases}$$

Set $\psi(t) \stackrel{\text{def}}{=} \int_t^\infty f(s)ds$. Clearly, $\varphi = \psi$ outside U. □

Lemma 8.10. *Let φ a nonincreasing function on $\mathbb{R}_+$ with $\lim_{t\to+\infty}\varphi(t) = 0$. Suppose that the integral operator with kernel function*

$$(x,y) \mapsto \varphi(\max\{x,y\})\chi_\Delta(x)\chi_\Delta(y)$$

belongs to $S_{1/2}$ *for a measurable subset* Δ *of* $\mathbb{R}_+$. *Then* $\varphi' = 0$ *almost everywhere on* Δ.

Proof. The result follows from Lemmas 8.4 and 8.9. □

Theorem 8.11. *Let* ψ *be a function with bounded variation on* $[0, 1]$. *Suppose that* $Q_\psi^{[0,1]} \in S_{1/2}$. *Then* $\psi' = 0$ *almost everywhere on* $[0, 1]$.

Proof. Again, we may assume that ψ is real. We may also make the assumption that ψ is continuous at 0 and at 1, and $\psi(t) = \max\{\psi(t^-), \psi(t^+)\}$ for any $t \in (0, 1)$. With any nondegenerate closed interval $I \subset [0, 1]$ we associate the function $\varphi_I : I \to \mathbb{R}$ defined by $\varphi_I(x) = \sup\{\psi(t) : t \in I \quad \text{and} \quad t \geq x\}$. Set

$$\Delta(I) \stackrel{\text{def}}{=} \{x \in I : \varphi_I(x) = \psi(x)\}.$$

Clearly, $\Delta(I)$ is closed. By Lemma 8.10, $\psi' = \varphi_I' = 0$ almost everywhere on $\Delta(I)$.

Set $E_- = \{x \in (0, 1) : \psi'(x) < 0\}$. Let $a \in E_-$. Clearly, $a \in \Delta(I)$ if I is small enough and $a \in I$. Consequently,

$$E_- \subset \bigcup_{n=2}^{\infty} \left(\bigcup_{k=1}^{n-1} \Delta \left(\left[\frac{k-1}{n}, \frac{k+1}{n} \right] \right) \right).$$

We have shown that Lemma 8.10 implies $\boldsymbol{m}(\Delta(I) \cap E_-) = 0$ for every I. Consequently, $\boldsymbol{m}(E_-) = 0$. Thus, we have proved that $\psi' \geq 0$ almost everywhere. It remains to apply this result to $-\psi$. □

The following fact is an immediate consequence of Theorem 8.11.

Corollary 8.12. *Suppose that* $Q_\psi \in S_{1/2}$. *Then* $\psi' = 0$ *almost everywhere on any interval* $I \subset \mathbb{R}_+$ *on which* ψ *is of bounded variation.* □

9. Sturm–Liouville theory and $p = 1/2$

If φ is real, then Q_φ is self-adjoint, so its singular values are the absolute values of its eigenvalues. Hence, we next study the eigenvalues and eigenfunctions. For simplicity we consider only the case of symbols φ which vanish on $(1, \infty)$; thus it does not matter whether we consider Q_φ on $L^2(\mathbb{R}_+)$ or $Q_\varphi^{[0,1]}$ on $L^2[0, 1]$.

Suppose that $\varphi \in C^1[0, 1]$ and that φ is real. Let λ be a non-zero eigenvalue of $Q_\varphi^{[0,1]}$ and $g \in L^2[0, 1]$ a corresponding eigenfunction,

$$\lambda g(x) = \varphi(x) \int_0^x g(y)\, dy + \int_x^1 \varphi(y) g(y)\, dy, \quad 0 \leq x \leq 1. \tag{9.1}$$

The right hand side is a continuous function of x; hence, $g \in C[0,1]$ and (9.1) holds for every x (and not just a.e.). By (9.1) again, $g \in C^1[0,1]$ with

$$\lambda g'(x) = \varphi'(x) \int_0^x g(y)\,dy. \tag{9.2}$$

Define $G(x) = \int_0^x g(y)\,dy$. Then (9.2) can be written as the system

$$\begin{aligned} G'(x) &= g(x) \\ g'(x) &= \lambda^{-1}\varphi'(x)G(x) \end{aligned} \tag{9.3}$$

and we have, using (9.1) with $x = 1$, the boundary conditions

$$\begin{aligned} G(0) &= 0, \\ g(1) &= \lambda^{-1}\varphi(1)G(1). \end{aligned} \tag{9.4}$$

Conversely, any solution of (9.3) with the boundary conditions (9.4) satisfies (9.2) and (9.1), so the problem of finding the singular values of Q_φ reduces to finding the $\lambda \neq 0$ for which (9.3) and (9.4) have a solution. Note that (9.3) can be written as a Sturm–Liouville problem

$$\lambda G''(x) = \varphi'(x)G(x). \tag{9.5}$$

If $g(x_0) = G(x_0) = 0$ for some $x_0 \in [0,1]$, then (9.3) shows, by the standard uniqueness theorem, that g vanishes identically, a contradiction. In particular, since $G(0) = 0$, we have $g(0) \neq 0$, and we may normalize the eigenfunction g by $g(0) = 1$.

For every $\lambda \neq 0$, (9.3) has a unique solution (G_λ, g_λ) with $G_\lambda(0) = 0$, $g_\lambda(0) = 1$. It thus follows that all non-zero eigenvalues of Q_φ are simple, and that $\lambda \neq 0$ is an eigenvalue if and only if

$$g_\lambda(1) = \lambda^{-1}\varphi(1)G_\lambda(1). \tag{9.6}$$

Example. Let $\varphi(x) = 1 - x$, $x \in [0,1]$, and $\varphi(x) = 0$, $x > 1$. Then (9.3) gives $G''(x) = -\lambda^{-1}G(x)$, and we find the solutions $g_\lambda(x) = \cos \lambda^{-1/2}x$, $\lambda > 0$, and $g_\lambda(x) = \cosh |\lambda|^{-1/2}x$, $\lambda < 0$.

Since $\varphi(1) = 0$, condition (9.6) is simply $g_\lambda(1) = 0$, and the non-zero eigenvalues are given by $\cos \lambda^{-1/2} = 0$ or $\lambda^{-1/2} = (n + \frac{1}{2})\pi$, $n = 0, 1, \ldots$. ($\lambda < 0$ is impossible in this case; in other words, Q_φ is a positive operator, as is also seen by Theorem 4.1.) Hence, the non-zero eigenvalues are $\{(n + \frac{1}{2})^{-2}\pi^{-2}\}_{n=0}^\infty$ and the singular values are $s_n = \pi^{-2}(n + \frac{1}{2})^{-2}$, $n \geq 0$.

The behaviour $s_n(Q_\varphi) \asymp (n+1)^{-2}$ found in the above example holds for all smooth φ on $[0,1]$ by Sturm–Liouville theory, as will be seen in Theorem 9.3. Hence, for smooth φ with compact support, we have $Q_\varphi \in S_{1/2,\infty}$ but nothing better.

Let $BV[0,1]$ denote the Banach space of functions on $[0,1]$ with bounded variation, with the seminorm $\|\varphi\|'_{BV} = \int_0^1 |d\varphi|$ and the norm $\|\varphi\|_{BV} = \|\varphi\|'_{BV} + \sup|\varphi|$.

The following result is essentially the same as Theorem 8.1. However, we use in this section a different approach based on the study of eigenvalues of Sturm–Liouville operators.

Theorem 9.1. *If* $\varphi \in BV[0,1]$, *and* $\varphi = 0$ *on* $(1,\infty)$, *then* $Q_\varphi \in \boldsymbol{S}_{1/2,\infty}$ *and* $\|Q_\varphi\|_{\boldsymbol{S}_{1/2,\infty}} \le C\|\varphi\|_{BV}$. *More precisely,*

$$s_n(Q_\varphi) \le C_1\|\varphi\|_{BV}(n+1)^{-2}, \quad n \ge 0, \tag{9.7}$$

and

$$s_n(Q_\varphi) \le C_2\|\varphi\|'_{BV} n^{-2}, \quad n \ge 1. \tag{9.8}$$

Proof. Note that (9.8) follows from (9.7) since a symbol φ constant on $[0,1]$ yields a rank one operator Q_φ.

We use methods from Sturm–Liouville theory, and begin by making some simplifications.

(i) Replacing φ by a regularization φ_ε such that $\varphi_\varepsilon \to \varphi$ in $L^2[0,1]$ and thus $Q_{\varphi_\varepsilon} \to Q_\varphi$ in $\boldsymbol{S}_2$ as $\varepsilon \to 0$, we see that we may assume $\varphi \in C^1[0,1]$.

(ii) Considering real and imaginary parts separately, we may assume that φ is real, and thus Q_φ self-adjoint.

(iii) Subtracting a constant times $\chi_{[0,1]}$, which yields a rank 1 operator, we may assume that $\varphi(1) = 0$.

(iv) By homogeneity, we may assume that $\|\varphi\|'_{BV} = \int_0^1 |\varphi'| \le 1$ and show that then $\|Q_\varphi\|_{\boldsymbol{S}_{1/2,\infty}} \le C$.

(v) Using

$$\begin{aligned}\varphi(x) &= -\int_x^1 \varphi'(y)\,dy \\ &= -\int_x^1 \max\{\varphi'(y),0\}\,dy - \int_x^1 \min\{\varphi'(y),0\}\,dy \\ &= \varphi_1(x) - \varphi_2(x),\end{aligned}$$

and the corresponding decomposition $Q_\varphi = Q_{\varphi_1} - Q_{\varphi_2}$, we may also assume that $\varphi' \le 0$ and thus $\varphi \ge 0$. By Theorem 4.1, Q_φ is then a positive operator.

(vi) Similarly, writing $\varphi = 2\varphi_1 - \varphi_2$ with $\varphi_2 = 1 - x$ and $\varphi_1 = (\varphi + \varphi_2)/2$, we may further assume that $\varphi' \le -1/2$ on $[0,1]$.

Let $\lambda > 0$ and let, as above, (G_λ, g_λ) be the solution to (9.3) with $g_\lambda(0) = 1$, $G_\lambda(0) = 0$. Thus λ is an eigenvalue if and only if (9.6) holds, i.e., by (iii), if and only if $g_\lambda(1) = 0$.

Write $\omega = \lambda^{-1/2}$ and express $(g_\lambda, \omega G_\lambda)$ in polar coordinates

$$\begin{aligned} g_\lambda(x) &= R_\omega(x)\cos\Theta_\omega(x), \\ \omega G_\lambda(x) &= R_\omega(x)\sin\Theta_\omega(x), \end{aligned} \tag{9.9}$$

where $R_\omega(x) = \sqrt{g^2 + \omega^2 G^2} > 0$ and Θ_ω is continuous with $\Theta_\omega(0) = 0$. Note that λ is an eigenvalue if and only if $\cos\Theta_\omega(1) = 0$, i.e. $\Theta_\omega(1) = n\pi + \pi/2$ for some integer n.

Since $R_\omega(x) > 0$, R_ω and Θ_ω belong to $C^1[0,1]$, and (9.9) and (9.3) yield

$$\lambda^{1/2} R_\omega^2 \Theta_\omega' = gG' - Gg' = g^2 - \lambda^{-1}\varphi' G^2 = R_\omega^2(\cos^2\Theta_\omega - \varphi'\sin^2\Theta_\omega)$$

and thus

$$\Theta_\omega' = \omega(\cos^2\Theta_\omega - \varphi'\sin^2\Theta_\omega). \tag{9.10}$$

In particular, since $\varphi' < 0$ by (vi), $\Theta_\omega' > 0$ and thus $\Theta_\omega(1) > 0$.

Now suppose $0 < \omega < \nu$ and consider the corresponding functions Θ_ω and Θ_ν. We claim that

$$\Theta_\omega(x) < \Theta_\nu(x), \quad 0 < x \le 1. \tag{9.11}$$

Indeed, since $\Theta_\omega(0) = 0 = \Theta_\nu(0)$ and, by (9.10), $\Theta_\omega'(0) = \omega < \nu = \Theta_\nu'(0)$, (9.11) holds in $(0, \delta)$ for some $\delta > 0$. Hence, if (9.11) fails, there exists some $x_1 \in (0, 1]$ such that $\Theta_\omega(x) < \Theta_\nu(x)$ for $0 < x < x_1$ but $\Theta_\omega(x_1) = \Theta_\nu(x_1)$. This would imply $\Theta_\omega'(x_1) \ge \Theta_\nu'(x_1)$; on the other hand, then

$$\cos^2\Theta_\omega(x_1) - \varphi'\sin^2\Theta_\omega(x_1) = \cos^2\Theta_\nu(x_1) - \varphi'\sin^2\Theta_\nu(x_1) > 0,$$

recalling (vi), and (9.10) would yield $\Theta_\omega'(x_1) < \Theta_\nu'(x_1)$, a contradiction.

From (9.11) follows in particular that the function $\omega \mapsto \Theta_\omega(1)$ is strictly increasing. Hence, there is for each integer $n \ge 0$ at most one value of ω, ω_n say, such that $\Theta_{\omega_n}(1) = n\pi + \pi/2$, and thus a corresponding eigenvalue $\lambda_n = \omega_n^{-2}$. (The solution to (9.4) depends continuously on ω, with $\Theta_0(1) = 0$ and $\Theta_\omega(1) \to \infty$ as $\omega \to \infty$, so ω_n exists for every $n \ge 1$, but we do not need that.) Integrating (9.10) we obtain by (iv)

$$\Theta_\omega(1) = \int_0^1 \Theta_\omega'(x)\,dx \le \omega \int_0^1 (1 + |\varphi'(x)|)\,dx \le 2\omega$$

and thus $2\omega_n \ge \Theta_{\omega_n}(1) = n\pi + \pi/2$, which yields

$$\lambda_n = \omega_n^{-2} \le 4\pi^{-2}(n + \tfrac{1}{2})^{-2}. \qquad \square$$

Considering again functions on the whole half-line $\mathbb{R}_+$, we now can prove an endpoint result corresponding to Theorem 5.1.

Theorem 9.2. *If $\varphi \in Y_{1/2}$, then $Q_\varphi \in S_{1/2,\infty}$.*

Proof. Define $A_n^{(k)}$ and $A^{(k)}$ as in (3.3) and (3.4), but now for all integers k. For $k \geq 1$, (3.7) holds for every p, and thus $\varphi \in Y_{1/2} \subset X_{1/2}$ implies that

$$\sum_{k\geq 1} \|\mathcal{P}_{A^{(k)}} Q_\varphi\|_{S_{1/2}}^{1/2} \leq \sum_{k\geq 1} C2^{-pk/2} < \infty.$$

Moreover, by symmetry, $\|\mathcal{P}_{A^{(-k)}} Q_\varphi\|_{S_{1/2}} = \|\mathcal{P}_{A^{(k)}} Q_\varphi\|_{S_{1/2}}$ so

$$\sum_{k\leq -1} \|\mathcal{P}_{A^{(k)}} Q_\varphi\|_{S_{1/2}}^{1/2} < \infty$$

too. It follows that

$$Q_\varphi - \mathcal{P}_{A^{(0)}} Q_\varphi = \sum_{k\neq 0} \mathcal{P}_{A^{(k)}} Q_\varphi \in S_{1/2}. \tag{9.12}$$

Next, $\mathcal{P}_{A^{(0)}} Q_\varphi$ is the direct sum of $\mathcal{P}_{A_n^{(0)}} Q_\varphi$, $n \in \mathbb{Z}$, which act in the orthogonal spaces $L^2[2^n, 2^{n+1}]$. By translation invariance, Lemma 2.1 and Theorem 9.1, with $\varphi_n(x) = 2^n \varphi(2^n x + 2^n)$,

$$\begin{aligned}\|\mathcal{P}_{A_n^{(0)}} Q_\varphi\|_{S_{1/2,\infty}} &= \|Q_\varphi^{[2^n,2^{n+1}]}\|_{S_{1/2,\infty}} = \|Q_{\varphi_n}^{[0,1]}\|_{S_{1/2,\infty}} \leq C\|\varphi_n\|_{BV[0,1]} \\ &\leq C' 2^n \int_{2^n}^\infty |d\varphi|.\end{aligned}$$

By Theorem 5.3, we thus have

$$\sum_{n\in\mathbb{Z}} \left(\|\mathcal{P}_{A_n^{(0)}} Q_\varphi\|_{S_{1/2,\infty}}\right)^{1/2} < \infty$$

and it follows from Lemma 7.6 (or as in the proof of Lemma 9.5 below) that $\mathcal{P}_{A^{(0)}} Q_\varphi \in S_{1/2,\infty}$. By (9.12), $Q_\varphi \in S_{1/2,\infty}$ too. □

Theorem 9.2 is the best possible; for any reasonably smooth φ, the singular numbers $s_n(Q_\varphi)$ decrease like n^{-2} but not faster. More precisely, we have the following very precise result. Recall that a function in $Y_{1/2}$ has locally bounded variation and thus is a.e. differentiable.

Theorem 9.3. *Let $\varphi \in Y_{1/2}$. Then*

$$n^2 s_n(Q_\varphi) \to \pi^{-2}\|\varphi'\|_{L^{1/2}} = \pi^{-2}\left(\int_0^\infty |\varphi'(x)|^{1/2}\,dx\right)^2 < \infty \quad \text{as } n \to \infty. \tag{9.13}$$

Equivalently,

$$\varepsilon^{1/2}|\{n : s_n(Q_\varphi) > \varepsilon\}| \to \pi^{-1}\int_0^\infty |\varphi'(x)|^{1/2}\,dx < \infty \quad \text{as } \varepsilon \to 0. \tag{9.14}$$

In particular, $n^2 s_n(Q_\varphi) \to 0$ as $n \to \infty$ if and only if $\varphi' = 0$ a.e.

Proof. Note first that by the Cauchy–Schwarz inequality and (5.1),

$$\int_0^\infty |\varphi'(x)|^{1/2}\,dx \le \sum_{n\in\mathbb{Z}} 2^{n/2}\left(\int_{2^n}^{2^{n+1}} |\varphi'(x)|\,dx\right)^{1/2} \le \|\varphi\|_{Y_{1/2}}^{1/2} < \infty. \tag{9.15}$$

For smooth and positive symbols on a finite interval, (9.13) follows by standard Sturm–Liouville theory, see [LS, §1.2 with the transformation in §1.1]. Indeed, much more refined asymptotics of s_n can be given [LS, Chapter 5].

We present here another proof that applies in the general case. We prove a sequence of lemmas. The first implies that (9.13) and (9.14) are equivalent.

Lemma 9.4. *For any bounded operator T on a Hilbert space,*

$$\limsup_{\varepsilon\to 0} \varepsilon^{1/2}|\{n : s_n > \varepsilon\}| = \big(\limsup_{n\to\infty}(n^2 s_n)\big)^{1/2}$$

and similarly with lim inf *instead of* lim sup *on both sides.*

Proof. If $\limsup \varepsilon^{1/2}|\{n : s_n > \varepsilon\}| < a$ for some $a > 0$, then for all small ε, $|\{n : s_n > \varepsilon\}| < a\varepsilon^{-1/2}$. Taking $\varepsilon = a^2(n+1)^{-2}$, we see that for large n, $s_n \le \varepsilon$, and thus $(n+1)^2 s_n \le a^2$, so $\limsup_{n\to\infty} n^2 s_n \le a^2$. The converse is similar, and the second part follows similarly by reversing the inequalities. □

Lemma 9.5. *If $T_1, \dots, T_N$ are bounded operators on Hilbert spaces $H_1, \dots, H_N$, then*

$$\big(\limsup_{n\to\infty} n^2 s_n(T_1 \oplus \cdots \oplus T_N)\big)^{1/2} \le \sum_{k=1}^N \big(\limsup_{n\to\infty} n^2 s_n(T_k)\big)^{1/2}$$

and

$$\big(\liminf_{n\to\infty} n^2 s_n(T_1 \oplus \cdots \oplus T_N)\big)^{1/2} \ge \sum_{k=1}^N \big(\liminf_{n\to\infty} n^2 s_n(T_k)\big)^{1/2}.$$

Proof. The singular numbers $s_n(T_1 \oplus \cdots \oplus T_N)$ consist of all $s_i(T_j)$, rearranged into a single nonincreasing sequence. Hence,

$$|\{n : s_n(T_1 \oplus \cdots \oplus T_N) > \varepsilon\}| = \sum_{j=1}^N |\{n : s_n(T_j) > \varepsilon\}|$$

and the result follows by Lemma 9.4. □

For arbitrary sums we have the following estimate.

Lemma 9.6. *If T and U are bounded operators in a Hilbert space, and $0 < \delta < 1$, then*

$$\limsup_{n\to\infty} n^2 s_n(T+U) \le (1-\delta)^{-2} \limsup_{n\to\infty} n^2 s_n(T) + \delta^{-2} \limsup_{n\to\infty} n^2 s_n(U), \quad (9.16)$$

$$\liminf_{n\to\infty} n^2 s_n(T+U) \ge (1-\delta)^{2} \liminf_{n\to\infty} n^2 s_n(T) - \delta^{-2} \limsup_{n\to\infty} n^2 s_n(U). \quad (9.17)$$

Proof. By (2.1), $s_n(T+U) \le s_{[(1-\delta)n]}(T) + s_{[\delta n]}(U)$, and (9.16) follows, together with

$$\liminf_{n\to\infty} n^2 s_n(T+U) \le (1-\delta)^{-2} \liminf_{n\to\infty} n^2 s_n(T) + \delta^{-2} \limsup_{n\to\infty} n^2 s_n(U). \quad (9.18)$$

Replacing here T by $T+U$ and U by $-U$, we obtain (9.17) by rearrangement. □

Letting $\delta \to 0$ in (9.17) and (9.18), we obtain the following result by Fan [GK1].

Lemma 9.7. *If T and U are bounded operators in a Hilbert space, $\lim_{n\to\infty} n^2 s_n(T)$ exists and $n^2 s_n(U) \to 0$ as $n \to \infty$, then $\lim_{n\to\infty} n^2 s_n(T+U) = \lim_{n\to\infty} n^2 s_n(T)$.* □

Lemma 9.8. *The set of $\varphi \in Y_{1/2}$ such that* (9.13) *holds is a closed set.*

Proof. Suppose that $\varphi_k \to \varphi$ in $Y_{1/2}$ and that (9.13) holds for each φ_k. By Lemma 9.6 and Theorem 9.2, for every k and $0 < \delta < 1$,

$$\begin{aligned}\limsup_{n\to\infty} n^2 s_n(Q_\varphi) &\le (1-\delta)^{-2} \limsup_{n\to\infty} n^2 s_n(Q_{\varphi_k}) + \delta^{-2} \limsup_{n\to\infty} n^2 s_n(Q_{\varphi-\varphi_k}) \\ &\le (1-\delta)^{-2}\pi^{-2} \|\varphi_k'\|_{L^{1/2}} + C\delta^{-2} \|\varphi-\varphi_k\|_{Y_{1/2}}\end{aligned} \quad (9.19)$$

and similarly

$$\liminf_{n\to\infty} n^2 s_n(Q_\varphi) \ge (1-\delta)^{2}\pi^{-2} \|\varphi_k'\|_{L^{1/2}} - C\delta^{-2} \|\varphi-\varphi_k\|_{Y_{1/2}}. \quad (9.20)$$

Moreover, by (9.15), $\|(\varphi-\varphi_k)'\|_{L^{1/2}} \le \|\varphi-\varphi_k\|_{Y_{1/2}} \to 0$ as $k \to \infty$, and so $\|\varphi_k'\|_{L^{1/2}} \to \|\varphi'\|_{L^{1/2}}$. Letting first $k \to \infty$ and then $\delta \to 0$ in (9.19) and (9.20), we obtain (9.13). □

Lemma 9.9. *If φ is linear on a finite interval I, then*

$$n^2 s_n(Q_\varphi^I) \to \pi^{-2}\left(\int_I |\varphi'(x)|^{1/2}\,dx\right)^2 \quad \textit{as } n \to \infty.$$

Proof. Let $\varphi(x) = \alpha + \beta x$ with α, β complex numbers. Suppose first that $I = [0, 1]$. By the example at the beginning of the section and homogeneity,

$$s_n(Q^I_{-\beta+\beta x}) = |\beta|\pi^{-2}(n+\tfrac{1}{2})^{-2}$$

so $n^2 s_n(Q^I_{-\beta+\beta x}) \to \pi^{-2}|\beta|$ as $n \to \infty$. Since $Q^I_{\alpha+\beta x} - Q^I_{-\beta+\beta x} = Q^I_{\alpha+\beta}$ is a rank one operator, Lemma 9.7 (or, more simply, $s_{n+1}(Q^I_{-\beta+\beta x}) \le s_n(Q^I_{\alpha+\beta x}) \le s_{n-1}(Q^I_{-\beta+\beta x})$) yields

$$n^2 s_n(Q^I_\varphi) \to \pi^{-2}|\beta| = \pi^{-2}\Big(\int_I |\varphi'(x)|^{1/2}\,dx\Big)^2. \tag{9.21}$$

If $I = [0, a]$, we have by Lemma 2.1 and (9.21)

$$n^2 s_n(Q^I_\varphi) = n^2 s_n(Q^{[0,1]}_{\varphi_a}) \to \pi^{-2}a^2|\beta| = \pi^{-2}\Big(\int_I |\varphi'(x)|^{1/2}\,dx\Big)^2,$$

and the general case follows by translation invariance. □

Completion of the proof of Theorem 9.3.

Step 1. *φ is piecewise linear on* $[0, 1]$ *and* $\varphi = 0$ *on* $(1, \infty)$. Let $0 = t_0 < t_1 < \cdots < t_N = 1$ be such that φ is linear on every $I_i = [t_{i-1}, t_i]$, $i = 1, \ldots, N$. Let $H_i = L^2(I_i)$, so $L^2[0, 1] = H_1 \oplus \cdots \oplus H_N$, and let $P_i : L^2[0, 1] \to H_i$ denote the orthogonal projection.

Since each $P_i Q_\varphi P_j$, $i \ne j$, has rank 1, $Q_\varphi - \sum_{i=1}^N P_i Q_\varphi P_i$ has finite rank and by Lemma 9.7 (or directly), it suffices to consider $\sum_{i=1}^N P_i Q_\varphi P_i = Q^{I_1}_\varphi \oplus \cdots \oplus Q^{I_N}_\varphi$. By Lemma 9.9,

$$n^2 s_n(Q^{I_i}_\varphi) \to \pi^{-2}\Big(\int_{t_{i-1}}^{t_i} |\varphi'(x)|^{1/2}\,dx\Big)^2,$$

and thus Lemma 9.5 yields

$$\Big(\lim_{n\to\infty} n^2 s_n\Big(\sum_{i=1}^N P_i Q_\varphi P_i\Big)\Big)^{1/2} = \sum_{i=1}^N \Big(\lim_{n\to\infty} n^2 s_n(Q^{I_i}_\varphi)\Big)^{1/2} = \pi^{-1}\int_0^1 |\varphi'(x)|^{1/2}\,dx,$$

which proves (9.13).

Step 2. *φ is absolutely continuous on* $[0, 1]$ *and* $\varphi = 0$ *on* $(1, \infty)$. Approximate φ' by step functions h_n such that $\|\varphi' - h_n\|_{L^1[0,1]} < 1/n$, and let

$$\psi_n(x) = \begin{cases} \varphi(0) + \int_0^x h_n(y)\,dy, & x \le 1, \\ 0, & x > 1. \end{cases}$$

Then (9.13) holds for each ψ_n by Step 1, and $\psi_n \to \varphi$ in $BV[0, 1]$ and thus in $Y_{1/2}$, see Corollary 5.5, so (9.13) holds by Lemma 9.8.

Step 3. *φ has bounded variation on* $[0, 1]$, $\varphi = 0$ *on* $(1, \infty)$ *and φ is singular, i.e., $\varphi' = 0$ almost everywhere.* We may assume that φ is right-continuous. Then $\varphi(x) = \varphi(0) + \int_0^x d\mu$ for some singular complex measure μ supported on $[0, 1]$. Given any $\varepsilon > 0$, there thus exists a sequence of intervals $(I_i)_1^\infty$ in $[0, 1]$ such that

$\sum_{i=1}^{\infty} |I_i| < \varepsilon$ and $|\mu|\left([0,1] \setminus \bigcup_{i=1}^{\infty} I_i\right) = 0$. Let N be a positive integer such that $|\mu|\left([0,1] \setminus \bigcup_{i=1}^{N} I_i\right) < \varepsilon$.

We may assume that each I_i is closed, and by combining any two of $I_1, \dots, I_N$ that overlap, we may assume that $I_1, \dots, I_N$ are disjoint. The complement $[0,1] \setminus \bigcup_{i=1}^{N} I_i$ is also a finite disjoint union of intervals, say $\bigcup_{j=1}^{M} J_j$.

For each interval I, Theorem 9.1 and Lemma 2.1 yield

$$\sup_{n \geq 1} n^2 s_n(Q_\varphi^I) \leq C|I| \|\varphi\|'_{BV(I)} \leq C|I|\,|\mu|(I). \tag{9.22}$$

Moreover, as in Step 1 of the proof,

$$Q_\varphi = Q_\varphi^{I_1} \oplus \cdots \oplus Q_\varphi^{I_N} \oplus Q_\varphi^{J_1} \oplus \cdots \oplus Q_\varphi^{J_M} + R,$$

where R has finite rank. Hence, by Lemma 9.5, (9.22) and the Cauchy–Schwarz inequality,

$$\begin{aligned}
\Big(\limsup_{n\to\infty} n^2 s_n(Q_\varphi)\Big)^{1/2} &\leq \sum_{i=1}^{N} \Big(\limsup_{n\to\infty} n^2 s_n(Q_\varphi^{I_i})\Big)^{1/2} + \sum_{j=1}^{M} \Big(\limsup_{n\to\infty} n^2 s_n(Q_\varphi^{J_j})\Big)^{1/2} \\
&\leq C \sum_{i=1}^{N} \big(|I_i|\,|\mu|(I_i)\big)^{1/2} + C \sum_{j=1}^{M} \big(|J_j|\,|\mu|(J_j)\big)^{1/2} \\
&\leq C\Big(\sum_{i=1}^{N} |I_i|\Big)^{1/2} \Big(\sum_{i=1}^{N} |\mu|(I_i)\Big)^{1/2} \\
&\quad + C\Big(\sum_{j=1}^{M} |J_j|\Big)^{1/2} \Big(\sum_{j=1}^{M} |\mu|(J_j)\Big)^{1/2} \\
&\leq C\varepsilon^{1/2}\big(|\mu|[0,1]\big)^{1/2} + C \cdot 1 \cdot \varepsilon^{1/2}.
\end{aligned}$$

The result $n^2 s_n(Q_\varphi) \to 0$ follows by letting $\varepsilon \to 0$.

Step 4. *φ has bounded variation on* $(0, a)$ *and* $\varphi = 0$ *on* (a, ∞) *for some* $a > 0$. By Lemma 2.1, it suffices to consider the case $a = 1$. We can decompose $\varphi = \varphi_{\mathrm{a}} + \varphi_{\mathrm{s}}$ on $[0, 1]$, with φ_{a} absolutely continuous and φ_{s} singular; let $\varphi_{\mathrm{a}} = \varphi_{\mathrm{s}} = 0$ on $(1, \infty)$. By Steps 2 and 3,

$$n^2 s_n(Q_{\varphi_{\mathrm{a}}}) \to \pi^{-2} \|\varphi'_{\mathrm{a}}\|_{1/2} = \pi^{-2} \|\varphi'\|_{1/2}$$

and $n^2 s_n(Q_{\varphi_{\mathrm{s}}}) \to 0$, and the result follows by Lemma 9.7.

Step 5. *$\varphi \in Y_{1/2}$ is arbitrary.* Define, for $N \geq 1$,

$$\varphi_N(x) = \begin{cases} \varphi(1/N) - \varphi(N), & 0 < x \leq 2^{-N}, \\ \varphi(x) - \varphi(N), & 2^{-N} < x \leq 2^N, \\ 0, & 2^N < x. \end{cases}$$

It is easily seen that each φ_N has bounded variation and that $\|\varphi - \varphi_N\|_{Y_{1/2}} \to 0$ as $N \to \infty$, cf. (5.1). Thus the result follows by Step 4 and Lemma 9.8. □

As corollaries, we obtain new proofs of some results from §8.

Corollary 9.10. *If I is a finite interval and φ has bounded variation on I, then*

$$n^2 s_n(Q_\varphi^I) \to \pi^{-2}\left(\int_I |\varphi'(x)|^{1/2}\,dx\right)^2 \quad \text{as } n \to \infty.$$

Proof. By translation invariance, we may assume $I = [0, a]$. Then, defining $\varphi = 0$ outside I, we have $\varphi \in Y_{1/2}$ by Corollary 5.5, and the result follows by Theorem 9.3. □

Corollary 9.11. *If φ has locally bounded variation and $Q_\varphi \in S_{1/2}$, then $\varphi' = 0$ a.e.*

Proof. If $0 < a < b < \infty$, then φ has bounded variation on $[a, b]$, and since

$$n^2 s_n\left(Q_\varphi^{[a,b]}\right) \le n^2 s_n(Q_\varphi) \to 0,$$

Corollary 9.10 yields $\int_a^b |\varphi'|^{1/2} = 0$. Hence, $\varphi' = 0$ a.e. □

Corollary 9.12. *If φ is locally absolutely continuous and $Q_\varphi \in S_{1/2}$, then $\varphi = 0$.* □

Remark. More generally, in the last two corollaries, $S_{1/2}$ can be replaced by any Schatten–Lorentz space $S_{1/2,q}$ with $q < \infty$.

10. More on $p = 1$

Although $\varphi \in X_1$ does not imply $Q_\varphi, Q_\varphi^+ \in S_1$, the corresponding weak results holds. There is, however, a striking difference between Q_φ and Q_φ^+; as is shown in (ii) and (iii) below, for every $\varphi \in X_1$ not a.e. equal to 0, $ns_n \to 0$ for Q_φ but not for Q_φ^+. (Note that Theorem 6.5 implies that nothing can be said about the rate of convergence of $ns_n(Q_\varphi)$ to 0. In particular, if $q < \infty$, then $\varphi \in X_1$ does not imply that $Q_\varphi \in S_{1,q}$.)

Theorem 10.1. *If $\varphi \in X_1$ then the following hold:*

(i) $Q_\varphi, Q_\varphi^+, Q_\varphi^- \in S_{1,\infty}$.

(ii) $ns_n(Q_\varphi) \to 0$ *as* $n \to \infty$.

(iii) $ns_n(Q_\varphi^+) = ns_n(Q_\varphi^-) \to \pi^{-1}\int_0^\infty |\varphi(x)|\,dx$ *as* $n \to \infty$.

Proof. Since $X_1 \subset X_\infty$, Q_φ^+ is bounded by Theorem 3.1. By Theorem 4.2,

$$(Q_\varphi^+)^* Q_\varphi^+ = Q_\Phi$$

where $\Phi(x) = \int_x^\infty |\varphi(y)|^2\,dy$. By (1.5), $x^{1/2}\Phi(x)^{1/2} \in L^1(dx/x)$, so $\Phi \in Y_{1/2}$ by Theorem 5.3, and hence, $Q_\Phi \in S_{1/2,\infty}$ by Theorem 9.2. Consequently, $Q_\varphi^+ \in S_{1,\infty}$. The same holds for $Q_\varphi^- = (Q_\varphi^+)^*$ and $Q_\varphi = Q_\varphi^+ + Q_\varphi^-$.

Moreover, $s_n(Q_\Phi) = s_n(Q_\varphi^+)^2$, and thus (iii) follows from Theorem 9.3 applied to Φ.

For (ii), we observe that $\varphi \mapsto Q_\varphi$ thus is a bounded linear map $X_1 \to S_{1,\infty}$, and that the set of C^1 functions with compact support is dense in X_1 and mapped (by Theorem 9.2) into the closed subspace $S_{1,\infty}^0 = \{T \in S_{1,\infty} : ns_n(T) \to 0 \text{ as } n \to \infty\}$ of $S_{1,\infty}$. Hence, $Q_\varphi \in S_{1,\infty}^0$ for every $\varphi \in X_1$. □

Remark. Note that (i) and (iii) were earlier obtained in [EEH] for a more general class of operators. Moreover, in [EEH] the authors also consider the same operators on $L^p(\mathbb{R}_+)$ and obtain similar results for approximation numbers. See also related results in [NaS]. We also mention here [GK1, Remark IV.8.3] and [GK2, Theorem III.2.4], where similar asymptotic formulas are given for abstract Volterra operators with trace class imaginary parts.

Remark. For the related operators $\mathbf{Q}_\mu^+$, we similarly obtain that if $x\mu(x,\infty) \in L^{1/2}(dx/x)$, then $\mathbf{Q}_\mu^+ \in S_{1,\infty}$, and

$$ns_n(\mathbf{Q}_\mu^+) \to \pi^{-1} \int_0^\infty \left(\frac{d\mu}{dx}\right)^{1/2} dx \quad \text{as } n \to \infty,$$

where $\frac{d\mu}{dx}$ is the Radon–Nikodym derivative of the absolutely continuous component of μ. In particular, for such μ, $ns_n(\mathbf{Q}_\mu^+) \to 0$ if and only if μ is singular.

We saw earlier that $\varphi \in X_1$ is not enough to insure that Q_φ is in the trace class. Furthermore the previous theorem shows that if $\varphi \in X_1$ then neither Q_φ^+ nor Q_φ^- will be in the trace class. However, the combination of size and regularity results for singular numbers given in the previous theorem does insure that these operators have a well defined Dixmier trace. Because of the recent interest in the Dixmier trace we digress briefly to record this observation. For more about the Dixmier trace and its uses we refer to IV.2.β of [C].

Let ℓ^∞ be the space of bounded sequences indexed by non-negative integers and let c_1 be the closed subspace consisting of sequences $\{a_n\}$ for which $\lim_n a_n$ exists. It follows from the Hahn–Banach theorem that the functional $\lim(\cdot)$ which is defined on c_1 has a positive continuous extension, $\lim_\omega(\cdot)$, to all of ℓ^∞. By saying $\lim_\omega(\cdot)$ is positive we mean that if $a_n \geq 0$ for $n = 0, 1, 2, \ldots$ then $\lim_\omega(\{a_n\}) \geq 0$. This extension is not unique and we are using the subscript ω to denote the particular choice. It was noted by Dixmier in [D] that $\lim_\omega(\cdot)$ can also be selected to have the following scaling property:

$$\lim_\omega (a_0, a_0, a_1, a_1, a_2, a_2, \dots) = \lim_\omega (a_0, a_1, a_2, \dots).$$

A simple proof is in [C]. (Although the scaling is important for the general theory it has no role in our discussion.)

Consider now the operator ideal $\boldsymbol{S}_\Omega \supset \boldsymbol{S}_{1,\infty}$ that consists of the operators T on Hilbert space such that

$$\|T\|_{\boldsymbol{S}_\Omega} \stackrel{\text{def}}{=} \sup_{n\geq 0} \frac{\sum_{k=0}^n s_k(T)}{\sum_{k=0}^n \frac{1}{k+1}} < \infty. \tag{10.1}$$

Suppose that $T \in \boldsymbol{S}_\Omega$. For a fixed choice of $\lim_\omega(\cdot)$ we define a Dixmier trace, $\operatorname{trace}_\omega(\cdot)$, as follows. For *positive* $T \in \boldsymbol{S}_\Omega$ set

$$\operatorname{trace}_\omega(T) = \lim_\omega\left(\left\{\frac{1}{\log(n+2)}\sum_{k=0}^n \lambda_k(T)\right\}\right).$$

Here the λ_k are the (necessarily non-negative) eigenvalues of the positive operator T arranged in decreasing order. Although perhaps not obvious at first glance, it is not difficult to see that, in fact, if T_1 and T_2 are two positive operators in $\boldsymbol{S}_\Omega$ then $\operatorname{trace}_\omega(T_1 + T_2) = \operatorname{trace}_\omega(T_1) + \operatorname{trace}_\omega(T_2)$. A proof of this is also in [C]. Using this fact, the functional $\operatorname{trace}_\omega(\cdot)$ can be extended uniquely by linearity to all of $T \in \boldsymbol{S}_\Omega$. For $T \in \boldsymbol{S}_\Omega$ the value of $\operatorname{trace}_\omega(T)$ need not be independent of ω. However, there are certain operators for which $\operatorname{trace}_\omega(T)$ is independent of ω. Such operators are defined to be *measurable*. In this case we will write $\operatorname{trace}_{\mathrm{D}}(T)$ for this common value and refer to it as *the* Dixmier trace of T.

Corollary 10.2.

(i) *If $\varphi \in X_1$, then the operators $|Q_\varphi^+|$ and $|Q_\varphi^-|$ are measurable and*

$$\operatorname{trace}_{\mathrm{D}}(|Q_\varphi^+|) = \operatorname{trace}_{\mathrm{D}}(|Q_\varphi^-|) = \frac{1}{\pi}\int_0^\infty |\varphi(x)|\,dx.$$

(ii) *If $\varphi \in Y_{1/2}$, then $|Q_\varphi|^{1/2}$ is measurable and*

$$\operatorname{trace}_{\mathrm{D}}(|Q_\varphi|^{1/2}) = \frac{1}{\pi}\int_0^\infty |\varphi'(x)|^{1/2}\,dx.$$

(iii) *If $\varphi \in X_1$, then Q_φ is measurable and*

$$\operatorname{trace}_{\mathrm{D}}(Q_\varphi) = 0.$$

(iv) *If $\varphi \in X_1$, then Q_φ^+ and Q_φ^- are measurable and*

$$\operatorname{trace}_{\mathrm{D}}(Q_\varphi^+) = \operatorname{trace}_{\mathrm{D}}(Q_\varphi^-) = 0.$$

Proof. We start with (i). From the very definitions $s_n(|Q_\varphi^+|) = s_n(Q_\varphi^+)$ and hence, the previous theorem gives the asymptotic behavior of $\{s_n(|Q_\varphi^+|)\}_{n\geq 0}$. Those asymptotics, together with the fact that $|Q_\varphi^+|$ is a positive operator, insure that $|Q_\varphi^+|$ is measurable and has the indicated Dixmier trace. A similar argument applies to $|Q_\varphi^-|$ and, after noting that $s_n\left(|Q_\varphi|^{1/2}\right) = s_n(Q_\varphi)^{1/2}$ and taking note of Theorem 9.3, to part (ii).

We now consider (iii). By Theorem 10.1, we have $\lim_{n\to\infty} ns_n(Q_\varphi) = 0$. Also $s_n(Q_\varphi) = s_n(Q_\varphi^*)$. Thus by (2.1), $\lim_{n\to\infty} ns_n((Q_\varphi + Q_\varphi^*)/2) = 0$. We now use the spectral projection to write $\frac{1}{2}(Q_\varphi + Q_\varphi^*)$ as a difference of two positive operators $\frac{1}{2}(Q_\varphi + Q_\varphi^*)_\pm$ and note that we will have

$$\lim_{n\to\infty} ns_n\left(\tfrac{1}{2}(Q_\varphi + Q_\varphi^*)_\pm\right) = 0.$$

Arguing similarly with the skew-adjoint part of Q_φ, we realize Q_φ as a linear combination of four positive operators each of which have singular numbers which tend to zero more rapidly than n^{-1}. Those positive operators are certainly measurable and have Dixmier trace zero. The result we want now follows by the linearity of $\mathrm{trace}_{\mathrm{D}}(\cdot)$.

For (iv) we first pick and fix a choice $\mathrm{trace}_\omega(\cdot)$. Assume for the moment that φ is real, supported in $[0, 1]$ and in L^2. $R_\varphi \overset{\text{def}}{=} Q_\varphi^+ - Q_\varphi^{+*} = Q_\varphi^+ - Q_\varphi^-$ has real anti-symmetric kernel. Thus it is normal and its eigenvalues are imaginary and symmetric. Hence, $\mathrm{i}R_\varphi$ is symmetric and its positive and negative parts, $(\mathrm{i}R_\varphi)_\pm$ are unitarily equivalent. Thus

$$\mathrm{trace}_\omega(R_\varphi) = -\mathrm{i}\ \mathrm{trace}_\omega((\mathrm{i}R_\varphi)_+) + \mathrm{i}\ \mathrm{trace}_\omega((\mathrm{i}R_\varphi)_-) = 0.$$

Taking note of the fact that $\lim_\omega(\cdot)$ is continuous on ℓ^∞ and of the norm estimates implicit in the previous theorem we see that we can extend this result by linearity and continuity and conclude that $\mathrm{trace}_\omega(R_\varphi) = 0$ for all $\varphi \in X_1$. Now we use the fact that ω was arbitrary to conclude $\mathrm{trace}_{\mathrm{D}}(R_\varphi) = 0$. By linearity this result together with the result in (iii) yields (iv). □

For a function φ defined on a finite or infinite interval I, we define the standard and L^p moduli of continuity by

$$\begin{aligned} \omega_\varphi^{(\infty)}(h; I) &\overset{\text{def}}{=} \sup\{|\varphi(x) - \varphi(y)| : x, y \in I,\ |x - y| \leq h\}, \\ \omega_\varphi^{(p)}(h; I) &\overset{\text{def}}{=} \sup_{0\leq s\leq h}\left(\int_{I\cap(I-s)} |\varphi(x+s) - \varphi(x)|^p\,dx\right)^{1/p}, \quad 1 \leq p < \infty, \end{aligned} \tag{10.2}$$

where $0 < h \leq |I|$ and $I - s = \{x - s : x \in I\} = \{x : x + s \in I\}$. It follows easily from Minkowski's inequality that

$$\omega_\varphi^{(p)}(h; I) \leq 2\omega_\varphi^{(p)}(h/2; I), \quad 1 \leq p \leq \infty. \tag{10.3}$$

Note further that for a finite interval I,

$$\omega_\varphi^{(p)}(h; I) \leq |I|^{1/p-1/q}\omega_\varphi^{(q)}(h; I), \quad p < q \leq \infty. \tag{10.4}$$

We often omit I from the notation.

An alternative L^p modulus of continuity is defined by

$$\widetilde{\omega}_\varphi^{(p)}(h; I) \overset{\text{def}}{=} \Big((2h)^{-1} \iint\limits_{\substack{x,y\in I \\ |x-y|<h}} |\varphi(x) - \varphi(y)|^p \, dx \, dy \Big)^{1/p}.$$

This is equivalent to $\omega_\varphi^{(p)}(h; I)$ defined above by the following lemma, which probably is well-known to some experts.

Lemma 10.3. *Let $1 \le p < \infty$. Then, with C_p depending on p only,*

$$\widetilde{\omega}_\varphi^{(p)}(h; I) \le \omega_\varphi^{(p)}(h; I) \le C_p \widetilde{\omega}_\varphi^{(p)}(h; I).$$

Proof. The left hand inequality follows by

$$\begin{aligned} \big(\widetilde{\omega}_\varphi^{(p)}(h; I)\big)^p &= \frac{1}{h} \iint\limits_{\substack{x,y\in I \\ 0<y-x<h}} |\varphi(y) - \varphi(x)|^p \, dx \, dy \\ &= \frac{1}{h} \int_0^h \int_{x \in I \cap (I-s)} |\varphi(x+s) - \varphi(x)|^p \, dx \, dy \le \big(\omega_\varphi^{(p)}(h; I)\big)^p. \end{aligned}$$

For the converse, we assume for convenience that $I = [0, 1]$. The result then follows for every finite I by a linear change of variables, and for infinite I by considering $I \cap [-n, n]$ and letting $n \to \infty$. Thus $I = [0, 1]$ and $I \cap (I - s) = [0, 1 - s]$.

Let $\varphi_s(x) = \varphi(x + s)$. Assume first that $h \le 1/2$. Then, for $0 \le s, t \le h$, by Minkowski's inequality,

$$\|\varphi - \varphi_s\|^p_{L^p[0,1/2]} \le C_p \|\varphi - \varphi_t\|^p_{L^p[0,1/2]} + C_p \|\varphi_s - \varphi_t\|^p_{L^p[0,1/2]}.$$

Averaging over $t \in [0, h]$ we find

$$\begin{aligned} &\|\varphi - \varphi_s\|^p_{L^p[0,1/2]} \\ &\quad \le \frac{C_p}{h} \int_0^h \int_0^{1/2} \big(|\varphi(x) - \varphi(x+t)|^p + |\varphi(x+s) - \varphi(x+t)|^p \big) \, dx \, dt \\ &\quad \le \frac{C_p}{h} \iint\limits_{\substack{x,y\in [0,1] \\ 0<y-x<h}} |\varphi(y) - \varphi(x)|^p \, dx \, dy = C_p \big(\widetilde{\omega}_\varphi^{(p)}(h; [0, 1])\big)^p. \end{aligned}$$

A similar argument, now taking $s - h \le t \le s$, yields the same estimate for $\|\varphi - \varphi_s\|^p_{L^p[1/2-s,1-s]}$, and summing we find

$$\|\varphi - \varphi_s\|_{L^p[0,1-s]} \le C_p \widetilde{\omega}_\varphi^{(p)}(h; [0, 1]) \tag{10.5}$$

for every $0 \le s \le h$, which proves the result for $h \le 1/2$.

If $1/2 < h \le 1$, the result follows from the case $h \le 1/2$ and (10.3). □

For simplicity, we state the following lemma for $I = [0, 1]$ only.

Lemma 10.4. *Let $1 \le p < \infty$. If $\varphi \in L^p[0,1]$ and $0 < t \le 1$, there exists a decomposition $\varphi = \varphi_0 + \varphi_1$ with*

$$\|\varphi_0\|_{L^p[0,1]} \le C_p \omega_\varphi^{(p)}(t) \quad \text{and} \quad \|\varphi_1\|'_{BV[0,1]} \le C_p t^{-1} \omega_\varphi^{(p)}(t).$$

In other words, the Peetre K-functional (see [BL]), can be estimated by

$$K(t, \varphi; L^p[0,1], BV'[0,1]) \le C_p \omega_\varphi^{(p)}(t), \quad 0 < t \le 1.$$

Proof. Take $\varphi_1(x) = \frac{1}{t}\int_{(1-t)x}^{(1-t)x+t} \varphi(y)\,dy$ and $\varphi_0 = \varphi - \varphi_1$. Then φ_1 is absolutely continuous, and thus

$$\begin{aligned}
\|\varphi_1\|'_{BV} &= \int_0^1 |\varphi_1'(x)|\,dx = \frac{1-t}{t}\int_0^1 \big|\varphi\big((1-t)x+t\big) - \varphi\big((1-t)x\big)\big|\,dx \\
&= \frac{1}{t}\int_0^{1-t} |\varphi(y+t) - \varphi(y)|\,dy \\
&\le \frac{1}{t}\Big(\int_0^{1-t} |\varphi(y+t) - \varphi(y)|^p\,dy\Big)^{1/p} \le \frac{1}{t}\omega_\varphi^{(p)}(t).
\end{aligned}$$

Moreover, using Hölder's inequality again,

$$|\varphi_0(x)|^p = \left|\frac{1}{t}\int_{(1-t)x}^{(1-t)x+t} \big(\varphi(x) - \varphi(y)\big)\,dy\right|^p \le \frac{1}{t}\int_{(1-t)x}^{(1-t)x+t} |\varphi(x) - \varphi(y)|^p\,dy$$

and thus

$$\int_0^1 |\varphi_0(x)|^p \le \frac{1}{t}\iint_{\substack{x,y\in[0,1]\\ |y-x|<t}} |\varphi(y) - \varphi(x)|^p\,dx\,dy = 2\big(\omega_\varphi^{(p)}(t)\big)^p. \qquad \square$$

Theorem 10.5. *If I is a finite interval and $\varphi \in L^2(I)$, then*

$$s_n(Q_\varphi^I) \le C\frac{|I|^{1/2}}{n}\omega_\varphi^{(2)}\Big(\frac{|I|}{n}\Big) \le C\frac{|I|}{n}\omega_\varphi^{(\infty)}\Big(\frac{|I|}{n}\Big), \quad n \ge 1.$$

Proof. By a linear change of variables, we may assume that $I = [0, 1]$, cf. Lemma 2.1. Then, using the decomposition given by Lemma 10.4 with $t = 1/n$, (2.1), Theorem 10.1 and Theorem 9.1, we find, for $n \ge 1$,

$$\begin{aligned}
s_{2n-1}(Q_\varphi) &\le s_{n-1}(Q_{\varphi_0}) + s_n(Q_{\varphi_1}) \\
&\le Cn^{-1}\|\varphi_0\|_{L^2} + Cn^{-2}\|\varphi_1\|'_{BV} \le Cn^{-1}\omega_\varphi^{(2)}(1/n),
\end{aligned}$$

and the result follows, using (10.3) and (10.4). $\square$

In particular, we see that a Dini condition implies $Q_\varphi \in S_1$.

Corollary 10.6. *If $\varphi \in L^2[0,1]$ is such that $\int_0^1 \omega_\varphi^{(2)}(t)\frac{dt}{t} < \infty$, in particular if $\int_0^1 \omega_\varphi^{(\infty)}(t)\,\frac{dt}{t} < \infty$, then $Q_\varphi^{[0,1]} \in \boldsymbol{S}_1$.*

Proof. Theorem 3.1 shows that Q_φ is bounded, and Theorem 10.5 yields

$$\sum_{n=2}^{\infty} s_n(Q_\varphi) \le C \sum_{n=2}^{\infty} \frac{1}{n}\,\omega_\varphi^{(2)}\left(\frac{1}{n}\right) \le C \int_0^1 \omega_\varphi^{(\infty)}(t)\,\frac{dt}{t}. \qquad \square$$

By a simple change of variables, Corollary 10.6 applies to other finite intervals too. Moreover, for functions φ on $\mathbb{R}_+$, we have the following corresponding sufficient conditions for $Q_\varphi \in \boldsymbol{S}_1$.

Theorem 10.7. *If $\varphi \in X_1$ and*

$$\sum_{n=-\infty}^{\infty} 2^{n/2} \int_0^{2^n} \omega_\varphi^{(2)}\big(t;\,[2^n, 2^{n+1}]\big)\frac{dt}{t} < \infty$$

then $Q_\varphi \in \boldsymbol{S}_1$.

Proof. Let $I_n = [2^n, 2^{n+1}]$. Then Theorem 10.5 yields

$$\begin{aligned}\|Q_\varphi^{I_n}\|_{\boldsymbol{S}_1} &= \sum_{k=0}^{\infty} s_k(Q_\varphi^{I_n}) \le 2\|Q_\varphi^{I_n}\|_{\boldsymbol{S}_2} + C|I_n|^{1/2}\sum_{k=2}^{\infty}\frac{1}{k}\omega_\varphi^{(2)}\left(\frac{2^n}{k};\,I_n\right)\\ &\le C2^{n/2}\|\varphi\|_{L^2(I_n)} + C2^{n/2}\int_0^{2^n}\omega_\varphi^{(2)}(t;\,I_n)\,\frac{dt}{t}\end{aligned}$$

and the result follows by Theorem 6.4 and (1.3). $\square$

Corollary 10.8. *If $\varphi \in X_1$ and*

$$\sum_{n=-\infty}^{\infty} 2^{n} \int_0^{2^n} \omega_\varphi^{(\infty)}\big(t;\,[2^n, 2^{n+1}]\big)\,\frac{dt}{t} < \infty$$

then $Q_\varphi \in \boldsymbol{S}_1$. $\square$

Note that for the functions φ_N considered in Theorem 6.5, the estimate of the singular numbers in Theorem 10.5 is sharp (within a constant factor) and the estimates of the $\boldsymbol{S}_1$ norm implicit in Corollary 10.6, Theorem 10.7 and Corollary 10.8 are of the right order.

We do not know whether the condition in Theorem 10.7 is necessary, but we will give a related necessary condition using the L^1 modulus of continuity in §15.

We have in these applications of Theorem 10.5 considered $\boldsymbol{S}_1$ only, but the same arguments apply to $\boldsymbol{S}_p$ for other p too. In particular, Theorem 10.7 extends as follows (see the remark after Theorem 6.4).

Theorem 10.9. *Let $1/2 < p \le 1$. If $\varphi \in X_p$ and*

$$\sum_{n=-\infty}^{\infty} 2^{n(1-p/2)} \int_0^{2^n} \left(\omega_\varphi^{(2)}(t; [2^n, 2^{n+1}])\right)^p t^{p-2}\, dt < \infty,$$

then $Q_\varphi \in S_p$. □

Note also the following immediate consequence of Theorem 10.5.

Corollary 10.10. *If I is a finite interval and φ satisfies a Hölder (Lipschitz) condition $|f(x) - f(y)| \le C|x - y|^\alpha$ for $x, y \in I$, where $0 < \alpha \le 1$, then $Q_\varphi \in S_{1/(1+\alpha),\infty}$ and thus $Q_\varphi \in S_p$ for every $p > 1/(1+\alpha)$.* □

11. Averaging projection

In this section we study properties of the averaging projection onto the set of operators of the form Q_ψ. Let us first define the averaging projection on S_2. Let T be an operator on $L^2(\mathbb{R}_+)$ of class S_2 with kernel function $k = k_T \in L^2((\mathbb{R}_+)^2)$. We define the function φ on $\mathbb{R}_+$ by

$$\varphi(x) = \frac{1}{2x}\left(\int_0^x k(x,t)dt + \int_0^x k(s,x)\, ds\right), \quad x > 0. \tag{11.1}$$

We define the averaging projection $\mathcal{Q}$ on S_2 by

$$\mathcal{Q}T \stackrel{\text{def}}{=} Q_\varphi.$$

It is not hard to see that if $Q_\psi \in S_2$, then $\mathcal{Q}Q_\psi = Q_\psi$. It is also easy to see that $\|\mathcal{Q}T\|_{S_2} \le \|T\|_{S_2}$ for any $T \in S_2$, and so $\mathcal{Q}$ is the orthogonal projection of S_2 onto the set of operators of the form Q_ψ.

We will show in this section that $\mathcal{Q}$ is a bounded linear operator on S_p for $1 < p \le 2$. This allows us to define by duality the projection $\mathcal{Q}$ on the classes S_p for $2 \le p < \infty$. We also show that $\mathcal{Q}$ is unbounded on S_1 but it has weak type (1,1), i.e., $s_n(\mathcal{Q}T)(1+n) \le C\|T\|_{S_1}$. Finally, we use this result to show that $\mathcal{Q}$ maps the Matsaev ideal into the set of compact operators.

Theorem 11.1. *Let $1 < p \le 2$. Then $\mathcal{Q}$ is a bounded projection on S_p.*

Proof. Let T be an integral operator in S_p with kernel function k and let φ be defined by (11.1). We have to show that $\varphi \in X_p$ (see the definition in the Introduction). We can identify in a natural way the dual space X_p^* with the space $Z_{p'}$ of functions f on $\mathbb{R}_+$ such that

$$\sum_{n\in\mathbb{Z}} 2^{-np'/2}\left(\int_{2^n}^{2^{n+1}} |f(x)|^2\, dx\right)^{p'/2} < \infty$$

with respect to the pairing

$$(\varphi, f) = \int_0^\infty \varphi(x) f(x)\,dx. \tag{11.2}$$

Here $p' = p/(p-1)$. Suppose that f is a function on $(0, \infty)$. Define the function ψ on $\mathbb{R}_+$ by $\psi(x) = \frac{f(x)}{2x}$, $x > 0$. It is straightforward to see from the definition of the X_p spaces that $f \in X_p^*$ if and only if $\psi \in X_{p'}$ and the norm of f in X_p^* and the norm of ψ in $X_{p'}$ are equivalent. It is also easy to see that for $1 < p < \infty$ the space X_p is reflexive.

Let us show that if f is a bounded function in X_p^* with compact support in $(0, \infty)$, then

$$(\varphi, f) = \operatorname{trace} T Q_\psi. \tag{11.3}$$

We have

$$\begin{aligned}
\operatorname{trace} T Q_\psi &= \iint_{\mathbb{R}_+^2} k(x, y)\psi(\max\{x, y\})\,dxdy \\
&= \int_0^\infty \psi(x)\left(\int_0^x k(x, t)dt + \int_0^x k(s, x)ds\right) dx \\
&= \int_0^\infty 2x\psi(x)\varphi(x)\,dx = \int_0^\infty \varphi(x) f(x)\,dx.
\end{aligned}$$

It follows that

$$\begin{aligned}
\sup\{|(\varphi, f)| : f \in X_p^*,\ \|f\|_{X_p^*} \le 1\} &\le C\|T\|_{S_p} \sup\{\|Q_\psi\|_{S_{p'}} : \|\psi\|_{X_{p'}} \le 1\} \\
&\le C\|T\|_{S_p}
\end{aligned}$$

by Theorem 3.3. It follows that $\varphi \in X_p$, and again by Theorem 3.3, $\mathcal{Q}T \in \boldsymbol{S}_p$. □

Theorem 11.1 allows us to define for $1 < p < 2$ the adjoint operator $\mathcal{Q}^*$ on $\boldsymbol{S}_{p'}$. Since $\mathcal{Q}$ is an orthogonal projection on $\boldsymbol{S}_2$, $\mathcal{Q}$ is a self-adjoint operator on $\boldsymbol{S}_2$. We denote the adjoint operator $\mathcal{Q}^*$ on $\boldsymbol{S}_{p'}$ by the same symbol $\mathcal{Q}$.

Thus we can consider the projection $\mathcal{Q}$ on any class $\boldsymbol{S}_p$ with $1 < p < \infty$. It is easy to show that if T is an integral operator with kernel function k of class $\boldsymbol{S}_p$, $2 < p < \infty$, then $\mathcal{Q}T = Q_\varphi$, where φ is defined by (11.1) and $Q_\varphi \in \boldsymbol{S}_p$. We are going to prove that for any $T \in \boldsymbol{S}_p$, $2 < p < \infty$, the operator $\mathcal{Q}T$ has the form Q_φ for a function $\varphi \in X_p$.

Theorem 11.2. *Let T be an operator of class $\boldsymbol{S}_p$, $2 < p < \infty$. Then there exists a function $\varphi \in X_p$ such that $\mathcal{Q}T = Q_\varphi$.*

Proof. Let $\mathfrak{X}_p$ be the space of operators of the form Q_φ with $\varphi \in X_p$. Clearly, $\mathfrak{X}_p$ is a Banach space with norm

$$\|Q_\varphi\|_{\mathfrak{X}_p} = \|\varphi\|_{X_p}.$$

It follows from Theorems 11.1 and 3.3 that for $T \in S_2$

$$\|\mathcal{Q}T\|_{\mathfrak{X}_p} \leq C\|\mathcal{Q}T\|_{S_p} \leq C\|T\|_{S_p}.$$

Since S_2 dense in S_p, it follows that $\mathcal{Q}T \in \mathfrak{X}_p$ for any $T \in S_p$. □

We consider now the behavior of $\mathcal{Q}$ on S_1. It follows from Theorem 11.1 that if $T \in S_1$, then $\mathcal{Q}T \in S_p$ for any $p > 1$. The next result shows that $\mathcal{Q}T$ does not have to be in S_1 but it has to be in $S_{1,\infty}$.

Theorem 11.3. (i) *There exists an operator T in S_1 such that $\mathcal{Q}T \notin S_1$.*

(ii) *$\mathcal{Q}$ has weak type $(1,1)$, i.e., $\mathcal{Q}$ maps S_1 into $S_{1,\infty}$, i.e.,*

$$s_n(\mathcal{Q}T) \leq C(1+n)^{-1}\|T\|_{S_1}, \quad T \in S_1.$$

Lemma 11.4.

$$\mathcal{Q}S_1 = \mathfrak{X}_1 \stackrel{\text{def}}{=} \{Q_\varphi : \varphi \in X_1\}.$$

Let us first deduce Theorem 11.3 from Lemma 11.4.

Proof of Theorem 11.3. (i) is an immediate consequence of Lemma 11.4 and the Example following Theorem 6.5. (ii) also follows immediately from Lemma 11.4 and Theorem 10.1. □

Proof of Lemma 11.4. Let us first show that $\mathcal{Q}S_1 \subset \mathfrak{X}_1$. Let $T \in S_1$ and $\mathcal{Q}T = Q_\varphi$. We have to prove that $\varphi \in X_1$. Consider the space Z_∞^0 that consists of functions f on $\mathbb{R}_+$ such that

$$\lim_{n\to\pm\infty}\left(2^{-n}\int_{2^n}^{2^{n+1}}|f(x)|^2\,dx\right) = 0.$$

It is not difficult to see that $(Z_\infty^0)^* = X_1$ with respect to the pairing (11.2). As in the proof of Theorem 11.1 we define the function ψ by $\psi(x) = \frac{f(x)}{2x}$, $x > 0$. It follows from (11.3) that

$$|(\varphi, f)| \leq C\|T\|_{S_1}\|Q_\psi\| \leq C\|T\|_{S_1}\|\psi\|_{X_\infty^0} \leq C\|T\|_{S_1}\|f\|_{Z_\infty^0},$$

and so φ determines a continuous linear functional on Z_∞^0. Hence, $\varphi \in X_1$.

To prove that $\mathcal{Q}S_1 = \mathfrak{X}_1$, we consider the operator $A : S_1 \to X_1$ defined by $AT = \varphi$, where φ is the function on $\mathbb{R}_+$ such that $\mathcal{Q}T = Q_\varphi$. We have to show that A maps S_1 onto X_1. Consider the conjugate operator $A^* : X_1^* \to \mathcal{B}(L^2(\mathbb{R}_+))$.

It is easy to see that with respect to the pairing (11.2) the space X_1^* can be identified with the space Z_∞ that consists of functions f on $\mathbb{R}_+$ such that

$$\sup_{n\in\mathbb{Z}} 2^{-n/2}\left(\int_{2^n}^{2^{n+1}} |f(x)|^2\,dx\right)^{1/2} < \infty.$$

Consider the operator $J : X_1^* \to X_\infty$ defined by $(Jf)(x) = \frac{f(x)}{2x}$, $x > 0$. It is easy to see that J maps isomorphically X_1^* onto X_∞.

It can easily be verified that $A^*f = Q_{Jf}$. It follows from Theorem 3.1 that $\|A^*f\| \geq C\|f\|_{X_1^*}$. It follows that A maps $\boldsymbol{S}_1$ onto X_1. □

Remark. In [Pel3] metric properties of the averaging projection $\mathcal{P}$ onto the space of Hankel matrices were studied. In particular, it was shown in [Pel3] that $\mathcal{P}\boldsymbol{S}_1 \subset \boldsymbol{S}_{1,2}$. However, it turns out that the averaging projection $\mathcal{Q}$ onto the operators Q_φ has different properties. Theorem 11.3 shows that $\mathcal{Q}\boldsymbol{S}_1 \subset \boldsymbol{S}_{1,\infty}$ but it follows from Lemma 11.4 and the remark preceding Theorem 6.6 that $\mathcal{Q}\boldsymbol{S}_1 \not\subset \boldsymbol{S}_{1,q}$ for any $q < \infty$.

Recall that the Matsaev ideal $\boldsymbol{S}_\omega$ consists of the operators T on Hilbert space such that

$$\|T\|_{\boldsymbol{S}_\omega} \stackrel{\text{def}}{=} \sum_{n\geq 0} \frac{s_n(T)}{1+n} < \infty.$$

It is easy to see that $\boldsymbol{S}_p \subset \boldsymbol{S}_\omega$ for any $p < \infty$.

Consider now the operator ideal $\boldsymbol{S}_\Omega$ defined by (10.1). It is easy to see that $\boldsymbol{S}_{1,\infty} \subset \boldsymbol{S}_\Omega$. It is well known (see [GK1]) that $\boldsymbol{S}_\omega^* = \boldsymbol{S}_\Omega$ with respect to the pairing

$$\{T, R\} = \operatorname{trace} TR, \quad T \in \boldsymbol{S}_\omega, \quad R \in \boldsymbol{S}_\Omega. \tag{11.4}$$

Theorem 11.5. *The averaging projection $\mathcal{Q}$ defined on $\boldsymbol{S}_2$ extends to a bounded linear operator from $\boldsymbol{S}_\omega$ to the space of compact operators. If $T \in \boldsymbol{S}_\omega$, then $\mathcal{Q}T = Q_\varphi$ for a function φ in X_∞^0.*

Proof. Let us prove that $\mathcal{Q}$ extends to a bounded operator from $\boldsymbol{S}_\omega$ to the space of compact operators. The proof of the second part of the theorem is the same as the proof of Theorem 11.2. Since the finite rank operators are dense in $\boldsymbol{S}_\omega$, it is sufficient to show that $\mathcal{Q}$ extends to a bounded operator from $\boldsymbol{S}_\omega$ to $\mathcal{B}(L^2(\mathbb{R}_+))$.

Let $T \in \boldsymbol{S}_2$ and $R \in \boldsymbol{S}_1$. By Theorem 11.3, $\mathcal{Q}R \in \boldsymbol{S}_{1,\infty} \subset \boldsymbol{S}_\Omega$. We have

$$\{\mathcal{Q}T, R\} = \{T, \mathcal{Q}R\},$$

and so

$$\begin{aligned}|\{\mathcal{Q}T, R\}| &\leq C\|T\|_{\boldsymbol{S}_\omega}\|\mathcal{Q}R\|_{\boldsymbol{S}_\Omega}\\ &\leq C\|T\|_{\boldsymbol{S}_\omega}\|\mathcal{Q}R\|_{\boldsymbol{S}_{1,\infty}} \leq C\|T\|_{\boldsymbol{S}_\omega}\|R\|_{\boldsymbol{S}_1}\end{aligned}$$

by Theorem 11.3. Since $S_1^* = \mathcal{B}(L^2(\mathbb{R}_+))$ with respect to the pairing (11.4), it follows that $\|\mathcal{Q}T\| \leq C\|T\|_{S_\omega}$, and so $\mathcal{Q}$ extends to a bounded linear operator from S_ω to $\mathcal{B}(L^2(\mathbb{R}_+))$. □

12. Finite rank

We say that φ is a *step function* if there exist finitely many numbers $0 = x_0 < x_1 < \dots < x_N < \infty$ such that φ is a.e. constant on each interval (x_{i-1}, x_i), and zero on (x_N, ∞). The number of steps of φ then is the smallest possible N in this definition.

There is a natural correspondence between operators Q_φ where the symbol φ is a step function, with given $x_1 < \dots < x_N$, and matrices of the form $\{a_{\max\{i,j\}}\}$. We need a simple result for such matrices, but will not pursue their study further.

Lemma 12.1. *If $a_1, \dots, a_n$ are complex numbers, then the matrix $\{a_{\max\{i,j\}}\}_{1\leq i,j\leq n}$ has determinant $a_n \prod_{i=1}^{n-1}(a_i - a_{i+1})$.*

Proof. Denote this determinant by $D(a_1, \dots, a_n)$. Subtracting the last row from all others, we see that $D(a_1, \dots, a_n) = a_n D(a_1 - a_n, \dots, a_{n-1} - a_n)$, and the result follows by induction. □

Theorem 12.2. *Q_φ has finite rank if and only if φ is a step function. In this case, the rank of Q_φ equals the number of steps of φ.*

Proof. If φ is a step function with N steps, we have $\varphi = \sum_{i=1}^N a_i \chi_{(0,x_i)}$ for some a_i and $x_i > 0$, and thus Q_φ is a linear combination of N rank one operators.

Conversely, suppose that $\operatorname{rank}(Q_\varphi) = M < \infty$. Suppose that $n > M$ and that $z_1 < \dots < z_n$ are Lebesgue points of φ. If $h > 0$ and $f_{z,h} = h^{-1}\chi_{(z,z+h)}$, then the matrix $(\langle Q_\varphi f_{z_i,h}, f_{z_j,h}\rangle)_{ij}$ has rank at most $M < n$ and thus its determinant vanishes. As $h \to 0$, as shown in the proof of Theorem 4.1, $\langle Q_\varphi f_{z_i,h}, f_{z_j,h}\rangle \to \varphi(\max\{z_i, z_j\}) = \varphi(z_{\max\{i,j\}})$, and thus the determinant of $\big(\varphi(z_{\max\{i,j\}})\big)_{ij}$ vanishes too. By Lemma 12.1, this implies that either $\varphi(z_i) = \varphi(z_{i+1})$ for some $i < n$ or $\varphi(z_n) = 0$.

Consequently, if $z_1 < \dots < z_n$ are Lebesgue points of φ such that $\varphi(z_i) \neq \varphi(z_{i+1})$ for $i < n$ and $\varphi(z_n) \neq 0$, then $n \leq M$. Choose such a sequence $z_1 < \dots < z_n$ with n maximal. If $z \in (z_i, z_{i+1}) \cap \operatorname{Leb}(\varphi)$ for some $i < n$, then either $\varphi(z) = \varphi(z_i)$ or $\varphi(z) = \varphi(z_{i+1})$, since n is maximal. Moreover, for the same reason, if $\varphi(z) = \varphi(z_i)$, then $\varphi(z') = \varphi(z_i)$ for every $z' \in (z_i, z) \cap \operatorname{Leb}(\varphi)$, and if $\varphi(z) = \varphi(z_{i+1})$, then $\varphi(z') = \varphi(z_{i+1})$ for every $z' \in (z, z_{i+1}) \cap \operatorname{Leb}(\varphi)$. Together with similar arguments for the intervals $(0, z_1)$ and (z_n, ∞), which we leave to the reader, this easily shows that φ is a step function with at most n steps. □

13. A class of integral operators on $L^2(\mathbb{R})$

In this section we associate with the operator Q_φ on $L^2(\mathbb{R}_+)$ an integral operator on $L^2(\mathbb{R})$ and we study these operators.

For a function $\varphi \in L^2_{\mathrm{loc}}(\mathbb{R}_+)$ we define the function $\varphi^\heartsuit$ on $\mathbb{R}$ by

$$\varphi^\heartsuit(t) \stackrel{\mathrm{def}}{=} 2\varphi(e^{2t})e^{2t}, \quad t \in \mathbb{R}; \tag{13.1}$$

this defines a one-to-one correspondence between $L^2_{\mathrm{loc}}(\mathbb{R}_+)$ and $L^2_{\mathrm{loc}}(\mathbb{R})$. With a function ψ on $L^2(\mathbb{R})$ we associate the function $\check{\psi}$ on $\mathbb{R} \times \mathbb{R}$ defined by

$$\check{\psi}(s,t) = \psi(\max\{s,t\})e^{-|s-t|}, \quad s,t \in \mathbb{R}, \tag{13.2}$$

and denote by K_ψ the integral operator on $L^2(\mathbb{R})$ with kernel function $\check{\psi}$ (if it makes sense):

$$(K_\psi f)(s) = \int_{\mathbb{R}} \psi(\max\{s,t\})e^{-|s-t|} f(t)\,dt.$$

Theorem 13.1. *Let $\varphi \in L^2_{\mathrm{loc}}(\mathbb{R}_+)$. Then the operators Q_φ and $K_{\varphi^\heartsuit}$ are unitarily equivalent.*

Theorem 13.1 certainly means that the boundedness of one of the operators implies the boundedness of the other one.

Proof. Consider the unitary operator $U : L^2(\mathbb{R}_+) \to L^2(\mathbb{R})$ defined as follows $(Uf)(t) = \sqrt{2} f(e^{2t})e^t$. It remains to observe that $K_{\varphi^\heartsuit} U = U Q_\varphi$. □

We can identify $L^2(\mathbb{R}_+)$ with the subspace of $L^2(\mathbb{R})$ which consists of the functions vanishing on $(-\infty, 0)$. We can now extend in a natural way the operator of triangular projection $\mathcal{P}$ to act on the space of operators on $L^2(\mathbb{R})$ by defining it in the same way as it has been done in §2. We keep the same notation, $\mathcal{P}$, for this extension. We put $K_\psi^+ \stackrel{\mathrm{def}}{=} \mathcal{P}K_\psi$ and $K_\psi^- \stackrel{\mathrm{def}}{=} K_\psi - \mathcal{P}K_\psi^+$.

It is easily seen from the proof of Theorem 13.1 that the operators Q_φ^+ and Q_φ^- are unitarily equivalent to the operators $K_{\varphi^\heartsuit}^+$ and $K_{\varphi^\heartsuit}^-$ respectively.

It is easy to see that, for any $p > 0$,

$$\varphi \in X_p \Leftrightarrow \sum_{n=-\infty}^{\infty} \|\varphi^\heartsuit\|^p_{L^2[n,n+1]} < \infty \tag{13.3}$$

(and correspondingly for X_∞ and X_∞^0) and, using Theorem 5.1 (vii),

$$\varphi \in Y_p \Leftrightarrow \sum_{n=-\infty}^{\infty} \|\varphi^\heartsuit\|^p_{BV[n,n+1]} < \infty. \tag{13.4}$$

(In (13.3) and (13.4), the intervals $[n, n+1]$ can be replaced by any partition of $\mathbb{R}$ into intervals of the same length.) We can thus translate results from the preceding sections to K_ψ, for example as follows.

Theorem 13.2. *Let $\psi \in L^2_{\mathrm{loc}}(\mathbb{R})$. The following are equivalent:*

(i) *K_ψ is bounded on $L^2(\mathbb{R})$;*

(ii) *K^+_ψ is bounded on $L^2(\mathbb{R})$;*

(iii) $\sup_{n\in\mathbb{Z}} \int_n^{n+1} |\psi(x)|^2 \, dx < \infty$.

Proof. The theorem is a direct consequence of Theorems 3.1 and 13.1. □

Similarly we find from Theorems 3.3, 6.2, and 5.1, respectively, the following three theorems.

Theorem 13.3. *Let $\psi \in L^2_{\mathrm{loc}}(\mathbb{R})$. If $1 < p < \infty$, the following are equivalent:*

(i) *$K_\psi \in S_p$;*

(ii) *$K^+_\psi \in S_p$;*

(iii) $\sum_{n\in\mathbb{Z}} \|\psi\|^p_{L^2[n,n+1]} < \infty$. □

Theorem 13.4. *If $K_\psi \in S_1$, then $\sum_{n\in\mathbb{Z}} \|\psi\|_{L^2[n,n+1]} < \infty$ and thus $\psi \in L^1(\mathbb{R})$. Moreover, then* trace $K_\psi = \int_{\mathbb{R}} \psi(x)\, dx$. □

Theorem 13.5. *If $1/2 < p \le 1$ and $\sum_{n\in\mathbb{Z}} \|\psi\|^p_{BV[n,n+1]} < \infty$, then $K_\psi \in S_p$.* □

The following two results involving the modulus of continuity also can be obtained by changes of variables in the corresponding Theorems 10.5 and 10.9, using (10.3) and Lemma 10.3, but the details are involved and we prefer to imitate the proofs.

Theorem 13.6. *If $\psi \in L^2(\mathbb{R})$ has support on* $[0, 1]$, *then*

$$s_n(K_\psi) \le C \frac{1}{n} \omega^{(2)}_\psi \left(\frac{1}{n}\right) + C \frac{1}{n^2} \|\psi\|_{L^2}, \quad n \ge 1.$$

Proof. We interpolate using Lemma 10.4 as in the proof of Theorem 10.5. □

Theorem 13.7. *Let $1/2 < p \le 1$. If $\sum_{n\in\mathbb{Z}} \|\psi\|^p_{L^2[n,n+1]} < \infty$ and*

$$\sum_{n=-\infty}^{\infty} \int_0^1 (\omega^{(2)}_\psi(t; [n, n+1]))^p t^{p-2} \, dt < \infty,$$

then $K_\psi \in S_p$.

Proof. We argue as in the proofs of Theorems 10.7 and 10.9, using Theorem 13.6. □

We denote by $\mathcal{F}$ the Fourier transformation on $L^2(\mathbb{R}^n)$, which is a unitary operator defined by (2.3) for $f \in L^2(\mathbb{R}^n) \bigcap L^1(\mathbb{R}^n)$. Let T be the integral operator with kernel function $k \in L^2(\mathbb{R}^2)$. Denote by R the integral operator with kernel function $\mathcal{F}k$. The following lemma has a straightforward verification.

Lemma 13.8. $\mathcal{F}T\mathcal{F} = R$. □

Note that Lemma 13.8 is similar to Lemma 2.3.

Corollary 13.9. *Let* $p > 0$. *Then* $\|T\|_{S_p} = \|R\|_{S_p}$. □

Denote by Z be the integral operator with kernel function $(x, y) \mapsto (\mathcal{F}k)(x, -y)$. By Lemma 2.3, T is unitarily equivalent to Z. Indeed, the equality $\mathcal{F}T\mathcal{F} = R$ implies $\mathcal{F}T\mathcal{F}^{-1} = Z$.

Lemma 13.10. *If* $\psi \in L^1(\mathbb{R})$, *then* $\breve{\psi} \in L^1(\mathbb{R}^2)$, *and*

$$(\mathcal{F}\breve{\psi})(x, y) = (\mathcal{F}\psi)(x + y)\left(\frac{1}{1 - 2\pi i x} + \frac{1}{1 - 2\pi i y}\right). \tag{13.5}$$

Proof. The inclusion $\breve{\psi} \in L^1(\mathbb{R}^2)$ is obvious. We have

$$\begin{aligned}
(\mathcal{F}\breve{\psi})(x, y) &= \int_{\mathbb{R}^2} \psi(\max\{s, t\})e^{-|s-t|}e^{-2\pi i s x - 2\pi i t y}\, ds\, dt \\
&= \int_{\mathbb{R}} \left(\psi(s)e^{-s}e^{-2\pi i s x} \int_{-\infty}^{s} e^{t-2\pi i t y} dt\right) ds \\
&\quad + \int_{\mathbb{R}} \left(\psi(t)e^{-t}e^{-2\pi i t y} \int_{-\infty}^{t} e^{s-2\pi i s x} ds\right) dt \\
&= \int_{\mathbb{R}} \frac{\psi(s)e^{-2\pi i s(x+y)}\, ds}{1 - 2\pi i y} + \int_{\mathbb{R}} \frac{\psi(t)e^{-2\pi i t(x+y)}\, dt}{1 - 2\pi i x} \\
&= \left(\frac{1}{1 - 2\pi i x} + \frac{1}{1 - 2\pi i y}\right)(\mathcal{F}\psi)(x + y).
\end{aligned}$$

□

Consider the functions

$$\breve{\psi}_+(s, t) \stackrel{\text{def}}{=} \chi_{\{(s,t):s>t\}}\breve{\psi}(s, t) \quad \text{and} \quad \breve{\psi}_-(s, t) \stackrel{\text{def}}{=} \chi_{\{(s,t):s<t\}}\breve{\psi}(s, t). \tag{13.6}$$

It can easily be seen from the proof of Lemma 13.10 that

$$(\mathcal{F}\breve{\psi}_+)(x, y) = \frac{(\mathcal{F}\psi)(x + y)}{1 - 2\pi i y} \quad \text{and} \quad (\mathcal{F}\breve{\psi}_-)(x, y) = \frac{(\mathcal{F}\psi)(x + y)}{1 - 2\pi i x}. \tag{13.7}$$

It is easy to verify that if ψ is a tempered distribution on $\mathbb{R}$ (see §2), we can define tempered distributions $\check{\psi}$, $\check{\psi}_+$, and $\check{\psi}_-$ by (13.2) and (13.6); the formal definitions are by duality and analogous to (2.4)–(2.6). It is easy to check that formulas (13.5) and (13.7) also hold for tempered distributions ψ.

As a corollary to Theorem 13.2, we have the following lemma.

Lemma 13.11. *Let $\psi \in L^2_{\text{loc}}(\mathbb{R})$. Suppose that the operator K_ψ is bounded on $L^2(\mathbb{R})$. Then ψ determines a tempered distribution.* □

Theorem 13.12. *Let $0 < p < \infty$. Suppose that $\varphi \in L^2_{\text{loc}}(\mathbb{R}_+)$ and $\varphi^\heartsuit$ is defined by* (13.1). *The following are equivalent:*

(i) $Q_\varphi \in \boldsymbol{S}_p$;

(ii) $K_{\varphi^\heartsuit} \in \boldsymbol{S}_p$;

(iii) *the integral operator on $L^2(\mathbb{R})$ with kernel*

$$(x, y) \mapsto (\mathcal{F}\varphi^\heartsuit)(x + y)\left(\frac{1}{1 - 2\pi \mathrm{i} x} + \frac{1}{1 - 2\pi \mathrm{i} y}\right)$$

belongs to $\boldsymbol{S}_p$.

Proof. The theorem is a consequence of Theorem 13.1, Lemma 13.8 and Lemma 13.10. □

Note that if for $p > 2$ we have a tempered distribution in (iii) rather than a function, by the integral operator we mean the operator determined by this tempered distribution (see §2).

In the same way one can prove the following result.

Theorem 13.13. *Let $0 < p < \infty$ and let $\varphi \in L^2_{\text{loc}}(\mathbb{R}_+)$. The following are equivalent:*

(i) $Q^+_\varphi \in \boldsymbol{S}_p$;

(ii) $Q^-_\varphi \in \boldsymbol{S}_p$;

(iii) $K^+_{\varphi^\heartsuit} \in \boldsymbol{S}_p$;

(iv) $K^-_{\varphi^\heartsuit} \in \boldsymbol{S}_p$;

(v) *the integral operator on $L^2(\mathbb{R})$ with the kernel*

$$(x, y) \mapsto \frac{(\mathcal{F}\varphi^\heartsuit)(x + y)}{1 - 2\pi \mathrm{i} x}$$

belongs to $\boldsymbol{S}_p$. □

It is straightforward to show that if $p > 0$ and the integral operator on $L^2(\mathbb{R})$ with kernel function $\frac{h(x+y)}{x+\alpha}$ belongs to $\boldsymbol{S}_p$ for some $\alpha \in \mathbb{C} \setminus \mathbb{R}$, then it belongs to $\boldsymbol{S}_p$ for any $\alpha \in \mathbb{C} \setminus \mathbb{R}$. Let us show that such an integral operator can belong to $\boldsymbol{S}_p$ for $p \le 1$ only if it is zero.

Consider the operator $E : \mathcal{D}(\mathbb{R}^2) \to \mathcal{D}(\mathbb{R}_+ \times \mathbb{R}_+)$ defined by the following formula $(Ef)(s,t) \stackrel{\text{def}}{=} \frac{1}{2}(st)^{-1/2} f(\frac{1}{2}\log s, \frac{1}{2}\log t)$. Clearly, E is an isomorphism. Consequently, the conjugate operator E' is an isomorphism from $\mathcal{D}'(\mathbb{R}_+ \times \mathbb{R}_+)$ onto $\mathcal{D}'(\mathbb{R}^2)$. Clearly, $(E'\Phi)(x,y) = 2\Phi(e^{2x}, e^{2y})e^x e^y$ if $\Phi \in L^1_{\text{loc}}(\mathbb{R}_+ \times \mathbb{R}_+)$. Put $2\Phi(e^{2x}, e^{2y})e^x e^y \stackrel{\text{def}}{=} (E'\Phi)(x,y)$ for $\Phi \in \mathcal{D}'(\mathbb{R}_+ \times \mathbb{R}_+)$.

Theorem 13.14. *Let $\Phi \in \mathcal{D}'(\mathbb{R}_+ \times \mathbb{R}_+)$. Then Φ determines a bounded operator on $L^2(\mathbb{R}_+)$ if and only if the distribution $2\Phi(e^{2x}, e^{2y})e^x e^y$ determines a bounded operator on $L^2(\mathbb{R})$. Moreover, these two operators are unitarily equivalent operators.*

Proof. It suffices to note that

$$\langle 2\Phi(e^{2x}, e^{2y})e^x e^y, \sqrt{2}e^y f(e^{2y})\overline{\sqrt{2}e^x g(e^{2x})}\rangle = \langle \Phi(s,t), f(t)\overline{g(s)}\rangle$$

for any $f, g \in \mathcal{D}(\mathbb{R}_+ \times \mathbb{R}_+)$, and the map $h \mapsto \sqrt{2}e^x h(e^{2x})$ is a unitary operator from $L^2(\mathbb{R}_+)$ onto $L^2(\mathbb{R})$. □

Theorem 13.15. *Let $h \in \mathcal{D}'(\mathbb{R})$. Suppose that the distribution $\frac{h(x+y)}{1-2\pi ix}$ determines an operator on $L^2(\mathbb{R})$ of class $\boldsymbol{S}_1$. Then $h = 0$.*

Proof. By Lemma 2.2, $\frac{h(x+y)}{1-2\pi ix} \in \mathcal{S}'(\mathbb{R}^2)$. Consequently, $h(x+y) \in \mathcal{S}'(\mathbb{R}^2)$, whence $h \in \mathcal{S}'(\mathbb{R})$. Thus, there exists a distribution $\varphi \in \mathcal{D}'(\mathbb{R}_+)$ such that $\varphi^\heartsuit \in \mathcal{S}'(\mathbb{R})$ and $\mathcal{F}\varphi^\heartsuit = h$ (the operation $\varphi \mapsto \varphi^\heartsuit$ defined in (13.1) extends in an obvious way to distributions φ). Lemma 2.3 and formula (13.7) imply that Q_φ^- and thus also Q_φ^+ belongs to $\boldsymbol{S}_1$. By Theorem 2.4, $\varphi \in L^2_{\text{loc}}(\mathbb{R}_+)$. Thus $\varphi = 0$ by Theorem 6.6. □

We are going to prove now that for $p > 1/2$ if the integral operator with kernel function $h(x+y)\left(\frac{1}{x+\alpha} + \frac{1}{y+\beta}\right)$ belongs to $\boldsymbol{S}_p$ for some $\alpha, \beta \in \mathbb{C}\setminus\mathbb{R}$ with $\alpha+\beta \notin \mathbb{R}$, then it belongs to $\boldsymbol{S}_p$ for any $\alpha, \beta \in \mathbb{C} \setminus \mathbb{R}$. We will also show that this is not true for $p \le 1/2$.

Lemma 13.16. *The function*

$$(x, y) \mapsto \frac{1}{x+y+\mathrm{i}}\, \chi_{[0,1]}(x), \quad x, y \in \mathbb{R},$$

is a Schur multiplier of $\boldsymbol{S}_p$ for any $p > 0$.

Proof. First we prove that the function

$$(x, y) \mapsto \frac{1}{x+y+\mathrm{i}}\, \chi_{[0,1]}(x)\chi_{\mathbb{R}\setminus[-2,2]}(y), \quad x, y \in \mathbb{R}, \tag{13.8}$$

is a Schur multiplier of $\boldsymbol{S}_p$. We have

$$\frac{1}{x+y+\mathrm{i}}\, \chi_{[0,1]}(x)\chi_{\mathbb{R}\setminus[-2,2]}(y) = \sum_{n\geq 0}(-1)^n \frac{(x+\mathrm{i})^n}{y^{n+1}}\chi_{[0,1]}(x)\chi_{\mathbb{R}\setminus[-2,2]}(y).$$

Clearly, the p-multiplier norm of the n-th summand is bounded by $2^{-n/2}$. Consequently, the function (13.8) is a Schur multiplier of $\boldsymbol{S}_p$. It remains to prove that the function

$$(x, y) \mapsto \frac{1}{x+y+\mathrm{i}}\, \chi_{[0,1]}(x)\chi_{[-2,2]}(y), \quad x, y \in \mathbb{R},$$

is a Schur multiplier of $\boldsymbol{S}_p$. For any $(\xi, \eta) \in [0,1] \times [-2,2]$ we can expand the function $\frac{1}{x+y+\mathrm{i}}$ in a Taylor series in a neighborhood of (ξ, η). It follows easily that for a sufficiently small $\varepsilon > 0$ the function

$$(x, y) \mapsto \frac{1}{x+y+\mathrm{i}}\, \chi_{[0,1]\cap[\xi-\varepsilon,\xi+\varepsilon]}(x)\chi_{[-2,2]\cap[\eta-\varepsilon,\eta+\varepsilon]}(y), \quad x, y \in \mathbb{R},$$

is a Schur multiplier of $\boldsymbol{S}_p$. It remains to choose a finite subcover of $[0,1] \times [-2,2]$ that consists of rectangles of the form $[\xi-\varepsilon, \xi+\varepsilon] \times [\eta-\varepsilon, \eta+\varepsilon]$. □

Remark. In the same way we can prove that the function

$$(x, y) \mapsto \frac{1}{x+y+\alpha}\, \chi_{[\xi,\eta]}(x), \quad x, y \in \mathbb{R},$$

is a Schur multiplier of $\boldsymbol{S}_p$ for any $p > 0$ for any $\alpha \in \mathbb{C} \setminus \mathbb{R}$ and for any $\xi, \eta \in \mathbb{R}$.

Corollary 13.17. *Let $\alpha, \beta, \gamma \in \mathbb{C}$ and $\gamma \notin \mathbb{R}$. Then the function*

$$(x, y) \mapsto \frac{(x+\alpha)(y+\beta)}{x+y+\gamma}\chi_{[0,1]}(x), \quad x, y \in \mathbb{R},$$

is a Schur multiplier of $\boldsymbol{S}_p$ for any $p > 0$.

Proof. We have

$$\frac{(x+\alpha)(y+\beta)}{x+y+\gamma} = (x+\alpha)\left(1 + \frac{\beta-\gamma-x}{x+y+\gamma}\right).$$

It remains to note that

$$\chi_{[0,1]}(x), \quad (x+\alpha)\chi_{[0,1]}(x), \quad (\beta-\gamma-x)\chi_{[0,1]}(x), \quad \text{and} \quad \frac{1}{x+y+\gamma}\chi_{[0,1]}(x)$$

are Schur multipliers of $\boldsymbol{S}_p$. □

Corollary 13.18. *Let $p > 0$ and let α, $\beta \in \mathbb{C}\setminus\mathbb{R}$ such that $\alpha+\beta \notin \mathbb{R}$. Suppose that the integral operator on $L^2(\mathbb{R})$ with kernel function*

$$(x, y) \mapsto h(x+y)\left(\frac{1}{x+\alpha}+\frac{1}{y+\beta}\right), \quad x, y \in \mathbb{R},$$

belongs to $\boldsymbol{S}_p$. Then the integral operator with kernel function

$$(x, y) \mapsto h(x+y)\chi_{[0,1]}(x), \quad x, y \in \mathbb{R},$$

belongs to $\boldsymbol{S}_p$. □

Theorem 13.19. *Let $p > 1/2$ and let α_0, $\beta_0 \in \mathbb{C}\setminus\mathbb{R}$ such that $\alpha_0+\beta_0 \notin \mathbb{R}$. Suppose that the integral operator on $L^2(\mathbb{R})$ with kernel function*

$$(x, y) \mapsto h(x+y)\left(\frac{1}{x+\alpha_0}+\frac{1}{y+\beta_0}\right), \quad x, y \in \mathbb{R},$$

belongs to $\boldsymbol{S}_p$. Then the integral operator with kernel function

$$(x, y) \mapsto h(x+y)\left(\frac{1}{x+\alpha}+\frac{1}{y+\beta}\right), \quad x, y \in \mathbb{R},$$

also belongs to $\boldsymbol{S}_p$ for any α, $\beta \in \mathbb{C}\setminus\mathbb{R}$.

Proof. By Corollary 13.18, the integral operator with kernel $h(x+y)\chi_{[0,1]}(x)$ belongs to $\boldsymbol{S}_p$. Obviously, for any $n \in \mathbb{Z}$,

$$\|h(x+y)\chi_{[0,1]}(x)\|_{\boldsymbol{S}_p} = \|h(x+y)\chi_{[n,n+1]}(x)\|_{\boldsymbol{S}_p}$$

(as usual we write $\|k\|_{\boldsymbol{S}_p}$ for the $\boldsymbol{S}_p$ norm (or quasi-norm) of the integral operator with kernel k). Consequently,

$$\left\|h(x+y)\left(\frac{1}{x+\alpha}-\frac{1}{x+\alpha_0}\right)\chi_{[n,n+1]}(x)\right\|_{\boldsymbol{S}_p} \le C(1+|n|)^{-2}.$$

It is now clear that the integral operator with kernel function

$$(x, y) \mapsto h(x+y)\left(\frac{1}{x+\alpha}-\frac{1}{x+\alpha_0}\right), \quad x, y \in \mathbb{R},$$

belongs to $\boldsymbol{S}_p$ for $p > 1/2$. Similarly, we prove that the integral operator with kernel function

$$(x, y) \mapsto h(x+y)\left(\frac{1}{y+\beta}-\frac{1}{y+\beta_0}\right), \quad x, y \in \mathbb{R},$$

belongs to $\boldsymbol{S}_p$ for $p > 1/2$. □

Theorem 13.20. *Let $\varphi \in L^2_{\text{loc}}(\mathbb{R}_+)$, $a \in \mathbb{R} \setminus \{0\}$, and $p > 1/2$. Put*

$$\varphi_{[a]}(t) \stackrel{\text{def}}{=} \varphi(t^a)t^{a-1}.$$

Then $Q_\varphi \in \boldsymbol{S}_p$ if and only if $Q_{\varphi_{[a]}} \in \boldsymbol{S}_p$.

Proof. Recall that $\varphi^\heartsuit(t) = 2\varphi(e^{2t})e^{2t}$ and $\varphi^\heartsuit_{[a]}(t) = 2\varphi_{[a]}(e^{2t})e^{2t} = 2\varphi(e^{2at})e^{2at}$. Consequently, $\varphi^\heartsuit_{[a]}(t) = \varphi^\heartsuit(at)$. By Theorem 13.12, $Q_{\varphi_{[a]}} \in \boldsymbol{S}_p$ if and only if the integral operator on $L^2(\mathbb{R})$ with kernel

$$(x, y) \mapsto (\mathcal{F}\varphi^\heartsuit)\left(\frac{x+y}{a}\right)\left(\frac{1}{1-2\pi \mathrm{i}x} + \frac{1}{1-2\pi \mathrm{i}y}\right), \qquad x, y \in \mathbb{R},$$

belongs to $\boldsymbol{S}_p$, and thus if and only if the integral operator on $L^2(\mathbb{R})$ with kernel

$$(x, y) \mapsto (\mathcal{F}\varphi^\heartsuit)(x+y)\left(\frac{1}{1-2\pi \mathrm{i}ax} + \frac{1}{1-2\pi \mathrm{i}ay}\right), \qquad x, y \in \mathbb{R},$$

belongs to $\boldsymbol{S}_p$. By Theorem 13.19, this holds if and only if the integral operator on $L^2(\mathbb{R})$ with kernel

$$(x, y) \mapsto (\mathcal{F}\varphi^\heartsuit)(x+y)\left(\frac{1}{1-2\pi \mathrm{i}x} + \frac{1}{1-2\pi \mathrm{i}y}\right), \qquad x, y \in \mathbb{R},$$

belongs to $\boldsymbol{S}_p$. It remains to apply Theorem 13.12 once more. □

Corollary 13.21. *Let $\psi \in L^2_{\text{loc}}(\mathbb{R})$, $a \in \mathbb{R} \setminus \{0\}$, and $p > 1/2$. Define $\psi_a(t) \stackrel{\text{def}}{=} \psi(at)$. Then $K_\psi \in \boldsymbol{S}_p$ if and only if $K_{\psi_a} \in \boldsymbol{S}_p$.* □

Remark. Theorem 13.20 and its corollary do not generalize to the case $p \le 1/2$. Indeed, if φ is the characteristic function of an interval, then $Q_\varphi \in \boldsymbol{S}_p$ for any $p > 0$ but if $a \ne 1$, then $Q_{\varphi_{[a]}} \notin \boldsymbol{S}_p$ by Corollary 8.12. It follows from the proof above that Theorem 13.19 too does not extend to $p \le 1/2$.

Remark. Note that if $\varphi \in X_p$, then $\varphi_{[a]}(t) \in X_p$ for any $a \in \mathbb{R} \setminus \{0\}$ and any $p > 0$. Moreover, if $\varphi \in Y_p$, then $\varphi_{[a]}(t) \in Y_p$ for any $a \in \mathbb{R} \setminus \{0\}$ and any $p > 0$. Indeed, let $A > 1$. It is easy to see that (1.3) is equivalent to the condition

$$\sum_{n\in\mathbb{Z}} A^{np/2}\left(\int_{A^n}^{A^{n+1}} |\varphi(x)|^2\,dx\right)^{p/2} < \infty,$$

while the condition in Theorem 5.3 (vii) is equivalent to

$$\sum_{n\in\mathbb{Z}} \|x\varphi(x)\|^p_{BV[A^n, A^{n+1}]} < \infty,$$

which easily implies the above assertions.

Theorem 13.22. *Let $\alpha, \beta \in \mathbb{C} \setminus \mathbb{R}$ and let $p > 0$. Then the integral operator on $L^2(\mathbb{R})$ with kernel function*

$$(x, y) \mapsto h(x+y)\left(\frac{1}{x+\alpha} + \frac{1}{y+\beta}\right), \quad x, y \in \mathbb{R},$$

belongs to $\boldsymbol{S}_p$ if and only if convolution with the function $h(x)(x+\alpha+\beta)$ is an operator from $L^2(\mathbb{R}, (1+x^2)\,dx)$ to $L^2(\mathbb{R}, (1+x^2)^{-1}\,dx)$ of class $\boldsymbol{S}_p$.

Proof. Clearly, the integral operator on $L^2(\mathbb{R})$ with kernel function

$$(x, y) \mapsto h(x+y)\left(\frac{1}{x+\alpha} + \frac{1}{y+\beta}\right), \quad x, y \in \mathbb{R},$$

belongs to $\boldsymbol{S}_p$ if and only if so does the integral operator with kernel function

$$(x, y) \mapsto h(x-y)\frac{x-y+\alpha+\beta}{(x+\alpha)(y-\beta)}, \quad x, y \in \mathbb{R}.$$

To complete the proof, it remains to observe that multiplication by $(x-\beta)^{-1}$ is an isomorphism from $L^2(\mathbb{R})$ onto $L^2(\mathbb{R}, (1+x^2)\,dx)$ and multiplication by $(x+\alpha)^{-1}$ is an isomorphism from $L^2(\mathbb{R}, (1+x^2)^{-1}\,dx)$ onto $L^2(\mathbb{R})$. □

Corollary 13.23. *Let $\alpha, \beta, \gamma \in \mathbb{C} \setminus \mathbb{R}$ such that $\alpha + \beta \notin \mathbb{R}$ and let $p > 1/2$. Then the integral operator with kernel function*

$$(x, y) \mapsto h(x+y)\left(\frac{1}{x+\alpha} + \frac{1}{y+\beta}\right), \quad x, y \in \mathbb{R},$$

belongs to $\boldsymbol{S}_p$ if and only if convolution with the function $h(x)(x+\gamma)$ is an operator from $L^2(\mathbb{R}, (1+x^2)\,dx)$ to $L^2(\mathbb{R}, (1+x^2)^{-1}\,dx)$ of class $\boldsymbol{S}_p$.

Proof. It suffices to apply Theorem 13.19. □

Remark. In the same way we can prove that the following statements are equivalent for any $\alpha \in \mathbb{C} \setminus \mathbb{R}$ and for any $p > 0$:

(i) the integral operator on $L^2(\mathbb{R})$ with kernel function

$$(x, y) \mapsto k(x+y)(x+\alpha)^{-1}, \quad x, y \in \mathbb{R},$$

belongs to $\boldsymbol{S}_p$;

(ii) the integral operator on $L^2(\mathbb{R})$ with kernel function

$$(x, y) \mapsto k(x+y)(y+\alpha)^{-1}, \quad x, y \in \mathbb{R},$$

belongs to $\boldsymbol{S}_p$;

(iii) convolution with k is an operator from $L^2(\mathbb{R})$ to $L^2(\mathbb{R}, (1+x^2)^{-1}\,dx)$ of class $\boldsymbol{S}_p$;

(iv) convolution with k is an operator from $L^2(\mathbb{R}, (1+x^2)\,dx)$ to $L^2(\mathbb{R})$ of class $\boldsymbol{S}_p$.

Let us repeat that Theorem 13.15 implies that the integral operator with kernel function $k(x+y)(x+\alpha)^{-1}$ can be a nonzero operator in $\boldsymbol{S}_p$ only if $p > 1$.

14. Matrix representation

Let φ be a function in $L^2_{\text{loc}}(\mathbb{R})$ such that $\varphi(x+1) = \varphi(x)$, $x \in \mathbb{R}$. Consider the operators $Q_\varphi^{[0,1]}$ and $Q_\varphi^{[0,2]}$ on $L^2[0,1]$ and $L^2[0,2]$ respectively. Obviously,

$$\|Q_\varphi^{[0,1]}\|_{\boldsymbol{S}_p} \le \|Q_\varphi^{[0,2]}\|_{\boldsymbol{S}_p}, \quad 0 < p < \infty. \tag{14.1}$$

Obviously,

$$\|Q_\varphi^{[0,2]}\| \ge \|\mathcal{P}_{[1,2]\times[0,1]}Q_\varphi^{[0,2]}\| = \|\varphi\|_{L^2[0,1]}. \tag{14.2}$$

It is also easy to see that

$$\|Q_\varphi^{[0,2]}\|_{\boldsymbol{S}_p} \le C(p)\big(\|Q_\varphi^{[0,1]}\|_{\boldsymbol{S}_p} + \|\varphi\|_{L^2[0,1]}\big), \quad 0 < p < \infty, \tag{14.3}$$

where $C(p)$ is a constant that may depend only on p.

Theorem 14.1. *Let* $0 < p < \infty$. *Suppose that* φ *is a function in* $L^2_{\text{loc}}(\mathbb{R})$ *such that* $\varphi(x) = \varphi(x+1)$ *and* $\psi(x) \stackrel{\text{def}}{=} \varphi(1-x)$. *Then*

$$C_1(p)\big(\|Q_\varphi^{[0,1]}\|_{\boldsymbol{S}_p} + \|Q_\psi^{[0,1]}\|_{\boldsymbol{S}_p}\big) \le \|Q_\varphi^{[0,2]}\|_{\boldsymbol{S}_p} \le C_2(p)\big(\|Q_\varphi^{[0,1]}\|_{\boldsymbol{S}_p} + \|Q_\psi^{[0,1]}\|_{\boldsymbol{S}_p}\big).$$

Proof. To prove the left inequality, consider the integral operator K on $L^2[0,1]$ with kernel function

$$(x, y) \mapsto \varphi(\min\{x, y\}).$$

Clearly, the operators K and $Q_\psi^{[0,1]}$ are unitarily equivalent, and therefore we have $\|K\|_{\boldsymbol{S}_p} = \|Q_\psi^{[0,1]}\|_{\boldsymbol{S}_p}$ for any $p > 0$. Note that

$$\varphi(\min\{x, y\}) + \varphi(\max\{x, y\}) = \varphi(x) + \varphi(y).$$

Hence, $K + Q_\varphi^{[0,1]}$ is the integral operator with kernel function $(x, y) \mapsto \varphi(x) + \varphi(y)$. Thus $\|Q_\varphi^{[0,1]} + K\|_{\boldsymbol{S}_p} \le C(p)\|\varphi\|_{L^2}$. Now the left inequality is obvious. To prove the right inequality, we have to show that

$$\|\varphi\|_{L^2[0,1]} \le C(p)\big(\|Q_\varphi^{[0,1]}\|_{\boldsymbol{S}_p} + \|Q_\psi^{[0,1]}\|_{\boldsymbol{S}_p}\big).$$

Clearly,

$$(K + Q_\varphi^{[0,1]})1 = \varphi + \int_0^1 \varphi(t)\,dt.$$

It follows that

$$\left\|\left(K+Q_{\varphi}^{[0,1]}\right)1\right\|_{L^2[0,1]}^2=\|\varphi\|_{L^2[0,1]}^2+3\left|\int_0^1\varphi(t)\,dt\right|^2\geq\|\varphi\|_{L^2[0,1]}^2.$$

Thus,

$$\left\|K+Q_{\varphi}^{[0,1]}\right\|_{S_p}\geq\left\|K+Q_{\varphi}^{[0,1]}\right\|\geq\|\varphi\|_{L^2[0,1]},$$

and so

$$\begin{aligned}\|\varphi\|_{L^2[0,1]}&\leq\left\|\left(K+Q_{\varphi}^{[0,1]}\right)\right\|_{S_p}\\&\leq C(p)\left(\|K\|_{S_p}+\left\|Q_{\varphi}^{[0,1]}\right\|_{S_p}\right)\\&=C(p)\left(\left\|Q_{\varphi}^{[0,1]}\right\|_{S_p}+\left\|Q_{\psi}^{[0,1]}\right\|_{S_p}\right).\end{aligned}$$

□

Corollary 14.2. *Under the hypotheses of Theorem* 14.1

$$\left\|Q_{\psi}^{[0,2]}\right\|_{S_p}\leq C(p)\left\|Q_{\varphi}^{[0,2]}\right\|_{S_p}.$$

□

Theorem 14.3. *Let* $p>0$ *and* $a\in\mathbb{R}$. *If* φ *is a function satisfying the hypotheses of Theorem* 14.1 *and* $\psi(x)\stackrel{\text{def}}{=}\varphi(x-a)$, *then*

$$\left\|Q_{\psi}^{[0,2]}\right\|_{S_p}\leq C(p)\left\|Q_{\varphi}^{[0,2]}\right\|_{S_p}.$$

Proof. Clearly, it is sufficient to consider the case $a\in(0,1)$. Then $Q_{\psi}^{[0,1]}$ is by a translation unitarily equivalent to $Q_{\varphi}^{[1-a,2-a]}$, and thus

$$\left\|Q_{\psi}^{[0,1]}\right\|_{S_p}=\left\|Q_{\varphi}^{[1-a,2-a]}\right\|_{S_p}\leq\left\|Q_{\varphi}^{[0,2]}\right\|_{S_p}.$$

The result follows by (14.3) and (14.2). □

Let ϕ be a function on the unit circle $\mathbb{T}$. Put $\mathbb{T}_+\stackrel{\text{def}}{=}\{\zeta\in\mathbb{T}:\operatorname{Im}\zeta\geq0\}$, $\mathbb{T}_-\stackrel{\text{def}}{=}\{\zeta\in\mathbb{T}:\operatorname{Im}\zeta<0\}$ and

$$k_\phi(\zeta,\tau)\stackrel{\text{def}}{=}\begin{cases}\phi(\zeta^2), & \zeta\bar{\tau}\in\mathbb{T}_+,\\ \phi(\tau^2), & \zeta\bar{\tau}\in\mathbb{T}_-,\end{cases}\qquad(\zeta,\tau)\in\mathbb{T}^2. \tag{14.4}$$

It is easy to see that the functions ϕ and k_ϕ are equimeasurable. In particular, $\|\phi\|_{L^2(\mathbb{T})}=\|k_\phi\|_{L^2(\mathbb{T}^2)}$. Note also that if ϕ is continuous on $\mathbb{T}$, then k_ϕ is continuous on $\mathbb{T}^2$. Let $\phi\in L^2(\mathbb{T})$. Denote by K_ϕ the integral operator on $L^2(\mathbb{T})$ with kernel function k_ϕ.

Theorem 14.4. *Let $p > 0$. Suppose that $\phi \in L^2(\mathbb{T})$ and $\varphi(t) \stackrel{\text{def}}{=} \phi(e^{2\pi i t})$, $t \in \mathbb{R}$. Then*

$$C_1(p)\|Q_\varphi^{[0,2]}\|_{S_p} \le \|K_\phi\|_{S_p} \le C_2(p)\|Q_\varphi^{[0,2]}\|_{S_p},$$

where $C_1(p)$ and $C_2(p)$ may depend only on p.

Proof. Consider the integral operator K on $L^2[0,1]$ with kernel function $k \in L^2([0,1]^2)$ defined by

$$k(x,y) = \begin{cases} \varphi(2\max\{x,y\}), & |x-y| \le 1/2, \\ \varphi(2\min\{x,y\}), & |x-y| > 1/2. \end{cases}$$

It is easy to see that K is unitarily equivalent to K_ϕ. For $\alpha, \beta = 0, 1$ we consider the integral operator $K^{(\alpha,\beta)}$ with kernel function

$$(x,y) \mapsto k(x,y)\chi_{[\alpha/2,(1+\alpha)/2]}(x)\chi_{[\beta/2,(1+\beta)/2]}(y).$$

Using the substitution $(x,y) \mapsto (2x,2y)$, we find that $2\|K^{(0,0)}\|_{S_p} = \|Q_\varphi^{[0,1]}\|_{S_p}$. In a similar way we can obtain $2\|K^{(1,1)}\|_{S_p} = \|Q_\varphi^{[0,1]}\|_{S_p}$. Let $\psi(t) \stackrel{\text{def}}{=} \varphi(1-t)$. It is also easy to see that $2\|K^{(0,1)}\|_{S_p} = 2\|K^{(1,0)}\|_{S_p} = \|Q_\psi^{[0,1]}\|_{S_p}$. Hence,

$$\begin{aligned}\frac{1}{4}\left(\|Q_\varphi^{[0,1]}\|_{S_p} + \|Q_\psi^{[0,1]}\|_{S_p}\right) &= \frac{1}{4}\sum_{\alpha=0}^{1}\sum_{\beta=0}^{1}\|K^{(\alpha,\beta)}\|_{S_p} \\ &\le \|K_\phi\|_{S_p} \le C(p)\sum_{\alpha=0}^{1}\sum_{\beta=0}^{1}\|K^{(\alpha,\beta)}\|_{S_p} \\ &= C(p)\left(\|Q_\varphi^{[0,1]}\|_{S_p} + \|Q_\psi^{[0,1]}\|_{S_p}\right).\end{aligned}$$

It remains to apply Theorem 14.1. □

We denote by $\hat{f}(n)$ denote the nth Fourier coefficient of a function f in $L^1(\mathbb{T})$. For convenience we put

$$\hat{f}(n+1/2) \stackrel{\text{def}}{=} 0, \quad n \in \mathbb{Z}.$$

For a function k in $L^1(\mathbb{T}^2)$ we denote by $\{\hat{k}(m,n)\}_{(m,n)\in\mathbb{Z}^2}$ the sequence of its Fourier coefficients.

Let ϕ be a function on $\mathbb{T}$. Put

$$k_\phi^+(\zeta,\tau) \stackrel{\text{def}}{=} \begin{cases} \phi(\zeta^2), & \zeta\bar{\tau} \in \mathbb{T}_+, \\ 0, & \zeta\bar{\tau} \in \mathbb{T}_-, \end{cases}$$

and

$$k_\phi^-(\zeta,\tau) \stackrel{\text{def}}{=} \begin{cases} 0, & \zeta\bar{\tau} \in \mathbb{T}_+, \\ \phi(\tau^2), & \zeta\bar{\tau} \in \mathbb{T}_-. \end{cases}$$

Clearly, $k_\phi^+(\zeta,\tau) = k_\phi^-(\tau,\zeta)$ and $k_\phi(\zeta,\tau) = k_\phi^+(\zeta,\tau) + k_\phi^-(\zeta,\tau)$, where k_ϕ is defined by (14.4).

Lemma 14.5. *Let $\phi \in L^1(\mathbb{T})$. Then for $(m,n) \in \mathbb{Z}^2$*

$$\hat{k}_\phi^+(m,n) = \begin{cases} \frac{1}{2}\hat{\phi}(\frac{m}{2}), & n = 0, \\ \frac{\mathrm{i}}{\pi n}\hat{\phi}(\frac{m+n}{2}), & m, n \text{ are odd}, \\ 0, & \text{otherwise}. \end{cases}$$

Proof. Let us first observe that for any $\zeta \in \mathbb{T}$ and $n \in \mathbb{Z}$ we have

$$\int_{\{\tau\in\mathbb{T}:\zeta\bar{\tau}\in\mathbb{T}_+\}} \tau^{-n}\,dm(\tau) = \begin{cases} \frac{1}{2}, & n = 0, \\ \frac{\mathrm{i}}{\pi n}\zeta^{-n}, & n \text{ is odd}, \\ 0, & \text{otherwise}. \end{cases}$$

It follows that

$$\int_{\mathbb{T}} k_\phi^+(\zeta,\tau)\zeta^{-m}\tau^{-n}\,dm(\zeta) = \int_{\{\tau\in\mathbb{T}:\zeta\bar{\tau}\in\mathbb{T}_+\}} \phi(\zeta^2)\zeta^{-m}\tau^{-n}\,dm(\tau)$$

$$= \begin{cases} \frac{1}{2}\phi(\zeta^2)\zeta^{-m}, & n = 0, \\ \frac{\mathrm{i}}{\pi n}\phi(\zeta^2)\zeta^{-m-n}, & n \text{ is odd}, \\ 0, & \text{otherwise}. \end{cases}$$

It remains to integrate the last identity in ζ. □

Corollary 14.6. *Let $\phi \in L^1(\mathbb{T})$. Then*

$$\hat{k}_\phi^-(m,n) = \begin{cases} \frac{1}{2}\hat{\phi}(\frac{n}{2}), & m = 0, \\ \frac{\mathrm{i}}{\pi m}\hat{\phi}(\frac{m+n}{2}), & m, n \text{ are odd}, \\ 0, & \text{otherwise}. \end{cases}$$

Proof. It suffices to observe that $k_\phi^-(\zeta,\tau) = k_\phi^+(\tau,\zeta)$. □

Corollary 14.7. *Let $\phi \in L^1(\mathbb{T})$. Then*

$$\hat{k}_\phi(m,n) = \begin{cases} \frac{1}{2}\hat{\phi}\left(\frac{m}{2}\right), & n = 0, \\ \frac{1}{2}\hat{\phi}\left(\frac{n}{2}\right), & m = 0, \\ \hat{\phi}(0), & m = n = 0, \\ \frac{\mathrm{i}}{\pi}\left(\frac{1}{m} + \frac{1}{n}\right)\hat{\phi}\left(\frac{m+n}{2}\right), & m,\ n \text{ are odd}, \\ 0, & \text{otherwise}. \end{cases}$$

Proof. It suffices to observe that $k_\phi(\zeta,\tau) = k_\phi^+(\zeta,\tau) + k_\phi^+(\tau,\zeta)$. □

Theorem 14.8. *Suppose that $\varphi(t) = \sum_{k\in\mathbb{Z}} a_k e^{2\pi \mathrm{i}kt}$ and $\sum_{k\in\mathbb{Z}} |a_k|^2 < \infty$, and let $p > 0$. Then $Q_\varphi^{[0,2]} \in \boldsymbol{S}_p$ if and only if the matrix*

$$\left\{ a_{m+n+1}\left(\frac{1}{m+\frac{1}{2}} + \frac{1}{n+\frac{1}{2}}\right)\right\}_{m,n\in\mathbb{Z}} \tag{14.5}$$

belongs to $\boldsymbol{S}_p$.

Here we identify operators on $\ell^2(\mathbb{Z})$ with their matrices with respect to the standard orthonormal basis of $\ell^2(\mathbb{Z})$.

Proof. Consider the function ϕ on $\mathbb{T}$ defined by $\phi(z) = \sum_{n\in\mathbb{Z}} a_n z^n$. By Theorem 14.4, $Q_\varphi^{[0,2]} \in \boldsymbol{S}_p$ if and only if $K_\phi \in \boldsymbol{S}_p$. It easy to see that the operator K_ϕ belongs to $\boldsymbol{S}_p$ if and only if the matrix $\{\hat{k}_\phi(m,n)\}_{m,n\in\mathbb{Z}}$ belongs to $\boldsymbol{S}_p$. Corollary 14.7 implies that $\hat{k}_\phi(m,n) \neq 0$ only if mn is odd or $mn = 0$. Hence, it is easy to check that $\{\hat{k}_\phi(m,n)\}_{m,n\in\mathbb{Z}} \in \boldsymbol{S}_p$ if and only if $\{\hat{k}_\phi(2m+1,2n+1)\}_{m,n\in\mathbb{Z}} \in \boldsymbol{S}_p$. It remains to note that

$$\hat{k}_\phi(2m+1,2n+1) = \frac{\mathrm{i}}{\pi}\left(\frac{1}{2m+1} + \frac{1}{2n+1}\right)a_{m+n+1}$$

by Corollary 14.7. □

Clearly, the same reasoning shows that $Q_\varphi^{[0,2]}$ is bounded on $L^2[0,1]$ if and only if the matrix (14.5) is bounded. The following result shows that the boundedness of (14.5) is equivalent to its membership of $\boldsymbol{S}_p$, $p > 1$.

Theorem 14.9. *Let $\{a_k\}_{k\in\mathbb{Z}}$ be a two-sided sequence of complex numbers and let*

$$A \stackrel{\text{def}}{=} \left\{ a_{m+n+1}\left(\frac{1}{m+\frac{1}{2}} + \frac{1}{n+\frac{1}{2}}\right)\right\}_{m,n\in\mathbb{Z}}.$$

Suppose that $p > 1$. The following are equivalent:

(i) *A is a bounded operator on $\ell^2(\mathbb{Z})$;*

(ii) *$A \in \boldsymbol{S}_p$;*

(iii) *$\{a_k\}_{k\in\mathbb{Z}} \in \ell^2(\mathbb{Z})$.*

Proof. Suppose that A is bounded. Then the sequence

$$\left\{a_{n+1}\left(2+\frac{1}{n+\frac{1}{2}}\right)\right\}_{n\in\mathbb{Z}}$$

belongs to $\ell^2(\mathbb{Z})$ which implies (iii). It is clear that (iii) is equivalent to the fact that $\varphi \in L^2[0,1]$. By Theorem 3.3, $Q_\varphi^{[0,2]} \in \boldsymbol{S}_p$, and so by Theorem 14.8, $A \in \boldsymbol{S}_p$. The implication (ii) $\Rightarrow$ (i) is trivial. □

Related results, that matrices of a roughly similar sort are bounded if and only if they are in certain $\boldsymbol{S}_p$ can be found in [W1].

Remark. Note that the following identities hold:

$$\frac{\mathrm{i}}{\pi}\sum_{m,n\in\mathbb{Z}} a_{m+n+1}\left(\frac{1}{m+\frac{1}{2}}+\frac{1}{n+\frac{1}{2}}\right)\zeta^m\tau^n = \frac{\phi(\zeta)-\phi(\tau)}{\sqrt{\zeta\tau}},$$

$$\frac{\mathrm{i}}{\pi}\sum_{m,n\in\mathbb{Z}} \frac{a_{m+n+1}}{n+\frac{1}{2}}\zeta^m\tau^n = \frac{\phi(\zeta)}{\sqrt{\zeta\tau}},$$

$$\frac{\mathrm{i}}{\pi}\sum_{m,n\in\mathbb{Z}} \frac{a_{m+n+1}}{m+\frac{1}{2}}\zeta^m\tau^n = -\frac{\phi(\tau)}{\sqrt{\zeta\tau}},$$

where $\sqrt{\zeta\tau}$ is chosen so that $\bar{\tau}\sqrt{\zeta\tau} \in \mathbb{T}_+$ (the series converge in $L^2(\mathbb{T}^2)$).

Indeed, it suffices to note that by Corollary 14.7,

$$\frac{\mathrm{i}}{\pi}\sum_{m,n\in\mathbb{Z}} a_{m+n+1}\left(\frac{1}{2m+1}+\frac{1}{2n+1}\right)\zeta^{2m+1}\tau^{2n+1} = \frac{1}{2}\big(k_\phi(\zeta,\tau)-k_\phi(\zeta,-\tau)\big),$$

by Lemma 14.5,

$$\frac{\mathrm{i}}{\pi}\sum_{m,n\in\mathbb{Z}} \frac{a_{m+n+1}}{2n+1}\zeta^{2m+1}\tau^{2n+1} = \frac{1}{2}\big(k_\phi^+(\zeta,\tau)-k_\phi^+(\zeta,-\tau)\big),$$

and by Corollary 14.6,

$$\frac{\mathrm{i}}{\pi}\sum_{m,n\in\mathbb{Z}} \frac{a_{m+n+1}}{2m+1}\zeta^{2m+1}\tau^{2n+1} = \frac{1}{2}\big(k_\phi^-(\zeta,\tau)-k_\phi^-(\zeta,-\tau)\big).$$

Remark. Note that if $p > 1/2$, then

$$A \stackrel{\text{def}}{=} \left\{ a_{m+n+1} \left(\frac{1}{m+\frac{1}{2}} + \frac{1}{n+\frac{1}{2}} \right) \right\}_{m,n\in\mathbb{Z}} \in \boldsymbol{S}_p$$

if and only if

$$B \stackrel{\text{def}}{=} \left\{ a_{m+n} \left(\frac{1}{m+\frac{1}{2}} + \frac{1}{n+\frac{1}{2}} \right) \right\}_{m,n\in\mathbb{Z}} \in \boldsymbol{S}_p. \tag{14.6}$$

Indeed, put $\psi(t) \stackrel{\text{def}}{=} e^{2\pi i t}\varphi(t)$. By Theorem 14.8, it suffices to prove that $Q_\varphi^{[0,2]} \in \boldsymbol{S}_p$ implies $Q_\psi^{[0,2]} \in \boldsymbol{S}_p$. This follows from Theorem 7.3.

Note however that for $p \le 1/2$ this is not true. Indeed, if $a_0 = 1$ and $a_n = 0$ for $n \ne 0$ (in other words, $\varphi(t) = 1$, $t \in \mathbb{R}$), then it is easy to see that A is the zero matrix, and so it belongs to $\boldsymbol{S}_p$ for any $p > 0$. On the other hand, the matrix B has nonzero entries $-(n^2-1/4)^{-1}$ for $m = -n$, and so it belongs to $\boldsymbol{S}_p$ only for $p > 1/2$. The situation is similar in the case where the restriction $\varphi\big|[0,1)$ is the characteristic function of an interval in which case $A \in \boldsymbol{S}_p$ for any $p > 0$ but $B \in \boldsymbol{S}_p$ only for $p > 1/2$, see Theorem 8.11 and Corollary 9.11.

Suppose now that $p > 1/2$ and consider the following submatrices of the matrix B defined by (14.6):

$$B_1 = \left\{ a_{m+n} \left(\frac{1}{m+\frac{1}{2}} + \frac{1}{n+\frac{1}{2}} \right) \right\}_{m,n\ge 0},$$

$$B_2 = \left\{ a_{m+n} \left(\frac{1}{m+\frac{1}{2}} + \frac{1}{n+\frac{1}{2}} \right) \right\}_{m\ge 0,\, n<0},$$

$$B_3 = \left\{ a_{m+n} \left(\frac{1}{m+\frac{1}{2}} + \frac{1}{n+\frac{1}{2}} \right) \right\}_{m<0,\, n\ge 0},$$

and

$$B_4 = \left\{ a_{m+n} \left(\frac{1}{m+\frac{1}{2}} + \frac{1}{n+\frac{1}{2}} \right) \right\}_{m,n<0}.$$

Clearly, $B \in \boldsymbol{S}_p$ if and only if all matrices B_j, $1 \le j \le 4$, belong to $\boldsymbol{S}_p$.
It is direct that $B_1 \in \boldsymbol{S}_p$ if and only if the matrix

$$\left\{ a_{j+k} \frac{j+k+1}{(j+1)(k+1)} \right\}_{j,k\ge 0} \tag{14.7}$$

belongs to $\boldsymbol{S}_p$. Matrices of the form

$$\{a_{j+k}(1+j)^{\alpha}(1+k)^{\beta}\}_{j,k\geq 0} \tag{14.8}$$

are called *weighted Hankel matrices.* It was proved in [Pel2] that if $\alpha > -1/2$, $\beta > -1/2$, and $0 < p \leq 1$, the matrix (14.8) belongs to $\boldsymbol{S}_p$ if and only if the function $\sum_{n\geq 0} a_n z^n$ belongs to the Besov class $B_{p\,p}^{1/p+\alpha+\beta}$ of functions on the unit circle $\mathbb{T}$. More recent results on Schatten class properties of weighted Hankel matrices are in [RW] and [W2]. However, in the case of interest, the weighted Hankel matrix (14.7) for $\alpha = \beta = -1$, no characterization of such matrices of class $\boldsymbol{S}_p$ is known. In the next section we obtain some necessary conditions for the matrix (14.7) to belong to $\boldsymbol{S}_1$.

It is also easy to see that $B_4 \in \boldsymbol{S}_p$ if and only if the weighted Hankel matrix

$$\left\{a_{-(j+k+2)}\frac{j+k+1}{(j+1)(k+1)}\right\}_{j,k\geq 0} \tag{14.9}$$

belongs to $\boldsymbol{S}_p$.

It can also be easily shown that $B_2 \in \boldsymbol{S}_p$ if and only if $B_3 \in \boldsymbol{S}_p$ and this is equivalent to the fact that the *weighted Toeplitz matrix*

$$\left\{a_{j-k-1}\frac{j-k}{(1+j)(1+k)}\right\}_{j,k\geq 0} \tag{14.10}$$

belongs to $\boldsymbol{S}_p$.

Summarizing the above, we can state the following result.

Theorem 14.10. *Let $1/2 < p \leq 1$ and let φ be a function in $L^2[0,1]$ of the form $\varphi(t) = \sum_{n\in\mathbb{Z}} a_n e^{2\pi i n t}$, $t \in [0,1]$. Then $Q_\varphi^{[0,1]} \in \boldsymbol{S}_p$ if and only if the matrices* (14.7), (14.9), *and* (14.10) *belong to $\boldsymbol{S}_p$.* □

In the next section we use the results above to obtain necessary conditions for the nuclearity of operators Q_φ.

Let us consider now the family of functions $\{F_\lambda\}_{\lambda\in\mathbb{C}}$ on $[0,1)^2$ defined for a function $\varphi \in L^1[0,1]$ by

$$F_\lambda(t,s) \stackrel{\text{def}}{=} \varphi\left(\max\{s,t\}\right) e^{-2\pi\lambda i|s-t|} + \varphi\left(\min\{s,t\}\right) e^{2\pi\lambda i|s-t|-2\pi\lambda i}.$$

Clearly, $F_\lambda \in L^1([0,1)^2)$. We identify $[0,1)^2$ via the map $(s,t) \mapsto (e^{2\pi i s}, e^{2\pi i t})$ with $\mathbb{T}^2$ and we can consider the Fourier coefficients of functions on $[0,1)^2$.

Theorem 14.11. *Suppose that $\lambda \notin \mathbb{Z}$. Then*

$$\widehat{F_\lambda}(m,n) = \frac{1-e^{-2\pi i\lambda}}{2\pi i}\left(\frac{1}{\lambda-m}+\frac{1}{\lambda-n}\right)\hat{\varphi}(m+n). \tag{14.11}$$

Proof. We have

$$\widehat{F_\lambda}(m,n) \stackrel{\text{def}}{=} \iint\limits_{[0,1)\times[0,1)} F_\lambda(s,t)e^{-2\pi i m s-2\pi i n t}\,ds\,dt = \iint\limits_{t\geq s} + \iint\limits_{t\leq s}.$$

Let us compute the first integral:

$$\iint\limits_{t\ge s} = \int_0^1 \left(\int_0^t \varphi(t) e^{2\pi i\lambda(s-t)} e^{-2\pi i m t - 2\pi i n s}\, ds \right) dt$$

$$+ \int_0^1 \left(\int_s^1 \varphi(s) e^{2\pi i\lambda(t-s) - 2\pi i\lambda} e^{-2\pi i m t - 2\pi i n s}\, dt \right) ds$$

$$= \int_0^1 \varphi(t) e^{-2\pi i\lambda t - 2\pi i m t} \left(\int_0^t e^{2\pi i s(\lambda - n)}\, ds \right) dt$$

$$+ \int_0^1 \varphi(s) e^{-2\pi i\lambda s - 2\pi i\lambda - 2\pi i n s} \left(\int_s^1 e^{2\pi i t(\lambda - m)}\, dt \right) ds$$

$$= \int_0^1 \varphi(t) e^{-2\pi i\lambda t - 2\pi i m t} \frac{e^{2\pi i t(\lambda-n)} - 1}{2\pi i(\lambda - n)} dt$$

$$+ \int_0^1 \varphi(s) e^{-2\pi i\lambda s - 2\pi i\lambda - 2\pi i n s} \frac{e^{2\pi i\lambda} - e^{2\pi i s(\lambda-m)}}{2\pi i(\lambda - m)} ds$$

$$= \frac{\hat{\varphi}(m+n)}{2\pi i(\lambda - n)} - \frac{1}{2\pi i(\lambda - n)} \int_0^1 \varphi(t) e^{-2\pi i\lambda t - 2\pi i m t}\, dt$$

$$+ \frac{1}{2\pi i(\lambda - m)} \int_0^1 \varphi(s) e^{-2\pi i\lambda s - 2\pi i n s}\, ds - \frac{e^{-2\pi i\lambda}\hat{\varphi}(m+n)}{2\pi i(\lambda - m)}.$$

Similarly,

$$\iint\limits_{t\le s} = \frac{\hat{\varphi}(m+n)}{2\pi i(\lambda - m)} - \frac{1}{2\pi i(\lambda - m)} \int_0^1 \varphi(s) e^{-2\pi i\lambda s - 2\pi i n s}\, ds$$

$$+ \frac{1}{2\pi i(\lambda - n)} \int_0^1 \varphi(t) e^{-2\pi i\lambda t - 2\pi i m t}\, dt - \frac{e^{-2\pi i\lambda}\hat{\varphi}(m+n)}{2\pi i(\lambda - n)}$$

which implies (14.11). □

Theorem 14.12. *Let $\lambda \in \mathbb{C}$. The integral operator with kernel function F_λ is bounded on $L^2[0, 1]$ if and only if $\varphi \in L^2([0, 1))$.*

Proof. Clearly, the integral operator with kernel function F_λ belongs to S_2 if $\varphi \in L^2([0, 1))$. Suppose now that the integral operator with kernel function F_λ is bounded. If $\lambda \in \mathbb{Z}$, then

$$F_\lambda(s, t) = \left(\varphi(\max\{s, t\}) e^{-4\pi i\lambda \max\{s,t\}} + \varphi(\min\{s, t\}) e^{-4\pi i\lambda \min\{s,t\}}\right) e^{2\pi i\lambda(s+t)}$$

$$= \left(\varphi(s) e^{-4\pi i\lambda s} + \varphi(t) e^{-4\pi i\lambda t}\right) e^{2\pi i\lambda(s+t)}.$$

Consequently, the boundedness of this operator implies that $\varphi \in L^2[0, 1]$. If $\lambda \notin \mathbb{Z}$, then the boundedness of the integral operator with kernel function F_λ implies that

$$\sum_{m\in\mathbb{Z}} \left|\widehat{F}_\lambda(m, 0)\right|^2 < +\infty$$

and by Theorem 14.11 we obtain

$$\sum_{m\in\mathbb{Z}} |\hat{\varphi}(m)|^2 \left|\frac{2\lambda - m}{\lambda - m}\right|^2 < +\infty,$$

whence $\varphi \in L^2[0, 1]$. □

Lemma 14.13. *Let $w, a \in \mathbb{C}$ and $w \neq 1$. Let $p > \frac{1}{2}$. Then the function*

$$(s, t) \mapsto (w - e^{a|s-t|})^{-1} \chi_\Delta(s) \chi_\Delta(t), \quad s, t \in \mathbb{R},$$

is a Schur multiplier of $\boldsymbol{S}_p$ if Δ is an interval of sufficiently small length.

Proof. Clearly, it suffices to consider the case $p < 1$. Note that $w - e^{a|s-t|}$ is a Schur multiplier of $\boldsymbol{S}_p(L^2(\Delta))$ by Theorem 7.3, since

$$w - e^{a|s-t|} = w - e^{a\max\{s,t\}} e^{-a\min\{s,t\}}.$$

We have to prove that this multiplier is an isomorphism of $\boldsymbol{S}_p(L^2(\Delta))$ if Δ has sufficiently small length. For $\omega \in L^\infty(\Delta^2)$ we put

$$\|\omega\|_{\mathfrak{M}_p(\Delta)} \stackrel{\text{def}}{=} \sup \|\omega k\|_{\boldsymbol{S}_p(L^2(\Delta))},$$

where the *supremum* is taken over all integral operators with kernel $k \in L^2(\Delta^2)$ such that $\|k\|_{\boldsymbol{S}_p} = 1$. Here by $\|k\|_{\boldsymbol{S}_p}$ we mean the $\boldsymbol{S}_p$ norm (quasi-norm if $p < 1$) of the integral operator with kernel function k. Obviously, it suffices to prove the inequality

$$\|e^{a\max\{s,t\}} e^{-a\min\{s,t\}} - 1\|_{\mathfrak{M}_p(\Delta)} < |w - 1|$$

provided the length of Δ is sufficiently small. Theorem 7.7 implies that, for any $x_0 \in \Delta$,

$$\lim_{|\Delta|\to 0} \|e^{a(\max\{s,t\}-x_0)} - 1\|_{\mathfrak{M}_p(\Delta)} = 0 \quad \text{and} \quad \lim_{|\Delta|\to 0} \|e^{-a(\min\{s,t\}-x_0)} - 1\|_{\mathfrak{M}_p(\Delta)} = 0.$$

Hence, the desired inequality is obvious. □

Theorem 14.14. *Suppose that $\lambda \notin \mathbb{Z}$ and $p > 1/2$. Then $Q_\varphi^{[0,2]} \in \boldsymbol{S}_p$ if and only if the integral operator with kernel function F_λ belongs to $\boldsymbol{S}_p$.*

Proof. Suppose that $Q_\varphi^{[0,2]} \in \boldsymbol{S}_p$. Then the integral operators with kernel functions $\varphi(\max\{s, t\})$ and $\varphi(\min\{s, t\})$ belong to $\boldsymbol{S}_p(L^2[0, 1])$ (see Theorem 14.1). Note that

$e^{2\pi i\lambda|s-t|} = e^{2\pi i\lambda(2\max\{s,t\}-s-t)}$. It follows now from Theorem 7.3 that the integral operator with kernel function F_λ belongs to $\boldsymbol{S}_p$.

Suppose now that the integral operator with kernel function F_λ belongs to $\boldsymbol{S}_p$. We have to prove that $Q_\varphi^{[0,2]} \in \boldsymbol{S}_p$. By Theorem 14.12, $\varphi \in L^2[0,1]$, and so it suffices to show that $Q_\varphi^{[0,1]} \in \boldsymbol{S}_p$. By Lemma 14.13, we can choose a positive number δ such that the function $(e^{2\pi i\lambda} - e^{4\pi i\lambda|s-t|})^{-1}$ belongs to $\mathfrak{M}_p(\Delta)$ for any interval Δ of length less than δ. We can represent the interval $[0,1)$ in the form $\bigcup_{j=1}^N \Delta_j$, where the Δ_j are pairwise disjoint intervals with lengths less than δ. Clearly,

$$F_\lambda(s,t) = \varphi(\max\{s,t\})e^{-2\pi\lambda i|s-t|} + (\varphi(s)+\varphi(t) - \varphi(\max\{s,t\})e^{2\pi\lambda i|s-t|-2\pi\lambda i}.$$

Let $s,\ t \in \Delta_j$. Then

$$\varphi(\max\{s,t\}) = \frac{F_\lambda(s,t) - (\varphi(s)+\varphi(t))e^{2\pi\lambda i|s-t|-2\pi\lambda i}}{e^{-2\pi\lambda i|s-t|} - e^{2\pi\lambda i|s-t|-2\pi\lambda i}} =$$

$$\frac{\left(F_\lambda(s,t) - (\varphi(s)+\varphi(t))e^{2\pi\lambda i(\max\{s,t\}-\min\{s,t\}-1)}\right)}{e^{2\pi\lambda i} - e^{4\pi\lambda i|s-t|}} e^{2\pi\lambda i(1+\max\{s,t\}-\min\{s,t\})}.$$

Theorem 7.3 and Lemma 14.13 imply that the integral operator with kernel function

$$(s,t) \mapsto \varphi(\max\{t,s\})\chi_{\Delta_j}(t)\chi_{\Delta_j}(s), \quad s,\tau \in \mathbb{R},$$

belongs to $\boldsymbol{S}_p$. To complete the proof, it remains to observe that the kernel function

$$(s,t) \mapsto \varphi(\max\{s,t\}) - \sum_{j=1}^N \varphi(\max\{s,t\})\chi_{\Delta_j}(s)\chi_{\Delta_j}(t)$$

determines a finite rank operator. □

Theorem 14.14 implies that if $p > 1/2$, $\lambda_1,\ \lambda_2 \notin \mathbb{Z}$, and $F_{\lambda_1} \in \boldsymbol{S}_p$, then $F_{\lambda_2} \in \boldsymbol{S}_p$. This can also be easily deduced from the following elementary fact: if $x \in \ell^2(\mathbb{Z})$ and $y \in \ell^p(\mathbb{Z})$ with $p \le 2$, then $\{x_{m+n}\, y_n\}_{m,n\in\mathbb{Z}} \in \boldsymbol{S}_p$.

15. Necessary conditions for $Q_\varphi \in \boldsymbol{S}_1$

In this section we obtain various necessary conditions for $Q_\varphi \in \boldsymbol{S}_1$.

Theorem 15.1. *Let $\{a_n\}_{n\ge0}$ be a sequence in ℓ^2. If the matrix*

$$\Gamma = \left\{a_{j+k}\left(\frac{1}{j+\frac12} + \frac{1}{k+\frac12}\right)\right\}_{j,k\ge0}$$

belongs to $\boldsymbol{S}_1$, then the function $\sum_{n\ge0}\log(2+n)a_n z^n$ belongs to the Hardy class H^1.

We need the following well-known lemma (see e.g., [Pel1]).

Lemma 15.2. *Suppose that the matrix $\{a_{jk}\}_{j,k\geq 0}$ belongs to $\boldsymbol{S}_1$. Then the function $\sum_{n\geq 0}\left(\sum_{j=0}^{n} a_{j\,n-j}\right)z^n$ belongs to the Hardy class H^1.*

Proof. It is sufficient to prove this when the matrix has rank one in which case the result is an immediate consequence of the fact that $H^2 \cdot H^2 \subset H^1$. □

Lemma 15.3. *Let $m \in \mathbb{Z}_+$ and let*

$$\beta_n \stackrel{\text{def}}{=} \sum_{j=0}^{n} \frac{1}{j+\frac{1}{2}}. \tag{15.1}$$

Then there exists $d \in \mathbb{R}$ such that

$$|\beta_n - \log(2+n) - d| \leq C\frac{1}{1+n}.$$

Proof. We use the following well known fact (see, for example, [Z, Ch. I, (8.9)])

$$\left|\sum_{j=1}^{n} \frac{1}{j} - \log n - \gamma\right| \leq C \cdot n^{-1}, \tag{15.2}$$

where γ is the Euler constant. We have

$$\sum_{j=0}^{n} \frac{1}{j+\frac{1}{2}} = 2\sum_{j=0}^{n} \frac{1}{2j+1} = 2\sum_{j=1}^{2n+1} \frac{1}{j} - \sum_{j=1}^{n} \frac{1}{j},$$

and so by (15.2),

$$\left|\sum_{j=0}^{n} \frac{1}{j+\frac{1}{2}} - 2\log(2n+1) + \log n - \gamma\right| \leq C \cdot n^{-1}$$

which implies the result. □

Proof of Theorem 15.1. By Lemma 15.2, we have

$$\sum_{m\geq 0} a_m \left(\sum_{j=0}^{m}\left(\frac{1}{j+\frac{1}{2}} + \frac{1}{(m-j+\frac{1}{2})}\right)\right)z^m = 2\sum_{m\geq 0} a_m \left(\sum_{j=0}^{m} \frac{1}{j+\frac{1}{2}}\right)z^m \in H^1.$$

Since $\{a_n\}_{n\geq 0} \in \ell^2$, it is not hard to check that $\sum_{n\geq 0} \log(2+n) a_n z^n \in H^1$ by Lemma 15.3. □

Theorem 15.4. *Let $\varphi \in L^2[0,b]$ and $\varphi(x) = \sum_{n\in\mathbb{Z}} a_n e^{2\pi i n x/b}$. If $Q_\varphi^{[0,b]} \in \boldsymbol{S}_1$, then the functions $\sum_{n\geq 0} a_n \log(2+n) z^n$ and $\sum_{n\geq 0} a_{-n} \log(2+n) z^n$ in the unit disc $\mathbb{D}$ belong to the Hardy class H^1.*

Proof. Without loss of generality we may assume that $b = 1$. By Theorem 14.8, the matrices

$$\Gamma = \left\{ a_{j+k+1} \left(\frac{1}{j+\frac{1}{2}} + \frac{1}{k+\frac{1}{2}} \right) \right\}_{j,k\geq 0}$$

and

$$\Gamma = \left\{ a_{-(j+k+1)} \left(\frac{1}{j+\frac{1}{2}} + \frac{1}{k+\frac{1}{2}} \right) \right\}_{j,k\geq 0}$$

belong to $\boldsymbol{S}_1$. The result follows now from Theorem 15.1. □

Theorem 15.5. *Let I be a compact interval in $(0, \infty)$ and let φ be a function in $L^1_{\mathrm{loc}}(\mathbb{R}_+)$ such that $Q_\varphi \in \boldsymbol{S}_1$. If*

$$a_n = \int_I \varphi(x) e^{-2\pi i n x/|I|}\, dx, \quad n \in \mathbb{Z},$$

then the functions $\sum_{n\geq 0} a_n \log(2+n) z^n$ and $\sum_{n\geq 0} a_{-n} \log(2+n) z^n$ belong to the Hardy class H^1.

Proof. Since I is separated away from 0, it follows that $\varphi \big| I \in L^2(I)$. We can now apply a translation and reduce the result to Theorem 15.4. □

Corollary 15.6. *Under the hypotheses of either Theorem* 15.4 *or* 15.5 *the following holds:*

(i) $|a_n| \leq C(\log(2+|n|))^{-1}$, $n \in \mathbb{Z}$;

(ii) *suppose that $\{n_k\}_{k\geq 0}$ is an Hadamard lacunary sequence of positive integers, i.e.,*

$$\inf_{k\geq 0} \frac{n_{k+1}}{n_k} > 1,$$

then

$$\sum_{k\geq 0} |a_{n_k}|^2 (\log(1+n_k^2))^2 < \infty \quad \textit{and} \quad \sum_{k\geq 0} |a_{-n_k}|^2 (\log(1+n_k^2))^2 < \infty.$$

Proof. (i) follows immediately from Theorem 15.5 and the obvious fact that the Fourier coefficients of an H^1 function are bounded. Finally, (ii) is an immediate consequence of Theorem 15.5 and Paley's inequality (see [Z, v. 2, Ch. XII, (7.8)]). □

Note that if I is a compact interval in $(0, \infty)$, the restrictions of function in X_1 to I fill the space $L^2(I)$, and so the sequence of Fourier coefficients $\{a_n\}_{n\in\mathbb{Z}}$ can be an arbitrary sequence in ℓ^2. Thus Corollary 15.6 also shows that the condition $\varphi \in X_1$ is not sufficient for $Q_\varphi \in \boldsymbol{S}_1$.

Now we are going to use Theorem 13.12 to obtain another necessary condition for $Q_\varphi \in S_1$. We denote by $\mathfrak{H}^1$ the Stein–Weiss space of functions f in $L^1(\mathbb{R})$ such that $\mathcal{F}^{-1}(\chi_{\mathbb{R}_+}\mathcal{F}f) \in L^1(\mathbb{R})$, where $\mathcal{F}$ is Fourier transformation.

Theorem 15.7. *Let $h \in L^2_{\mathrm{loc}}(\mathbb{R})$. Suppose that the integral operator on $L^2(\mathbb{R})$ with kernel function*

$$(x,y) \mapsto h^{\spadesuit}(x,y) \stackrel{\text{def}}{=} h(x+y)\left(\frac{1}{x+\mathrm{i}} + \frac{1}{y-\mathrm{i}}\right), \quad x, y \in \mathbb{R},$$

belongs to S_1. Then the Fourier transform of the function $h(x)\log(1+x^2)$ belongs to the Stein–Weiss space $\mathfrak{H}^1$.

Proof. Clearly, the integral operator with kernel function $h^{\spadesuit}\chi_{[0,+\infty)^2}$ belongs to S_1. Put

$$g(x) \stackrel{\text{def}}{=} \int_{\mathbb{R}} h^{\spadesuit}(t, x-t)\chi_{[0,+\infty)^2}(t, x-t)\,dt.$$

We have

$$g(x) = \begin{cases} h(x)\int_0^x \left(\frac{1}{t+\mathrm{i}} + \frac{1}{x-t-\mathrm{i}}\right) dt = h(x)\log(1+x^2), & x > 0, \\ 0, & x < 0. \end{cases}$$

It follows from Theorem 6.3 that $\mathcal{F}g \in L^1(\mathbb{R})$. In the same way it can be shown that the Fourier transform of the function $h(x)\log(1+x^2)\chi_{\mathbb{R}_-}(x)$ belongs to $L^1(\mathbb{R})$. This implies the result. □

Corollary 15.8. *Let $h \in L^2_{\mathrm{loc}}(\mathbb{R})$ and let $a, b \in \mathbb{C} \setminus \mathbb{R}$ such that $a + b \notin \mathbb{R}$. Suppose that the integral operator on $L^2(\mathbb{R})$ with kernel function*

$$h^{\spadesuit}_{a,b}(x,y) \stackrel{\text{def}}{=} h(x+y)\left(\frac{1}{x+a} + \frac{1}{y+b}\right)$$

belongs to S_1. Then the Fourier transform of the function $h(x+c)\log(1+x^2)$ belongs to $\mathfrak{H}^1$ for any $c \in \mathbb{R}$.

Proof. It is obvious that the integral operators on $L^2(\mathbb{R})$ with kernel functions $h(x+y+c)\left(\frac{1}{x+a+c} + \frac{1}{y+b}\right)$ belong to S_1. Consequently, by Theorem 13.19, the integral operator on $L^2(\mathbb{R})$ with kernel function $h(x+y+c)\left(\frac{1}{x+\mathrm{i}} + \frac{1}{y-\mathrm{i}}\right)$ belongs to S_1. It remains to apply Theorem 15.7. □

Corollary 15.9. *Suppose that h, a and b satisfy the hypotheses of Corollary 15.8. Then $h(x)\log|x| \to 0$ as $|x| \to \infty$.* □

Now Corollary 15.9 and Theorem 13.12 imply the following theorem.

Theorem 15.10. *Let $\varphi \in L^2_{\mathrm{loc}}(\mathbb{R}_+)$. Suppose that $Q_\varphi \in \boldsymbol{S}_1$. Then*

$$\log|x| \int_{\mathbb{R}_+} \varphi(t) t^{x\mathrm{i}}\, dt \to 0 \quad as \quad |x| \to \infty. \tag{15.3}$$

□

Note that it follows from Theorem 6.2 that $\varphi \in L^1(\mathbb{R}_+)$, and so the integral in (15.3) is well defined. It is easy to see that if φ is an arbitrary L^2 function supported on a compact subset of $(0, \infty)$, then $\varphi \in X_1$. However, φ does not have to satisfy (15.3), and so Theorem 15.10 also implies that the condition $\varphi \in X_1$ is not sufficient for $Q_\varphi \in \boldsymbol{S}_1$.

We conclude this section with necessary conditions on the L^1 modulus of continuity of symbols. If f is a function on $\mathbb{T}$, then its L^1 modulus of continuity $\omega_f^{(1)}$ is defined by, in analogy with (10.2),

$$\omega_f^{(1)}(t) \stackrel{\text{def}}{=} \sup_{\zeta \in \mathbb{T},\ |1-\zeta|<t} \int_{\mathbb{T}} |f(\zeta\tau) - f(\tau)|\, d\boldsymbol{m}(\tau), \quad t > 0.$$

The following result is possibly known to experts. We were not able to find a reference, and we prove it here.

Theorem 15.11. *Let $f \in L^1(\mathbb{T})$ and let*

$$g(z) = \sum_{n \in \mathbb{Z}} \frac{\hat{f}(n)}{\log(|n| + 2)} z^n.$$

Then $g \in L^1(\mathbb{T})$ and

$$\lim_{t \to 0} \omega_g^{(1)}(t) \log \frac{1}{t} = 0. \tag{15.4}$$

Consider the function $\mathbf{h}$ on $\mathbb{T}$ defined by

$$\mathbf{h}(z) \stackrel{\text{def}}{=} \sum_{n \in \mathbb{Z}} (\log(|n| + 2))^{-1} z^n.$$

It is well-known (see, for example, [Z, Ch. V, (1.5)]) that the series converges for $z \in \mathbb{T} \setminus \{1\}$, $\mathbf{h} \geq 0$ and $\mathbf{h} \in L^1(\mathbb{T})$. We define the function h on $\mathbb{R}$ by

$$h(x) \stackrel{\text{def}}{=} \mathbf{h}(e^{\mathrm{i}x}) = (\log 2)^{-1} + 2 \sum_{n \geq 1} (\log(n + 2))^{-1} \cos nx. \tag{15.5}$$

Then h is continuously differentiable on $\mathbb{R} \setminus 2\pi\mathbb{Z}$, see [Z, Ch. V, Miscellaneous theorems and examples, 7].

We use the following notation. Let φ and ψ be nonvanishing functions on an interval $(0, \alpha)$. We write

$$\varphi \stackrel{0}{\sim} \psi, \quad \text{if } \lim_{x \to 0} \frac{\varphi(x)}{\psi(x)} = 1.$$

Lemma 15.12. *Let h be the function defined by* (15.5). *Then*

$$h(x) \overset{0}{\sim} \frac{\pi}{x(\log x)^2} \tag{15.6}$$

and

$$h'(x) \overset{0}{\sim} -\frac{\pi}{x^2(\log x)^2}. \tag{15.7}$$

Proof. (15.6) is proved in [Z, Ch. V, (2.17)]. Let us prove (15.7). Using Abel's transformation, we obtain

$$\begin{aligned} h(x) &= \sum_{n\geq 0} \big((\log(n+2))^{-1} - (\log(n+3))^{-1}\big) \frac{\sin(n+\frac{x}{2})}{\sin\frac{x}{2}} \\ &= \cot\frac{x}{2}\Big(\sum_{n\geq 0} \big((\log(n+2))^{-1} - (\log(n+3))^{-1}\big)\sin nx\Big) \\ &\quad + \sum_{n\geq 0} \big((\log(n+2))^{-1} - (\log(n+3))^{-1}\big)\cos nx. \end{aligned}$$

Consequently,

$$\begin{aligned} h'(x) &= -\frac{1}{2\sin^2\frac{x}{2}}\Big(\sum_{n\geq 0} \big((\log(n+2))^{-1} - (\log(n+3))^{-1}\big)\sin nx\Big) \\ &\quad + \cot\frac{x}{2}\Big(\sum_{n\geq 0} \big((\log(n+2))^{-1} - (\log(n+3))^{-1}\big) n\cos nx\Big) \\ &\quad - \sum_{n\geq 0} \big((\log(n+2))^{-1} - (\log(n+3))^{-1}\big) n \sin nx \overset{\text{def}}{=} \Sigma_1 + \Sigma_2 + \Sigma_3. \end{aligned}$$

It remains to observe that

$$\Sigma_1 \overset{0}{\sim} -\frac{\pi}{x^2(\log x)^2}$$

by [Z, Ch. V, (2.13)], while

$$\Sigma_2 \overset{0}{\sim} -\frac{2\pi}{x^2(\log x)^3},$$

and

$$\Sigma_3 \overset{0}{\sim} -\frac{1}{x(\log x)^2}$$

by [Z, Ch. V, (2.18)]. □

Corollary 15.13. *The following inequality holds*

$$\int_{\mathbb{T}} |\mathbf{h}(\zeta\tau) - \mathbf{h}(\tau)|\, dm(\tau) \le C \left(\log \frac{3}{|\zeta - 1|} \right)^{-1}$$

for any $\zeta \in \mathbb{T}$.

Proof. It suffices to prove that

$$\int_{-\pi}^{\pi} |h(x+t) - h(x)|\, dx \le C \left(\log \frac{1}{t} \right)^{-1}$$

for sufficiently small positive t. We have

$$\begin{aligned}
\int_{-\pi}^{\pi} |h(x+t) - h(x)|\, dx &= \int_{|x| \le 2t} |h(x+t) - h(x)|\, dx \\
&\quad + \int_{2t \le |x| \le \pi} |h(x+t) - h(x)|\, dx \\
&\le 2 \int_{|x| \le 3t} |h(x)|\, dx + \int_{2t \le |x| \le \pi} |h(x+t) - h(x)|\, dx \\
&\stackrel{\text{def}}{=} 2I_1 + I_2.
\end{aligned}$$

Using Lemma 15.12, we obtain

$$I_1 \le C \int_0^{3t} \frac{1}{x(\log x)^2}\, dx \le C \left(\log \frac{1}{t} \right)^{-1}$$

and

$$I_2 \le Ct \int_{2t}^{\pi} \frac{dx}{x^2 (\log \frac{x}{10})^2} \le C \left(\log \frac{1}{t} \right)^{-2},$$

if $t > 0$ is sufficiently small. □

Proof of Theorem 15.11. Note that $g = f * \mathbf{h}$. Consequently, $g \in L^1(\mathbb{T})$. It follows easily from Corollary 15.13 that

$$\omega_g^{(1)}(t) \le C \|f\|_{L^1} \left(\log \frac{3}{t} \right)^{-1}, \quad 0 < t < 2.$$

The result follows now from the obvious fact that (15.4) holds for trigonometric polynomials f. □

For a function $f \in L^1(\mathbb{R})$, we defined the L^1 modulus of continuity $\omega_f^{(1)}$ in (10.2):

$$\omega_f^{(1)}(t) = \sup_{|s| \le t} \int_{\mathbb{R}} |f(x+s) - f(x)|\, dx, \quad t > 0.$$

In fact this definition can be extended to functions f not necessarily in $L^1(\mathbb{R})$. It is sufficient to assume that

$$\int_{\mathbb{R}} |f(x+s) - f(x)|\, dx < \infty, \quad s \in \mathbb{R}.$$

In a similar way we can prove the following analog of Theorem 15.11.

Theorem 15.14. *Let $f \in L^1(\mathbb{R})$. Then there exists a function $g \in L^1(\mathbb{R})$ such that*

$$(\mathcal{F} f)(x) = (\mathcal{F} g(x)) \log(|x| + 2), \quad x \in \mathbb{R},$$

and

$$\lim_{t \to 0} \omega_g^{(1)}(t) \log \frac{1}{t} = 0$$

Proof. Indeed, let

$$\mathfrak{h}(x) \stackrel{\text{def}}{=} \int_{\mathbb{R}} (\log(2 + |t|))^{-1} e^{-2\pi i t x}\, dt = 2 \int_0^\infty (\log(2 + t))^{-1} \cos(2\pi t x)\, dt.$$

Then $\mathfrak{h}$ is an even positive continuously differentiable function on $\mathbb{R} \setminus \{0\}$. We can repeat the above reasoning to prove that

$$\mathfrak{h}(x) \stackrel{0}{\sim} \frac{1}{2x(\log x)^2}$$

and

$$\mathfrak{h}'(x) \stackrel{0}{\sim} -\frac{1}{2x^2(\log x)^2}.$$

Moreover, $|\mathfrak{h}(x)| \le C \cdot x^{-2}$ and $|\mathfrak{h}'(x)| \le C \cdot x^{-2}$ everywhere. These estimates allow us to obtain the inequality

$$\int_{\mathbb{R}} |\mathfrak{h}(x+t) - \mathfrak{h}(x)|\, dx \le C \left(\log \frac{1}{t}\right)^{-1}$$

for $t \in (0, \frac{1}{2})$ and repeat the reasoning in the proof of Theorem 15.11. □

Let us introduce some more notation. Set $\mathbb{C}_+ \stackrel{\text{def}}{=} \{z \in \mathbb{C} : \operatorname{Im} z > 0\}$ and $\mathbb{C}_- \stackrel{\text{def}}{=} \{z \in \mathbb{C} : \operatorname{Im} z < 0\}$. Let f be a function in $L^1_{\text{loc}}(\mathbb{R})$ such that

$$\int_{\mathbb{R}} \frac{|f(t)|}{1 + |t|}\, dt < +\infty.$$

Consider the Cauchy transform of f defined by

$$(\mathcal{C}f)(\zeta) \stackrel{\text{def}}{=} \frac{1}{2\pi \mathrm{i}} \int_{\mathbb{R}} \frac{f(t)\,dt}{t-\zeta}, \quad \operatorname{Im}\zeta \neq 0.$$

It is well known that $(\mathcal{C}f)\big|\mathbb{C}_+$ and $(\mathcal{C}f)\big|\mathbb{C}_-$ are holomorphic functions of bounded characteristic in $\mathbb{C}_+$ and $\mathbb{C}_-$ respectively, and so they have finite angular boundary values almost everywhere on $\mathbb{R}$. Set

$$f_+(x) \stackrel{\text{def}}{=} \lim_{y\to 0+} (\mathcal{C}f)(x+\mathrm{i}y) \quad \text{and} \quad f_-(x) \stackrel{\text{def}}{=} -\lim_{y\to 0-} (\mathcal{C}f)(x+\mathrm{i}y).$$

By the Privalov theorem (see [Pr] for the case of a rectifiable Jordan curve), $f = f_+ + f_-$ almost everywhere on $\mathbb{R}$. If $f \in L^1(\mathbb{R}) + L^2(\mathbb{R})$, then

$$(\mathcal{C}f)(z) = \int_0^{\infty} (\mathcal{F}f)(t) e^{2\pi \mathrm{i} tz}\,dt, \quad z \in \mathbb{C}_+,$$

and

$$(\mathcal{C}f)(z) = -\int_{-\infty}^{0} (\mathcal{F}f)(t) e^{2\pi \mathrm{i} tz}\,dt, \quad z \in \mathbb{C}_-.$$

Note that f_+ does not have to be in $L^1(\mathbb{R})$ for an arbitrary function $f \in L^1(\mathbb{R})$. In fact, if $f \in L^1(\mathbb{R})$, then $f_+ \in L^1(\mathbb{R})$ if and only if f belongs to the Stein–Weiss space $\mathfrak{H}^1$.

Theorem 15.15. *Let $f \in \mathfrak{H}^1$. Suppose that there exists a function $g \in L^1(\mathbb{R})$ such that $(\mathcal{F}f)(x) = (\mathcal{F}g)(x)\log(1+x^2)$ for all $x \in \mathbb{R}$. Then*

$$\lim_{t\to 0} \omega_{g_+}^{(1)}(t) \log\frac{1}{t} = 0$$

and

$$\lim_{t\to 0} \omega_{g_-}^{(1)}(t) \log\frac{1}{t} = 0.$$

Note, however, that the assumptions of Theorem 15.15 do not imply that $g_+ \in L^1(\mathbb{R})$ or $g_- \in L^1(\mathbb{R})$.

We need some auxiliary facts. Let $\mathcal{M}(\mathbb{R})$ be the space of finite Borel measures on $\mathbb{R}$.

Lemma 15.16. *Let $f \in L^1(\mathbb{R})$. Suppose that $f'' \in \mathcal{M}(\mathbb{R})$ (in the distributional sense). Then $\mathcal{F}f \in L^1(\mathbb{R})$ and*

$$\|\mathcal{F}f\|_{L^1(\mathbb{R})} \le C\sqrt{\|f\|_{L^1(\mathbb{R})}\|f''\|_{\mathcal{M}(\mathbb{R})}}.$$

Proof. The result follows from the obvious inequality:

$$|(\mathcal{F}f)(x)| \le \min\left\{\|f\|_{L^1}, \frac{\|f''\|_{\mathcal{M}(\mathbb{R})}}{4\pi^2 x^2}\right\}, \quad x \in \mathbb{R}. \qquad \square$$

Corollary 15.17. *Let $f \in L^1(\mathbb{R})$. Suppose that* $\operatorname{supp} \mathcal{F} f$ *is bounded above. Then*

$$\omega^{(1)}_{f_+}(t) \le C \cdot t, \quad t > 0.$$

Proof. It suffices to construct a function $g_s \in L^1(\mathbb{R})$ such that $\|g_s\|_{L^1(\mathbb{R})} \le C|s|$ and $f_+(x+s) - f_+(x) = (f * g_s)(x)$ for all $x \in \mathbb{R}$. Suppose that $\operatorname{supp} f \subset (-\infty, M]$, where $M > 0$. We may take a function g_s such that

$$(\mathcal{F} g_s)(t) = \begin{cases} 0, & t \le 0, \\ e^{2\pi i s t} - 1, & t \in [0, M], \\ e^{2\pi i s(2M-t)} - 1, & t \in [M, 2M], \\ 0, & t \ge 2M. \end{cases}$$

Clearly, $f_+(x+s) - f_+(x) = (f * g_s)(x)$ for all $x \in \mathbb{R}$. The inequality $\|g_s\|_{L^1(\mathbb{R})} \le C|s|$ follows from Lemma 15.16 (with C depending on M). □

Lemma 15.18. *Set* $\rho(t) \stackrel{\text{def}}{=} 2\log(2+|t|) - \log(1+t^2)$. *Then* $\mathcal{F}\rho \in L^1$.

Proof. It suffices to observe that ρ is even, $\lim_{t\to\infty} \rho(t) = 0$, ρ has two continuous derivatives on $(0, \infty)$, and

$$\int_0^\infty t|\rho''(t)|\, dt < \infty;$$

this implies that $\rho(x) = -\int_{|x|}^\infty \rho'(t)\, dt$ and

$$\rho(x) = \int_{|x|}^\infty (t - |x|)\rho''(t)\, dt = \int_0^\infty \rho''(t)(t - |x|)_+\, dt = \int_0^\infty \rho''(t) t \mathcal{F} K_t(x)\, dt,$$

where K_t is the Fejér kernel with $\|K_t\|_{L^1} = 1$. □

Lemma 15.19. *Suppose that φ is an even positive function in $C^2(\mathbb{R})$ such that $\varphi(x) = \log(1+x^2)$ for sufficiently large $|x|$. Then $\mathcal{F}(\varphi^{-1}) \in L^1$.*

Proof. See the proof of the previous lemma. □

Proof of Theorem 15.15. We prove the first equality (the proof of the second one is the same). Let ψ be a function in $L^1(\mathbb{R})$ such that $\operatorname{supp} \mathcal{F}\psi$ is compact and $\operatorname{supp} \mathcal{F}\psi = 1$ in a neighborhood of 0. Then $f = f * \psi + (f - f * \psi)$. The Fourier transform of the first summand has a compact support while the support of the Fourier transform of the second summand does not contain 0. Thus it is sufficient to consider two cases.

Case 1, $\operatorname{supp} \mathcal{F} f$ is compact. The result follows from Corollary 15.17.

Case 2, $0 \notin \operatorname{supp} \mathcal{F} f$. Clearly,

$$(\mathcal{F} f_+)(x) = (\mathcal{F} g_+)(x) \log(1+x^2), \quad x \in \mathbb{R}.$$

Let $\delta > 0$ be such that $\mathcal{F} f = 0$ on $[-\delta, \delta]$, and let φ be an even positive function in $C^2(\mathbb{R})$ such that $\varphi(x) = \log(1+x^2)$ for $|x| \ge \delta$. By Lemma 15.19, $1/\varphi = \mathcal{F}\Phi$ for some $\Phi \in L^1(\mathbb{R})$. Hence,

$$\mathcal{F} g_+ = \mathcal{F} f_+/\varphi = \mathcal{F}(f_+ * \Phi),$$

which implies that $g_+ \in L^1(\mathbb{R})$. Moreover, if ρ is as in Lemma 15.18, and thus $\rho = \mathcal{F} F$ with $F \in L^1(\mathbb{R})$, then

$$2(\mathcal{F} g_+)(x)\log(2+|x|) = (\mathcal{F} g_+)(x)\rho(x) + (\mathcal{F} f_+)(x) = \mathcal{F}(g_+ * F + f_+)(x),$$

so $(\mathcal{F} g_+)(x)\log(2+|x|)$ is the Fourier transform of an L^1-function. It remains to apply Theorem 15.14. □

Theorem 15.20. *Let φ be a function in $L^2[0,b]$ such that $Q_\varphi^{[0,b]} \in \boldsymbol{S}_1$ and let $\varphi(x) = \sum_{n\in\mathbb{Z}} a_n e^{2\pi i n x/b}$. If*

$$\phi_+(\zeta) \stackrel{\text{def}}{=} \sum_{n\ge 0} a_n \zeta^n \quad \text{and} \quad \phi_-(\zeta) \stackrel{\text{def}}{=} \sum_{n<0} a_n \zeta^n, \tag{15.8}$$

then

$$\lim_{t\to 0} \omega_{\phi_+}^{(1)}(t) \log\frac{1}{t} = 0 \quad \text{and} \quad \lim_{t\to 0} \omega_{\phi_-}^{(1)}(t) \log\frac{1}{t} = 0. \tag{15.9}$$

Proof. The result follows immediately from Theorems 15.4 and 15.11. □

Theorem 15.21. *Let I be a compact interval in $(0,\infty)$ and let φ be a function in $L^1_{\text{loc}}(\mathbb{R}_+)$ such that $Q_\varphi \in \boldsymbol{S}_1$. If*

$$a_n = \int_I \varphi(x) e^{-2\pi i n x/|I|}\,dx, \quad n \in \mathbb{Z},$$

and ϕ_+ and ϕ_- are defined by (15.8), *then* (15.9) *holds.*

Proof. The result is an immediate consequence of Theorem 15.20. □

Recall that for a function $\varphi \in L^2_{\text{loc}}(\mathbb{R}_+)$ the function $\varphi^\heartsuit$ is defined by (13.1). Note that if $Q_\varphi \in \boldsymbol{S}_1$, then by Theorem 6.2, $\varphi \in L^1(\mathbb{R}_+)$ and thus $\varphi^\heartsuit \in L^1(\mathbb{R})$.

Theorem 15.22. *Let φ be a function in $L^2_{\text{loc}}(\mathbb{R}_+)$ such that $Q_\varphi \in \boldsymbol{S}_1$. Then*

$$\lim_{t\to 0} \omega_{(\varphi^\heartsuit)_+}^{(1)}(t) \log\frac{1}{t} = 0 \quad \text{and} \quad \lim_{t\to 0} \omega_{(\varphi^\heartsuit)_-}^{(1)}(t) \log\frac{1}{t} = 0.$$

In particular,

$$\lim_{t\to 0} \omega_{\varphi^\heartsuit}^{(1)}(t) \log\frac{1}{t} = 0.$$

Proof. The result follows from Theorem 13.12, Corollary 15.8, and Theorem 15.15. □

This result should be compared to Theorems 10.7 and 13.7. In particular, if φ has compact support in $\mathbb{R}_+$, we see that a Dini condition on the L^2 modulus of continuity is sufficient for $Q_\varphi \in \boldsymbol{S}_1$, while the slightly weaker condition $\lim_{t\to 0} \omega_\varphi^{(1)}(t) \log \frac{1}{t} = 0$ on the L^1 modulus of continuity is necessary.

Theorem 15.23. *Let φ be a function in $L^2_{\mathrm{loc}}(\mathbb{R}_+)$ such that $Q_\varphi \in \boldsymbol{S}_1$. Then*

$$\lim_{a\to 1} \int_{\mathbb{R}_+} |\varphi(ax) - \varphi(x)|\, dx \cdot \log \frac{1}{|a-1|} = 0.$$

Proof. By Theorem 15.22, we have

$$\lim_{t\to 0} \int_{\mathbb{R}} \left|\varphi(e^{2s+2t})e^{2s+2t} - \varphi(e^{2s})e^{2s}\right| ds \cdot \log \frac{1}{t} = 0.$$

Substituting $e^{2s} = x$ and $e^{2t} = a$, we obtain

$$\lim_{a\to 1} \int_{\mathbb{R}_+} |a\varphi(ax) - \varphi(x)|\, dx \cdot \log \frac{1}{|a-1|} = 0.$$

It remains to observe that by Theorem 6.2, $\varphi \in L^1(\mathbb{R}_+)$ and obviously,

$$\lim_{a\to 1} |a-1| \cdot \log \frac{1}{|a-1|} = 0. \qquad \square$$

16. Dilation of symbols

Let φ be a function in $L^2_{\mathrm{loc}}(\mathbb{R})$ such that $\varphi(x+1) = \varphi(x)$, $x \in \mathbb{R}$. For $a > 0$ we define the function φ_a on $[0, 1]$ by $\varphi_a(x) \stackrel{\mathrm{def}}{=} \varphi(ax)$ for $x \in [0, 1)$. We are going to obtain in this section upper and lower estimates for $\left\| Q_{\varphi_a}^{[0,1]} \right\|_{\boldsymbol{S}_p}$.

Note that we can extend φ_a to $\mathbb{R}$ as a 1-periodic function on $\mathbb{R}$. Using an obvious estimate, see (14.1)–(14.3),

$$C_1\Big(\big\| Q_{\varphi_a}^{[0,1]} \big\|_{\boldsymbol{S}_p} + \|\varphi_a\|_{L^2[0,1]}\Big) \le \big\| Q_{\varphi_a}^{[0,2]} \big\|_{\boldsymbol{S}_p} \le C_2\Big(\big\| Q_{\varphi_a}^{[0,1]} \big\|_{\boldsymbol{S}_p} + \|\varphi_a\|_{L^2[0,1]}\Big), \tag{16.1}$$

we can reduce the estimation of $\|Q_{\varphi_a}^{[0,1]}\|_{\boldsymbol{S}_p}$ to that of $\|Q_{\varphi_a}^{[0,2]}\|_{\boldsymbol{S}_p}$. We can consider the Fourier coefficients of φ_a defined by

$$\hat{\varphi}_a(n) \stackrel{\mathrm{def}}{=} \int_0^1 \varphi_a(t) e^{-2\pi i n t}\, dt, \qquad n \in \mathbb{Z}.$$

Theorem 16.1. *Let φ be a 1-periodic function in $L^2_{\mathrm{loc}}(\mathbb{R})$ and let $a > 0$. Suppose that φ has bounded variation on $[0, 1]$. Then*

$$\|Q_{\varphi_a}^{[0,2]}\|_{\boldsymbol{S}_1} \le C(\varphi) \log(2 + a)$$

and

$$\|Q_{\varphi_a}^{[0,2]}\|_{S_p} \le C(\varphi)(1+a)^{1/p-1}, \quad 1/2 < p < 1.$$

Proof. The result follows from Theorem 7.7. □

Theorem 16.2. *Let φ be a nonconstant 1-periodic function in $L^2_{\text{loc}}(\mathbb{R})$. Then for $a \ge 1$*

$$\|Q_{\varphi_a}^{[0,1]}\|_{S_1} \ge C(\varphi)\log(2+a).$$

Proof. It follows from (16.1) that it is sufficient to prove that

$$\|Q_{\varphi_a}^{[0,2]}\|_{S_1} \ge C(\varphi)\log(2+a). \tag{16.2}$$

First we consider the case where a is an integer. There exists an integer $k \in \mathbb{Z} \setminus \{0\}$ such that $\hat{\varphi}(k) \ne 0$. By Corollary 15.6,

$$|\hat{\varphi}_a(l)| \le C(\log(2+|l|))^{-1}\|Q_{\varphi_a}\|_{S_1}.$$

Substituting $l = ak$, we obtain (16.2), since $\hat{\varphi}_a(ak) = \hat{\varphi}(k)$.

Let now a be an arbitrary number in $(1, \infty)$. For any $\sigma \in [1, 2]$ there exists $k_\sigma \in \mathbb{Z} \setminus \{0\}$ such that $\hat{\varphi}_\sigma(k_\sigma) \ne 0$. Consequently, there exists a neighborhood U_σ of σ such that $\hat{\varphi}_\tau(k_\sigma) \ne 0$ for any $\tau \in U_\sigma$. The first part of the proof allows us to obtain the required estimate for any $a > 1$ such that $a/N \in U_\sigma$ for some positive integer N. To complete the proof, we can choose a finite subcover U_{σ_j} of $[1, 2]$. □

Theorem 16.3. *Let φ be a nonconstant 1-periodic function in $L^2_{\text{loc}}(\mathbb{R})$ and let $0 < p < 1$. Then for $a \ge 1$*

$$\left\|Q_{\varphi_a}^{[0,1]}\right\|_{S_p} \ge C(\varphi)a^{1/p-1}.$$

Proof. It suffices to consider the case when a is an even integer. The general case may be reduced to this special case in the same way as in the proof of Theorem 16.2. With any kernel k on the square $[0, 1)^2$ and any integer $n \ge 1$ we associate the kernel $k^{[n]}$ defined by

$$k^{[n]}(x, y) = n^2 \int_{\frac{j}{n}}^{\frac{j+1}{n}} \int_{\frac{l}{n}}^{\frac{l+1}{n}} k(t, s)\, dt\, ds, \quad \text{if } x \in \left[\frac{j}{n}, \frac{j+1}{n}\right) \text{ and } y \in \left[\frac{l}{n}, \frac{l+1}{n}\right).$$

Clearly, $\|k^{[n]}\|_{S_p} \le \|k\|_{S_p}$ for any positive p (recall that $\|k\|_{S_p}$ means the S_p-norm (or quasinorm) of the integral operator with kernel k).

Suppose that

$$\int_0^1 \varphi(t)\, dt \ne 2\int_0^1 t\varphi(t)\, dt.$$

Put $k_n(x, y) \stackrel{\text{def}}{=} \varphi(n \max\{x, y\})$ for $x, y \in [0, 1)$. Clearly,

$$\left\|k_{2n}^{[n]} - k_{2n}^{[2n]}\right\|_{S_p} \le 2^{1/p}\|k_{2n}\|_{S_p}.$$

It is not hard to see that on $\left[\frac{j}{n}, \frac{j+1}{n}\right) \times \left[\frac{l}{n}, \frac{l+1}{n}\right)$

$$k_{2n}^{[n]} = \begin{cases} \int_0^1 \varphi(t)\,dt, & j \neq l, \\ \frac{1}{2}\int_0^1 \varphi(t)\,dt + \int_0^1 t\varphi(t)\,dt, & j = l. \end{cases}$$

Next, on $\left[\frac{j}{2n}, \frac{j+1}{2n}\right) \times \left[\frac{l}{2n}, \frac{l+1}{2n}\right)$

$$k_{2n}^{[2n]} = \begin{cases} \int_0^1 \varphi(t)\,dt, & j \neq l, \\ 2\int_0^1 t\varphi(t)\,dt, & j = l. \end{cases}$$

Thus, the kernel $k_{2n}^{[n]} - k_{2n}^{[2n]}$ vanishes outside the "diagonal" $\bigcup_{j=0}^{n-1}\left[\frac{j}{n}, \frac{j+1}{n}\right) \times \left[\frac{j}{n}, \frac{j+1}{n}\right)$. Clearly, for $x,\ y \in \left[0, \frac{n-1}{n}\right) \times \left[0, \frac{n-1}{n}\right)$ we have

$$k_{2n}^{[n]}(x, y) - k_{2n}^{[2n]}(x, y) = k_{2n}^{[n]}\left(x + \frac{1}{n}, y + \frac{1}{n}\right) - k_{2n}^{[2n]}\left(x + \frac{1}{n}, y + \frac{1}{n}\right).$$

Consequently,

$$\|k_{2n}^{[n]} - k_{2n}^{[2n]}\|_{S_p} = n^{1/p}\|(k_{2n}^{[n]} - k_{2n}^{[2n]})\chi_{[0,\frac{1}{n})\times[0,\frac{1}{n})}\|_{S_p}.$$

We have

$$k_{2n}^{[n]} - k_{2n}^{[2n]} = \frac{1}{2}\int_0^1 \varphi(t)\,dt - \int_0^1 t\varphi(t)\,dt$$

on $\left(\left[0, \frac{1}{2n}\right) \times \left[0, \frac{1}{2n}\right)\right) \cup \left(\left[\frac{1}{2n}, \frac{1}{n}\right) \times \left[\frac{1}{2n}, \frac{1}{n}\right)\right)$ and

$$k_{2n}^{[n]} - k_{2n}^{[2n]} = \int_0^1 t\varphi(t)\,dt - \frac{1}{2}\int_0^1 \varphi(t)\,dt$$

on $\left(\left[0, \frac{1}{2n}\right) \times \left[\frac{1}{2n}, \frac{1}{n}\right)\right) \cup \left(\left[\frac{1}{2n}, \frac{1}{n}\right) \times \left[0, \frac{1}{2n}\right)\right)$. Now it is easy to see that

$$\|k_{2n}^{[n]} - k_{2n}^{[2n]}\|_{S_p} = cn^{1/p-1}$$

for some nonzero c, since $\int_0^1 \varphi(t)\,dt \neq 2\int_0^1 t\varphi(t)\,dt$.

Suppose now that φ is an arbitrary nonconstant 1-periodic function. It suffices to prove that there exists $b \in \mathbb{R}$ such that $\int_0^1 \varphi(t-b)\,dt \neq 2\int_0^1 t\varphi(t-b)\,dt$. Suppose that

$$\int_0^1 \varphi(t-b)\,dt = 2\int_0^1 t\varphi(t-b)\,dt, \quad b \in \mathbb{R}. \tag{16.3}$$

Let h be the 1-periodic function such that $h(t) = 2t - 1$ for $t \in [0, 1)$. Clearly, $\hat{h}(n) \neq 0$ for $n \neq 0$. Thus it follows from (16.3) that φ is constant. □

Acknowledgement

This research was begun by the first and third authors. The two other authors joined them during the Conference on Function Spaces, Interpolation Theory, and related topics in honour of Jaak Peetre on his 65th birthday at Lund University, Sweden, August 17–22, 2000, where three of the authors were present and several of the results were proved. We thank Jonathan Arazy and Vladimir Maz'ya for helpful comments.

References

[A] A. B. Aleksandrov, Toeplitz Schur multipliers of $S_p(L^2(G))$, Preprint, IHES, 2000.

[AO] A. B. Alekseev and V. L. Oleinik, Estimates for widths of the unit ball of $\overset{\circ}{L}{}^l_p$ in $L_p(\mu)$ (Russian), in: Probl. Math. Phys. 7, Leningrad University, 1974, 3–7.

[BL] J. Bergh and J. Löfström, Interpolation spaces, Springer-Verlag, Berlin–Heidelberg–New York, 1976.

[B] M. Sh. Birman, A proof of the Fredholm trace formula as an application of a simple embedding for kernels of integral operators of trace class in $L^2(\mathbb{R}^m)$, Preprint, Dept. Math., Linköping University, LITH-MAT-R-89-30, 1989/90.

[BS] M. Sh. Birman and M. Z. Solomyak, Spectral theory of selfadjoint operators in Hilbert space, Math. Appl. (Soviet Series), D. Reidel Publishing Co., Dordrecht, 1987.

[C] A. Connes, Noncommutative geometry, Academic Press, San Diego, CA, 1994.

[D] J. Dixmier, Existence de traces non normales (French), C. R. Acad. Sci. Paris Sér. A-B, 262 (1966), A1107–A1108.

[EEH] D. E. Edmunds, W. D. Evans, and D. J. Harris, Two-sided estimates for the approximation numbers of certain Volterra integral operators, Studia Math. 124 (1997), 59-80.

[ES] D. E. Edmunds and V. D. Stepanov, On the singular numbers of certain Volterra integral operators, J. Funct. Anal. 134 (1) (1995), 222–246.

[GK1] I. C. Gohberg and M. G. Krein, Vvedenie v teoriyu lineinykh nesamosopryazhyonnykh operatorov (Russian), Nauka, Moscow, 1965; English transl.: Introduction to the theory of linear nonselfadjoint operators in Hilbert space, Amer. Math. Soc., Providence, R.I., 1969.

[GK2] I. C. Gohberg and M. G. Krein, Teoriya Vol'terrovykh ooperatorov v gil'bertovom prostranstve i eyo prilozheniya (Russian), Nauka, Moscow, 1967; English transl.: Theory and applications of Volterra operators in Hilbert space, Amer. Math. Soc., Providence, R.I., 1970.

[K] N. J. Kalton, Linear operators on L^p for $0 < p < 1$, Trans. Amer. Math. Soc. 259 (1980), 319–355.

[LS] B. M. Levitan and I. S. Sargsjan, Operatory Shturma–Liuvillya i Diraka (Russian), Nauka, Moscow, 1988; English transl.: Sturm–Liouville and Dirac operators, Kluwer, Dordrecht, 1991.

[MV] V. G. Maz'ya and I. E. Verbitsky, The Schrödinger operator on the energy space: boundedness and compactness criteria, Preprint, 2000.

[NaS] K. Naimark and M. Solomyak, Regular and pathological eigenvalue behavior for the equation $\lambda u'' = Vu$ on the semiaxis, Integral Equations Operator Theory 20 (1994), 335–349.

[NeS] J. Newman and M. Solomyak, Two-sided estimates on singular values for a class of integral operators on the semi-axis, J. Funct. Anal. 151 (1997), 504-530.

[No] K. Nowak, Schatten ideal behavior of a generalized Hardy operator, Proc. Amer. Math. Soc. 118 (1993), 479–483.

[OP] V. L. Oleinik and B. S. Pavlov, On criteria of boundedness and complete continuity of certain imbedding operators (Russian), in: Probl. Math. Phys. 4: Spectral Theory. Wave Processes (Russian), Leningrad University, 1970, 112–117.

[Pee] J. Peetre, New thoughts on Besov spaces, Duke Univ. Press., Durham, N.C., 1976.

[Pel1] V. V. Peller, Hankel operators of class $\mathbf{S}_p$ and their applications (rational approximation, Gaussian processes, the problem of majorizing operators) (Russian), Mat. Sb. 41 (1980), 538–581; English transl.: Math. USSR Sb. 41 (1982), 443–479.

[Pel2] V. V. Peller, A description of Hankel operators of class $\mathbf{S}_p$ for $p > 0$, an investigation of the rate of rational approximation, and other applications (Russian), Mat. Sb. 122 (1983), 481–510; English transl.: Math. USSR Sb. 50 (1985), 465–494.

[Pel3] V. V. Peller, Metric properties of the averaging projection onto the set of Hankel matrices (Russian), Dokl. Akad Nauk SSSR 278 (1984), 271–285; English transl.: Soviet Math. Dokl. 30 (1984), 362–368.

[Pr] I. I. Privalov, Granichnye svoistva analiticheskikh funktsii (Russian) [Boundary properties of analytic functions], 2nd ed., GITTL, Moscow–Leningrad, 1950; German transl.: Randeigenschaften analytischer Funktionen, VEB Deutscher Verlag, Berlin, 1956.

[RW] R. Rochberg and Z. Wu, A new characterization of Dirichlet type spaces and applications. Illinois J. Math. 37 (1993), 101–122.

[R] S. Yu. Rotfel'd, On singular numbers of the sum of compact operators (Russian), in: Prob. Math. Phys. 3: Spectral Theory (Russian), Leningrad University, 1968, 81–87.

[Sch] L. Schwartz, Théorie des distributions (French), Hermann, Paris, 1966.

[St] V. D. Stepanov, On the lower bounds for Schatten-von Neumann norms of certain Volterra integral operators, J. London Math. Soc. (2) 61 (2000), 905–922.

[W1] Z. Wu, Hankel and Toeplitz operators on Dirichlet spaces, Integral Equations Operator Theory **15** (1992), no. 3, 503–525.

[W2] Z. Wu, A class of bilinear forms on Dirichlet type spaces, J. London Math. Soc. (2) 54 (1996), no. 3, 498–514.

[Z] A. Zygmund, Trigonometric series, 2nd ed., Cambridge University Press, London–New York, 1968.

Alexei B. Aleksandrov, St-Petersburg Branch, Steklov Institute of Mathematics, Fontanka 27, 191011 St-Petersburg, Russia

E-mail: alex@pdmi.ras.ru

Svante Janson, Department of Mathematics, Uppsala University, 751 06 Uppsala, Sweden

E-mail: svante@math.uu.se
URL: http://www.math.uu.se/~svante/

Vladimir V. Peller, Department of Mathematics, Michigan State University, East Lansing, MI 48824, U.S.A.

E-mail: peller@math.msu.edu
URL: http://www.mth.msu/People/Faculty/peller_v.html

Richard Rochberg, Department of Mathematics, Washington University, St. Louis, MO 63130, U.S.A.

E-mail: rr@math.wustl.edu

Invariant symbolic calculi and eigenvalues of invariant operators on symmetric domains

Jonathan Arazy and Harald Upmeier *

Dedicated to Jaak Peetre on the occasion of his 65th anniversary

Abstract. We study the structure of invariant symbolic calculi $\mathcal{A}$ in the context of weighted Bergman spaces on symmetric domains $D = G/K$ and the eigenvalues of the associated link transforms $\mathcal{A}'\mathcal{A}$. We parametrize all such calculi by K-invariants maps which have very simple description. We also introduce and study the properties of the fundamental function $\mathfrak{a}_{\mathcal{A}}(\lambda)$ associated with an invariant symbolic calculus $\mathcal{A}$. Our main result is the formula for the eigenvalues of the associated link transform $\mathcal{A}'\mathcal{A}$:

$$\widetilde{\mathcal{A}'\mathcal{A}}(\underline{\lambda}) = \frac{\mathfrak{a}_{\mathcal{A}}(\underline{\lambda})\,\overline{\mathfrak{a}_{\mathcal{A}}(\overline{\underline{\lambda}})}}{\mathfrak{a}_{\mathcal{T}}(\underline{\lambda})},$$

where $\mathcal{T}$ is the Toeplitz calculus.

2000 Mathematics Subject Classification: 32M15, 46E22, 47B35, 81T70

0. Introduction

Let $D \equiv G/K$ be a hermitian symmetric domain in $\mathbb{C}^d$ and let $\mathcal{H}$ be a Hilbert space of holomorphic functions on D with reproducing kernel $K(z, w)$, which is invariant under an irreducible projective representation U of G. An *invariant symbolic calculus* $\mathcal{A}$ is a linear map $b \mapsto \mathcal{A}_b$ from a G-invariant subspace Dom($\mathcal{A}$) of functions ("symbols") on D into the space Op($\mathcal{H}$) of operators on $\mathcal{H}$ which intertwines the natural actions of G on symbols and operators:

$$U(g)\mathcal{A}_b U(g)^{-1} = \mathcal{A}_{b \circ g^{-1}}, \quad \forall g \in G,\ \forall b \in \mathrm{Dom}(\mathcal{A}).$$

*Both authors were supported by a grant from the German–Israeli Foundation (GIF), I-415-023.06/95. Part of this work was carried over while the authors visited the Erwin Schrödinger Institute for Mathematical Physics (ESI) in Vienna.

The *adjoint* of $\mathcal{A}$ is the map $\mathcal{A}' : \mathrm{Op}(\mathcal{H}) \to$ {functions on D} defined by

$$\langle \mathcal{A}'(T), b\rangle_{L^2(D,\mu_0)} = \langle T, \mathcal{A}_b\rangle_{S_2}, \quad \forall T \in \mathrm{Dom}(\mathcal{A}'),\ \forall b \in \mathrm{Dom}(\mathcal{A}),$$

where μ_0 is the G-invariant measure on D and S_2 is the Hilbert–Schmidt class. The operator $\mathcal{B} := \mathcal{A}'\mathcal{A} : \mathrm{Dom}(\mathcal{A}) \to$ {functions on D} is the *link transform* associated with $\mathcal{A}$. It is *G-invariant:* $\mathcal{B}(f \circ g) = (\mathcal{B}f) \circ g$ for all $g \in G$ and $f \in \mathrm{Dom}(\mathcal{B})$, and is therefore diagonalized by the *exponential functions* $\{e_{\underline{\lambda}}\}$ in $\mathrm{Dom}(\mathcal{B})$:

$$\mathcal{B}e_{\underline{\lambda}} = \widetilde{\mathcal{B}}(\underline{\lambda})\, e_{\underline{\lambda}}.$$

The link transform $\mathcal{B} = \mathcal{A}'\mathcal{A}$ associated with the invariant symbolic calculus $\mathcal{A}$ maps the *active symbol* b of $\mathcal{A}_b$ into its *passive symbol*: $\mathcal{B}(b) = \mathcal{A}'(\mathcal{A}_b)$. For instance, the link transform associated with the *Toeplitz calculus* $\mathcal{T}$ is the well-known *Berezin transform* $\mathcal{B} := \mathcal{T}'\mathcal{T}$, which plays a central role in quantization on symmetric domains.

The link transform and its eigenvalues reflect characteristic features of the underlying quantization procedure. For example, on the cotangent bundle $\mathbb{C}^n = T^*(\mathbb{R}^n)$ the well-known *Weyl calculus* $\mathcal{W}$ is unitary (i.e. $\mathcal{W}'\mathcal{W} = I$), whereas the Toeplitz (or, anti-Wick) calculus $\mathcal{T}$ yields the contraction semi-group of the Laplace operator: $\mathcal{T}'\mathcal{T} = \exp(\beta\Delta)$ for some $\beta > 0$ depending on the underlying Hilbert space. Even more interesting are quantizations of curved symmetric spaces, such as the unit disk and its higher dimensional generalizations (Cartan and Siegel domains).

In this paper we develop a unified approach to compute the eigenvalues $\widetilde{\mathcal{B}}(\underline{\lambda})$ of the link transforms $\mathcal{B}$. This approach is based on a new factorization technique and on a parametrization of the invariant symbolic calculi by K-invariant operators on $\mathcal{H}$ which have very simple structure. Let K_o be the reproducing kernel at the base point $o \equiv K \in G/K$ of D. The formula for the eigenvalues of $\mathcal{B} := \mathcal{A}'\mathcal{A}$ is expressed in terms of the *fundamental function* $\mathfrak{a}_{\mathcal{A}}(\underline{\lambda}) := \langle \mathcal{A}_{e_{\underline{\lambda}}}(K_o), K_o\rangle_{\mathcal{H}}$ of the calculus $\mathcal{A}$ via

$$\widetilde{\mathcal{B}}(\underline{\lambda}) = \frac{\mathfrak{a}_{\mathcal{A}}(\underline{\lambda})\,\overline{\mathfrak{a}_{\mathcal{A}}(\overline{\underline{\lambda}})}}{\mathfrak{a}_{\mathcal{T}}(\underline{\lambda})}.$$

Our approach gives also new proof for the known results concerning the eigenvalues of the Berezin transforms in the context of the weighted Bergman spaces over symmetric domains. For the flat case $D = \mathbb{C}^d$ and the associated weighted Fock spaces $\mathcal{F}_\nu$, we also show that the general approach developed here not only clarifies the relationship between the standard calculi (Toeplitz, Weyl, and Wick calculi) in a very satisfactory manner, but also enables one to construct entirely new invariant symbolic calculi which, nevertheless, can be fully analyzed by closed formulas.

The organization of the paper is as follows. In Section 1 we give first the necessary background on symmetric domains $D = G/K$, Jordan theory, and invariant Hilbert spaces $\mathcal{H}$ of holomorphic functions on D. Subsections 1.2 and 1.3 give the background on the Fock spaces on $\mathbb{C}^d$ and on the weighted Bergman spaces over symmetric tube domains respectively. In particular, we give explicit formulas for the reproducing kernels and the exponential functions in these setups. In Section 2 we introduce and study the notions of invariant symbolic calculus ("quantization") and the associated link transform. We establish the one-to-one correspondence between the invariant symbolic calculi and the K-invariant operators on $\mathcal{H}$. The main examples (Toeplitz and Weyl calculi, as well as the calculi associated with the projections onto the K-irreducible subspaces) are discussed. Of special importance is the study of a real-analytic reflection ψ of $D = G/K$ and the induced involution on the Lie algebra $\mathfrak{A}$ of spectral parameters. The main result of this section is Theorem 2.23 which provides a new general formula for the eigenvalues of the link transforms. In Section 3 we introduce the Wick calculus in a general setting via the sesqui-holomorphic extension of real-analytic functions. We also introduce and study the properties of the *fundamental function* associated with an invariant symbolic calculus. Our main result (Theorem 3.12) provides a simple formula for the eigenvalues of the link transform in terms of this fundamental function. The remaining two sections are devoted to applications of this general formula, for weighted Fock spaces over $\mathbb{C}^d$ (Section 4) and weighted Bergman spaces over symmetric tube domains (Section 5). In the flat case we find unexpected relations between the classical functional calculi on $\mathbb{C}^d$, and present a new class of calculi which are still explicitly solvable. For symmetric tube domains, we present a new proof for the spectral analysis of the Berezin transform.

It should be remarked that our formulas for the eigenvalues of link transforms can be translated into corresponding formulas describing the link transforms in terms of canonical sets of generators of the ring of invariant differential operators on D (via the Harish-Chandra transform).

The general theory of invariant symbolic calculi on symmetric domains developed here may be applied in a much wider setting. For example – in the context of real symmetric domains our approach led to some definite results, particularly in the rank-one case (see [AU01]). It is possible and very interesting to develop our theory in the context of the invariant Hilbert spaces of holomorphic functions on symmetric Siegel domains, which are associated with the Wallach parameter but are not part of the holomorphic discrete series. The discrete Wallach points are of special interest (and most difficult to handle), since they are the only points which survive when the rank of the domains becomes infinite. The integral formulas for the invariant inner products associated with the Wallach set obtained in [AU97], [AU98] and [AU99] will be very useful in this goal. Finally, our theory can also be developed in the context of NA-invariant and vector-valued Hilbert spaces of holomorphic functions on symmetric domains.

1. Invariant Hilbert spaces of holomorphic functions on symmetric domains

1.1. The general framework

Let D be an irreducible hermitian symmetric domain in $\mathbb{C}^d$ with a distinguished base point $o \in D$ (called the "origin"). Then D can be realized as the quotient $D = G/K$, where $G := \mathrm{Iso}(D) \subseteq \mathrm{Aut}(D)$ is the real Lie group of all biholomorphic Riemannian isometries of D, and $K := \{g \in G;\, g(o) = o\}$ is the maximal compact subgroup of G. Our main applications are in the cases where D is either $\mathbb{C}^d$, or irreducible symmetric tube domain (i.e. a symmetric Siegel domain of type I), see Subsections 1.2 and 1.3 bellow. Nevertheless, we prefer to develop our theory in the general setup, so as to allow future applications (for instance, to general symmetric hermitian spaces). [He78] and [He84] are our general references for symmetric spaces and semi-simple Lie groups. [Hu63], [Gi64] and [FK94] are the general references for analysis on symmetric domains, and [Lo77], [Up87] are the references to the analysis on symmetric domains from the Jordan-theoretic point of view. [UU96] and [Un98] are general references for pseudo-differential analysis on symmetric cones and quantization respectively.

It is known that for every $z \in D$ there is a unique *symmetry* $s_z \in G$ (i.e. $s_z \circ s_z = 1_G$, the unit of G) for which z is the unique fixed point. Moreover,

$$s_z = g \circ s_o \circ g^{-1}, \quad \forall g \in G,\ g(o) = z. \tag{1.1}$$

The *Cartan involution* induced by s_o, $\Theta(g) := s_o \circ g \circ s_o$, $g \in G$ gives rise to the *Iwasawa decomposition*

$$G = NAK \tag{1.2}$$

in which A and N are maximal abelian and maximal nilpotent subgroups of G respectively. NA is a maximal solvable subgroup of G, and the evaluation map $NA \ni g \mapsto g(o) \in D$ is a surjective Riemannian isometry. Thus, for every $z \in D$ there exists a unique element $g_z \in NA$ for which $g_z(o) = z$. In what follows we shall use the important map $\psi : D \to D$ which is defined by

$$\psi(z) := g_z^{-1}(o). \tag{1.3}$$

We shall show that ψ is a real-analytic diffeomorphism of period 2 of D whose unique fixed point is the origin o. It is known that any pair of points in D can be joined by a unique geodesic line. Also, since the elements of G act on D as Riemannian isometries, they permute the geodesic lines in D. In particular, the symmetry s_z maps each geodesic line through z into itself, reverses its orientation and preserves distances. Given $w \in D$, let z be the mid-point along the geodesic line between o and z. We denote $\varphi_w := s_z$. Thus

$$\varphi_w \circ \varphi_w = 1_G, \quad \varphi_w(o) = w, \quad \text{and} \quad \varphi_w(w) = o.$$

Let $\mathfrak{A}$ be the Lie algebra of A. It is isomorphic to $\mathbb{R}^r$, where r is the *rank* of D. Given $g \in G$ with Iwasawa decomposition $g = n\,a\,k$ (with $n \in N, a \in A, k \in K$) let $\mathfrak{A}(g) \in \mathfrak{A}$ be the unique element for which $\exp \mathfrak{A}(g) = a$. Then $\mathfrak{A}(n_1 g k_1) = \mathfrak{A}(g)$ for all $n_1 \in N, k_1 \in K$, and $\mathfrak{A}(a_1 a_2) = \mathfrak{A}(a_1) + \mathfrak{A}(a_2)$ for all $a_1, a_2 \in A$. Let $\mathfrak{A}^{*\mathbb{C}} \equiv \mathbb{C}^r$ be the complexification of the dual of $\mathfrak{A}$ and let $\underline{\rho} \in \mathfrak{A}^*$ be the half sum of the positive roots. The *exponential functions*

$$e_{\underline{\lambda}}(z) := \exp\langle \mathfrak{A}(g) | \underline{\lambda} + \underline{\rho} \rangle, \quad g \in G, \ \ g(o) = z, \tag{1.4}$$

are N-invariant functions on D which are the eigenfunctions of NA

$$e_{\underline{\lambda}}(h(z)) = e_{\underline{\lambda}}(h(o))\, e_{\underline{\lambda}}(z), \quad \forall h \in NA, \ \ \forall z \in D. \tag{1.5}$$

Let us define

$$\mathcal{X}_{\underline{\lambda}} = \overline{\operatorname{span}}\{e_{\underline{\lambda}} \circ g;\, g \in G\} \tag{1.6}$$

where the closure is taken in the topology of uniform convergence on compact subsets of D. It is known that $\operatorname{span}\{\mathcal{X}_{\underline{\lambda}};\, \underline{\lambda} \in \mathbb{C}^r\}$ is dense in $C^\infty(D)$ in that topology. The *spherical function* associated with the exponential function $e_{\underline{\lambda}}$ is

$$\phi_{\underline{\lambda}}(z) := \int_K e_{\underline{\lambda}}(k(z))\, dk.$$

Then $\phi_{\underline{\lambda}} \in \mathcal{X}_{\underline{\lambda}}$ is K-invariant, $\phi_{\underline{\lambda}}(o) = e_{\underline{\lambda}}(o) = 1$, and for every $f \in \mathcal{X}_{\underline{\lambda}}$

$$\int_K f(k(z))\, dk = f(o)\, \phi_{\underline{\lambda}}(z), \quad \forall z \in D. \tag{1.7}$$

Let W be the *Weyl group* of D. It is a subgroup of $\mathrm{GL}(\mathbb{C}^r)$ which contains the permutation group, and moreover

$$\phi_{\underline{\lambda}} = \phi_{\underline{\lambda}'} \quad \text{if and only if there exists } w \in W \text{ so that } \underline{\lambda} = w(\underline{\lambda}'). \tag{1.8}$$

It follows that

$$\mathcal{X}_{\underline{\lambda}} = \mathcal{X}_{\underline{\lambda}'} \quad \text{if and only if there exists } w \in W \text{ so that } \underline{\lambda} = w(\underline{\lambda}').$$

We denote by $\mathrm{Diff}(D)^G$ the algebra of *G-invariant differential operators* on D (i.e. differential operators T on D so that $T(f \circ g) = T(f) \circ g$ for all $g \in G$ and all $f \in C^\infty(D)$). A fundamental property of the exponential and spherical functions is that they are *joint eigen-functions for* $\mathrm{Diff}(D)^G$:

$$T(\phi_{\underline{\lambda}}) = \widetilde{T}(\underline{\lambda})\, \phi_{\underline{\lambda}}, \quad T(e_{\underline{\lambda}}) = \widetilde{T}(\underline{\lambda})\, e_{\underline{\lambda}}, \quad \forall T \in \mathrm{Diff}(D)^G, \ \forall \underline{\lambda} \in \mathbb{C}^r. \tag{1.9}$$

A fundamental result of Harish-Chandra is that the eigen-value map $T \mapsto \widetilde{T}(\underline{\lambda})$ (called the *Harish-Chandra transform*) is an algebra isomorphism of $\mathrm{Diff}(D)^G$ onto the algebra $\mathbb{C}[x_1, \ldots, x_r]^W$ of W-invariant polynomials in $x_1, \ldots, x_r$. In particular, $\mathrm{Diff}(D)^G$ is commutative.

Using standard tools from spectral theory, this result extends to more general (bounded or unbounded) G-invariant operators on D (see [E98]). Thus, if T is a

G-invariant operator, defined on a G-invariant subspace $\mathcal{X}$ of functions on D, then (1.9) holds whenever $e_{\underline{\lambda}} \in \mathcal{X}$, and the generalized Harish-Chandra transform $T \mapsto \widetilde{T}(\cdot)$ is injective. Thus T is uniquely determined by its eigenvalues $\widetilde{T}(\underline{\lambda})$, and they can be computed by

$$\widetilde{T}(\underline{\lambda}) = T(e_{\underline{\lambda}})(o) = T(\phi_{\underline{\lambda}})(o). \tag{1.10}$$

In what follows we shall use (1.10) with the exponential function, since they are much simpler than the spherical functions, and obey (1.5).

Let μ_0 be the (unique up to a multiplicative constant) *G-invariant measure* on D, i.e.

$$\int_D f\, d\mu_0 = \int_G f(g(o))\, dg.$$

The normalization of μ_0 will be fixed in the definition (1.21) below. For any Borel measure μ on D we define a function $\tilde{\mu}$ by

$$\mathrm{Dom}(\tilde{\mu}) := \{\underline{\lambda} \in \mathbb{C}^r;\, e_{\underline{\lambda}} \in L^1(D, \mu)\}$$

and

$$\tilde{\mu}(\underline{\lambda}) := \int_D e_{\underline{\lambda}}(z)\, d\mu(z), \quad \forall \underline{\lambda} \in \mathrm{Dom}(\tilde{\mu}).$$

Notice that if μ is K-invariant then $\tilde{\mu}$ is its (spherical) *Fourier transform* and

$$\tilde{\mu}(\underline{\lambda}) := \int_D \phi_{\underline{\lambda}}(z)\, d\mu(z), \quad \forall \underline{\lambda} \in \mathrm{Dom}(\tilde{\mu}).$$

In particular, if f is a measurable function on D then $\tilde{f} := \widetilde{f\, d\mu_0}$, i.e.

$$\tilde{f}(\underline{\lambda}) := \int_D e_{\underline{\lambda}}(z)\, f(z)\, d\mu_0(z). \tag{1.11}$$

Again, if f is K-invariant then $\tilde{f}$ is its (spherical) Fourier transform,

$$\tilde{f}(\underline{\lambda}) := \int_D \phi_{\underline{\lambda}}(z)\, f(z)\, d\mu_0(z) \quad \forall \underline{\lambda} \in \mathrm{Dom}(\tilde{f}).$$

Next, if μ is a K-invariant measure on D then the *convolution operator*

$$(C_\mu f)(z) := \int_D f(g(w))\, d\mu(w), \quad \text{where } g \in G \text{ and } g(o) = z,$$

is well-defined (on an appropriate domain) and G-invariant. Its Harish-Chandra transform is the Fourier transform of μ:

$$\widetilde{C_\mu}(\underline{\lambda}) = C_\mu(e_{\underline{\lambda}})(o) = \int_D e_{\underline{\lambda}}\, d\mu = \tilde{\mu}(\underline{\lambda}).$$

Let $\mathcal{H}$ be a Hilbert space of holomorphic functions on D on which the point evaluation functionals $\mathcal{H} \ni f \mapsto f(w)$, $w \in D$, are continuous, and let $K(z, w) =$

$K_w(z)$ be the *reproducing kernel* of $\mathcal{H}$. Let

$$k_w(z) := \frac{K_w(z)}{\|K_w\|} = \frac{K(z,w)}{K(w,w)^{1/2}}$$

be the normalized kernel at the point $w \in D$. For convenience we adopt the normalization

$$K(o,o) = 1.$$

We assume that G acts on $\mathcal{H}$ by means of an irreducible *projective representation* U of the form:

$$U(g)(f)(z) := j(g^{-1},z)\, f(g^{-1}(z)), \quad \forall g \in G, \ \forall z \in D. \tag{1.12}$$

Thus, each operator $U(g)$ is isometric on $\mathcal{H}$, the function $j(g^{-1},z)$ is holomorphic in z for all $g \in G$, and

$$j(g_2^{-1} \circ g_1^{-1}, z) = c(g_1,g_2)\, j(g_2^{-1}, g_1^{-1}(z))\, j(g_1^{-1}, z), \quad \forall g_1, g_2 \in G, \ \forall z \in D, \tag{1.13}$$

where $c(g_1,g_2)$ is a unimodular constant depending only on g_1, g_2. This is clearly equivalent to

$$U(g_1 \circ g_2) = c(g_1,g_2)\, U(g_1)\, U(g_2), \quad \forall g_1, g_2 \in G. \tag{1.14}$$

The relationship (1.13) yields also

$$|j(g_2^{-1} \circ g_1^{-1}, z)| = |j(g_2^{-1}, g_1^{-1}(z))|\, |j(g_1^{-1}, z)|, \quad \forall g_1, g_2 \in G, \ \forall z \in D.$$

In particular,

$$|j(k_1 \circ k_2, o)| = |j(k_1,o)|\, |j(k_2,o)|, \quad \forall k_1, k_2 \in K.$$

This fact and the compactness of K yield

$$|j(k,o)| = 1, \quad \forall k \in K. \tag{1.15}$$

Also, (1.13) and the fact that $j(1_G, z) \equiv 1$ lead to $c(g, g^{-1}) = 1$, i.e.

$$j(g^{-1}, g(z))\, j(g,z) = 1, \quad \forall g \in G, \ z \in D. \tag{1.16}$$

This implies that $U(g^{-1}) = U(g)^{-1}$ for all $g \in G$, and that $c(g_1,g_2)\, c(g_2^{-1}, g_1^{-1}) = 1$ for all $g_1, g_2 \in G$. Thus the natural action of G on operators on $\mathcal{H}$,

$$\pi(g)(T) := U(g)\, T\, U(g)^{-1}, \quad g \in G,$$

is a genuine representation, i.e. $\pi(g_1 \circ g_2) = \pi(g_1)\, \pi(g_2)$ for all $g_1, g_2 \in G$.

The fact that the operators $U(g)$, $g \in G$, are unitary on $\mathcal{H}$ is reflected in the transformation formula of the reproducing kernel:

$$j(g,z)\, K(g(z), g(w))\, \overline{j(g,w)} = K(z,w), \quad \forall g \in G, \ \forall z, w \in D, \tag{1.17}$$

a fact which can also be written as $(U(g) \otimes \overline{U(g)})(K) = K$. Also, (1.17) can be written as

$$U(g)(K_w) = \overline{j(g,w)}\, K_{g(w)}, \quad \forall g \in G,\ \forall w \in D. \tag{1.18}$$

Notice that since $K(o,o) = 1$, (1.17) implies in particular that

$$|j(g,o)|^{-2} = K(z,z), \quad \text{where } g \in G,\ \ g(o) = z. \tag{1.19}$$

Of particular interest is the case where $\mathcal{H}$ is the *weighted Bergman space*

$$\mathcal{H} := L^2_a(D,\mu) = L^2(D,\mu) \cap \{\text{holomorphic functions}\}$$

with respect to a K-invariant absolutely continuous measure μ on D. In this case we require that the measure μ transforms under G via

$$d\mu(g(z)) = |j(g,z)|^2\, d\mu(z), \quad \forall g \in G. \tag{1.20}$$

Thus U extends to a projective representation of $L^2(D,\mu)$. The transformation rule (1.20) of the measure μ and (1.15) yield easily that the measure μ_0 defined by

$$d\mu_0(z) := |j(g,o)|^{-2}\, d\mu(z) = K(z,z)\, d\mu(z), \quad \text{where } g \in G \text{ and } g(o) = z \tag{1.21}$$

is well-defined (independent of the particular $g \in G$ satisfying $g(o) = z$), and is G-*invariant*, i.e.

$$d\mu_0(g(z)) = d\mu_0(z), \quad \forall g \in G.$$

1.2. Weighted Fock spaces on $\mathbb{C}^d$

For $\nu > 0$ consider on $\mathbb{C}^d$ the probability measure

$$d\mu_\nu(z) = \left(\frac{\nu}{\pi}\right)^d e^{-\nu|z|^2}\, dm(z),$$

where $dm(z)$ is Lebesgue measure. The *weighted Fock space*

$$\mathcal{F}_\nu = \mathcal{F}_\nu(\mathbb{C}^d) = L^2_a(\mathbb{C}^d, \mu_\nu) := L^2(\mathbb{C}^d, \mu_\nu) \cap \{\text{holomorphic functions}\}$$

has a reproducing kernel

$$K^{(\nu)}(z,w) = e^{\nu(z,w)}.$$

Let G be the semi-direct product of the unitary group $U(d)$ and the translation group $T := \{g_w;\ w \in \mathbb{C}^d\} \equiv \mathbb{C}^d$, where $g_w(z) = z + w$. Notice that in the Iwasawa decomposition of G we have $K = U(d)$, $A = \{1_G\}$, $N = T$, and that $U(d)$ normalizes T:

$$k\, g_w\, k^{-1} = g_{k(w)}, \quad \forall\, k \in U(d),\ \forall\, w \in \mathbb{C}^d.$$

G acts isometrically on $L^2(\mathbb{C}^d, \ \mu_\nu)$ by the rule

$$U^{(\nu)}(g)\, f(z) = j(g^{-1}, z)\, f(g^{-1}(z)),$$

where

$$j(g^{-1}, z) = \exp\{-\frac{\nu}{2}\,|g(0)|^2 + \nu\langle z, g(0)\rangle\}$$
$$= K^{(\nu)}(z, g(0))/K^{(\nu)}(g(0), g(0)) = k^{(\nu)}_{g(0)}(z).$$

Here $k^{(\nu)}_w(z) := K^{(\nu)}(z, w)/K^{(\nu)}(w, w)^{1/2}$. Indeed, for $g \in G$

$$d\mu_\nu(g(z)) = |j(g, z)|^2\, d\mu_\nu(z), \tag{1.22}$$

i.e. (1.20) holds. To prove (1.22) let us write $g = g_w\, k$, where $k \in U(d)$ and $g_w \in T$. Then

$$\begin{aligned} d\mu_\nu(g(z)) &= (\nu/\pi)^d\ e^{-\nu\,|k\,z+w|^2}\ dm(k\,z + w) \\ &= (\nu/\pi)^d\ e^{-\nu\,|z|^2}\ e^{-2\nu\ \mathrm{Re}\langle z, k^*(w)\rangle - \nu\,|w|^2}\ dm(z) \\ &= \left|e^{-\nu\,\langle z, k^*(w)\rangle\ -\frac{1}{2}\ \nu|w|^2}\right|^2 d\mu_\nu(z) \\ &= \left|e^{\nu\,\langle z, g^{-1}(0)\rangle\ -\frac{\nu}{2}\ |g^{-1}(0)|^2}\right|^2 d\mu_\nu(z) = |j(g, z)|^2\, d\mu_\nu(z). \end{aligned}$$

The action $U^{(\nu)}$ of G is an irreducible *projective representation* (see (1.14)) since, by elementary calculations, one obtains (1.13) with the unimodular constant

$$c(g_1, g_2) := \exp\left\{i\ \nu\ \mathrm{Im}\langle g_1(0), g_1'(0)(g_2(0))\rangle\right\}.$$

Let us define

$$d\mu_0(w) = \left(\frac{\nu}{\pi}\right)^d\ dm(w).$$

Then μ_0 is clearly G-invariant. The reason for the particular normalization is the following relationship between μ_0 and μ_ν

$$d\mu_0(z) = |j(g, 0)|^{-2}\, d\mu_\nu(z), \quad g \in G,\ g(0) = z,$$

in accordance with (1.21). The ring of G-invariant differential operators is simply the polynomial ring $\mathbb{C}[\Delta]$, where

$$\Delta = \langle\partial|\partial\rangle = \sum_{j=1}^{d} \partial_j\ \overline{\partial}_j$$

is the Euclidean Laplacian (properly normalized). The exponential functions

$$e_{a,b}(z) = e^{\langle z|b\rangle}\ e^{\langle a|z\rangle}, \quad a, b \in \mathbb{C}^d \tag{1.23}$$

are eigen-functions of Δ:

$$\Delta(e_{a,b}) = \lambda\ e_{a,b} \quad \text{with } \lambda = \langle a|b\rangle$$

as well as of the translation subgroup T:

$$e_{a,b} \circ g_w = e_{a,b}(w)\, e_{a,b}, \quad \forall w \in \mathbb{C}^d.$$

Notice that $e_{a,b}$ is bounded if and only if $b = -a$, and in this case $e_{a,-a}(z) = e^{2i\,\mathrm{Im}\langle a|z\rangle}$ and $\lambda = \langle a, -a\rangle = -\|a\|^2$. The map $\psi(w) = g_w^{-1}(0)$ is simply $\psi(w) = -w = s_0(w)$ and the exponential functions transform under ψ according to the rule

$$e_{a,b} \circ \psi = e_{-a,-b}. \tag{1.24}$$

Since $\overline{e_{a,b}} = e_{b,a}$, it follows that

$$\overline{e_{a,b} \circ \psi} = e_{-b,-a}. \tag{1.25}$$

In this case the spherical function ϕ_λ associated with $e_{a,b}$ ($\lambda = \langle a|b\rangle$) is calculated explicitly

$$\phi_\lambda(z) = \int_{U(d)} e_{a,b}(k(z))\, dk = \sum_{\ell=0}^{\infty} \frac{1}{(d)_\ell} \frac{\lambda^\ell \|z\|^{2\ell}}{\ell!} = {}_0F_1(d;\ \lambda\,\|z\|^2).$$

1.3. Weighted Bergman spaces over symmetric tube domains

Let X be an irreducible Euclidean Jordan algebra of dimension d with unit element e, and let $\Omega = \{x^2;\ X \ni x \text{ invertible}\}$ be the associated symmetric cone. The *triple product* $\{x, y, z\} = (xy)z + (yz)x - (zx)y$ extends to the complexification $Z = X^{\mathbb{C}} = X \oplus iX$, and Z carries the structure of a *JB^*-algebra* with product $zw = \{z, e, w\}$, unit e, and involution $z^* = \{e, z, e\}$ (see [Up87]). The *tube domain* associated with Ω is

$$T(\Omega) := X + i\Omega = \left\{z \in Z;\ \frac{z - z^*}{2i} \in \Omega\right\}.$$

It is well known that $T(\Omega)$ is an irreducible hermitian symmetric domain (symmetric Siegel domain of type I). Thus with respect to the Bergman metric the holomorphic symmetry at ie, $s_{ie}(z) = -z^{-1}$, is an isometry, and the group $G := \mathrm{Aut}\,(T(\Omega))$ of all biholomorphic automorphisms of $T(\Omega)$ acts on it transitively. Let $G = NAK$ be the *Iwasawa decomposition* with respect to the Cartan involution $g \mapsto s_{ie}\, g\, s_{ie}$. Thus $K = \{g \in G;\ g(ie) = ie\}$ is a maximal compact subgroup of G and $T(\Omega) \equiv G/K \equiv NA$, with A maximal abelian and N maximal nilpotent subgroups of G. Since NA acts on $T(\Omega)$ simply transitively, for any $z \in T(\Omega)$ there exists a unique element $g_z \in NA$ so that $g_z(ie) = z$.

Let $\mathrm{N}(z)$ and $\mathrm{tr}(z) = \langle z|e\rangle$ be the *determinant* ("*norm*") and *trace* polynomials (defined on X via the spectral theorem, and extended linearly to the complexification $Z = X^{\mathbb{C}}$). Fix a frame $\{e_j\}_{j=1}^r$ of pairwise orthogonal primitive idempotents in X, where r is the *rank* of X (also, the rank of $T(\Omega)$). Thus $e = \sum_{j=1}^r e_j$, and Z has a *Peirce decomposition*

$$Z = \sum_{1 \le i \le j \le r}^{\oplus} Z_{i,j}$$

relative to $\{e_j\}_{j=1}^r$ (see [Lo77], [FK94], and [Up87]). For $1 \le k \le r$ let N_k be the determinant polynomial of the JB^*-sub-algebra

$$Z_k := \sum_{1\le i\le j\le k}^{\oplus} Z_{i,j}$$

whose unit is $u_k = \sum_{j=1}^k e_j$. It is known that the *characteristic multiplicity*

$$a = \dim_{\mathbb{C}} Z_{i,j}, \quad 1 \le i < j \le r$$

is independent of the frame and of the pair (i, j) with $i < j$. Let P be the orthogonal projection from Z onto Z_k and extend N_k to all of Z via $\mathrm{N}_k(z) = \mathrm{N}_k(P_k(z))$. Clearly, $\mathrm{N}_r = \mathrm{N}$.

The *conical function* associated with $\boldsymbol{s} = (s_1, s_2, \ldots, s_r) \in \mathbb{C}^r$ is defined on $T(\Omega)$ via

$$\mathrm{N}_{\boldsymbol{s}}(z) := \mathrm{N}_1(z)^{s_1-s_2}\,\mathrm{N}_2(z)^{s_2-s_3}\ldots\mathrm{N}_r(z)^{s_r}, \quad z \in T(\Omega).$$

Notice that if $\boldsymbol{s} \in \mathbb{N}^r$ and $\boldsymbol{s} \ge 0$ in the sense that $s_1 \ge s_2 \ge \cdots \ge s_r \ge 0$ then $\mathrm{N}_{\boldsymbol{s}}(z)$ is a polynomial.

The *Gindikin–Koecher Gamma function* is defined for $\boldsymbol{s} \in \mathbb{C}^r$ with $\mathrm{Re}(s_j) > (j-1)\frac{a}{2}$ for all j by the absolutely convergent integral

$$\Gamma_\Omega(\boldsymbol{s}) = \int_\Omega e^{-\operatorname{tr}(x)}\,\mathrm{N}_{\boldsymbol{s}}(x)\,d\mu_\Omega(x)$$

where $d\mu_\Omega(x) = \mathrm{N}(x)^{-\frac{d}{r}}dx$ is the (unique up to a constant multiple) measure on Ω which is invariant under $\mathrm{GL}(\Omega) := \{g \in \mathrm{GL}(X); G(\Omega) = \Omega\}$. It is known (see [Gi64] and [FK94]) that

$$\Gamma_\Omega(\boldsymbol{s}) = (2\pi)^{\frac{d-r}{2}} \prod_{j=1}^r \Gamma\left(s_j - (j-1)\frac{a}{2}\right). \tag{1.26}$$

This allows to extend $\Gamma_\Omega(\boldsymbol{s})$ to an entire meromorphic function on $\mathbb{C}^r$. Let us denote

$$\tau(z, w) := \frac{z - w^*}{2i}, \quad z, w \in Z \text{ and } \tau(z) := \tau(z, z) = \mathrm{Im}(z).$$

The functions $\mathrm{N}_{\boldsymbol{s}} \circ \tau$ are *joint eigenfunctions* of the group NA:

$$\mathrm{N}_{\boldsymbol{s}}(\tau(g(z))) = \mathrm{N}_{\boldsymbol{s}}(\tau(g(ie)))\,\mathrm{N}_{\boldsymbol{s}}(\tau(z)), \quad \forall\, g \in NA. \tag{1.27}$$

See [UU94]. Thus the exponential functions in this context are given by

$$e_{\underline{\lambda}}(z) := \mathrm{N}_{\underline{\lambda}+\underline{\rho}}(\tau(z, z)), \quad z \in T(\Omega), \tag{1.28}$$

where $\rho := (\frac{1}{2}((j-1)a+1))_{j=1}^r$ is the half-sum of the positive roots. Property (1.27) allows the derivation of the fundamental formula

$$\frac{1}{\Gamma_\Omega(s)} \int_\Omega e^{-(\frac{z}{i}|x)} \mathrm{N}_s(x) d\mu_\Omega(x) = \mathrm{N}_s\left(\left(\frac{z}{i}\right)^{-1}\right) = \mathrm{N}^*_{-s^*}\left(\frac{z}{i}\right) \tag{1.29}$$

valid for all $s \in \mathbb{C}^r$ with $\mathrm{Re}(s_j) > (j-1)\frac{a}{2}$ for all j, and all $z \in T(\Omega)$. Here, $s^* = (s_r, s_{r-1}, s_{r-2}, \dots, s_1)$ and $\mathrm{N}^*_{\boldsymbol{\alpha}}$ are the conical functions associated with the frame in reverse order $\{e_r, e_{r-1}, e_{r-2}, \dots, e_1\}$.

The *Wallach set* $W(D)$ is the set of all $\nu \in \mathbb{C}$ for which the function

$$K^{(\nu)}(z, w) := \mathrm{N}(\tau(z, w))^{-\nu}, \quad z, w \in T(\Omega) \tag{1.30}$$

is *positive definite*. It is known (see [Be75], [Gi75], [VR76], [W79], [La86], [La87], [FK90]) that

$$W(D) = \left\{0, \frac{a}{2}, 2\frac{a}{2}, \dots, (r-1)\frac{a}{2}\right\} \cup \left((r-1)\frac{a}{2}, \infty\right).$$

For $\nu \in W(D)$ let $\mathcal{H}_\nu$ be the Hilbert space of analytic functions on $T(\Omega)$ whose reproducing kernel is $K^{(\nu)}(z, w)$. Let $p = 2\frac{d}{r} = (r-1)a + 2$ be the *genus* of $T(\Omega)$. Then (with J denoting the Jacobian)

$$(J g(z))^{\nu/p} K^{(\nu)}(g(z), g(w)) \overline{Jg(w)}^{\nu/p} = K^{(\nu)}(z, w), \quad \forall g \in G, \ \forall z, w \in T(\Omega).$$

Thus G acts isometrically on $\mathcal{H}_\nu$ via

$$U^{(\nu)}(g) f = (J(g^{-1}))^{\nu/p} \cdot (f \circ g^{-1}), \quad g \in G, \ f \in \mathcal{H}_\nu,$$

and this action is an irreducible *projective representation* of G which becomes a unitary representation when $\nu \in W(T(\Omega)) \cap \frac{1}{2}\mathbb{N}$. Thus, in the notation of Subsection 1.1,

$$j(g, z) := (Jg(z))^{\frac{\nu}{p}}, \quad \forall g \in G, \ \forall z \in T(\Omega).$$

If $\nu > p - 1$ then $\mathcal{H}_\nu$ is the *weighted Bergman space*

$$\mathcal{H}_\nu = L^2_a(T(\Omega), \mu_\nu) = L^2(T(\Omega), \mu_\nu) \cap \{\text{holomorphic functions}\}$$

where

$$d\mu_\nu(z) := a(\nu)\ \mathrm{N}(\tau(z))^{\nu-p} dm(z),$$

$dm(z)$ being the Lebesgue measure, and

$$a(\nu) := \frac{\Gamma_\Omega(\nu)}{(4\pi)^d\, \Gamma_\Omega(\nu - \frac{d}{r})}.$$

The measure μ_ν transforms according to the rule

$$d\mu_\nu(g(z)) = |Jg(z)|^{\frac{2\nu}{p}} d\mu_\nu(z), \tag{1.31}$$

in accordance with (1.20). In particular, the measure $\mathrm{N}(\tau(z))^{-p}\, dm(z)$ is the unique, up to a multiplicative factor, *G-invariant measure* on $T(\Omega)$. It will be convenient for us to use the following normalization for the G-invariant measure:

$$d\mu_0(z) := a(\nu)\ \mathrm{N}(\tau(z))^{-p}\ dm(z).$$

With this normalization, the measures μ_0 and μ_ν are related by

$$d\mu_0(z) = |Jg(ie)|^{-\frac{2\nu}{p}}\ d\mu_\nu(z), \quad g \in G,\ g(ie) = z,$$

in accordance with (1.21).

2. Invariant symbolic calculi

2.1. Invariant symbolic calculi and covariant fields of operators

As in Section 1, let D be a hermitian symmetric domain in $\mathbb{C}^d$ and let $\mathcal{H}$ be a Hilbert space of holomorphic functions on D with reproducing kernel $K(z, w) = K_w(z)$, on which the group $G := \mathrm{Iso}(D)$ of all biholomorphic Riemannian isometries of D acts isometrically and irreducibly by means of a projective representation U with multiplier $j(g, z)$ (1.12).

Definition 2.1. An *invariant symbolic calculus* is a map $b \mapsto \mathcal{A}_b$ from a G-invariant subspace Dom $(\mathcal{A})$ of functions on D into the space $\mathrm{Op}(\mathcal{H})$ of closed operators on $\mathcal{H}$, such that

1. $\mathrm{span}\{\mathcal{X}_{\underline{\lambda}} \cap \mathrm{Dom}(\mathcal{A});\ \underline{\lambda} \in \mathbb{C}^r\} = \mathrm{span}\{e_{\underline{\lambda}} \circ g;\ g \in G, \underline{\lambda} \in \mathbb{C}^r\}$ is dense in $\mathrm{Dom}(\mathcal{A})$ in the topology of uniform convergence on compact subsets.

2. $\mathrm{span}\{K_w;\ w \in D\} \subset \mathrm{Dom}(\mathcal{A}_b),\quad \forall b \in \mathrm{Dom}(\mathcal{A})$;

3. $\mathcal{A}$ intertwines the natural group actions on functions and operators:

$$U(g)\mathcal{A}_b U(g)^{-1} = \mathcal{A}_{b \circ g^{-1}}, \quad \forall\, g \in G,\ \forall\, b \in \mathrm{Dom}\,(\mathcal{A}). \tag{2.1}$$

The function b is called the *active symbol* (or, strong symbol) of the operator $\mathcal{A}_b$. The invariant symbolic calculi appear naturally in Berezin's theory of *quantization* on symmetric domains, see [Be71], [Be72], [Be73], [Be74-1], [Be74-2], [Be75] and [Be78].

Remarks. (i) If the function 1 belongs to $\mathrm{Dom}(\mathcal{A})$ then $\mathcal{A}_1$ commutes with all the operators $U(g)$, $g \in G$. Since U is irreducible, $\mathcal{A}_1$ is a multiple of the identity operator. In this case we normalize $\mathcal{A}$ by requiring that

$$\mathcal{A}_1 = I.$$

(ii) We are vague here about the nature of the functions (symbols) in $\mathrm{Dom}(\mathcal{A})$ and the nature of the operators $\mathcal{A}_b$. In the applications the symbols are measurable (or even continuous) functions, and the operators are bounded.

(iii) One can study the more general setup in which $\mathcal{A}_b$ is an operator from $\mathcal{H}$ into another Hilbert space $\mathcal{L}$ of functions (holomorphic or not) on D which is invariant under an isometric action V of G. The intertwining property is then

$$V(g)\mathcal{A}_b U(g)^{-1} = \mathcal{A}_{b\circ g^{-1}} \quad \forall g \in G, \forall b \in \operatorname{Dom}(\mathcal{A}).$$

The space $\mathcal{L}$ need not be irreducible. For instance, let D be an irreducible hermitian symmetric domain (realized either as a Cartan or as a Siegel domain), let $\nu > p-1$ and let $\mathcal{L} := L^2(D, \mu_\nu)$ and $\mathcal{H} := L^2_a(D, \mu_\nu)$, the space of holomorphic functions in $L^2(D, \mu_\nu)$. $G = \operatorname{Aut}(D)$ acts isometrically on both $\mathcal{L}$ and $\mathcal{H}$ via $U^{(\nu)}(g)(f) := (J(g^{-1})(z))^{\nu/p} f \circ g^{-1}$. The study of *Hankel operators* and their generalizations [A96] fits in naturally here.

Another important case is when $\mathcal{H}$ and $\mathcal{L}$ are both Hilbert spaces of holomorphic functions on D on which G acts isometrically by means of (possibly different) irreducible projective representations U and V respectively. In this case one can also replace $\mathcal{L}$ by $\overline{\mathcal{L}} := \{\bar{f};\ f \in \mathcal{L}\}$ and replace V by $\overline{V}$, which is defined by $\overline{V}(f) := \overline{V(\bar{f})}$.

Notice that since $\operatorname{span}\{K_w;\ w \in D\}$ is dense in $\mathcal{H}$, $\mathcal{A}_b$ is determined by its action on the kernel functions K_w, $w \in D$. We define

$$A_b(z, w) := \mathcal{A}_b\,(K_w)(z)/K_w(z) = \frac{\langle \mathcal{A}_b K_w, K_z\rangle}{\langle K_w, K_z\rangle}. \tag{2.2}$$

Then $\mathcal{A}_b$ is completely determined by the function $A_b(z, w)$. Notice that the function $A_b(z, z)$ is the *Berezin symbol* of $\mathcal{A}_b$, and it determines the function $A_b(z, w)$, since the latter is sesqui-holomorphic in (z, w), i.e. holomorphic in z and anti-holomorphic in w, see Subsection 3.1. Also, the mapping $b \mapsto A_b$ (2.2) is *G-covariant*, i.e.

$$A_b(g(z), g(w)) = A_{b\circ g}(z, w), \quad \forall b \in \operatorname{Dom}(\mathcal{A}). \tag{2.3}$$

Indeed, this follows easily by (1.17) and (2.1).

Conversely, if $b \mapsto A_b$ is a map from a G-invariant space $\mathcal{X}$ of functions on D into functions $A_b(z, w)$ on $D \times D$, which are sesqui-holomorphic and satisfy (2.3), then we can define a map $\mathcal{X} \ni b \mapsto \mathcal{A}_b \in \operatorname{Op}(\mathcal{H})$ via $\mathcal{A}_b(K_w)(z) = K(z, w)A_b(z, w)$, and (2.1) will hold. Thus, *the invariant symbolic calculi are in one-to-one correspondence with the covariant maps into sesqui-holomorphic functions.* A similar statement is true for the version of the theory in which $\mathcal{A}_b : \mathcal{H} \to \mathcal{L}$.

Next, define an operator $\beta : \mathcal{X} = \operatorname{Dom}(\mathcal{A}) \to \{$holomorphic functions on $D\}$ by

$$\beta(b)(z) = A_b(z, o) = \mathcal{A}_b(K_o)(z)/K_o(z). \tag{2.4}$$

Then β is *K-invariant*, i.e.

$$\beta(b \circ k) = \beta(b) \circ k, \quad \forall k \in K.$$

Indeed, by (2.3) $\beta(b)(kz) = A_b(kz, o) = A_b(kz, ko) = A_{b\circ k}(z, o) = \beta(b \circ k)(z)$. Also, β determines A and $\mathcal{A}$ via

$$A_b(z, w) = \beta(b \circ g)(g^{-1}(z)) \quad \text{and} \quad \mathcal{A}_b(K_w)(z) = K(z, w)\, \beta(b \circ g)(g^{-1}(z)), \tag{2.5}$$

where $g \in G$ is any element for which $g(o) = w$. We therefore obtain

Corollary 2.1. (2.4) *and* (2.5) *establish a one-to-one correspondence between the invariant symbolic calculi* $b \mapsto \mathcal{A}_b$ *and the* K*-invariant maps* $b \mapsto \beta(b)$.

In what follows we shall assume that β is an *integral operator*:

$$\beta(b)(z) = \int_D F(z, w)\, b(w)\, d\mu_0(w),$$

where $F(z, w)$ is holomorphic in z and is K-invariant in the sense that

$$F(k(z), k(w)) = F(z, w), \quad \forall k \in K,\ \forall z, w \in D. \tag{2.6}$$

This is indeed the case for the interesting calculi $\mathcal{A}$ on the weighted Fock and Bergman spaces. Next, for each $\eta \in D$ define an operator B_η on $\mathcal{H}$ via its action on the reproducing kernel functions

$$B_\eta(K_w)(z) = F\big(g^{-1}(z), g^{-1}(\eta)\big)\, K(z, w), \tag{2.7}$$

where, again, g is any element of G with $g(o) = w$. It is well-known and easy to prove that the kernel functions $\{K_w\}_{w\in D}$ are linearly independent. Hence, the K-invariance (2.6) shows that B_η is well-defined (independent of the choice of g) on span$\{K_w;\, w \in D\}$. Under mild assumptions on F (or β) B_η is also a closed operator. This property can be derived also on the basis of the properties of the adjoint $\mathcal{A}'$ of $\mathcal{A}$, which will be studied in the next subsection (see (2.18)).

Proposition 2.2. *For every* $b \in \mathrm{Dom}(\mathcal{A})$,

$$\mathcal{A}_b = \int_D b(\eta)\, B_\eta\, d\mu_0(\eta), \tag{2.8}$$

where the integral converges in the weak operator topology on the dense subspace span$\{K_w;\, w \in D\}$.

Proof. It is enough to act on the reproducing kernel functions. Now (with $g \in G,\ g(o) = w$),

$$\begin{aligned}
\mathcal{A}_b(K_w)(z) &= A_b(z, w)\, K(z, w) = \beta(b \circ g)(g^{-1}(z))\, K(z, w) \\
&= \int_D F\left(g^{-1}(z), \xi\right) b(g(\xi))\, d\mu_0(\xi)\, K(z, w) \\
&= \int_D F\left(g^{-1}(z), g^{-1}(\eta)\right) b(\eta)\, d\mu_0(\eta)\, K(z, w) \\
&= \int_D b(\eta)\, B_\eta(K_w)(z)\, d\mu_0(\eta) = \left(\int_D b(\eta)\, B_\eta\, d\mu_0(\eta)\right)(K_w)(z). \qquad \square
\end{aligned}$$

Proposition 2.3. *The map $D \ni \eta \mapsto B_\eta \in \mathrm{Op}(\mathcal{H})$ is G-equivariant, namely*

$$U(g)^{-1} B_\eta U(g) = B_{g^{-1}(\eta)}, \quad \forall g \in G, \quad \forall \eta \in D. \tag{2.9}$$

Proof. Again, it is enough to act on the kernel functions $\{K_w\}_{w\in D}$. Now, for all $g \in G$,

$$\begin{aligned}(U(g)^{-1} B_\eta U(g) K_w)(z) &= \overline{j(g,w)} \left(U(g^{-1}) B_\eta K_{g(w)}\right)(z)\\ &= \overline{j(g,w)}\, j(g,z)\, (B_\eta K_{g(w)})(g(z))\\ &= j(g,z)\overline{j(g,w)}\, F((gh)^{-1}(g(z)), (gh)^{-1}(\eta))\, K(g(z), g(w))\end{aligned}$$

where $h \in G$ is any element for which $h(o) = w$. Using (1.30) we see that the last expression is

$$F\left(h^{-1}(z), h^{-1}(g^{-1}(\eta))\right) K(z,w) = B_{g^{-1}(\eta)}(K_w)(z).$$ □

Definition 2.2. A *covariant field of operators* on $\mathcal{H}$ is a family $\{B_\eta\}_{\eta\in D} \subseteq \mathrm{Op}(\mathcal{H})$ satisfying (2.9).

Corollary 2.4. *Given a covariant field of operators $\{B_\eta\}_{\eta\in D}$ on $\mathcal{H}$ define a map $b \mapsto \mathcal{A}_b$ via* (2.8). *Then* (2.1) *holds. Thus, the invariant symbolic calculi $\mathcal{A}$ are in one-to-one correspondence with the covariant fields of operators $\{B_\eta\}_{\eta\in D}$ via* (2.8).

Proposition 2.5. *Let $\{B_\eta\}_{\eta\in D}$ be a covariant field of operators on $\mathcal{H}$.*

(i) *The operator B_o is K-invariant, i.e.*

$$U(k)B_oU(k)^{-1} = B_o, \quad \forall k \in K. \tag{2.10}$$

(ii) *B_o determines all other operators B_η via*

$$B_\eta = U(g)\, B_o\, U(g)^{-1}, \quad g \in G,\ g(o) = \eta. \tag{2.11}$$

(iii) *Given a K-invariant operator B on $\mathcal{H}$, define $B_o := B$, and let B_η be defined by* (2.11). *Then $\{B_\eta\}_{\eta\in D}$ are well-defined (independent of the choice of $g \in G$ for which $g(o) = \eta$) and form a covariant field of operators. Thus,* (2.11) *establishes a one-to-one correspondence between the covariant fields of operators on $\mathcal{H}$ and the K-invariant operators on $\mathcal{H}$.*

Proof. (2.10) follows from (2.9): $U(k)B_o\, U(k)^{-1} = B_k(o) = B_o$ for all $k \in K$. Also (2.11) follows from (2.9). Finally, (iii) follows from the fact that U is a projective representation. □

Corollary 2.6. *The invariant symbolic calculi $\mathcal{A}$ are in one-to-one correspondence with the K-invariant operators B on $\mathcal{H}$, $B \leftrightarrow \mathcal{A}^B$, via*

$$\mathcal{A}_b^B = \int_G b(g(o))\, U(g)\, B\, U(g)^{-1}\, dg = \int_D b(\xi) B_\xi\, d\mu_0.$$

Moreover, $\mathcal{A}^B$ is given also by the following formulas

$$\mathcal{A}_b^B = \int_D b(\eta)\, U(g_\eta)\, B\, U(g_\eta)^{-1}\, d\mu_0(\eta) = \int_D b(\eta)\, U(\varphi_\eta)\, B\, U(\varphi_\eta)\, d\mu_0(\eta),$$

where $g_\eta \in NA$ satisfies $g_\eta(o) = \eta$, and $\varphi_\eta \in G$ is the symmetry at the geodesic mid-point between o and η.

Remark. It is possible to develop our theory so that $\mathcal{A}$ will extend to a certain class of distributions. Then, the K-invariant operator B which determines $\mathcal{A}$ is given by $B = \mathcal{A}_{\delta_0}$, and the associated covariant field of operators is given by $B_\xi = \mathcal{A}_{\delta_\xi}$.

The next class of examples of invariant symbolic calculi, related to irreducible representations of K, is described in the setting of symmetric tube domains (cf. Section 1.3). The minor notational changes needed for the weighted Fock spaces are obvious. Let $\{\mathcal{P}_{\mathbf{m}}\}_{\mathbf{m}\geq 0}$ be the irreducible K-invariant subspaces of $\mathcal{H}$. It is known that these subspaces are pairwise orthogonal and mutually K-inequivalent, and $\sum_{\mathbf{m}\geq 0} \mathcal{P}_{\mathbf{m}}$ is dense in $\mathcal{H}$ (see [Sch69]). Therefore Schur's lemma in representation theory implies that every K-invariant operator B on $\mathcal{H}$ leaves each $\mathcal{P}_{\mathbf{m}}$ invariant and $B|_{\mathcal{P}_{\mathbf{m}}} = b_{\mathbf{m}}\, I_{\mathcal{P}_{\mathbf{m}}}$. Thus

$$B = \sum_{\mathbf{m}\geq 0} b_{\mathbf{m}} P_{\mathbf{m}}$$

where $P_{\mathbf{m}}$ is the projection onto $\mathcal{P}_{\mathbf{m}}$ which annihilates all the spaces $\mathcal{P}_{\mathbf{n}}$ for $\mathbf{n} \neq \mathbf{m}$.

Corollary 2.7. *The invariant symbolic calculi are in one-to-one correspondence with families $\{b_{\mathbf{m}}\}_{\mathbf{m}\geq 0}$ of complex numbers, via*

$$\mathcal{A} \longleftrightarrow B = \sum_{\mathbf{m}\geq 0} b_{\mathbf{m}}\, P_{\mathbf{m}}.$$

Remark. B_o is bounded on $\mathcal{H}$ if and only if all the $\{B_\eta;\ \eta \in D\}$ are bounded, and $\|B_\eta\| = \|B_o\|$ for all $\eta \in D$. Moreover, $B_o = \sum_{\mathbf{m}\geq 0} b_{\mathbf{m}} P_{\mathbf{m}}$ is bounded if and only if $\{b_{\mathbf{m}}\}_{\mathbf{m}\geq 0}$ is bounded.

Definition 2.3. $\mathcal{A}^{\mathbf{m}} = \mathcal{A}^{P_{\mathbf{m}}}$, i.e. $\mathcal{A}^{\mathbf{m}}$ is the invariant symbolic calculus determined by the K-invariant operator $B = P_{\mathbf{m}}$. Let $A^{\mathbf{m}}$, $F^{\mathbf{m}}$, and $\beta^{\mathbf{m}}$ be the maps associated with $\mathcal{A}^{\mathbf{m}}$ in the manner described above.

The importance of the $\mathcal{A}^{\mathbf{m}}$'s is exhibited in the following simple fact.

Proposition 2.8. *Given a K-invariant operator $T = \sum_{\mathbf{m}\geq 0} t_{\mathbf{m}}\, P_{\mathbf{m}}$ on $\mathcal{H}$, the corresponding invariant symbolic calculus is given by $\mathcal{A}_b = \mathcal{A}_b^T = \sum_{\mathbf{m}\geq 0} t_{\mathbf{m}}\, \mathcal{A}_b^{\mathbf{m}}$.*

Definition 2.4. The covariant field of operators associated with $P_{\mathbf{m}}$ is

$$P_{\mathbf{m},\eta} = U(g)\, P_{\mathbf{m}}\, U(g)^{-1}, \quad g \in G,\ g(o) = \eta.$$

Definition 2.5. The reproducing kernel of $\mathcal{P}_{\mathbf{m}}$ with respect to the inner product of $\mathcal{H}$ is denoted by $K_{\mathbf{m}}(z, w) = K_{\mathbf{m}}^{\mathcal{H}}(z, w)$.

Thus

$$P_{\mathbf{m}}(f)(z) = \langle f,\ K_{\mathbf{m}}(\cdot, z)\rangle_{\mathcal{H}}.$$

Lemma 2.9. *Let* $\mathbf{m} \geq 0$ *and* $\eta \in D$. *Then for every* $g \in G$ *with* $g^{-1}(o) = \eta$ *we have*

$$P_{\mathbf{m},\eta}(K_w)(z) = j(g, z)\, K_{\mathbf{m}}(g(z), g(w))\, \overline{j(g, w)} \tag{2.12}$$

$$A_b^{\mathbf{m}}(z, w) = K(z, w)^{-1} \int_D b(\eta)\, K_{\mathbf{m}}(g_\eta^{-1}(z), g_\eta^{-1}(w))\, j(g_\eta^{-1}, z)\, \overline{j(g_\eta^{-1}, w)}\, d\mu_0(\eta) \tag{2.13}$$

where g_η *is the unique element of* NA *for which* $g_\eta(o) = \eta$. *In particular,*

$$\beta^{\mathbf{m}}(b)(z) = K(z, o)^{-1} \int_D b(\eta)\, K_{\mathbf{m}}(g_\eta^{-1}(z), g_\eta^{-1}(o))\, j(g_\eta^{-1}, z)\, \overline{j(g_\eta^{-1}, o)}\, d\mu_0(\eta) \tag{2.14}$$

and

$$F^{\mathbf{m}}(z, \eta) = K(z, o)^{-1} K_{\mathbf{m}}(g_\eta^{-1}(z), g_\eta^{-1}(o))\, j(g_\eta^{-1}, z)\, \overline{j(g_\eta^{-1}, o)}. \tag{2.15}$$

Proof. By definition of $P_{\mathbf{m},\eta}$ we have

$$\begin{aligned} P_{\mathbf{m},\eta}\,(K_w)(z) &= \left(U(g)^{-1}\, P_{\mathbf{m}}\, U(g)\, K_w\right)(z) \\ &= P_{\mathbf{m}}\,(U(g)\, K_w)\,(g(z))\, j(g, z) \\ &= \langle U(g)\, K_w,\ K_{\mathbf{m}}(\cdot, g(z))\rangle_{\mathcal{H}}\, j(g, z) \\ &= \langle K_w, K_{\mathbf{m}}(g(\cdot), g(z)\rangle_{\mathcal{H}}\, j(g, z)\, \overline{j(g, w)} \\ &= K_{\mathbf{m}}(g(z), g(w))\, j(g, z)\, \overline{j(g, w)}. \end{aligned}$$

This establishes (2.12). Next,

$$\begin{aligned} \mathcal{A}_b^{\mathbf{m}}(K_w)(z) &= \int_D b(\eta)\, P_{\mathbf{m},\eta}(K_w)(z)\, d\mu_0(\eta) \\ &= \int_D b(\eta)\, K_{\mathbf{m}}\,(g_\eta^{-1}(z), g_\eta^{-1}(w))\, j(g_\eta^{-1}, z)\, j(g_\eta^{-1}, w)\, d\mu_0(\eta). \end{aligned}$$

This implies (2.13). Finally, (2.14) and (2.15) are direct consequences of (2.13). □

Remark. In (2.13), (2.14) and (2.15) we can replace g_η by any other $g \in G$ for which $g(o) = \eta$.

In what follows we describe the basic examples of invariant symbolic calculi.

Example 2.1. The most important symbolic calculus is the *Toeplitz calculus* (called also "Toeplitz quantization", or "Toeplitz–Berezin quantization"). Let us consider the case where $\mathcal{H}$ is the Bergman space

$$\mathcal{H} = L_a^2(D, \mu) := L^2(D, \mu) \cap \{\text{holomorphic functions on } D\},$$

and μ is a measure on D satisfying (1.20). Let $P : L^2(D, \mu) \to \mathcal{H}$ be the orthogonal projection. The *Toeplitz operator* with an active (strong) symbol $b \in L^\infty(D, \mu)$ is the operator $\mathcal{T}_b : \mathcal{H} \to \mathcal{H}$ defined by

$$(\mathcal{T}_b f)(z) := P(bf)(z) = \int_D b(w) f(w) K(z, w)\, d\mu(w), \qquad f \in \mathcal{H},\ z \in D.$$

It is well known that (2.1) holds with $\mathcal{A} = \mathcal{T}$.

Example 2.2 Another important invariant symbolic calculus, the *Weyl calculus* $\mathcal{W}$, is defined in the general setting via

$$\mathcal{W}_b := \int_D b(\eta)\, U(s_\eta)\, d\mu_0(\eta),$$

where $s_\eta \in G$ is the symmetry at η and μ_0 is the G-invariant measure on D. Namely, the domain of definition $\mathrm{Dom}(\mathcal{W})$ of $\mathcal{W}$ consists of all measurable functions b on D for which the integral

$$\mathcal{W}_b(f)(z) := \int_D b(\eta)\, (U(s_\eta) f)(z)\, d\mu_0(\eta)$$

converges weakly in $\mathcal{H}$ for all $f \in \mathcal{H}$. Using (1.1), (1.14) and (1.16) one obtains

$$U(g)\, U(s_\eta)\, U(g)^{-1} = U(s_{g(\eta)}), \quad \forall g \in G,\ \forall \eta \in D,$$

and this implies that (2.1) holds with $\mathcal{A} = \mathcal{W}$.

2.2. The adjoint map and the link transform

The *adjoint* of an invariant symbolic calculus $b \mapsto \mathcal{A}_b$ is the map $\mathcal{A}'$ from $\mathrm{Op}(\mathcal{H})$, the space of closed operators on $\mathcal{H}$, to functions on D is defined via

$$\mathrm{Dom}(\mathcal{A}') := \{T \in \mathrm{Op}(\mathcal{H});\ T\mathcal{A}_b^* \in S_1(\mathcal{H})\ \ \forall b \in \mathrm{Dom}(\mathcal{A})\},$$

where $S_1(\mathcal{H})$ is the space of *trace class* operators on $\mathcal{H}$, and

$$\langle \mathcal{A}'(T), b\rangle_{L^2(D,\mu_0)} = \langle T, \mathcal{A}_b\rangle_{S_2} = \mathrm{trace}(T\, \mathcal{A}_b^*) \tag{2.16}$$

for all $T \in \mathrm{Dom}(\mathcal{A}')$ and for all $b \in \mathrm{Dom}(\mathcal{A})$. Here S_2 is the *Hilbert–Schmidt class*. The function $\mathcal{A}'(T)$ is called the *passive* (or, weak) *symbol* of T.

Proposition 2.10. (i) *For every* $T \in \mathrm{Dom}(\mathcal{A}')$,

$$\mathcal{A}'\big(U(g)\, T\, U(g)^{-1}\big) = \mathcal{A}'(T) \circ g^{-1}, \quad \forall\, g \in G. \tag{2.17}$$

(ii) *For every* $T \in \mathrm{Dom}(\mathcal{A}')$,

$$\mathcal{A}'(T)(\eta) = \langle T, B_\eta\rangle_{S_2}, \quad \forall \eta \in D. \tag{2.18}$$

Proof. (i) is a consequence of (2.1) and (2.16). From (2.8) we obtain, formally,

$$\langle \mathcal{A}'(T), b\rangle_{L^2(\mu_0)} = \left\langle T, \int_D b(\eta)\, B_\eta\, d\mu_0(\eta)\right\rangle_{S_2} = \int_D \langle T, B_\eta\rangle_{S_2} \overline{b(\eta)}\, d\mu_0(\eta),$$

and this yields (2.18). □

Example 2.3. (i) For the Toeplitz calculus $\mathcal{T}$ (see Example 2.1) we have $B_z = k_z \otimes k_z$, where $k_z := K_z/\|K_z\|$. Hence

$$\mathcal{T}'(T)(z) = \langle T\, k_z, k_z\rangle_{\mathcal{H}}$$

is the *Berezin symbol* of T.

(ii) For the Weyl calculus $\mathcal{W}$ (see Example 2.2) we have

$$\mathcal{W}'(T)(\eta) = \langle T, U(s_\eta)\rangle_{S_2},$$

where $s_\eta \in G$ is the symmetry at $\eta \in D$.

Remark. (2.17) implies that the function $\mathcal{A}'(I)$ satisfies $\mathcal{A}'(I) \circ g = \mathcal{A}'(I)$ for every $g \in G$. Hence $\mathcal{A}'(I)$ is a constant function. We therefore assume without loss of generality (by modifying the definition of $\mathcal{A}'$ if necessary) that

$$\mathcal{A}'(I)(z) = 1, \quad \forall z \in D.$$

We now associate with an invariant symbolic calculus $\mathcal{A}$ two linear transformations, namely

$$\mathcal{B} := \mathcal{A}'\, \mathcal{A} \quad \text{and} \quad \mathcal{Q} := \mathcal{A}\, \mathcal{A}',$$

acting on functions on D, and on operators on $\mathcal{H}$ respectively. The operator $\mathcal{B}$ is called the *link transform* associated with $\mathcal{A}$, because it maps the active symbol b of $\mathcal{A}_b$ to the passive symbol $\mathcal{B}\, b = \mathcal{A}'(\mathcal{A}_b)$ of $\mathcal{A}_b$. We call $\mathcal{Q}$ the corresponding "co-transform". It maps an operator S to the operator $\mathcal{Q}(S)$ whose active symbol is the passive symbol of S.

Example 2.4 The link transform associated with the Toeplitz calculus $\mathcal{T}$ on the Bergman space $\mathcal{H} = L^2_a(D, \mu)$ with respect to a K-invariant measure μ is the *Berezin transform* $\mathcal{B} = \mathcal{T}'\,\mathcal{T}$ associated with μ:

$$\mathcal{B}(b)(z) = \langle b\, k_z, k_z\rangle_{\mathcal{H}} = \int_D b(w)\, \mathcal{K}(z, w)\, d\mu_0(w),$$

where $d\mu_0(w) := K(w, w)\, d\mu(w)$ is the (properly normalized) G-invariant measure on D and

$$\mathcal{K}(z, w) = \frac{|K(z, w)|^2}{K(z, z)\, K(w, w)}$$

is a *G-invariant* kernel, i.e.: $\mathcal{K}(g(z), g(w)) = \mathcal{K}(z, w)$ for all $g \in G$. This exhibits $\mathcal{B}$ as the operator of *convolution with* μ:

$$(\mathcal{B}b)(z) = \int_D b \circ g \, d\mu, \quad g \in G, \; g(o) = z. \tag{2.19}$$

Also, in the case of the Toeplitz calculus the co-transform $\mathcal{Q} = \mathcal{T}\,\mathcal{T}'$ is given by

$$\mathcal{Q}(S) = \int_D \langle S\, k_\eta, k_\eta\rangle_{\mathcal{H}} \; k_\eta \otimes k_\eta \; d\mu_0(\eta).$$

The link transforms associated with the Weyl calculus $\mathcal{W}$ are given formally by

$$\mathcal{W}'\mathcal{W}(b)(\eta) = \int_D b(\xi)\langle U(s_\xi), U(s_\eta)\rangle_{S_2} \; d\mu_0(\xi)$$

and

$$\mathcal{W}\mathcal{W}'(T) = \int_D \langle T, U(s_\xi)\rangle_{S_2} \; U(s_\xi)\; d\mu_0(\xi).$$

Proposition 2.11. *Let $\mathcal{A}$ be an invariant symbolic calculus with adjoint $\mathcal{A}'$.*
(i) *The link transform $\mathcal{B} = \mathcal{A}'\,\mathcal{A}$ is given by*

$$(\mathcal{B}\, b)(\eta) = \int_D b(\xi)\; \langle B_\xi, B_\eta\rangle_{S_2} \; d\mu_0(\xi). \tag{2.20}$$

(ii) *The co-transform $\mathcal{Q} = \mathcal{A}\,\mathcal{A}'$ is given by*

$$\mathcal{Q}(S) = \int_D \langle S, B_\eta\rangle_{S_2} \; B_\eta \; d\mu_0(\eta). \tag{2.21}$$

Proof. (i) Using (2.8) and (2.18) we obtain

$$\begin{aligned}(\mathcal{B}b)(\eta) &= \mathcal{A}'\,(\mathcal{A}_b)(\eta) = \langle \mathcal{A}_b, B_\eta\rangle_{S_2} \\ &= \Big\langle \int_D b(\xi)\; B_\xi \; d\mu_0(\xi), B_\eta \Big\rangle_{S_2} = \int_D \langle B_\xi, B_\eta\rangle_{S_2}\; d\mu_0(\xi).\end{aligned}$$

(ii) Formula (2.21) is a consequence of (2.8) and (2.18). □

Corollary 2.12. *For $b_1, b_2 \in \operatorname{Dom}(\mathcal{A})$,*

$$\langle \mathcal{A}_{b_1}, \mathcal{A}_{b_2}\rangle_{S_2} = \iint_{D\times D} b_1(\xi)\; \overline{b_2(\eta)}\; \langle B_\xi, B_\eta\rangle_{S_2}\; d\mu_0(\xi)\; d\mu_0(\eta).$$

Proof. Using (2.20) we obtain

$$\begin{aligned}\langle \mathcal{A}_{b_1}, \mathcal{A}_{b_2}\rangle_{S_2} &= \langle \mathcal{B}\, b_1, b_2\rangle_{L^2(\mu_0)} = \int_D (\mathcal{B}b_1(\xi)\; \overline{b_2(\eta)}\; d\mu_0(\eta) \\ &= \iint_{D\times D} b_1(\xi)\; \overline{b_2(\eta)}\; \langle B_\xi, B_\eta\rangle_{S_2}\; d\mu_0(\xi)\; d\mu_0(\eta).\end{aligned}$$ □

Proposition 2.13. (i) *The link transform $\mathcal{B} = \mathcal{A}'\mathcal{A}$ is G-invariant, i.e.*

$$\mathcal{B}(b \circ g) = (\mathcal{B}\, b) \circ g, \quad \forall\, g \in G.$$

(ii) *The co-transform $\mathcal{Q} = \mathcal{A}\,\mathcal{A}'$ is G-invariant in the sense that*

$$U(g)\, \mathcal{Q}(S)\, U(g)^{-1} = \mathcal{Q}\big(U(g)\, S\, U(g)^{-1}\big), \quad \forall\, g \in G.$$

Proof. This follows from (2.1) and (2.17). □

Remark. If $1 \in \mathrm{Dom}(\mathcal{A})$ then $1 \in \mathrm{Dom}(\mathcal{A}'\mathcal{A})$ as well, and

$$\mathcal{A}'\mathcal{A}(1) = 1.$$

Indeed, this follows from the facts that $\mathcal{A}_1 = I$ and $\mathcal{A}'(I) = 1$.

Lemma 2.14. *Let $\xi, \eta \in D$. Then*

$$\langle B_\xi, B_\eta\rangle_{S_2} = \int_D\int_D F(g^{-1}(z), g^{-1}(\xi))\, F(g^{-1}(z), g^{-1}(\eta))\, |K(z,w)|^2\, d\mu(z)\, d\mu(w) \tag{2.22}$$

where $g \in G$ in the inner integral is an arbitrary element for which $g(o) = w$.

Proof. It is a well-known fact that for every operator T on $\mathcal{H} = L^2_a(D, \mu)$ for which the trace $\mathrm{tr}(T)$ is well defined (i.e., $T \in S_1$ or $T \geq 0$),

$$\mathrm{tr}(T) = \int_D \langle T(K_w), K_w\rangle_{\mathcal{H}}\, d\mu(w).$$

Therefore, using (2.7) we obtain

$$\langle B_\xi, B_\eta\rangle_{S_2} = \mathrm{tr}(B_\eta^* B_\xi) = \int_D \langle B_\xi(K_w), B_\eta(K_w)\rangle_{\mathcal{H}}\, d\mu(w)$$

$$= \int_D \int_D F(g^{-1}(z), g^{-1}(\xi))\, \overline{F(g^{-1}(z), g^{-1}(\eta))}\, |K(z,w)|^2\, d\mu(z)\, d\mu(w). \square$$

Definition 2.6. Given two K-invariant operators T and S on $\mathcal{H}$, let $\mathcal{A}^T$, $\mathcal{A}^S$ be the associated invariant symbolic calculi. The corresponding mixed link transform and co-transform are

$$\mathcal{B}^{T,S} := (\mathcal{A}^S)'\, \mathcal{A}^T \quad \text{and} \quad \mathcal{Q}^{T,S} := \mathcal{A}^T\, (\mathcal{A}^S)' \tag{2.23}$$

respectively.

It is clear that these two transforms are G-invariant, i.e.

$$\mathcal{B}^{T,S}(b \circ g) = (\mathcal{B}^{T,S}\, b) \circ g, \quad \forall\, b \in \mathrm{Dom}(\mathcal{B}^{T,S}),\ \forall\, g \in G$$

and

$$U(g)\, \mathcal{Q}^{T,S}(X)\, U(g)^{-1} = \mathcal{Q}^{T,S}\big(U(g)\, X\, U(g)^{-1}\big), \quad \forall g \in G,\ \forall X \in \mathrm{Dom}(\mathcal{Q}^{T,S}).$$

Our results extend to the context of the mixed link transforms and co-transforms.

2.3. The reflection ψ and the associated involution of $\mathfrak{A}^{*\mathbb{C}}$

In this subsection we study in detail the real-analytic reflection ψ of D at o mentioned above (1.3) and its connection to a natural involution on $\mathfrak{A}^{*\mathbb{C}}$. Recall that for any $z \in D$ there exists a unique element $g_z \in NA \subset G$ for which $g_z(o) = z$. The map $\psi : D \to D$ is defined by

$$\psi(z) = g_z^{-1}(o). \tag{2.24}$$

Proposition 2.15. *ψ is a real-analytic diffeomorphism of D satisfying $\psi(\psi(z)) = z$ for all $z \in D$. The unique fixed point of ψ is o.*

Proof. Let $w = \psi(z) = g_z^{-1}(o)$. Then $g_w(o) = w = g_z^{-1}(o)$ hence $g_w = g_z^{-1}$. Thus $\psi(w) = g_z(o) = z$, i.e. $\psi(\psi(z)) = z$. The fact that ψ is a real-analytic diffeomorphism follows from the fact that $NA \ni g \mapsto g(o) \in D$ is a real-analytic diffeomorphism. Finally, $\psi(z) = z$ if and only if $g_z \circ g_z = 1_G$. Since $g_z \in NA$, this is equivalent to $g_z = 1_G$, and so $z = g_z(o) = 1_G(o) = o$. □

In view of Proposition 2.15 we call ψ a *reflection* of D at o. Unless $D = \mathbb{C}^d$, ψ is not holomorphic, and thus not a member of G.

Lemma 2.16. *There exists an involution $\underline{\lambda} \mapsto \underline{\lambda}^*$ on the Lie algebra $\mathfrak{A}^{*\mathbb{C}} \equiv \mathbb{C}^r$ such that*

$$\int_D \overline{e_{\underline{\lambda}}(\xi)}\, f(\psi(\xi))\, d\mu_0(\xi) = \int_D e_{\underline{\lambda}^*}(\xi)\, f(\xi)\, d\mu_0(\xi) = \tilde{f}(\underline{\lambda}^*)$$

holds for all admissible functions f on D and $\underline{\lambda} \in \mathbb{C}^r$.

Proof. In the case of the Fock space on $\mathbb{C}^d$ (see Subsection 1.2 above) we have $g_\xi(z) = z + \xi$ and thus $\psi(\xi) = (g_\xi)^{-1}(0) = -\xi$. Hence, for $a, b \in \mathbb{C}^d$ with $\langle a, b\rangle = \lambda$ one can take $e_\lambda(z) = e_{a,b}(z) = \exp(\langle a, z\rangle + \langle z, b\rangle)$. Therefore one obtains

$$\int_{\mathbb{C}^d} \overline{e_{a,b}(\xi)}\, f(\psi(\xi))\, d\mu_0(\xi) = \int_{\mathbb{C}^d} e_{-b,-a}(\xi)\, f(\xi)\, d\mu_0(\xi). \tag{2.25}$$

Since $\langle -b, -a\rangle = \bar{\lambda}$ the involution $*$ satisfying (2.25) is

$$\lambda^* := \bar{\lambda}. \tag{2.26}$$

In the case of a symmetric tube domain $T(\Omega) \subset \mathbb{C}^d$ we shall prove in Lemma 2.17 below that

$$(\widetilde{f \circ \psi})(\underline{\lambda}) = \tilde{f}(-\underline{\lambda}). \tag{2.27}$$

Since $\overline{e_{\underline{\lambda}}(z)} = e_{\underline{\bar{\lambda}}}(z)$ in this case, we obtain

$$\int_{T(\Omega)} \overline{e_{\underline{\lambda}}(\xi)}\, f(\psi(\xi))\, d\mu_0(\xi) = \int_{T(\Omega)} e_{-\underline{\bar{\lambda}}}(\xi)\, f(\xi)\, d\mu_0(\xi).$$

Hence Lemma 2.16 holds with the involution

$$\underline{\lambda}^* = -\overline{\underline{\lambda}}\,. \tag{2.28}$$

□

It remains to prove (2.27) in the setting of symmetric tube domains.

Lemma 2.17. *In the context of symmetric tube domains $T(\Omega)$, the map ψ satisfies*

$$\mathrm{Det}_{\mathbb{R}}((d\psi)(z)) = \mathrm{N}_{2\underline{\rho}-2p}(\tau(z)).$$

Proof. The *evaluation map*

$$\varepsilon : NA \to T(\Omega), \quad \varepsilon(g) = g(ie)$$

is a real-analytic diffeomorphism whose differential

$$d\varepsilon(\mathbf{1}) : \mathfrak{a} \oplus \mathfrak{n} = T_{\mathbf{1}}(NA) \to Z = T_{ie}(T(\Omega))$$

is also an evaluation map

$$d\varepsilon(\mathbf{1})(X) = X(ie), \quad \forall\, X \in \mathfrak{a} \oplus \mathfrak{n}$$

which is a real-linear isomorphism of $\mathfrak{a} \oplus \mathfrak{n}$ onto Z. For every $g \in NA$ consider the inner automorphism

$$c_g(h) = ghg^{-1}, \quad h \in G$$

and its differential

$$\mathrm{Ad}(g) = dc_g(\mathbf{1}) : \mathfrak{a} \oplus \mathfrak{n} \to \mathfrak{a} \oplus \mathfrak{n}\,.$$

Let $j : NA \to NA$ be the inversion map $j(h) = h^{-1}$. We claim that for every $g \in NA$ we have

$$g^{-1} \circ \varepsilon \circ j \circ c_g = \psi \circ g \circ \varepsilon, \tag{2.29}$$

namely the following diagram is commutative

$$\begin{array}{ccccccc} NA & \xrightarrow{c_g} & NA & \xrightarrow{j} & NA & \xrightarrow{\varepsilon} & T(\Omega) \\ \varepsilon \downarrow & & & & & & \downarrow g^{-1} \\ T(\Omega) & \xrightarrow{g} & & T(\Omega) & & \xrightarrow{\psi} & T(\Omega) \end{array} \tag{2.30}$$

Indeed, for every $h \in NA$

$$\begin{aligned}(g^{-1} \circ \varepsilon \circ j \circ c_g)(h) &= g^{-1}(((ghg^{-1})^{-1})(ie)) \\ &= g^{-1}(g(h^{-1}(g^{-1}(ie)))) = h^{-1}(g^{-1}(ie)).\end{aligned}$$

Also,

$$(\psi \circ g \circ \varepsilon)(h) = \psi(g(h(ie))) = (gh)^{-1}(ie) = h^{-1}(g^{-1}(ie)),$$

and (2.29) is proved. Since $T_w(T(\Omega)) = Z$ for all $w \in T(\Omega)$ and $dj(\mathbf{1}) = -I$, we obtain from (2.30) by taking differentials the following commutative diagram

$$\begin{array}{ccccccc} \mathfrak{a}\oplus\mathfrak{n} & \xrightarrow{\mathrm{Ad}(g)} & \mathfrak{a}\oplus\mathfrak{n} & \xrightarrow{-I} & \mathfrak{a}\oplus\mathfrak{n} & \xrightarrow{d\varepsilon(\mathbf{1})} & Z \\ d\varepsilon(\mathbf{1})\downarrow & & & & & & \downarrow d(g^{-1})(ie) \\ Z & \xrightarrow{dg(ie)} & & Z & & \xrightarrow{d\psi(z)} & Z, \end{array} \tag{2.31}$$

where $z := g(ie)$. Identifying $\mathfrak{a}\oplus\mathfrak{n}$ and Z via the evaluation map $d\varepsilon(\mathbf{1})$, we obtain from (2.31) by taking determinants

$$\mathrm{Det}_{\mathbb{R}}(\mathrm{Ad}(g))\,\mathrm{Det}_{\mathbb{R}}(d(g^{-1})(ie)) = \mathrm{Det}_{\mathbb{R}}(dg(ie))\,\mathrm{Det}_{\mathbb{R}}(d\psi(z)).$$

Now the well-known modulus function of NA, determined in a general setting in [AU99], yields

$$\mathrm{Det}_{\mathbb{R}}\,\mathrm{Ad}(g) = \mathrm{N}_{2\underline{\rho}}(\tau(z)).$$

Since $\mathrm{Det}_{\mathbb{R}}(dg(ie)) = \mathrm{N}(\tau(z))^p$ and $\mathrm{Det}_{\mathbb{R}}(dg^{-1}(ie)) = \mathrm{N}(\tau(z)^{-1})^p = \mathrm{N}(\tau(z))^{-p}$, we obtain

$$\mathrm{Det}_{\mathbb{R}}(d\psi(z)) = \frac{\mathrm{Det}_{\mathbb{R}}(\mathrm{Ad}(g))}{\mathrm{N}(\tau(z))^{2p}} = \mathrm{N}_{2\underline{\rho}-2p}(\tau(z)). \qquad \square$$

Corollary 2.18. *The G-invariant measure μ_0 on $T(\Omega)$ transforms under ψ according to the rule*

$$d\mu_0(\psi(z)) = \mathrm{N}_{2\underline{\rho}}(\tau(z))\,d\mu_0(z).$$

Proof. (1.27) shows that

$$\mathrm{N}(\tau(\psi(z))) = \mathrm{N}(\tau(g_z^{-1}(ie))) = \mathrm{N}(\tau(g_z(ie)))^{-1} = \mathrm{N}(\tau(z))^{-1}.$$

Therefore we obtain

$$\begin{aligned} d\mu_0(\psi(z)) &= \mathrm{N}(\tau(\psi(z)))^{-p} dm(\psi(z)) \\ &= \mathrm{N}(\tau(z))^p\ \mathrm{N}_{2\underline{\rho}-2p}(\tau(z))\,dm(z) = \mathrm{N}_{2\underline{\rho}}(\tau(z))\,d\mu_0(z). \end{aligned} \qquad \square$$

Corollary 2.19. *For every $\underline{\lambda} \in \mathbb{C}^r$ and $z \in T(\Omega)$*

$$e_{\underline{\lambda}}(\psi(z)) = e_{-\underline{\lambda}-\underline{\rho}}(z).$$

Proof. For every $\underline{\alpha} \in \mathbb{C}^r$ and $g \in NA$ we get by (1.27)

$$1 = \mathrm{N}_{\underline{\alpha}}(\tau(ie)) = \mathrm{N}_{\underline{\alpha}}(\tau(g(g^{-1}(ie)))) = \mathrm{N}_{\underline{\alpha}}(\tau(g(ie)))\ \mathrm{N}_{\underline{\alpha}}(\tau(g^{-1}(ie))).$$

Thus

$$\mathrm{N}_{\underline{\alpha}}(\tau(g^{-1}(ie))) = \frac{1}{\mathrm{N}_{\underline{\alpha}}(\tau(g(ie)))} = \mathrm{N}_{-\underline{\alpha}}(\tau(g(ie))).$$

Putting $z = g(ie)$ and using (1.28), this implies

$$e_{\underline{\lambda}}(\psi(z)) = \mathrm{N}_{\underline{\lambda}+\underline{\rho}}(\tau(g^{-1}(ie))) = \mathrm{N}_{-\underline{\lambda}-\underline{\rho}}(\tau(z)) = e_{-\underline{\lambda}-2\underline{\rho}}(z). \qquad \square$$

Corollary 2.20. *Let f be a locally integrable function on $T(\Omega)$ and $\underline{\lambda} \in \mathbb{C}^r$. Then $\underline{\lambda} \in \mathrm{Dom}\,(\widetilde{f \circ \psi})$ if and only if $-\underline{\lambda} \in \mathrm{Dom}\,(\tilde{f})$ and*

$$\widetilde{f \circ \psi}\,(\underline{\lambda}) = \tilde{f}(-\underline{\lambda}).$$

Proof.

$$\begin{aligned}
(\widetilde{f \circ \psi})\,(\underline{\lambda}) &= \int_{T(\Omega)} f(\psi(z))\, e_{\underline{\lambda}}(z)\, d\mu_0(z) = \int_{T(\Omega)} f(w)\, e_{\underline{\lambda}}(\psi(w))\, d\mu_0(\psi(w)) \\
&= \int_{T(\Omega)} f(w)\, e_{-\underline{\lambda}-2\underline{\rho}}(w)\ \mathrm{N}_{2\underline{\rho}}(\tau(w))\, d\mu_0(w) \\
&= \int_{T(\Omega)} f(w)\, e_{-\underline{\lambda}}(w)\, d\mu_0(w) \\
&= \tilde{f}(-\underline{\lambda}). \qquad \square
\end{aligned}$$

2.4. Eigenvalues of the link transforms

Let $\mathcal{A}$ be an invariant symbolic calculus on $\mathcal{H} = L^2_a(D, \mu)$, and let β, F, $\mathcal{A}'$ and $\mathcal{B}$ be as in previous sections. Recall that the "exponential functions" $e_{\underline{\lambda}}$, $\underline{\lambda} \in \mathbb{C}^r$, are joint eigen-functions of the G-invariant operators on D, and that the K-averages of the $e_{\underline{\lambda}}$'s are the spherical functions $\phi_{\underline{\lambda}}(z) = \int_K e_{\underline{\lambda}}(k(z))\, dk$. In particular, if $\mathcal{B}$ is a link transform then for every $e_{\underline{\lambda}} \in \mathrm{Dom}\,(\mathcal{B})$

$$\mathcal{B}(e_{\underline{\lambda}}) = \tilde{\mathcal{B}}(\underline{\lambda})\, e_{\underline{\lambda}}, \quad \text{where } \tilde{\mathcal{B}}(\underline{\lambda}) = \mathcal{B}(e_{\underline{\lambda}})(o).$$

In what follows we shall use the notation

$${}_zF(w) := F(z, w), \quad \forall z, w \in D$$

as well as the notion of the transform $\tilde{f}$ of functions f on D (see (1.11)).

Proposition 2.21. *For $e_{\underline{\lambda}} \in \mathrm{Dom}\,(\mathcal{B})$,*

$$\tilde{\mathcal{B}}(\underline{\lambda}) = \int_D {}_z\tilde{F}(\underline{\lambda}) \left(\widetilde{{}_zF \circ \psi}\right)(\underline{\lambda})\, |K(z, o)|^2\, d\mu(z) \tag{2.32}$$

where ${}_z\tilde{F}(\underline{\lambda})$ and $\left(\widetilde{{}_zF \circ \psi}\right)(\underline{\lambda})$ are defined via (1.11).

Proof. Using (2.20) with $b = e_{\underline{\lambda}}$ and $\eta = o$, as well as (2.22) we find

$$\begin{aligned}\tilde{\mathcal{B}}(\underline{\lambda}) &= \int_D e_{\underline{\lambda}}(\xi)\, \langle B_\xi, B_o\rangle_{S_2}\, d\mu_0(\xi) \\ &= \int_D e_{\underline{\lambda}}(\xi) \left\{ \int_D \left(\int_D F(g_w^{-1}(z), g_w^{-1}(\xi))\, \overline{F(g_w^{-1}(z), g_w^{-1}(o))} \right.\right. \\ &\qquad\qquad \left.\left. |K(z,w)|^2\, d\mu(z) \right) d\mu(w) \right\} d\mu_0(\xi)\,.\end{aligned}$$

Letting $u = g_w^{-1}(z)$, then

$$K(z,w) = K(g_w(u), g_w(o)) = K(u,o)/j(g_w,u)\, \overline{j(g_w,o)}\,.$$

Using (1.20) and (1.21) with g_w, we obtain

$$\begin{aligned}&\tilde{\mathcal{B}}(\underline{\lambda}) \\ &= \int_D e_{\underline{\lambda}}(\xi) \int_D \left[\int_D {}_uF(g_w^{-1}(\xi))\, \overline{{}_uF(g_w^{-1}(o))}\, |K(u,o)|^2\, d\mu(u) \right] d\mu_0(w)\, d\mu_0(\xi).\end{aligned}$$

Interchanging the order of integration, we obtain

$$\begin{aligned}&\tilde{\mathcal{B}}(\underline{\lambda}) \\ &= \int_D |K(u,o)|^2\, d\mu(u) \int_D d\mu_0(w) \left[\int_D {}_uF(g_w^{-1}(\xi))\, e_{\underline{\lambda}}(\xi)\, d\mu_0(\xi) \right] \overline{{}_uF(\psi(w))}.\end{aligned}$$

The substitution $\eta = g_w^{-1}(\xi)$ and the fact that $e_{\underline{\lambda}}(\xi) = e_{\underline{\lambda}}(g_w(\eta)) = e_{\underline{\lambda}}(w)\, e_{\underline{\lambda}}(\eta)$ lead to

$$\begin{aligned}&\tilde{\mathcal{B}}(\underline{\lambda}) \\ &= \int_D |K(u,o)|^2\, d\mu(u) \left[\int_D d\mu_0(w)\, \overline{{}_uF(\psi(w))}\, e_{\underline{\lambda}}(w) \right] \int_D {}_uF(\eta)\, e_{\underline{\lambda}}(\eta)\, d\mu_0(\eta) \\ &= \int_D \widetilde{{}_uF}(\underline{\lambda})\, \widetilde{\overline{{}_uF} \circ \psi}(\underline{\lambda})\, |K(u,o)|^2\, d\mu(u).\end{aligned}$$

□

Formula (2.32) can be extended to the context of the link transform $\mathcal{B}^{T,S} = (\mathcal{A}^S)'\,\mathcal{A}^T$, see (2.23). Indeed, if F^T, F^S are the F-functions associated with $\mathcal{A}^T$ and $\mathcal{A}^S$ respectively then the proof of Proposition 2.21 yields also the following result.

Proposition 2.22. *The eigenvalues of the link transform $\mathcal{B}^{T,S}$ are given by*

$$\widetilde{\mathcal{B}^{T,S}}(\underline{\lambda}) = \int_D \widetilde{{}_zF^T}(\underline{\lambda})\, \widetilde{\overline{{}_zF^S} \circ \psi}(\underline{\lambda})\, |K(z,o)|^2\, d\mu(z).$$

Notice that, by definition,

$$\widetilde{{}_zF^T}(\underline{\lambda}) = \int_D e_{\underline{\lambda}}(\xi) F^T(z,\xi)\, d\mu_0(\xi) = \beta^T(e_{\underline{\lambda}})(z) = A^T_{e_{\underline{\lambda}}}(z,o) = A^T_{e_{\underline{\lambda}}}(K_o)(z)/K_o(z).$$

Also

$$\left(\widetilde{{}_zF^S \circ \psi}\right)(\underline{\lambda}) = \overline{\int_D \overline{e_{\underline{\lambda}}(\xi)}\, F(z, \psi(\xi))\, d\mu_0(\xi)}. \tag{2.33}$$

Using Lemma 2.16, (2.33) can be written as

$$\left(\widetilde{{}_zF^S \circ \psi}\right)(\underline{\lambda}) = \overline{\int_D e_{\underline{\lambda}^*}(\xi)\, F^S(z, \xi)\, d\mu_0(\xi)} = \overline{\mathcal{A}^S_{e_{\underline{\lambda}^*}}(K_o)(z)}.$$

Therefore Proposition 2.22 yields our first main result.

Theorem 2.23. *Let $\mathcal{A}^T$, $\mathcal{A}^S$ be two invariant symbolic calculi associated with the K-invariant operators T, S on $\mathcal{H} = L^2_a(D, \mu)$. Then the eigenvalues of the associated link transform $\mathcal{B}^{T,S} = (\mathcal{A}^S)' \mathcal{A}^T$ are given by*

$$\begin{aligned} \widetilde{\mathcal{B}^{T,S}}(\underline{\lambda}) &= \int_D \mathcal{A}^T_{e_{\underline{\lambda}}}(K_o)(z)\, \overline{\mathcal{A}^S_{e_{\underline{\lambda}^*}}(K_o)(z)}\, d\mu(z) \qquad (2.34) \\ &= \langle \mathcal{A}^T_{e_{\underline{\lambda}}}(K_o), \mathcal{A}^S_{e_{\underline{\lambda}^*}}(K_o)\rangle_{L^2(D,\mu)}. \end{aligned}$$

For weighted Bergman spaces over a symmetric tube domain $T(\Omega)$ we obtain in particular

Theorem 2.24. *The eigenvalues of the link transform $\mathcal{B}$ associated with the invariant symbolic calculus $\mathcal{A}$ on $L^2_a(T(\Omega), \mu_\nu)$ are given by*

$$\tilde{\mathcal{B}}(\underline{\lambda}) = \langle \mathcal{A}_{e_{\underline{\lambda}}}(K_{ie}), \mathcal{A}_{e_{-\overline{\underline{\lambda}}}}(K_{ie})\rangle_{L^2_a(T(\Omega),\mu_\nu)}.$$

3. The Wick calculus, the fundamental function, and eigenvalues of link transforms

3.1. Sesqui-holomorphic extension of real analytic functions and the Wick calculus

Let D, G, $\mathcal{H}$, $K(z, w)$, U, μ, μ_0 etc. be as in the previous sections. Recall that a *sesqui-holomorphic* function $f(z, w)$ on $D \times D$ is a function which is holomorphic in z and anti-holomorphic in w. The following result is well-known.

Lemma 3.1. *Every sesqui-holomorphic function $f(z, w)$ on $D \times D$ is determined by its restriction to the "diagonal" $f(z, z)$. Namely, if $f(z, z) = 0$ for all $z \in D$, then $f(z, w) = 0$ for all $z, w \in D$ as well.*

We proceed with the definition of the *Wick calculus* $\mathcal{E}$. We denote by $\mathrm{Dom}(\mathcal{E})$ the space of all real analytic functions $\varphi : D \to \mathbb{C}$ for which there exists a (unique) sesqui-holomorphic function E_φ on all of $D \times D$ satisfying

$$E_\varphi(z, z) = \varphi(z), \quad \forall z \in D.$$

We call the map $\varphi \mapsto E_\varphi$ (defined on Dom($\mathcal{E}$)) the *extension operator*. Note that Dom($\mathcal{E}$) is a linear space, and it contains all real analytic functions on D of the form $f(z) = f_1(z)\,\overline{f_2(z)}$ where f_1 and f_2 are holomorphic in D (since then $E_f(z, w) = f_1(z)\,\overline{f_2(w)}$). Moreover, Dom($\mathcal{E}$) contains all exponential functions $e_{\underline{\lambda}}$, $\underline{\lambda} \in C^r$. This can be obtained via case by case considerations (using (1.23) in the case of the Fock space or (1.28) in the case of the weighted Bergman spaces over symmetric tube domains), or via the general formula (1.4)) and the relationship between the complex structures of G and D.

We define a map $\mathcal{E}$: Dom($\mathcal{E}$) $\to$ Op(span$\{K_w;\ w \in D\}$) $\subset$ Op($\mathcal{H}$) via its action on the kernel functions

$$\mathcal{E}_\varphi(K_w)(z) := E_\varphi(z, w)\,K(z, w), \quad \forall z, w \in D.$$

Lemma 3.2. *$\mathcal{E}$ is an invariant symbolic calculus, i.e.* Dom($\mathcal{E}$) *is G-invariant and*

$$U(g)\,\mathcal{E}_\varphi\,U(g)^{-1} = \mathcal{E}_{\varphi\circ g^{-1}}, \quad \forall \varphi \in \mathrm{Dom}(\mathcal{E}),\ \forall g \in G.$$

Proof. Since G consists of biholomorphic automorphisms of D, it is clear that whenever $\varphi \in$ Dom($\mathcal{E}$) and $g \in G$ the function $(z, w) \mapsto E_\varphi(g(z), g(w))$ is sesqui-holomorphic in all of $D \times D$. Moreover, $E_\varphi(g(z), g(z)) = \varphi(g(z))$ for all $z \in D$. Thus, $\varphi \circ g \in$ Dom($\mathcal{E}$) and

$$E_{\varphi\circ g}(z, w) = E_\varphi(g(z), g(w)), \quad \forall z, w \in D.$$

Thus Dom($\mathcal{E}$) is G-invariant and E is G-covariant (cf. (2.3)). From this it is clear by the discussion following (2.3) that $\mathcal{E}$ is the invariant symbolic calculus associated with E. □

Definition 3.1. The invariant symbolic calculus $\mathcal{E}$ is called the *Wick calculus.*

The name "Wick calculus" is justified by the following result.

Lemma 3.3. *Let $\varphi \in$* Dom($\mathcal{E}$) *admit a factorization $\varphi(z) = \varphi_1(z)\,\overline{\varphi_2(z)}$, $z \in D$, where $\varphi_1, \varphi_2 \in \mathcal{H}$. Then for the Toeplitz calculus $\mathcal{T}$ we have*

$$\mathcal{E}_\varphi = \mathcal{T}_{\varphi_1}\,\mathcal{T}_{\overline{\varphi_2}}. \tag{3.1}$$

Proof. $\mathcal{T}_{\varphi_1}$, $\mathcal{T}_{\overline{\varphi_2}}$ are well-defined operators, at least on span$\{K_w;\ w \in D\}$, and it is well known and easy to prove that

$$\mathcal{T}_{\overline{\varphi_2}}(K_w) = \overline{\varphi_2(w)}\,K_w, \quad \forall w \in D,$$

whereas $\mathcal{T}_{\varphi_1}$ is the operator of multiplication by φ_1. Therefore we obtain for all $z, w \in D$

$$\begin{aligned}(\mathcal{T}_{\varphi_1}\,\mathcal{T}_{\overline{\varphi_2}})(K_w)(z) &= \varphi_1(z)\,\mathcal{T}_{\overline{\varphi_2}}(K_w)(z) = \varphi_1(z)\,\overline{\varphi_2(w)}\,K(z, w)\\ &= E_\varphi(z, w)\,K(z, w) = \mathcal{E}_\varphi(K_w)(z),\end{aligned}$$

and this yields (3.1). □

Remark. It follows from (3.1) that if $\varphi \in \mathrm{Dom}(\mathcal{E})$ admits a representation $\varphi = \sum_{j=1}^{n} f_j \overline{h_j}$ with $f_j, h_j \in \mathcal{H}$, then $\mathcal{E}_\varphi = \sum_{j=1}^{n} \mathcal{T}_{f_j} \mathcal{T}_{\overline{h_j}}$, regardless of the representation of φ. This does not look obvious at first glance.

3.2. The fundamental function associated with an invariant symbolic calculus

The following Lemma is a key result. It indicates the central role played by the Wick calculus in our theory.

Lemma 3.4. *Let $\mathcal{A}$ be an invariant symbolic calculus. Define*

$$\mathfrak{a}_{\mathcal{A}}(\underline{\lambda}) = \mathcal{A}_{e_{\underline{\lambda}}}(K_o)(o), \quad \forall e_{\underline{\lambda}} \in \mathrm{Dom}(\mathcal{A}).$$

Then

$$\mathcal{A}_{e_{\underline{\lambda}}}(K_w)(z) = \mathfrak{a}_{\mathcal{A}}(\underline{\lambda})\, \mathcal{E}_{e_{\underline{\lambda}}}(K_w)(z) = \mathfrak{a}_{\mathcal{A}}(\underline{\lambda})\, E_{e_{\underline{\lambda}}}(z, w)\, K(z, w) \tag{3.2}$$

for all $e_{\underline{\lambda}} \in \mathrm{Dom}(\mathcal{A})$ and $z, w \in D$. Thus the Berezin symbol $A_{e_{\underline{\lambda}}}$ of $\mathcal{A}_{e_{\underline{\lambda}}}$ is given by

$$A_{e_{\underline{\lambda}}}(z, w) = \mathfrak{a}_{\mathcal{A}}(\underline{\lambda})\, E_{e_{\underline{\lambda}}}(z, w), \quad z, w \in D.$$

Proof. Both sides of (3.2) are sesqui-holomorphic in (z, w). In view of Lemma 3.1 it suffices to prove (3.2) for $z = w$. Let $e_{\underline{\lambda}} \in \mathrm{Dom}(\mathcal{A})$ and $z \in D$. Let $g_z \in NA \subset G$ be the unique element for which $g_z(o) = z$. Using (1.18), (2.1), (1.5) and (1.19) we obtain

$$\begin{aligned}\mathcal{A}_{e_{\underline{\lambda}}}(K_z)(z) &= \frac{U(g_z^{-1})(\mathcal{A}_{e_{\underline{\lambda}}}(U(g_z)K_o))(o)}{j(g_z, o)\, \overline{j(g_z, o)}} \\ &= \frac{\mathcal{A}_{e_{\underline{\lambda}} \circ g_z}(K_o)(o)}{|j(g_z, o)|^2} = \mathcal{A}_{e_{\underline{\lambda}}}(K_o)(o)\, e_{\underline{\lambda}}(z)\, K(z, z).\end{aligned}$$

This completes the proof. □

Corollary 3.5. *The map $\mathcal{A} \mapsto \mathfrak{a}_{\mathcal{A}}$ is injective. Thus $\mathfrak{a}_{\mathcal{A}}$ completely determines $\mathcal{A}$.*

Proof. Since $\mathrm{span}\{K_w;\ w \in D\}$ is dense in $\mathcal{H}$ it suffices to show that the function $\mathfrak{a}_{\mathcal{A}}$ determines the action of $\mathcal{A}_b$ on the kernel functions for all $b \in \mathrm{Dom}(\mathcal{A})$. To this end we claim first that for every $b \in \mathcal{X}_{\underline{\lambda}} \cap \mathrm{Dom}(\mathcal{A})$ (see (1.6))

$$\mathcal{A}_b(K_w)(z) = \mathfrak{a}_{\mathcal{A}}(\underline{\lambda})\, \mathcal{E}_b(K_w)(z) = \mathfrak{a}_{\mathcal{A}}(\underline{\lambda})\, E_b(z, w)\, K(z, w), \quad \forall z, w \in D. \tag{3.3}$$

Indeed, it is enough to prove this for $b = e_{\underline{\lambda}} \circ g$ for some $g \in G$, and in this case (1.19) and (3.2) yield

$$\begin{aligned}\mathcal{A}_b(K_w)(z) &= \big(U(g^{-1})\, \mathcal{A}_{e_{\underline{\lambda}}}\, U(g)\big)(K_w)(z) \\ &= j(g, z)\, \overline{j(g, w)}\, \mathcal{A}_{e_{\underline{\lambda}}}(K_{g(w)})(g(z))\end{aligned}$$

$$= \mathfrak{a}_{\mathcal{A}}(\underline{\lambda})\, E_{e_{\underline{\lambda}}}(g(z), g(w))\, j(g,z)\, \overline{j(g,w)}\, K(g(z), g(w))$$
$$= \mathfrak{a}_{\mathcal{A}}(\underline{\lambda})\, E_{e_{\underline{\lambda}}\circ g}(z,w)\, K(z,w).$$

Since $\mathrm{span}\{\cup_{\underline{\lambda}\in\mathbb{C}^r} \mathcal{X}_{\underline{\lambda}} \cap \mathrm{Dom}(\mathcal{A})\}$ is dense in $\mathrm{Dom}(\mathcal{A})$, (3.3) shows that $\mathfrak{a}_{\mathcal{A}}$ determines $\mathcal{A}$. □

Definition 3.2. The function $\mathfrak{a}_{\mathcal{A}}$ is called the *fundamental function* associated with the calculus $\mathcal{A}$.

As will become clear below, the eigenvalues of the link transforms can be expressed conveniently via the fundamental functions of the associated calculi. Notice also that we clearly have

$$\mathfrak{a}_{\mathcal{E}}(\underline{\lambda}) = 1, \quad \forall \underline{\lambda} \in \mathbb{C}^r.$$

This follows easily from (3.2).

Let B be the K-invariant operator on $\mathcal{H}$ which determines $\mathcal{A}$ via Corollary 2.6, i.e. $\mathcal{A} = \mathcal{A}^B$. For simplicity we write also $\mathfrak{a}_{\mathcal{A}^B}(\underline{\lambda}) = \mathfrak{a}_B(\underline{\lambda})$. Let $\{B_\xi\}_{\xi\in D}$ be the covariant field of operators generated by B via (2.11). Let us define

$$\mathfrak{b}_{\mathcal{A}}(\xi) := B_\xi(K_o)(o) = \langle B_\xi(K_o), K_o\rangle_{\mathcal{H}}, \quad \xi \in D.$$

Lemma 3.6. (i) *The function $\mathfrak{b}_{\mathcal{A}}(\xi)$ is K-invariant in the sense that $\mathfrak{b}_{\mathcal{A}}(k(\xi)) = \mathfrak{b}_{\mathcal{A}}(\xi)$ for all $k \in K$ and $\xi \in D$.*

(ii) *$\mathfrak{a}_{\mathcal{A}}$ is the spherical Fourier transform of $\mathfrak{b}_{\mathcal{A}}$, i.e., for $e_{\underline{\lambda}} \in \mathrm{Dom}(\mathcal{A})$ we have*

$$\mathfrak{a}_{\mathcal{A}}(\underline{\lambda}) = \widetilde{\mathfrak{b}_{\mathcal{A}}}(\underline{\lambda}) = \int_D e_{\underline{\lambda}}(\xi)\, \mathfrak{b}_{\mathcal{A}}(\xi)\, d\mu_0(\xi). \tag{3.4}$$

(iii) *$\mathfrak{a}_{\mathcal{A}}$ is invariant under the Weyl group W:*

$$\mathfrak{a}_{\mathcal{A}}(w(\underline{\lambda})) = \mathfrak{a}_{\mathcal{A}}(\underline{\lambda}), \quad \forall w \in W. \tag{3.5}$$

(iv) *Let $\psi(\xi) = g_\xi^{-1}(o)$, see* (2.24) *and Proposition* 2.15. *Then*

$$\mathfrak{b}_{\mathcal{A}}(\xi) = B(K_{\psi(\xi)})(\psi(\xi))\, K(\psi(\xi), \psi(\xi)), \quad \forall \xi \in D.$$

Proof. (i) Let $\xi \in D$ and $k \in K$. Then $B_{k(\xi)} = U(k)\, B_\xi\, U(k)^{-1}$. Hence, (2.11) yields

$$\begin{aligned} \mathfrak{b}_{\mathcal{A}}(k(\xi)) &= \langle B_\xi U(k)^{-1}(K_o), U(k)^{-1}(K_o)\rangle_{\mathcal{H}} \\ &= \langle B_\xi(\overline{j(k^{-1},o)}K_o), \overline{j(k^{-1},o)}K_o\rangle_{\mathcal{H}} = \langle B_\xi(K_o), K_o\rangle_{\mathcal{H}} = \mathfrak{b}_{\mathcal{A}}(\xi), \end{aligned}$$

since $j(k^{-1}, o)$ is unimodular and $k^{-1}(o) = o$.

(ii) This follows from the definition of $\mathfrak{a}_{\mathcal{A}}$:

$$\mathfrak{a}_{\mathcal{A}}(\underline{\lambda}) = \mathcal{A}_{e_{\underline{\lambda}}}(K_o)(o) = \int_D e_{\underline{\lambda}}(\xi)\, B_\xi(K_o)(o)\, d\mu_0(\xi) = \int_D e_{\underline{\lambda}}(\xi)\, \mathfrak{b}_{\mathcal{A}}(\xi)\, d\mu_0(\xi).$$

(iii) This follows by (i), (ii) and (1.8). Indeed, let $w \in W$, then

$$\mathfrak{a}_{\mathcal{A}}(\underline{\lambda}) = \int_D \phi_{\underline{\lambda}}(\xi)\, \mathfrak{b}_{\mathcal{A}}(\xi)\, d\mu_0(\xi) = \int_D \phi_{w(\underline{\lambda})}(\xi)\, \mathfrak{b}_{\mathcal{A}}(\xi)\, d\mu_0(\xi) = \mathfrak{a}_{\mathcal{A}}(w(\underline{\lambda})).$$

(iv) Using (2.11) we obtain for all $\xi \in D$

$$\begin{aligned} \mathfrak{b}_{\mathcal{A}}(\xi) &= B_\xi(K_o)(o) = (U(g_\xi)\, B\, U(g_\xi^{-1}))(K_o)(o) \\ &= j(g_\xi^{-1}, o)\, B(\overline{j(g_\xi, o)}\, K_{\psi(\xi)})(\psi(\xi)) = B(K_{\psi(\xi)})(\psi(\xi))\, K(\psi(\xi), \psi(\xi)). \end{aligned}$$

□

The proof of Corollary 3.5 yields the following result.

Proposition 3.7. *Let $\mathfrak{a}$ be a W-invariant holomorphic function, defined on a W-invariant subset of $\mathbb{C}^r$. Then there exists a unique invariant symbolic calculus $\mathcal{A}$ on $\mathcal{H}$ for which $\mathfrak{a} = \mathfrak{a}_{\mathcal{A}}$.*

Proof. Notice first that since $e_{\underline{\lambda}} \in \mathrm{Dom}(\mathcal{E})$ for all $\underline{\lambda}$ we have $\mathcal{X}_{\underline{\lambda}} \subset \mathrm{Dom}(\mathcal{E})$ (cf. (1.6)) by the G-invariance of $\mathrm{Dom}(\mathcal{E})$. For $\underline{\lambda} \in \mathrm{Dom}(\mathfrak{a})$ and $b \in \mathcal{X}_{\underline{\lambda}}$ we define $\mathcal{A}_b$ on the kernel functions via (3.3), and extend it to $\mathrm{span}\{K_w;\, w \in D\}$ by linearity. To check the invariance (2.1) let $g \in G$. Then, for all $z, w \in D$,

$$\begin{aligned} (U(g)\, \mathcal{A}_b\, U(g^{-1}))(K_w)(z) &= j(g^{-1}, z)\, \overline{j(g^{-1}, w)}\, \mathcal{A}_b(K_{g^{-1}(w)})(g^{-1}(z)) \\ &= j(g^{-1}, z)\, \overline{j(g^{-1}, w)}\, \mathfrak{a}(\underline{\lambda})\, E_b(g^{-1}(z), g^{-1}(w))\, K(g^{-1}(z), g^{-1}(w)) \\ &= \mathfrak{a}(\underline{\lambda})\, E_{b\circ g^{-1}}(z, w)\, K(z, w) = \mathcal{A}_{b\circ g^{-1}}(K_w)(z). \end{aligned}$$

We extend $\mathcal{A}$ to $\mathrm{span}\{\cup_{\underline{\lambda}\in \mathrm{Dom}(\mathfrak{a})} \mathcal{X}_{\underline{\lambda}}\}$ by linearity, and define $\mathrm{Dom}(\mathcal{A})$ to be this space. Then $\mathrm{Dom}(\mathcal{A})$ is G-invariant and $\mathcal{A}$ is an invariant symbolic calculus. Finally, the uniqueness of $\mathcal{A}$ follows from (3.3) and the fact that $\mathcal{A}_b$ is determined by its action on the kernel functions K_w, $w \in D$. □

The function $\mathfrak{b}_{\mathcal{A}}(\xi)$ is real analytic and K-invariant on D. The next result shows that such functions are in one-to-one correspondence with the invariant symbolic calculi.

Proposition 3.8. *Let $\mathfrak{b}$ be a K-invariant real analytic function on D. Then there exists a unique covariant field of operators $\{B_\xi\}_{\xi\in D}$ on $\mathcal{H}$ so that $\mathfrak{b}(\xi) = B_\xi(K_o)(o)$ for all $\xi \in D$. Consequently, $\mathfrak{b} = \mathfrak{b}_{\mathcal{A}}$ where $\mathcal{A}$ is the invariant symbolic calculus generated by $\{B_\xi\}_{\xi\in D}$ via (2.8).*

Proof. Fix $\xi \in D$ and define the operator B_ξ in the following steps. First, we define for every $w \in D$

$$B_\xi(K_w)(w) = K(w, w)\, \mathfrak{b}(g^{-1}(\xi)), \quad \text{where } g \in G, g(o) = w.$$

The K-invariance of $\mathfrak{b}$ implies that the definition is independent of the particular g. The real analyticity of $\mathfrak{b}$ implies the real analyticity of the function $w \mapsto B_\xi(K_w)(w)$.

We assume the existence of a unique sesqui-holomorphic extension to $D \times D$, which is denoted by $B_\xi(K_w)(z)$. Now extend this operator by linearity to $\operatorname{span}\{K_w;\ w \in D\}$. The family of operators $\{B_\xi\}_{\xi \in D}$ constructed in this way is covariant. Indeed, let $\xi, w \in D$ and let $h \in G$ so that $h(o) = w$. Then for any $g \in G$,

$$\begin{aligned}(U(g^{-1})\, B_\xi\, U(g))(K_w)(w) &= |j(g,w)|^2\, B_\xi(K_{g(w)})(g(w)) \\ &= |j(g,w)|^2\, K(g(w), g(w))\, b((g \circ h)^{-1}(\xi)) \\ &= K(w,w)\, b(h^{-1}(g^{-1}(\xi))) = B_{g^{-1}}(K_w)(w).\end{aligned}$$

Hence $U(g^{-1})\, B_\xi\, U(g) = B_{g^{-1}(\xi)}$. Finally, the relationship $\mathfrak{b}(\xi) = B_\xi(K_o)(o)$ follows from the definition of B_ξ. □

Proposition 3.9. *Let $\mathcal{A}$ be an invariant symbolic calculus on $\mathcal{H}$ and let $\mathcal{B}$ be a G-invariant operator on D. Define the composition $\mathcal{A}\,\mathcal{B}$ via*

$$\begin{aligned}\operatorname{Dom}(\mathcal{A}\,\mathcal{B}) &= \{b \in \operatorname{Dom}(\mathcal{B});\ \mathcal{B}_b \in \operatorname{Dom}(\mathcal{A})\} \text{ and} \\ (\mathcal{A}\,\mathcal{B})_b &:= \mathcal{A}_{\mathcal{B}(b)}, \quad \forall b \in \operatorname{Dom}(\mathcal{A}\,\mathcal{B}).\end{aligned}$$

Then, (i) *$\mathcal{A}\,\mathcal{B}$ is an invariant symbolic calculus, i.e. it satisfies* (2.1).

(ii) *The fundamental function of $\mathcal{A}\,\mathcal{B}$ is*

$$\mathfrak{a}_{\mathcal{A}\mathcal{B}}(\underline{\lambda}) = \mathfrak{a}_{\mathcal{A}}(\underline{\lambda})\, \widetilde{\mathcal{B}}(\underline{\lambda}).$$

(iii) *Let $\mathcal{A} = \mathcal{U}\,|\mathcal{A}|$ be the polar decomposition of $\mathcal{A}$ with respect to inner products of $L^2(D, \mu_0)$ and $S_2(\mathcal{H})$ respectively, and let $\mathcal{B} = \mathcal{V}\,|\mathcal{B}|$ be the polar decomposition of $\mathcal{B}$ with respect to $L^2(D, \mu_0)$. Then $|\mathcal{A}|$, $|\mathcal{B}|$ and $\mathcal{V}$ are G-invariant operators on D, $\mathcal{U}$ is an invariant symbolic calculus on $\mathcal{H}$, and the polar decomposition of $\mathcal{A}\,\mathcal{B}$ is*

$$\mathcal{A}\,\mathcal{B} = (\mathcal{U}\,\mathcal{V})\,(|\mathcal{A}|\,|\mathcal{B}|),$$

namely $|\mathcal{A}\,\mathcal{B}| = |\mathcal{A}|\,|\mathcal{B}|$ and the partial isometry in the minimal polar decomposition of $\mathcal{A}\,\mathcal{B}$ is $\mathcal{U}\,\mathcal{V}$.

Proof. (i) Since $\operatorname{Dom}(\mathcal{B})$ is G-invariant, $\operatorname{Dom}(\mathcal{B}) \cap \mathcal{X}_{\underline{\lambda}} \neq \{0\}$ implies that $\mathcal{X}_{\underline{\lambda}} \subseteq \operatorname{Dom}(\mathcal{B})$ and $\mathcal{B}_{|\mathcal{X}_{\underline{\lambda}}} = \widetilde{\mathcal{B}}(\underline{\lambda})\, I_{|\mathcal{X}_{\underline{\lambda}}}$. Thus $\mathcal{B} : \operatorname{Dom}(\mathcal{B}) \to \operatorname{Dom}(\mathcal{B})$ and the composition $\mathcal{A}\,\mathcal{B}$ is defined in $\operatorname{Dom}(\mathcal{A}) \cap \operatorname{Dom}(\mathcal{B})$. Next, for every $g \in G$ and $b \in \operatorname{Dom}(\mathcal{A}) \cap \operatorname{Dom}(\mathcal{B})$,

$$\begin{aligned}U(g)(\mathcal{A}\,\mathcal{B})_b\, U(g^{-1}) &= U(g)\, \mathcal{A}_{\mathcal{B}(b)}\, U(g^{-1}) \\ &= \mathcal{A}_{\mathcal{B}(b) \circ g^{-1}} = \mathcal{A}_{\mathcal{B}(b \circ g^{-1})} = (\mathcal{A}\,\mathcal{B})_{b \circ g^{-1}}.\end{aligned}$$

(ii) Let $e_{\underline{\lambda}} \in \operatorname{Dom}(\mathcal{A}) \cap \operatorname{Dom}(\mathcal{B})$. Since $\mathcal{B}(e_{\underline{\lambda}}) = \widetilde{\mathcal{B}}(\underline{\lambda})\, e_{\underline{\lambda}}$, we obtain

$$\mathfrak{a}_{\mathcal{A}\mathcal{B}}(\underline{\lambda}) = (\mathcal{A}\,\mathcal{B})_{e_{\underline{\lambda}}}(K_o)(o) = \mathcal{A}_{\mathcal{B}(e_{\underline{\lambda}})}(K_o)(o) = \widetilde{\mathcal{B}}(\underline{\lambda})\, \mathcal{A}_{e_{\underline{\lambda}}}(K_o)(o) = \widetilde{\mathcal{B}}(\underline{\lambda})\, \mathfrak{a}_{\mathcal{A}}(\underline{\lambda}).$$

(iii) For the modulus of $\mathcal{A}\,\mathcal{B}$ we have $|\mathcal{A}\,\mathcal{B}|^2 = (\mathcal{A}\,\mathcal{B})'(\mathcal{A}\,\mathcal{B}) = \mathcal{B}'\,|\mathcal{A}|^2\,\mathcal{B} = |\mathcal{A}|^2\,|\mathcal{B}|^2$, since all invariant operators commute. Taking the square root and using the commutativity again, we obtain $|\mathcal{A}\,\mathcal{B}| = |\mathcal{A}|\,|\mathcal{B}|$ and

$$\mathcal{A}\,\mathcal{B} = \mathcal{U}\,|\mathcal{A}|\,\mathcal{V}\,|\mathcal{B}| = \mathcal{U}\,\mathcal{V}\,|\mathcal{A}|\,|\mathcal{B}| = \mathcal{U}\,\mathcal{V}\,|\mathcal{A}\,\mathcal{B}|.$$

Since $\mathcal{V}$ maps $\mathrm{Dom}(\mathcal{A}) \cap \mathrm{Dom}(\mathcal{B})$ isometrically onto itself and this space is the domain of definition of $\mathcal{U}$, we see that $\mathcal{U}\,\mathcal{V}$ is a partial isometry, whose kernel is the same as that of $|\mathcal{A}\,\mathcal{B}|$. This completes the proof. □

Every invariant symbolic calculus $\mathcal{A}$ on $\mathcal{H}$ can be factorized in many ways as $\mathcal{A} = \mathcal{A}_1\,\mathcal{B}$ with $\mathcal{A}_1$ invariant symbolic calculus and $\mathcal{B}$ a G-invariant operator on D. Besides the trivial factorization $\mathcal{A} = \mathcal{A}I$ and the factorization coming from the polar decomposition $\mathcal{A} = \mathcal{U}\,|\mathcal{A}|$, there is a canonical factorization in which $\mathcal{A}_1$ is the Wick calculus $\mathcal{E}$.

Proposition 3.10. *Let $\mathcal{A}$ be an invariant symbolic calculus on $\mathcal{H}$. Then*

$$\mathcal{A} = \mathcal{E}\,C_{\mathcal{A}}, \tag{3.6}$$

where $C_{\mathcal{A}}$ is the operator of convolution with $\mathfrak{b}_{\mathcal{A}}$:

$$(C_{\mathcal{A}}f)(z) = \int_D f(g(\xi))\,\mathfrak{b}_{\mathcal{A}}(\xi)\,d\mu_0(\xi), \quad \textit{where } g \in G,\ g(o) = z.$$

Proof. $C_{\mathcal{A}}$ is certainly G-invariant, hence Proposition 3.9 guarantees that $\mathcal{E}\,C_{\mathcal{A}}$ is an invariant symbolic calculus. Moreover, the eigenvalues of $C_{\mathcal{A}}$ are $\widetilde{C_{\mathcal{A}}}(\underline{\lambda}) = \widetilde{\mathfrak{b}_{\mathcal{A}}}(\underline{\lambda}) = \mathfrak{a}_{\mathcal{A}}(\underline{\lambda})$. Hence,

$$\mathfrak{a}_{\mathcal{E}\,C_{\mathcal{A}}}(\underline{\lambda}) = \mathfrak{a}_{\mathcal{E}}(\underline{\lambda})\,\widetilde{C_{\mathcal{A}}}(\underline{\lambda}) = \mathfrak{a}_{\mathcal{A}}(\underline{\lambda}).$$

since $\mathfrak{a}_{\mathcal{E}}(\underline{\lambda}) \equiv 1$. Thus Corollary 3.5 guarantees that (3.6) holds. □

Remark. Propositions 3.8 and 3.10 can be generalized to distributions. We outline this generalization briefly. Let $\mathfrak{b}$ be a K-invariant distribution on D, and let $C_{\mathfrak{b}}$ be the operator of convolution with $\mathfrak{b}$:

$$(C_{\mathfrak{b}}f)(z) := \langle f \circ g, \mathfrak{b}\rangle, \quad \text{where } g \in G,\ g(o) = z.$$

Then $\mathcal{A} := \mathcal{E}\,C_{\mathfrak{b}}$ is an invariant symbolic calculus on $\mathcal{H}$ in its canonical factorization. In the important case of the Dirac measure $\mathfrak{b} = \delta_o$ we obtain $\mathcal{A} = \mathcal{E}$ since $C_{\delta_o} = I$. However, this general approach leads to some open problems which will be discussed in a subsequent publication.

3.3. Eigenvalues of link transforms via the fundamental functions

Proposition 3.11. *Let T be a K-invariant operator on $\mathcal{H} = L^2_a(D, \mu)$ and let $\mathfrak{a}_T$ be the fundamental function associated with the calculus $\mathcal{A}^T$. Let T^* the adjoint of T as an operator on $\mathcal{H}$. Then for all admissible $\underline{\lambda} \in \mathbb{C}^r$ we have*

$$\mathfrak{a}_{T^*}(\underline{\lambda}) = \overline{\mathfrak{a}_T(\overline{\underline{\lambda}})}. \tag{3.7}$$

Proof. Let $\mathfrak{b}_T(\xi) = \mathfrak{b}_{\mathcal{A}^T}(\xi) = \langle T_\xi K_o, K_o\rangle_{\mathcal{H}}$ and define $\mathfrak{b}_{T^*}(\xi)$ similarly. Then, $\mathfrak{b}_{T^*}(\xi) = \overline{\mathfrak{b}_T(\xi)}$. Using the fact that $\overline{e_{\underline{\lambda}}(\xi)} = e_{\overline{\underline{\lambda}}}(\xi)$ we obtain by (3.4)

$$\begin{aligned}\mathfrak{a}_{T^*}(\underline{\lambda}) &= \int_D e_{\underline{\lambda}}(\xi)\, \mathfrak{b}_{T^*}(\xi)\, d\mu_0(\xi)\\ &= \int_D e_{\underline{\lambda}}(\xi)\, \overline{\mathfrak{b}_T(\xi)}\, d\mu_0(\xi) = \overline{\int_D e_{\overline{\underline{\lambda}}}(\xi)\, \mathfrak{b}_T(\xi)\, d\mu_0(\xi)} = \overline{\mathfrak{a}_T(\overline{\underline{\lambda}})}.\end{aligned}$$ □

Remark. In the case where $\mathcal{H}$ is either the weighted Bergman space over a symmetric tube domain $L^2_a(T(\Omega), \mu_\nu)$ or the weighted Fock space $\mathcal{F}_\nu$ we have also

$$\mathfrak{a}_{T^*}(\underline{\lambda}) = \overline{\mathfrak{a}_T(\underline{\lambda}^*)}. \tag{3.8}$$

Indeed, in the case of $L^2_a(T(\Omega), \mu_\nu)$ we have $\underline{\lambda}^* = -\overline{\underline{\lambda}}$ (see Lemma 2.16 and (2.28)), and (3.8) follows from (3.5) and (3.7). In the case of $\mathcal{F}_\nu$, if $a, b \in \mathbb{C}^d$ are so that $\langle a, b\rangle = \lambda$ then $e_{\lambda^*} = e_{-b,-a}$ (see (1.25) and (2.26)). Hence

$$\overline{\mathfrak{a}_T(\lambda^*)} = \overline{\int_{\mathbb{C}^d} e_{-b,-a}(\xi)\, \mathfrak{b}_T(\xi)\, d\mu_0(\xi)} = \int_{\mathbb{C}^d} e_{b,a}(-\xi)\, \overline{\mathfrak{b}T(\xi)}\, d\mu_0(\xi) = \mathfrak{a}_{T^*}(\lambda),$$

since $\mathfrak{a}_T(\xi) = \mathfrak{a}_T(-\xi)$ by Lemma 3.6.

We now combine the results of Subsections 2.3 and 3.2 to obtain our main result.

Theorem 3.12. *Let S, T be K-invariant operators on $\mathcal{H} = L^2_a(D, \mu)$ and let $\mathfrak{a}_S$, $\mathfrak{a}_T$ be the fundamental functions of the associated invariant symbolic calculi $\mathcal{A}^S$ and $\mathcal{A}^T$ respectively. Let $\mathcal{B}^{S,T} = (\mathcal{A}^T)'\,\mathcal{A}^S$ be the corresponding link transform. Then the eigenvalues of $\mathcal{B}^{S,T}$ are expressed in terms of the fundamental functions in the following way*

$$\widetilde{\mathcal{B}^{S,T}}(\underline{\lambda}) = \frac{\mathfrak{a}_S(\underline{\lambda})\, \overline{\mathfrak{a}_T(\underline{\lambda}^*)}}{\mathfrak{a}_{\mathcal{T}}(\underline{\lambda})}, \quad \forall \underline{\lambda} \in \mathrm{Dom}(\mathcal{B}^{S,T}), \tag{3.9}$$

where $\mathfrak{a}_{\mathcal{T}}(\underline{\lambda})$ is the fundamental function associated with the Toeplitz calculus $\mathcal{T}$ and $\underline{\lambda} \mapsto \underline{\lambda}^*$ is the involution whose existence is guaranteed by Lemma 2.16.

Proof. We know by Theorem 2.23 that

$$\widetilde{\mathcal{B}^{S,T}}(\underline{\lambda}) = \int_D \mathcal{A}^S_{e_{\underline{\lambda}}}(K_o)(z)\, \overline{\mathcal{A}^T_{\underline{\lambda}^*}(K_o)(z)}\, d\mu(z).$$

However, Lemma 3.4 says that

$$\mathcal{A}^S_{e_{\underline{\lambda}}}(K_o)(z) = \mathfrak{a}_S(\underline{\lambda})\, E_{e_{\underline{\lambda}}}(z, o)\, K(z, o), \quad \mathcal{A}^T_{e_{\underline{\lambda}^*}}(K_o)(z) = \mathfrak{a}_T(\underline{\lambda}^*)\, E_{e_{\underline{\lambda}^*}}(z, o)\, K(z, o).$$

Thus,

$$\widetilde{\mathcal{B}^{S,T}}(\underline{\lambda}) = \mathfrak{a}_S(\underline{\lambda})\, \overline{\mathfrak{a}_T(\underline{\lambda}^*)}\, c(\underline{\lambda}), \tag{3.10}$$

where

$$c(\underline{\lambda}) = \int_D E_{e_{\underline{\lambda}}}(z, o)\,\overline{E_{e_{\underline{\lambda}*}}(z, o)}\,|K(z, o)|^2\, d\mu(z)$$

is independent of the particular calculi $\mathcal{A}^S$, $\mathcal{A}^T$. Let us apply (3.10) with $\mathcal{A}^S = \mathcal{A}^T = \mathcal{T}$. This yields

$$c(\underline{\lambda}) = \frac{\widetilde{\mathcal{T}'\mathcal{T}}(\underline{\lambda})}{\mathfrak{a}_{\mathcal{T}}(\underline{\lambda})\,\overline{\mathfrak{a}_{\mathcal{T}}(\underline{\lambda}^*)}}.$$

In the case of the Toeplitz calculus we have $\mathcal{T} = \mathcal{A}^{P_o}$, where $P_o := \langle\cdot, K_o\rangle_{\mathcal{H}} K_o$, the projection on $\mathbb{C}K_o$, is a self adjoint operator. Hence $\mathfrak{a}_{\mathcal{T}}(\underline{\lambda}^*) = \mathfrak{a}_{\mathcal{T}}(\underline{\lambda})$, and

$$\widetilde{\mathcal{T}'\mathcal{T}}(\underline{\lambda}) = \langle \mathcal{T}_{e_{\underline{\lambda}}}, P_o\rangle_{S_2(\mathcal{H})} = \mathcal{T}_{e_{\underline{\lambda}}}(K_o)(o) = \mathfrak{a}_{\mathcal{T}}(\underline{\lambda}).$$

Therefore

$$c(\underline{\lambda}) = \frac{1}{\mathfrak{a}_{\mathcal{T}}(\underline{\lambda})}.$$

Using this fact, Proposition 3.11 and (3.10) we obtain (3.9). □

Remark. In the case where $\mathcal{H}$ is either the weighted Bergman space over a symmetric tube domain $L^2_a(T(\Omega), \mu_\nu)$ or the weighted Fock space $\mathcal{F}_\nu$ we can write (3.9) also as

$$\widetilde{\mathcal{B}^{S,T}}(\underline{\lambda}) = \frac{\mathfrak{a}_S(\underline{\lambda})\,\overline{\mathfrak{a}_{\mathcal{T}}(\overline{\underline{\lambda}})}}{\mathfrak{a}_{\mathcal{T}}(\underline{\lambda})}, \quad \forall \underline{\lambda} \in \mathrm{Dom}(\mathcal{B}^{S,T}),$$

Indeed, this follows from (3.9) and (3.8).

Theorem 3.12 is a substantial improvement of Theorem 2.23, since instead of the inner product of the functions $\mathcal{A}^T_{e_{\underline{\lambda}}}(K_o)(z)$, $\mathcal{A}^S_{e_{\underline{\lambda}*}}(K_o)(z)$ one needs only to compute the product $\mathfrak{a}_{\mathcal{T}}(\underline{\lambda})\,\overline{\mathfrak{a}_{S^*}(\underline{\lambda})} = \mathcal{A}^T_{e_{\underline{\lambda}}}(K_o)(o)\,\mathcal{A}^S_{e_{\underline{\lambda}*}}(K_o)(o)$. Theorem 3.12 involves the fundamental function of the Toeplitz calculus $\mathfrak{a}_{\mathcal{T}}(\underline{\lambda})$, which is known in the cases under consideration. Indeed, in the case where $\mathcal{H}$ is the Fock space $\mathcal{F}_\nu$ we have

$$\mathfrak{a}_{\mathcal{T}}(\lambda) = e^{\frac{\lambda}{\nu}},$$

and in the case where $\mathcal{H}$ is the weighted Bergman space $L^2_a(T(\Omega), \mu_\nu)$ we have, by [UU94] (see also [Be78]),

$$\mathfrak{a}_{\mathcal{T}}(\underline{\lambda}) = \frac{\Gamma_\Omega(\underline{\lambda} + \underline{\rho} + \nu - \frac{d}{r})\,\Gamma_\Omega(-\underline{\lambda} + \underline{\rho} + \nu - \frac{d}{r})}{\Gamma_\Omega(\nu - \frac{d}{r})\,\Gamma_\Omega(\nu)} \tag{3.11}$$

We will give a new proof of this result in Section 5.

We close by observing that the results in this section yield also interesting information on the fundamental function of the unitary part of an invariant symbolic calculus $\mathcal{A}$. Let $\mathcal{A} = \mathcal{U}(\mathcal{A}'\mathcal{A})^{\frac{1}{2}}$ be the minimal polar decomposition of $\mathcal{A}$ (thus $\ker(\mathcal{A}) = \ker(\mathcal{U})$). We know that $\mathcal{U}$ itself is also an invariant symbolic calculus. Thus for $e_{\underline{\lambda}} \in \mathrm{Dom}(\mathcal{A})$ we have

$$\mathcal{A}_{e_{\underline{\lambda}}} = \mathcal{U}_{(\mathcal{A}'\mathcal{A})^{\frac{1}{2}} e_{\underline{\lambda}}} = \left(\widetilde{\mathcal{A}'\mathcal{A}}(\underline{\lambda})\right)^{\frac{1}{2}} \mathcal{U}_{e_{\underline{\lambda}}}.$$

From this it follows that

$$\mathfrak{a}_{\mathcal{A}}(\underline{\lambda}) = \left(\widetilde{\mathcal{A}'\mathcal{A}}(\underline{\lambda})\right)^{\frac{1}{2}} \mathfrak{a}_{\mathcal{U}}(\underline{\lambda}) = \left(\frac{\mathfrak{a}_{\mathcal{A}}(\underline{\lambda})\, \overline{\mathfrak{a}_{\mathcal{A}}(\overline{\underline{\lambda}})}}{\mathfrak{a}_{\mathcal{T}}(\underline{\lambda})}\right)^{\frac{1}{2}} \mathfrak{a}_{\mathcal{U}}(\underline{\lambda}).$$

Thus, if $\mathfrak{a}_{\mathcal{A}}(\underline{\lambda}) \neq 0$ then

$$\mathfrak{a}_{\mathcal{U}}(\underline{\lambda}) = \left(\frac{\mathfrak{a}_{\mathcal{A}}(\underline{\lambda})}{\overline{\mathfrak{a}_{\mathcal{A}}(\overline{\underline{\lambda}})}}\right)^{\frac{1}{2}} \mathfrak{a}_{\mathcal{T}}(\underline{\lambda})^{\frac{1}{2}} = \left(\frac{\mathfrak{a}_{\mathcal{A}}(\underline{\lambda})}{\mathfrak{a}_{\mathcal{A}^*}(\underline{\lambda})}\right)^{\frac{1}{2}} \mathfrak{a}_{\mathcal{T}}(\underline{\lambda})^{\frac{1}{2}}.$$

In particular, if $T = T^*$ then

$$\mathfrak{a}_{\mathcal{U}}(\underline{\lambda}) = \mathfrak{a}_{\mathcal{T}}(\underline{\lambda})^{\frac{1}{2}}.$$

4. Application to weighted Fock spaces

In this section we apply the general theory developed above to study various important invariant symbolic calculi in the context of the weighted Fock spaces $\mathcal{F}_\nu$ over $\mathbb{C}^d$ introduced in Subsection 1.2 above. After describing the standard calculi of Wick, Weyl and anti-Wick (Toeplitz) type in the following subsection, we construct new families of calculi, related to the "k - homogeneous" projections in the next subsection, and give a detailed numerical analysis of their spectral properties. This yields a rather surprising interrelationship between different calculi.

4.1. The Toeplitz, Weyl and Wick calculi

Proposition 2.21 yields the following result.

Theorem 4.1. *Let $\mathcal{A}$ be an invariant symbolic calculus on the weighted Fock space $\mathcal{F}_\nu = L^2_a(\mathbb{C}^d, \mu_\nu)$. Then the eigenvalues of the associated link transform $\mathcal{B} = \mathcal{A}'\mathcal{A}$ are given by*

$$\tilde{\mathcal{B}}(\lambda) = \int_{\mathbb{C}^d} \mathcal{A}_{e_{a,b}}(1)(z)\, \overline{\mathcal{A}_{e_{-b,-a}}(1)(z)}\, d\mu_\nu(z) = \langle \mathcal{A}_{e_{a,b}}(1),\ \mathcal{A}_{e_{-b,-a}}(1)\rangle_{\mathcal{F}_\nu} \quad (4.1)$$

where $a, b \in \mathbb{C}^d$ are arbitrary vectors for which $\lambda = \langle a|b\rangle$.

Proof. With the notation as in the previous section, we have

$$
\begin{aligned}
{}_z\tilde{F}(\lambda) &= \int_{\mathbb{C}^d} F(z,w)\, e_{a,b}(w)\, d\mu_0(w) = \beta(e_{a,b})(z) \\
&= \mathcal{A}_{e_{a,b}}(K_o)(z)/K_o(z) = \mathcal{A}_{e_{a,b}}(1)(z),
\end{aligned}
$$

since $K_o(w) \equiv 1$. Similarly,

$$
\begin{aligned}
{}_z\widetilde{\overline{F \circ \psi}}(\lambda) &= \int_{\mathbb{C}^d} \overline{F(z,\psi(w))}\, e_{a,b}(w)\, d\mu_o(w) = \overline{\int_{\mathbb{C}^d} F(z,\xi)\, \overline{e_{a,b}(\psi(\xi))}\, d\mu_0(\psi(\xi))} \\
&= \overline{\int_{\mathbb{C}^d} F(z,\xi)\, e_{-b,-a}(\xi)\, d\mu_0(\xi)} = \overline{\mathcal{A}_{e_{-b,-a}}(1)(z)}.
\end{aligned}
$$

These results imply (4.1) via (2.34). □

The extension of Theorem 4.1 to the link transform $\mathcal{B}^{T,S}$ is a consequence of Proposition 2.22.

Theorem 4.2. *Let $\mathcal{A}^T$, $\mathcal{A}^S$ be invariant symbolic calculi generated by the K-invariant operators T, S on holomorphic functions on $\mathbb{C}^d$. Then the eigenvalues of the associated link transform $\mathcal{B}^{T,S} = (\mathcal{A}^S)'\,\mathcal{A}^T$ are*

$$
\widetilde{\mathcal{B}^{T,S}}(\lambda) = \int_{\mathbb{C}^d} \mathcal{A}^T_{e_{a,b}}(1)(z)\overline{\mathcal{A}^S_{e_{-b,-a}}(1)(z)}\, d\mu_\nu(z) = \langle \mathcal{A}^T_{e_{a,b}}(1),\ \mathcal{A}^S_{e_{-b,-a}}(1)\rangle_{\mathcal{F}_\nu}
$$

where $a, b \in \mathbb{C}^d$ are arbitrary vectors for which $\langle a|b\rangle = \lambda$.

We turn now to some important examples to illustrate the scope of our theory.

Example 4.1. *Toeplitz calculus.* Let $\mathcal{T}_b = \mathcal{T}_b^{(\nu)}$ be the Toeplitz operator with symbol b on $\mathcal{F}_\nu = L^2_a(\mathbb{C}^d, \mu_\nu)$. Then

$$
\mathcal{T}_{e_{a,b}}(1)(z) = e^{\frac{\lambda}{\nu}}\, e^{\langle z|b\rangle},
$$

and in particular

$$
\mathfrak{a}_{\mathcal{T}}(\lambda) = e^{\frac{\lambda}{\nu}}.
$$

Indeed, if $\lambda = \langle a, b\rangle$ then

$$
\mathcal{T}_{e_{a,b}}(1)(z) = \int_{\mathbb{C}^d} e^{\langle w|b\rangle}\, e^{\nu\langle \frac{a}{\nu}+z|w\rangle}\, d\mu_\nu(w) = e^{\frac{\lambda}{\nu}}\, e^{\langle z|b\rangle}.
$$

Similarly,

$$
\mathcal{T}_{e_{-b,-a}}(1)(z) = e^{\frac{\overline{\lambda}}{\nu}}\, e^{-\langle z|a\rangle}.
$$

Hence

$$
\int_{\mathbb{C}^d} \mathcal{T}_{e_{a,b}}(1)(z)\, \overline{\mathcal{T}_{e_{-b,-a}}(1)(z)}\, d\mu_\nu(z) = e^{\frac{2\lambda}{\nu}} \int_{\mathbb{C}^d} e^{\langle z|b\rangle}\, e^{-\langle a|z\rangle}\, d\mu_\nu(z) = e^{\frac{\lambda}{\nu}}.
$$

Therefore the eigenvalues of the *Berezin transform* $\mathcal{B} = \mathcal{T}'\,\mathcal{T}$, i.e.

$$\mathcal{B}(f)(z) = \int_{\mathbb{C}^d} \frac{f(w)\,|K^{(\nu)}(z,w)|^2}{K^{(\nu)}(z,z)}\,d\mu_\nu(w) = \int_{\mathbb{C}^d} f(w)\,e^{-\nu|z-w|^2}\,d\mu_0(w)$$

are given by

$$\tilde{\mathcal{B}}(\lambda) = e^{\frac{\lambda}{\nu}}, \quad \lambda \in \mathbb{C}.$$

From here one concludes that

$$\mathcal{B} = e^{\frac{\Delta}{\nu}}.$$

We remark that in the case of the Berezin transform one can compute the eigenvalues directly. Indeed, if $\lambda = \langle a, b\rangle$ then

$$\tilde{\mathcal{B}}(\lambda) = \mathcal{B}(e_{a,b})(0) = \int_{\mathbb{C}^d} e_{a,b}(w)d\mu_\nu(w) = \int_{\mathbb{C}^d} e^{\langle w|b\rangle + \langle a|w\rangle} d\mu_\nu(w) = e^{\frac{\lambda}{\nu}}.$$

We calculate now the standard functions of our theory in the case of the Toeplitz calculus $\mathcal{T}$. First

$$T_b(z,w) = \mathcal{T}_b(K_w)(z)/K_w(z) = \int_{\mathbb{C}^d} b(\xi)\frac{K(z,\xi)\,K(\xi,w)}{K(z,w)}\,d\mu_\nu(\xi).$$

Hence

$$\beta^{\mathcal{T}}(b)(z) = T_b(z,0) = \int_{\mathbb{C}^d} b(\xi)\frac{K(z,\xi)}{K(\xi,\xi)}\,d\mu_0(\xi)$$

and therefore

$$F^{\mathcal{T}}(z,\xi) = \frac{K(z,\xi)}{K(\xi,\xi)} = e^{\nu\langle z-\xi|\xi\rangle}.$$

From this we conclude that

$$T_\eta(K_w)(z) = \frac{K(z,\eta)\,K(\eta,w)}{K(\eta,\eta)},$$

and therefore $T_\eta f = \langle f, k_\eta\rangle_\nu k_\eta$ for all $f \in \mathcal{F}_\nu$. In particular $T_0 = k_0 \otimes k_0 = 1 \otimes 1$, i.e. $T_0 f = \langle f, 1\rangle 1$. This is the K-invariant operator which determines $\mathcal{T}$. Also, it follows that

$$\mathfrak{b}_{\mathcal{T}}(\xi) = \langle T_\xi k_0, k_0\rangle = |\langle k_\xi, 1\rangle|^2 = e^{-\nu|\xi|^2}, \quad \forall \xi \in \mathbb{C}^d.$$

Example 4.2. *The Weyl calculus.* It is easy to see that the symmetry at w is given by $s_w(z) = -z + 2w$. Therefore the corresponding isometry of $L^2(\mathbb{C}^d, \mu_\nu)$ and $\mathcal{F}_\nu$ is

$$U(s_w)(f)(z) = f(-z+2w)\,e^{-2\nu|w|^2 + 2\nu\langle z|w\rangle}. \tag{4.2}$$

The Weyl calculus is defined by

$$\mathcal{W}_b = \int_{\mathbb{C}^d} b(w)\, U(s_w)\, \widetilde{d\mu_0}(w),$$

where $\widetilde{d\mu_0}(w) = 2^d\, d\mu_0(w) = \left(\frac{2\nu}{\pi}\right)^d\, dm(w)$. Namely,

$$\mathcal{W}_b(f)(z) = \left(\frac{2\nu}{\pi}\right)^d \int_{\mathbb{C}^d} b(w)\, f(-z+2w)\, e^{-2\nu\,|w|^2+2\nu\,(z|w)}\, dm(w).$$

It follows that for $a, b \in \mathbb{C}^d$ with $\lambda = \langle a, b\rangle$

$$\begin{aligned}\mathcal{W}_{e_{a,b}}(1)(z) &= \left(\frac{2\nu}{\pi}\right)^d \int_{\mathbb{C}^d} e^{(w|b)+(a|w)+2\nu\,(z|w)}\, e^{-2\nu\,|w|^2}\, dm(w)\\ &= \int_{\mathbb{C}^d} e^{(w|b)}\, K_{2\nu}\left(z+\frac{a}{2\nu},\, w\right) d\mu_{2\nu}(w) = e^{(z+\frac{a}{2\nu}|b)} = e^{\frac{\lambda}{2\nu}}\, e^{(z|b)}.\end{aligned}$$

In particular, if $\lambda = (a|b)$ then

$$a_{\mathcal{W}}(\lambda) = \mathcal{W}_{e_{a,b}}(1)(0) = e^{\frac{\lambda}{2\nu}}.$$

Similarly,

$$\mathcal{W}_{e_{-b,-a}}(1)(z) = e^{\frac{\bar{\lambda}}{2\nu}}\, e^{-(z|a)},$$

and hence,

$$\begin{aligned}\widetilde{\mathcal{W}'\mathcal{W}}(\lambda) &= \int_{\mathbb{C}^d} \mathcal{W}_{e_{a,b}}(1)(z)\, \overline{\mathcal{W}_{e_{-b,-a}}(1)(z)}\, d\mu_\nu(z)\\ &= e^{\frac{\lambda}{\nu}} \int_{\mathbb{C}^d} e^{(z|b)}\, e^{-(a|z)}\, d\mu_\nu(z) = e^{\frac{\lambda}{\nu}}\, e^{-\frac{(a|b)}{\nu}} = 1.\end{aligned}$$

Thus the eigenvalues of the link transform $\mathcal{W}'\mathcal{W}$ associated with the Weyl calculus are

$$\widetilde{\mathcal{W}'\mathcal{W}}(\lambda) \equiv 1, \quad \forall\, \lambda \in \mathbb{C}.$$

From this we deduce that

$$\mathcal{W}'\mathcal{W} = I,$$

namely $\mathcal{W}$ is an isometry from $L^2(\mathbb{C}^d, \mu_\nu)$ into the Hilbert–Schmidt operators $S_2(\mathcal{F}_\nu)$ on the weighted Fock space $\mathcal{F}_\nu = L^2_a(\mathbb{C}^d, \mu_\nu))$. This result is well-known, but its derivation by direct methods is more involved.

It is easy to compute the standard functions of our theory in the case of the Weyl calculus. First

$$\begin{aligned}W_b(z, w) &= \mathcal{W}_b(K_w)(z)/K_w(z)\\ &= e^{-2\nu(z|w)} \int_{\mathbb{C}^d} b(\xi)\, \frac{e^{2\nu\,(z|\xi)}\, e^{2\nu\,(\xi|w)}}{e^{2\nu\,(\xi|\xi)}}\, d\tilde{\mu}_0(\xi)\end{aligned}$$

$$= e^{-2\nu\langle z|w\rangle} \int_{\mathbb{C}^d} b(\xi)\, K^{(2\nu)}(z,\xi)\, K^{(2\nu)}(\xi,w)\, d\mu_{2\nu}(\xi).$$

Hence

$$\beta^{\mathcal{W}}(b)(z) = W_b(z,0) = \int_{\mathbb{C}^d} b(\xi)\, K^{(2\nu)}(z,\xi)\, d\mu_{2\nu}(\xi) = P^{(2\nu)}(b)(z).$$

Thus, with respect to the invariant measure $d\tilde{\mu}_0(\xi) = \left(\frac{2\nu}{\pi}\right)^d dm(\xi)$,

$$F^{\mathcal{W}}(z,\xi) = e^{2\nu\langle z|\xi\rangle - 2\nu|\xi|^2} = j(s_\xi, z).$$

Therefore we find easily that $W_\eta(K_w)(z) = K_w(s_\eta(z))\, j(s_\eta, z)$. Namely

$$W_\eta = U(s_\eta).$$

In particular, the K-invariant operator on $\mathcal{F}_\nu$ which determines $\mathcal{W}$ is

$$W_0 = U(s_0) = \sum_{k=0}^{\infty} (-1)^k\, P_k.$$

Moreover, using (4.2) we see that the fundamental function $\mathfrak{b}_{\mathcal{W}}(\xi)$ is given by

$$\mathfrak{b}_{\mathcal{W}}(\xi) = \langle W_\xi 1, 1\rangle = U(s_\xi(1)(0) = e^{-2\nu|\xi|^2}.$$

Example 4.3. *Wick calculus*. Let b be a real-analytic function with everywhere convergent Taylor expansion

$$b(z) = \sum_{\alpha,\beta\in\mathbb{N}^d} \frac{\partial^\alpha}{\alpha!}\frac{\overline{\partial}^\beta}{\beta!} b(0)\, z^\alpha \overline{z}^\beta.$$

We define the Wick calculus via

$$\mathcal{E}_b = \sum_{\alpha,\beta\in\mathbb{N}^d} \frac{\partial^\alpha}{\alpha!}\frac{\overline{\partial}^\beta}{\beta!} b(0)\, \mathcal{T}_{z^\alpha}\, \mathcal{T}^*_{z^\beta} \tag{4.3}$$

where $\mathcal{T}_{z^\alpha}$ is the Toeplitz operator with symbol z^α. Notice that this definition is consistent with the definition given in Subsection 3.1. Thus, if $b(z) = \sum_j f_j(z)\overline{g_j(z)}$ with f_j, g_j entire holomorphic functions, then the sesqui-holomorphic extension of b is $E_b(z,w) = \sum_j f_j(z)\overline{g_j(w)}$, and $\mathcal{E}_b = \sum_j \mathcal{T}_{f_j}\mathcal{T}^*_{g_j}$. In particular, if we use the Taylor series of b, we see that

$$\beta^{\mathcal{E}}(b)(z) = E_b(z,0) = \sum_{\alpha\in\mathbb{N}^d} \frac{\partial^\alpha}{\alpha!} b(0)\, z^\alpha.$$

In the special case of the symbol $e_{a,b}(z) = e^{\langle z|b\rangle + \langle a|z\rangle}$ we obtain $\mathcal{E}_{e_{a,b}}(1)(z) = E_{e_{a,b}}(z,0) = e^{\langle z|b\rangle}$, and similarly $\mathcal{E}_{e_{-b,-a}}(1)(z) = e^{-\langle z|a\rangle}$. Hence, if $\langle a,b\rangle = \lambda$

then

$$\int_{\mathbb{C}^d} \mathcal{E}_{e_{a,b}}(1)(z)\,\overline{\mathcal{E}_{e_{-b,-a}}(1)(z)}\,d\mu_\nu(z) = \int_{\mathbb{C}^d} e^{(z|b)}\,e^{-(a|z)}\,d\mu_\nu(z) = e^{-\frac{\lambda}{\nu}}.$$

Thus the eigenvalues of the link transform associated with the Wick calculus are

$$\widetilde{\mathcal{E}'\mathcal{E}}(\lambda) = e^{-\frac{\lambda}{\nu}}.$$

4.2. The calculi associated with the *k*-homogeneous projections

Let $\mathcal{P}_k$ denote the subspace of $\mathcal{F}_\nu$ consisting of all homogeneous polynomials of degree k. It has a reproducing kernel $K_k^{(}z, w) = K_k^{(\nu)}(z, w)$, where

$$K_k(z, w) = K_k^{(\nu)}(z, w) = \frac{\nu^k\,(z|w)^k}{k!}.$$

The orthogonal projection $P_k : \mathcal{F}_\nu \to \mathcal{P}_k$ is given by $P_k(f)(z) = \langle f,\ K_k(\cdot, z)\rangle_\nu$, and the covariant field of operators generated by P_k is $P_{k,\xi} = U(g_\xi)\,P_k\,U(g_\xi)^{-1}$.

Lemma 4.3. *For all $k \in \mathbb{N}$, $\xi \in \mathbb{C}^d$,*

$$\begin{aligned} P_{k,\xi}(K_w)(z) &= U(g_\xi) \otimes U(g_\xi)^*(K_k)(z, w) \\ &= j(g_\xi, z)\,K_k(z-\xi, w-\xi)\overline{j(g_\xi, w)} \\ &= e^{\nu\,(z|\xi)}\,K_k(z-\xi, w-\xi)e^{\nu\,(\xi|w)}\,e^{-\nu\,|\xi|^2}. \end{aligned} \tag{4.4}$$

Proof.

$$\begin{aligned} P_{k,\xi}(K_w)(z) &= P_k\,(U(g_{-\xi})\,K_w)(z-\xi)\,j(g_{-\xi}, z) \\ &= \langle U(g_{-\xi})\,K_w,\ K_k(\cdot, z-\xi)\rangle_\nu\,j(g_{-\xi}, z) \\ &= \langle K_w,\ U(g_\xi)\,K_k(\cdot, z-\xi)\rangle_\nu\,j(g_{-\xi}, z) \\ &= K_k(z-\xi,\ w-\xi)\,j(g_{-\xi}, z)\overline{j(g_{-\xi}, w)}. \end{aligned}$$ □

Lemma 4.4. *For any $\ell, j, n \in \mathbb{N}$ and $a, b \in \mathbb{C}^d$*

$$\int_{\mathbb{C}^d} |z|^{2\ell}\,(z|b)^j\,(a|z)^n\,d\mu_\nu(z) = \delta_{j,n}\,\frac{(a|b)^j}{\nu^{\ell+j}}\,\frac{(d+\ell+j-1)!\,j!}{(d+j-1)!}.$$

Proof. Integrating in polar coordinates (where $S = \{\xi \in \mathbb{C}^d;\ |\xi| = 1\}$ and σ is the $U(d)$-invariant probability measure on S) we find

$$\begin{aligned} &\int_{\mathbb{C}^d} |z|^{2\ell}\,(z|b)^j\,(a|z)^n\,d\mu_\nu(z) = \\ &\quad= \frac{2\nu^d}{(d-1)!}\int_0^\infty r^{2d-1}\,r^{2\ell}\,r^{j+n}\,e^{-\nu r^2}dr\,\int_S (\xi|b)^j\,(a|\xi)^n\,d\sigma(\xi) \\ &\quad= \delta_{j,n}\,\frac{\nu^d}{(d-1)!}\int_0^\infty t^{d+\ell+j-1}\,e^{-\nu t}\,dt\,\frac{j!\,(a,b)^j}{(d)_j} \end{aligned}$$

$$= \delta_{j,n} \frac{(d+\ell+j-1)!\, j!}{(d-1)!(d)_j} \frac{\langle a|b\rangle^j}{\nu^{\ell+j}}. \qquad \square$$

Theorem 4.5. *Let $k, \ell \in \mathbb{N}$ and write $\mathcal{B}^{k,\ell} := (\mathcal{A}^{P_\ell})' \mathcal{A}^{P_k}$. Then the eigenvalues of $\mathcal{B}^{k,\ell}$ are given by*

$$\widetilde{\mathcal{B}^{k,\ell}}(\lambda) = e^{\frac{\lambda}{\nu}} q_k(\frac{\lambda}{\nu})\, q_\ell(\frac{\lambda}{\nu}), \quad \forall \lambda \in \mathbb{C},$$

where

$$q_k(x) = \sum_{j=0}^{k} \binom{d+k-1}{d+j-1} \frac{x^j}{j!}. \tag{4.5}$$

Proof. Let $\lambda \in \mathbb{C}$ and let $a, b \in \mathbb{C}^d$ be so that $\langle a|b\rangle = \lambda$. Then, using Lemma 4.3, we have

$$\begin{aligned}
\tilde{\mathcal{B}}^{k,\ell}(\lambda) &= \big((\mathcal{A}^{P_\ell})' \mathcal{A}^{P_k} e_{a,b}\big)(0) = \langle \mathcal{A}^{P_k}_{e_{a,b}},\ P_\ell\rangle_{S_2} \\
&= \int_{\mathbb{C}^d} e_{a,b}(\xi)\, \langle P_{k,\xi},\ P_\ell\rangle_{S_2}\, d\mu_0(\xi) \\
&= \int_{\mathbb{C}^d} e_{a,b}(\xi) \left(\int_{\mathbb{C}^d} \int_{\mathbb{C}^d} P_{k,\xi}(K_w)(z)\, \overline{P_\ell(K_w)(z)}\, d\mu_\nu(z)\, d\mu_\nu(w) \right) d\mu_0(\xi) \\
&= \int_{\mathbb{C}^d} e^{\langle a|\xi\rangle} e^{\langle \xi|b\rangle} \left(\int_{\mathbb{C}^d} \int_{\mathbb{C}^d} e^{\nu\langle z|\xi\rangle} K_k(z-\xi,\ w-\xi)\, e^{\nu\langle \xi|w\rangle} \right. \\
&\qquad\qquad \left. K_\ell(w,z)\, d\mu_\nu(z)\, d\mu_\nu(w) \right) d\mu_\nu(\xi) \\
&= \int_{\mathbb{C}^d} e^{\langle a|\xi\rangle} e^{\langle \xi|b\rangle} \left(\int_{\mathbb{C}^d} \int_{\mathbb{C}^d} K_k(z, w-\xi) K_\ell(w, z+\xi) e^{-\nu\langle \xi|z\rangle}\, d\mu_\nu(z) \right. \\
&\qquad\qquad \left. e^{\nu\langle \xi|w\rangle}\, d\mu_\nu(w) \right) d\mu_\nu(\xi).
\end{aligned}$$

Interchanging the order of integration, the last integral becomes

$$\begin{aligned}
&\iint_{\mathbb{C}^d\times\mathbb{C}^d} \left[\int_{\mathbb{C}^d} \left\{ K_k(z, w-\xi)\, K_\ell(w, z+\xi)\, e^{\langle a|\xi\rangle} \right\} \right. \\
&\qquad\qquad \left. e^{\nu\langle \xi|w-z+\frac{b}{\nu}\rangle}\, d\mu_\nu(\xi) \right] d\mu_\nu(z)\, d\mu_\nu(w) \\
&\quad = \iint_{\mathbb{C}^d\times\mathbb{C}^d} K_k\left(z, z-\frac{b}{\nu}\right) K_\ell\left(w, w+\frac{b}{\nu}\right) e^{\langle a|w-z+\frac{b}{\nu}\rangle}\, d\mu_\nu(z)\, d\mu_\nu(w) \\
&\quad = e^{\frac{\lambda}{\nu}} I_k(a,b)\, I_\ell(a,b),
\end{aligned}$$

where

$$I_k(a,b) := \int_{\mathbb{C}^d} K_k\left(z, z+\frac{b}{\nu}\right) e^{\langle a|z\rangle}\, d\mu_\nu(z)$$

and we used the fact that $I_k(a,b) = I_k(-a,-b)$. To calculate $I_k(a,b)$ we expand $K_k(z, z+\frac{b}{\nu})$ and use Lemma 4.4,

$$\begin{aligned} I_k(a,b) &= \frac{\nu^k}{k!}\sum_{j=0}^{k}\binom{k}{j}\int_{\mathbb{C}^d} |z|^{2(k-j)}\,\frac{(z|b)^j}{\nu^j}\, e^{(a|z)}\, d\mu_\nu(z) \\ &= \nu^k \sum_{j=0}^{k}\frac{\nu^{-j}}{j!(k-j)!}\sum_{n=0}^{\infty}\frac{1}{n!}\int_{\mathbb{C}^d}|z|^{2(k-j)}(z|b)^j\,(a|z)^n\, d\mu_\nu(z) \\ &= \nu^k\sum_{j=0}^{k}\frac{\nu^{-j}}{(j!)^2(k-j)!}\,\frac{\lambda^j}{\nu^k}\,\frac{(d+k-1)!}{(d+j-1)!}\,j! \\ &= \sum_{j=0}^{k}\binom{d+k-1}{d+j-1}\frac{(\frac{\lambda}{\nu})^j}{j!} = q_k\left(\frac{\lambda}{\nu}\right). \end{aligned}$$ □

Remark 4.1. Notice that the polynomial q_k can be written also as

$$q_k(x) = \sum_{j=0}^{k}\frac{(d+j)_{k-j}}{(k-j)!}\,\frac{x^j}{j!}$$

and that

$$q_k(x) = \binom{k+d-1}{d-1}\sum_{j=0}^{k}\frac{(-k)_j}{(d)_j}\,\frac{(-x)^j}{j!} = \binom{k+d-1}{d-1}\,{}_1F_1(-k;d;-x).$$

Applying Theorem 4.5 we obtain the following result.

Theorem 4.6. *Let $T = \sum_{k=0}^{\infty} t_k P_k$, $S = \sum_{\ell=0}^{\infty} s_\ell P_\ell$ be K-invariant operators on holomorphic functions and let $\mathcal{A}^T$, $\mathcal{A}^S$ be the associated calculi. Let $\mathcal{B}^{T,S} = (\mathcal{A}^S)'\mathcal{A}^T$ be the corresponding link transform. Then*

$$\widetilde{\mathcal{B}^{T,S}}(\lambda) = e^{\frac{\lambda}{\nu}}\sum_{k=0}^{\infty} t_k\, q_k\left(\frac{\lambda}{\nu}\right)\sum_{\ell=0}^{\infty}\overline{s_\ell}\, q_\ell\left(\frac{\lambda}{\nu}\right).$$

Proof. We have

$$\begin{aligned} \widetilde{\mathcal{B}^{T,S}}(\lambda) &= \sum_{k,\ell\ge 0} t_k\,\overline{s_\ell}\,\widetilde{\mathcal{B}^{k,\ell}}(\lambda) \\ &= \sum_{k,\ell\ge 0} t_k\,\overline{s_\ell}\,e^{\frac{\lambda}{\nu}}\, q_k\left(\frac{\lambda}{\nu}\right)\, q_\ell\left(\frac{\lambda}{\nu}\right) = e^{\frac{\lambda}{\nu}}\sum_{k\ge 0} t_k\, q_k\left(\frac{\lambda}{\nu}\right)\sum_{\ell\ge 0}\overline{s_\ell}\, q_\ell\left(\frac{\lambda}{\nu}\right). \end{aligned}$$ □

The following example of a one-parameter family of calculi will play an important role in the sequel.

Example 4.4: Let $\alpha \in \mathbb{C}$, $\alpha \neq 1$ and set

$$t_k(\alpha) = (1-\alpha)^d \, \alpha^k \quad \text{and} \quad T(\alpha) = \sum_{k=0}^{\infty} t_k(\alpha) \, P_k,$$

i.e. $(T(\alpha)\, f)(z) = (1-\alpha)^d \; f(\alpha\, z)$. Let $\mathcal{A}^\alpha = \mathcal{A}^{T(\alpha)}$ be the associated calculus. Then Theorem 4.5 shows that the eigenvalues of $\mathcal{B}^{\alpha,\beta} = \mathcal{B}^{T(\alpha),T(\beta)}$ are given by

$$\widetilde{\mathcal{B}^{\alpha,\beta}}(\lambda) = e^{\frac{\lambda}{\nu}} \sum_{k\geq 0} \alpha^k \, q_k\left(\frac{\lambda}{\nu}\right) \sum_{\ell \geq 0} \overline{\beta^\ell} \, q_\ell\left(\frac{\lambda}{\nu}\right) \; (1-\alpha)^d (1-\overline{\beta})^d.$$

Claim: If $|\alpha| < 1$ then

$$\sum_{k=0}^{\infty} \alpha^k \, q_k(x) = \frac{1}{(1-\alpha)^d} \exp\left(\frac{\alpha}{1-\alpha}\, x\right). \tag{4.6}$$

Indeed, by absolute convergence we can interchange the order of summation and obtain

$$\begin{aligned}
\sum_{k=0}^{\infty} \alpha^k \, q_k(x) &= \sum_{j=0}^{\infty} \frac{x^j}{j!} \sum_{k=j}^{\infty} \frac{(d+j)_{k-j}}{(k-j)!}\, \alpha^k = \sum_{j=0}^{\infty} \frac{(\alpha x)^j}{j!} \sum_{\ell=0}^{\infty} (d+j)_\ell \, \frac{\alpha^\ell}{\ell!} \\
&= \sum_{j=0}^{\infty} \frac{(\alpha x)^j}{j!} \, (1-\alpha)^{-(d+j)} = \frac{1}{(1-\alpha)^d} \sum_{j=0}^{\infty} \frac{\left(\frac{\alpha x}{1-\alpha}\right)^j}{j!} \\
&= \frac{1}{(1-\alpha)^d} \exp\left(\frac{\alpha x}{1-\alpha}\right).
\end{aligned}$$

It follows that

$$\sum_{k=0}^{\infty} t_k(\alpha) \, q_k\,(\frac{\lambda}{\nu}) = \exp\left(\frac{\alpha}{1-\alpha}\,\frac{\lambda}{\nu}\right).$$

If also $|\beta| < 1$, then

$$\widetilde{\mathcal{B}^{\alpha,\beta}}(\lambda) = \exp\left\{\frac{1-\alpha\,\overline{\beta}}{(1-\alpha)(1-\overline{\beta})}\,\frac{\lambda}{\nu}\right\}, \tag{4.7}$$

and in particular,

$$\widetilde{\mathcal{B}^{\alpha,\alpha}}(\lambda) = \exp\left\{\frac{1-|\alpha|^2}{|1-\alpha|^2}\,\frac{\lambda}{\nu}\right\}. \tag{4.8}$$

By analytic continuation in α, $\overline{\beta}$ we obtain:

Proposition 4.7. *Let $\alpha, \beta \in \mathbb{C} \setminus \{1\}$. Then the eigenvalues of $\mathcal{B}^{\alpha,\beta}$ are given by* (4.7).

Corollary 4.8. *Let $\alpha, \beta \in \mathbb{C} \setminus \{1\}$ be so that $1 - \alpha\overline{\beta} = (1-\alpha)(1-\overline{\beta})$ (i.e. $\alpha + \overline{\beta} = 2\alpha\overline{\beta}$). Then $\mathcal{B}^{\alpha,\beta}$ coincides with the Berezin transform (namely, with the link transform of the Toeplitz calculus).*

Proof. We have $\widetilde{\mathcal{B}^{\alpha,\beta}}(\lambda) = e^{\frac{\lambda}{\nu}}$ for all $\lambda \in \mathbb{C}$. Since this is true for the Berezin transform $\mathcal{B}$ we conclude that $\mathcal{B}^{\alpha,\beta} = \mathcal{B}$. □

Corollary 4.9. *Let $\alpha, \beta \in \mathbb{C} \setminus \{1\}$ be so that $\alpha\overline{\beta} = 1$. Then $\mathcal{B}^{\alpha,\beta} = I$.*

Proof. We have $\widetilde{\mathcal{B}^{\alpha,\beta}}(\lambda) = 1$ for all $\lambda \in \mathbb{C}$. □

Corollary 4.10. *Let $\alpha \in \mathbb{C} \setminus \{1\}$ be so that $|\alpha| = 1$. Then $\mathcal{B}^{\alpha,\alpha} = I$. Consequently $\mathcal{A}^{\alpha}$ is an isometry.*

Proposition 4.11. *Let $\alpha \in \mathbb{C} \setminus \{1\}$ and let $\tau = \frac{1-|\alpha|^2}{|1-\alpha|^2} > 0$. Then $\mathcal{B}^{\alpha,\alpha} f = f * \mu_{\frac{\nu}{\tau}}$, i.e.*

$$(\mathcal{B}^{\alpha,\alpha} f)(z) = \int_{\mathbb{C}^d} f(g(w))\, d\mu_{\frac{\nu}{\tau}}(w).$$

Proof. The operator of convolution with $\mu_{\frac{\nu}{\tau}}$ is the Berezin transform for the Toeplitz calculus associated with $L^2_a(\mathbb{C}^d, \mu_{\frac{\nu}{\tau}})$. Hence its eigenvalues are $e^{\lambda/(\frac{\nu}{\tau})} = e^{\tau\frac{\lambda}{\nu}} = \widetilde{\mathcal{B}^{\alpha,\alpha}}(\lambda)$. □

The following general fact follows by elementary properties of intertwining operators.

Lemma 4.12. *Let $\mathcal{A}$ be an invariant symbolic calculus, considered formally as a map between Hilbert spaces $\mathcal{A} : L^2(D, \mu_0) \to S_2(\mathcal{H})$. Let*

$$\mathcal{A} = \mathcal{U}\mathcal{B}^{\frac{1}{2}} \quad (\mathcal{B} = \mathcal{A}'\mathcal{A}), \quad \text{and} \quad \mathcal{A} = \mathcal{Q}^{\frac{1}{2}}\mathcal{V} \quad (\mathcal{Q} = \mathcal{A}\mathcal{A}') \tag{4.9}$$

be the polar decompositions of $\mathcal{A}$, with $\mathcal{U}, \mathcal{V}$ minimal partial isometries (i.e. $\ker(\mathcal{U}) = \ker(\mathcal{A})$ *and* $\ker(\mathcal{V}) = \ker(\mathcal{A}')$*). Then*

$$\mathcal{B}^{\frac{1}{2}}(f \circ g) = (\mathcal{B}^{\frac{1}{2}} f) \circ g, \quad \forall\, g \in G, \ \forall\, f \in \operatorname{Dom}(\mathcal{B}^{\frac{1}{2}}). \tag{4.10}$$

$$\mathcal{Q}^{\frac{1}{2}}(\pi(g)(X)) = \pi(g)(\mathcal{Q}^{\frac{1}{2}}(X)), \quad \forall\, g \in G, \ \forall\, X \in \operatorname{Dom}(\mathcal{Q}^{\frac{1}{2}}), \tag{4.11}$$

where

$$\pi(g)(X) := U(g)\, X\, U(g)^{-1}, \quad \forall\, g \in G.$$

$\mathcal{U}$ and $\mathcal{V}$ are intertwining operators, i.e.

$$\pi(g)(\mathcal{U}_b) = \mathcal{U}_{b \circ g^{-1}}, \quad \pi(g)(\mathcal{V}_b) = \mathcal{V}_{b \circ g^{-1}}$$

for all $g \in G$ and b in the appropriate domains. Thus, $\mathcal{U}$ and $\mathcal{V}$ are invariant symbolic calculi.

Proof. (4.10) and (4.11) follow immediately from Proposition 2.13. Notice also that

$$\mathcal{A}'(\pi(g)\,X) = \mathcal{A}'(X)\circ g^{-1}, \quad \forall g \in G,\ \forall X \in \mathrm{Dom}(\mathcal{A}'), \tag{4.12}$$

The fact that

$$\pi(g)(\mathcal{A}_b) = \mathcal{A}_{b\circ g^{-1}}, \quad \forall\, g \in G,\ \forall\, X \in \mathrm{Dom}(\mathcal{A}),$$

clearly implies

$$\pi(g)(\mathcal{U}_f) = \mathcal{U}_{f\circ g^{-1}}, \quad \forall\, g \in G,\ \forall\, f \in \mathrm{Ran}(\mathcal{B}^{\frac{1}{2}}).$$

Hence, $\mathcal{U}$ is an invariant symbolic calculus with $\mathrm{Dom}(\mathcal{U}) = \mathrm{Ran}(\mathcal{B}^{\frac{1}{2}})$. Similarly, using the fact that $\mathcal{Q}^{\frac{1}{2}}$ is one-to-one on the range of $\mathcal{V}$, we see that $\pi(g)(\mathcal{V}_b) = \mathcal{V}_{b\circ g^{-1}}$ for all $g \in G$ and all $b \in \mathrm{Dom}(\mathcal{V})$. □

Theorem 4.13. *Let $k \in \mathbb{N}$ and let $\mathcal{A}^{P_k} = \mathcal{U}^{(k)}\,|\mathcal{A}^{P_k}|$, where $|\mathcal{A}^{P_k}| = (\mathcal{A}^{P_k\prime}\mathcal{A}^{P_k})^{\frac{1}{2}} = \mathcal{B}^{(k)\,\frac{1}{2}}$ be the polar decomposition* (4.9) *of the invariant symbolic calculus $\mathcal{A}^{P_k}$ associated with the projection P_k. Then*

$$\widetilde{(\mathcal{B}^{(k)})^{\frac{1}{2}}}(\lambda) = e^{\frac{\lambda}{2\nu}}\left|q_k\left(\frac{\lambda}{\nu}\right)\right|, \quad \forall\,\lambda \in (-\infty, 0],\ \textit{and}\ (\mathcal{B}^{(k)})^{\frac{1}{2}} = e^{\frac{\Delta}{2\nu}}\,|q_k|\left(\frac{\Delta}{\nu}\right). \tag{4.13}$$

as an operator on $L^2(\mathbb{C}^d, \mu_0) \equiv L^2(\mathbb{R}^{2d}, m)$. Moreover

$$\mathcal{U}^{(k)} = \mathcal{W}\,V_k, \quad \textit{where}\ \ V_k := \mathrm{sgn}(q_k)\left(\frac{\Delta}{\nu}\right) \tag{4.14}$$

and $\mathcal{W}$ is the Weyl calculus.

Proof. Let $a, b \in \mathbb{C}^d$ and denote $\lambda = \langle a|b\rangle$. We claim that for all $z, w \in \mathbb{C}^d$

$$\begin{aligned} A^{(k)}_{e_{a,b}}(z, w) &= \mathcal{A}^{(P_k)}_{e_{a,b}}(K_w)(z)/K_w(z) \\ &= \binom{d-1+k}{d-1}\,{}_1F_1\left(d+k;\ d;\ \frac{\lambda}{\nu}\right)\, e^{\langle z|b\rangle}\, e^{\langle a|w\rangle}. \end{aligned}$$

Here we use the standard notation ${}_1F_1\,(\alpha;\ \beta;\ x) = \sum_{n\geq 0}\frac{(\alpha)_n}{(\beta)_n}\frac{x^n}{n!}$. Since $A^{(k)}_{e_{a,b}}(z, w)$ is holomorphic in a and z and anti-holomorphic in b and w, it suffices to consider $a = b$ and $z = w$. In this case, we obtain from Lemma 4.3

$$\begin{aligned} \mathcal{A}^{P_k}_{e_{a,b}}(K_z)(z) &= \int_{\mathbb{C}^d} e^{\langle a|\xi\rangle}\, e^{\langle\xi|a\rangle}\, e^{\nu\,\langle z|\xi\rangle}\, K_k\,(z-\xi,\ z-\xi)\, e^{\nu\,\langle\xi|z\rangle}\, d\mu_\nu(\xi) \\ &= \frac{\nu^k}{k!}\int_{\mathbb{C}^d} e^{\nu\,\langle\eta+z|z+\frac{a}{\nu}\rangle}\, e^{\nu\,\langle z+\frac{a}{\nu}|\eta+z\rangle}\, |\eta|^{2k}\, d\mu_\nu\,(\eta+z) = \end{aligned}$$

$$= \frac{\nu^k}{k!} e^{\nu|z|^2} e^{(z|a)+(a|z)} \int_{\mathbb{C}^d} e^{(\eta|a)} e^{(a|\eta)} |\eta|^{2k} d\mu_\nu(\eta)$$

$$= \frac{\nu^k}{k!} e^{\nu|z|^2} e^{(z|a)+(a|z)} \sum_{m,n=0}^{\infty} \int_{\mathbb{C}^d} \frac{(\eta|a)^m}{m!} \frac{(a|\eta)^n}{n!} |\eta|^{2k} d\mu_\nu(\eta).$$

The last integral is evaluated by Lemma 4.4

$$\int_{\mathbb{C}^d} \frac{(\eta|a)^m}{m!} \frac{(a|\eta)^n}{n!} |\eta|^{2k} d\mu_\nu(\eta) = \delta_{m,n} \frac{(d+k+n-1)!}{(d-1)!\,(d)_n\, \nu^{k+n}} \frac{|a|^{2n}}{n!}. \tag{4.15}$$

Hence

$$\mathcal{A}^{P_k}_{e_{a,b}}(K_z)(z) = \binom{d+k-1}{d-1} \sum_{n=0}^{\infty} \frac{(d+k)_n}{(d)_n} \frac{|a|^{2n}}{n!\,\nu^n} e^{(z|a)} e^{(a|z)} e^{\nu|z|^2}$$

$$= \binom{d+k-1}{d-1} {}_1F_1\left(d+k;\ d;\ \frac{|a|^2}{\nu}\right) e^{(z|a)} e^{(a|z)} e^{\nu|z|^2}.$$

Since a sesqui-holomorphic function is determined by its value on the "diagonal", we see that

$$\mathcal{A}^{P_k}_{e_{a,b}}(K_w)(z) = \binom{d+k-1}{d-1} {}_1F_1\left(d+k;\ d;\ \frac{(a|b)}{\nu}\right) e^{(z|b)} e^{(a|w)} e^{\nu(z|w)},$$

and (4.15) is established.

Lemma 4.14. *For all $x \in \mathbb{C}$ and $k \in \mathbb{N}$*

$$e^x q_k(x) = \binom{d+k-1}{d-1} {}_1F_1(d+k;\ d;\ x).$$

Proof. Interchanging the order of summation, we get

$$e^x q_k(x) = \sum_{n=0}^{\infty} \frac{x^n}{n!} \sum_{j=0}^{k} \binom{d+k-1}{d+j-1} \frac{x^j}{j!} = \sum_{m=0}^{\infty} \sum_{j=0}^{k\wedge m} \binom{m}{j} \binom{d+k-1}{k-j} \frac{x^m}{m!}.$$

An easy combinatorial argument yields

$$\sum_{j=0}^{k\wedge m} \binom{m}{j} \binom{d+k-1}{k-j} = \binom{d+k+m-1}{k},$$

and thus

$$e^x q_k(x) = \sum_{m=0}^{\infty} \binom{d+k+m-1}{k} \frac{x^m}{m!}$$

$$= \sum_{j=0}^{k} \binom{d+k-1}{d-1} \sum_{m=0}^{\infty} \frac{(d+k)_m}{(d)_m} \frac{x^m}{m!}$$

$$= \binom{d+k-1}{d-1}\ {}_1F_1\,(d+k;\ d;\ x). \qquad \square$$

It follows from (4.15) and Lemma 4.14 that

$$A^{P_k}_{e_{a,b}}(z,w) = e^{\frac{\lambda}{\nu}} q_k\left(\frac{\lambda}{\nu}\right)\ e^{\langle z|b\rangle}\ e^{\langle a|w\rangle}. \tag{4.16}$$

Next, we know that

$$\widetilde{\mathcal{B}^{(k)}}(\lambda) = \widetilde{(\mathcal{A}^{P_k})'\,\mathcal{A}^{P_k}}(\lambda) = e^{\frac{\lambda}{\nu}}\ q_k\left(\frac{\lambda}{\nu}\right)^2, \quad \forall\,\lambda \in \mathbb{C}.$$

Considering $\mathcal{B}^{(k)\frac{1}{2}}$ as an operator on $L^2(\mathbb{C}^d, \mu_0)$, and knowing that the Plancherel measure is supported on $(-\infty, 0]$ (i.e. on the $e_{a,b}$ with $b = -a$), we conclude from

$$\widetilde{\mathcal{B}^{(k)}}^{\frac{1}{2}}(\lambda) = e^{\frac{\lambda}{2\nu}}\ \left|q_k\left(\frac{\lambda}{\nu}\right)\right|, \quad \forall\,\lambda \in (-\infty, 0]$$

that

$$\widetilde{\mathcal{B}^{(k)}}^{\frac{1}{2}} = e^{\frac{\Delta}{2\nu}}\ |q_k|\left(\frac{\Delta}{\nu}\right).$$

This establishes (4.13). Computing $A^{P_k}_{e_{a,b}}$ via the factorization $\mathcal{A}^{P_k} = \mathcal{U}^{(k)}\mathcal{B}^{(k)\frac{1}{2}}$, we obtain

$$A^{P_k}_{e_{a,b}}(z,w) = e^{\frac{\lambda}{2\nu}}\ \left|q_k\left(\frac{\lambda}{\nu}\right)\right|\ \frac{\mathcal{U}^{(k)}_{e_{a,b}}(K_w)(z)}{K_w(z)}.$$

Comparing this with (4.16) we conclude that

$$\frac{\mathcal{U}^{(k)}_{e_{a,b}}(K_w)(z)}{K_w(z)} = e^{\frac{\lambda}{2\nu}}\ \operatorname{sgn}\left(q_k\left(\frac{\lambda}{\nu}\right)\right)\ e^{\langle z|b\rangle}\ e^{\langle a|w\rangle}. \tag{4.17}$$

However, the Weyl transform $\mathcal{W}$ satisfies

$$\begin{aligned}
\mathcal{W}_{e_{a,b}}(K_w)(z) &= 2^d \int_{\mathbb{C}^d} e^{\langle \xi|b\rangle}\ e^{\langle a|\xi\rangle}\ U(s_\xi)(K_w)(z)\ d\mu_0(\xi)\\
&= \int_{\mathbb{C}^d} e^{\langle \xi|b\rangle}\ e^{\langle a|\xi\rangle}\ e^{\nu\langle -z+2\xi|w\rangle}\ e^{2\nu\langle z|\xi\rangle}\ d\mu_{2\nu}(\xi)\\
&= \int_{\mathbb{C}^d} e^{\langle \xi|b+2\nu w\rangle}\ e^{2\nu\langle z+\frac{a}{2\nu}|\xi\rangle}\ d\mu_{2\nu}(\xi)\ e^{-\nu\langle z|w\rangle}\\
&= e^{\frac{\lambda}{2\nu}}\ e^{\langle z|b\rangle}\ e^{\langle a|w\rangle}\ e^{\nu\langle z|w\rangle},
\end{aligned}$$

i.e.

$$W_{e_{a,b}}(z,w) = \frac{\mathcal{W}_{e_{a,b}}(K_w)(z)}{K_w(z)} = e^{\frac{\lambda}{2\nu}}\ e^{\langle z|b\rangle}\ e^{\langle a|w\rangle}. \tag{4.18}$$

Comparing (4.17) and (4.18) we conclude that

$$\mathcal{U}^{(k)}_{e_{a,b}} = \mathcal{W}_{V_k(e_{a,b})}, \quad \text{where } V_k := \operatorname{sgn}(q_k)\left(\frac{\Delta}{\nu}\right).$$

Standard approximation arguments yield that $\mathcal{U}^{(k)}_f = \mathcal{W}_{V_k(f)}$ for all admissible symbols f. Hence (4.14) is established, and the proof of Theorem 4.13 is complete. □

Theorem 4.15. *Let $\alpha \in \mathbb{C}\setminus\{1\}$ and let $\mathcal{A}^\alpha = \mathcal{U}^{(\alpha)}|\mathcal{A}^\alpha|$ be the polar decomposition of $\mathcal{A}^\alpha$, where $|\mathcal{A}^\alpha| = (\mathcal{B}^{\alpha,\alpha})^{\frac{1}{2}}$. Let $\alpha = \frac{z-1}{z+1}$ where $z = x+iy$. Then $|\mathcal{A}^\alpha| = \mathcal{B}^{\frac{x}{2}}$, where $\mathcal{B}$ is the Berezin transform associated with the Toeplitz calculus, and $\mathcal{U}^{(\alpha)} = \mathcal{A}^\beta$, where $\beta = \frac{iy-1}{iy+1}$.*

Proof. We know by (4.8) that

$$|\widetilde{\mathcal{A}^\alpha}|(\lambda) = \exp\left\{\frac{1}{2}\,\frac{1-|\alpha|^2}{|1-\alpha|^2}\,\frac{\lambda}{\nu}\right\} = \exp\left\{\frac{\lambda}{2\nu}\,x\right\}.$$

Also, $\tilde{\mathcal{B}}(\lambda) = e^{\frac{\lambda}{\nu}}$. Hence $|\widetilde{\mathcal{A}^\alpha}|(\lambda) = \widetilde{\mathcal{B}^{\frac{x}{2}}}(\lambda)$, and therefore $|\mathcal{A}^\alpha| = \mathcal{B}^{\frac{x}{2}}$. Next, using (4.6) and (4.16) we see that for any $a, b \in \mathbb{C}^d$ with $(a|b) = \lambda$ and any $z, w \in \mathbb{C}^d$ we have for $|\alpha| < 1$

$$\frac{\mathcal{A}^\alpha_{e_{a,b}}(K_w)(z)}{K_w(z)\, e^{(a|w)}\, e^{(z|b)}} = (1-\alpha)^d \sum_{k\geq 0} \alpha^k\, q_k\!\left(\frac{\lambda}{\nu}\right) e^{\frac{\lambda}{\nu}} = \exp\left\{\frac{1}{1-\alpha}\,\frac{\lambda}{\nu}\right\}.$$

By analytic continuation we conclude that

$$\frac{\mathcal{A}^\alpha_{e_{a,b}}(K_w)(z)}{K_w(z)\, e^{(a|w)}\, e^{(z|b)}} = \exp\left\{\frac{1}{1-\alpha}\,\frac{\lambda}{\nu}\right\}, \quad \forall\, \alpha \in \mathbb{C}\setminus\{1\}. \tag{4.19}$$

Since

$$|\mathcal{A}^\alpha|(e_{a,b}) = \left\{\frac{1}{2}\,\frac{1-|\alpha|^2}{|1-\alpha|^2}\,\frac{\lambda}{\nu}\right\} e_{a,b},$$

we obtain also

$$\frac{\mathcal{A}^\alpha_{e_{a,b}}(K_w)(z)}{K_w(z)\, e^{(a|w)}\, e^{(z|b)}} = \exp\left\{\frac{1}{2}\,\frac{1-|\alpha|^2}{|1-\alpha|^2}\,\frac{\lambda}{\nu}\right\} \frac{\mathcal{U}^{(\alpha)}_{e_{a,b}}(K_w)(z)}{K_w(z)\, e^{(a|w)}\, e^{(z|b)}}.$$

Therefore, by (4.19), we obtain

$$\frac{\mathcal{U}^{(\alpha)}_{e_{a,b}}(K_w)(z)}{K_w(z)\, e^{(a|w)}\, e^{(z|b)}} = \exp\left\{\left(\frac{1}{1-\alpha} - \frac{1}{2}\,\frac{1-|\alpha|^2}{|1-\alpha|^2}\right)\frac{\lambda}{\nu}\right\} = \exp\left\{\frac{1}{1-\beta}\,\frac{\lambda}{\nu}\right\} \tag{4.20}$$

with

$$\beta = -\frac{1-2\alpha+|\alpha|^2}{1-2\bar{\alpha}+|\alpha|^2} = \frac{iy-1}{iy+1}.$$

Comparing (4.20) and (4.19) we conclude that $\mathcal{U}^{(\alpha)} = \mathcal{A}^{\beta}$. □

Remark 4.2. In general, $\mathcal{U}^{(\alpha)} = \mathcal{A}^{\beta}$ is different from the Weyl calculus $\mathcal{W}$, and

$$\mathcal{U}^{(\alpha)} = \mathcal{W} \iff \beta = -1 \iff \alpha \in \mathbb{R} \setminus \{1\} \iff z \in \mathbb{R}.$$

Remark 4.3. Theorem 4.15 suggests another possible definition of the Wick calculus (4.3), namely

$$\mathcal{E} = \mathcal{W}\, \mathcal{B}^{-\frac{1}{2}} \tag{4.21}$$

where $\mathcal{B}$ is the Berezin transform and $\mathcal{W}$ is the Weyl calculus.

Indeed, using (4.18) we see that for all $a, b \in \mathbb{C}^d$ with $\lambda = \langle a|b\rangle$ and all $z, w \in \mathbb{C}^d$ we have

$$\frac{\left(\mathcal{W}\, \mathcal{B}^{-\frac{1}{2}}\right)_{e_{a,b}}(K_w)(z)}{K_w(z)\; e^{\langle a|w\rangle}\; e^{\langle z|b\rangle}} = e^{-\frac{\lambda}{2\nu}}\, \frac{\mathcal{W}_{e_{a,b}}(K_w)(z)}{K_w(z)\; e^{\langle a|w\rangle}\; e^{\langle z|b\rangle}} = 1.$$

On the other hand, using the definition (4.3) we see that $\mathcal{E}_{e_{a,b}} = \mathcal{T}_{e_{0,b}}\, \mathcal{T}^*_{e_{0,a}}$, and therefore

$$\frac{\mathcal{E}_{e_{a,b}}(K_w)(z)}{K_w(z)\; e^{\langle a|w\rangle}\; e^{\langle z|b\rangle}} = \frac{e^{\langle z|b\rangle}\; \mathcal{T}^*_{e_{0,a}}(K_w)(z)}{K_w(z)\; e^{\langle a|w\rangle}\; e^{\langle z|b\rangle}} = 1,$$

since for holomorphic symbols φ we have $\mathcal{T}^*_{\varphi}(K_w) = \overline{\varphi(w)}\, K_w$. Therefore (4.21) holds.

Remark 4.4. Theorem 4.15 suggests also to consider $\mathcal{E}$ as the limiting case of $\mathcal{A}^{\alpha}$ where $\alpha \to \infty$.

More precisely, using the parameter z instead of $\alpha = \frac{z-1}{z+1}$ and writing ${}^{(z)}\mathcal{A} = \mathcal{A}^{\alpha}$, we see that

$${}^{(x+iy)}\mathcal{A} = {}^{(iy)}\mathcal{A}\, (\mathcal{B}^{\frac{1}{2}})^{x}\,.$$

Comparing this with (4.21), we conclude that

$$\mathcal{E} = {}^{(-1)}\mathcal{A} = \mathcal{A}^{\infty}.$$

Summing up, the Toeplitz, Weyl and Wick calculi correspond to the points 0, -1, and ∞ respectively in the α-plane, and the points 0, 1, and -1 respectively in the z-planes.

The polynomials q_k (4.5) have an interesting orthogonality property.

Proposition 4.16. *The polynomials $\{q_k\}_{k=1}^{\infty}$ are orthogonal with respect to the probability measure*

$$d\rho(x) := \frac{1}{(d-1)!}\chi_{(-\infty,0)}(x)\, |x|^{d-1} e^{-|x|}\, dx,$$

where $\chi_{(-\infty,0)}(x)$ *is the indicator function of* $(-\infty,0)$. *Moreover,*

$$\|q_k\|^2_{L^2(\mathbb{R},\rho)} = \frac{(d)_k}{k!}.$$

Thus, the sequence of normalized polynomials $\{(k!/(d)_k)^{\frac{1}{2}}\, q_k(x)\}_{k=0}^{\infty}$ *is an orthonormal basis of* $L^2((-\infty,0),\rho)$.

Proof. Let $0 \le \ell < k$. Using the change of variables $t = -x$, we obtain

$$\begin{aligned}\int_{\mathbb{R}} x^\ell q_k(x)\,d\rho(x) &= \frac{(-1)^\ell (d)_k}{k!(d-1)!}\int_0^\infty t^\ell \left(\sum_{j=0}^{k} (-1)^j \binom{k}{j} \frac{t^j}{(d)_j}\right) t^{d-1} e^{-t}\,dt \\ &= \frac{(-1)^\ell (d)_k}{k!} \sum_{j=0}^{\ell} (-1)^j \binom{k}{j} \frac{(d)_{\ell+j}}{(d)_j} \\ &= \frac{(-1)^{\ell+d-1} (d)_k}{k!} \left(\frac{\partial}{\partial t}\right)^\ell \left(\sum_{j=0}^{k} \binom{k}{j} t^{d+j+\ell-1}\right)_{|t=-1} \\ &= \frac{(-1)^{\ell+d-1} (d)_k}{k!} \left(\frac{\partial}{\partial t}\right)^\ell \left(t^{d+\ell-1}(1+t)^k\right)_{|t=-1} = 0,\end{aligned}$$

since $k > \ell$. Thus, $\langle f, q_k\rangle_{L^2(\rho)} = 0$ for every polynomial f of degree at most $k-1$. In particular, $\{q_k\}_{k=0}^{\infty}$ are orthogonal in $L^2(\rho)$. The same calculations yield

$$\int_{\mathbb{R}} x^k q_k(x)\,d\rho(x) = \frac{(-1)^{k+d-1} (d)_k}{k!} \left(\frac{\partial}{\partial t}\right)^k \left(t^{d+k-1}(1+t)^k\right)_{|t=-1} = (d)_k.$$

Thus,

$$\begin{aligned}\int_{\mathbb{R}} q_k(x)^2\,d\rho(x) &= \sum_{\ell=0}^{k} \binom{d+k-1}{d+\ell-1} \frac{1}{\ell!} \int_{\mathbb{R}} x^\ell q_k(x)\,d\rho(x) \\ &= \frac{1}{k!}\int_{\mathbb{R}} x^k q_k(x)\,d\rho(x) = \frac{(d)_k}{k!}.\end{aligned}$$

□

We close this section with an example of invariant symbolic calculus which generalizes Example 4.4. It is more complicated but, nevertheless, explicitly solvable.

Example 4.5. Fix $2 \le n \in \mathbb{N}$ and $\alpha \in \mathbb{C} \setminus \{1\}$, and define

$$t_k = \begin{cases} \alpha^k & k \equiv 0 \pmod n \\ 0 & k \not\equiv 0 \pmod n \end{cases}$$

and $T = \sum_{k\geq 0} t_k\, P_k$. Let $\omega = e^{\frac{2\pi i}{n}}$ and $\alpha_\ell = \omega^\ell \alpha$. Then $\frac{1}{n}\sum_{\ell=0}^{n-1}(\alpha_\ell)^k = t_k$ for all k. Let $T^{(\ell)} = (1-\alpha_\ell)^d \sum_{k\geq 0}(\alpha_\ell)^k P_k$. Then $\frac{1}{n}\sum_{\ell=0}^{n-1} \frac{T^{(\ell)}}{(1-\alpha_\ell)^d} = T$, and hence $\frac{1}{n}\sum_{\ell=0}^{n-1} \frac{\mathcal{A}^{T^{(\ell)}}}{(1-\alpha_\ell)^d} = \mathcal{A}^T$. It follows that if $a, b \in \mathbb{C}^d$ and $\langle a|b\rangle = \lambda$ then for all $z, w \in \mathbb{C}^d$

$$\mathfrak{a}_T(\lambda) = \frac{\mathcal{A}^T_{e_{a,b}}(K_w)(z)}{K_w(z)\, e^{\langle a|w\rangle}\, e^{\langle z|b\rangle}} = \frac{1}{n}\sum_{\ell=0}^{n-1} \frac{1}{(1-\alpha_\ell)^d} \exp\left\{\frac{1}{1-\alpha_\ell}\frac{\lambda}{\nu}\right\}.$$

5. Application to Toeplitz calculus and the eigenvalues of the Berezin transform on symmetric tube domains

Let $T(\Omega)$ be a symmetric tube domain of rank r and genus p in $\mathbb{C}^d$ (see Subsection 1.3). Let $\nu > p-1$ and for each $b \in L^\infty(T(\Omega))$ let

$$\mathcal{T}_b = P^{(\nu)} M_b|_{L^2_a(T(\Omega),\mu_\nu)} \tag{5.1}$$

be the Toeplitz operator with symbol b on the weighted Bergman space $L^2_a(T(\Omega), \mu_\nu)$. Here $P^{(\nu)} : L^2(T(\Omega), \mu_\nu) \to L^2_a(T(\Omega), \mu_\nu)$ is the orthogonal projection and $M_b f = b f$ is the operator of multiplication by b. In this case the link transform is the Berezin transform $\mathcal{B}^{(\nu)} = \mathcal{T}'\mathcal{T}$. Our goal here is to give a new proof, via Theorem 2.23 of the following known result [UU94] (see also [Be78]). Here, for any $\underline{\lambda} = (\lambda_1, \ldots, \lambda_r) \in \mathbb{C}^r$ we denote

$$\underline{\lambda}^* = (\lambda_r, \lambda_{r-1}, \ldots, \lambda_1). \tag{5.2}$$

Also, $\underline{\rho} = (\rho_1, \rho_2, \ldots, \rho_r)$, the half sum of the positive roots, is given by

$$\rho_j = (j-1)\frac{a}{2} + \frac{1}{2}, \qquad j = 1, 2, \ldots, r, \tag{5.3}$$

and for simplicity we put also

$$t_\nu = \nu - \frac{p-1}{2}.$$

Theorem 5.1. *Let $\nu > p-1$. Then*

$$\mathrm{Dom}(\widetilde{\mathcal{B}^{(\nu)}}) = \{\underline{\lambda} \in \mathbb{C}^r;\ |\operatorname{Re}\lambda_j| < t_\nu\}$$

and for $\underline{\lambda} \in \mathrm{Dom}(\widetilde{\mathcal{B}^{(\nu)}})$,

$$\widetilde{\mathcal{B}^{(\nu)}}(\underline{\lambda}) = \frac{\Gamma_\Omega(\underline{\lambda} + \underline{\rho} + \nu - \frac{d}{r})\,\Gamma_\Omega(-\underline{\lambda}^* - \underline{\rho}^* + \nu)}{\Gamma_\Omega(\nu - \frac{d}{r})\,\Gamma_\Omega(\nu)} = \prod_{j=1}^r \frac{\Gamma(t_\nu + \lambda_j)\,\Gamma(t_\nu - \lambda_j)}{\Gamma(t_\nu + \rho_j)\,\Gamma(t_\nu - \rho_j)}. \tag{5.4}$$

In the derivation of (5.4) we shall use the following result, which is of independent interest. For $\underline{\alpha}, \underline{\beta}, \underline{\gamma} \in \mathbb{R}^r$ and $z, w \in T(\Omega)$ consider the integral

$$I_{\underline{\alpha},\underline{\beta},\underline{\gamma}}(z, w) = \int_{T(\Omega)} \mathrm{N}_{\underline{\alpha}}(\tau(z, \xi))\, \mathrm{N}_{\underline{\beta}}(\tau(\xi, \xi))\, \mathrm{N}_{\underline{\gamma}}(\tau(\xi, w))\, d\mu_0(\xi). \tag{5.5}$$

Definition 5.1. $\mathcal{D}$ is the set of all $(\underline{\alpha}, \underline{\beta}, \underline{\gamma})$ in $\mathbb{R}^r \times \mathbb{R}^r \times \mathbb{R}^r$ such that the integral (5.5) is absolutely convergent for all $(z, w) \in T(\Omega) \times T(\Omega)$, and the convergence is uniform on compact subsets of $T(\Omega) \times T(\Omega)$.

Theorem 5.2. *We have the inclusion*

$$\mathcal{D} \supseteq \Big\{(\underline{\alpha}, \underline{\beta}, \underline{\gamma}) \in \mathbb{R}^r \times \mathbb{R}^r \times \mathbb{R}^r;\ \alpha_j + \gamma_j < (j-1)\frac{a}{2} - p + 1,$$
$$\alpha_j + \beta_j + \gamma_j < (r-j)\frac{a}{2},\ \beta_j > \frac{d}{r} + (j-1)\frac{a}{2} \text{ for } 1 \le j \le r\Big\},$$

and for $(\underline{\alpha}, \underline{\beta}, \underline{\gamma}) \in \mathcal{D}$,

$$I_{\underline{\alpha},\underline{\beta},\underline{\gamma}}(z, w) = I_{\underline{\alpha},\underline{\beta},\underline{\gamma}}\ \mathrm{N}_{\underline{\alpha}+\underline{\beta}+\underline{\gamma}}(\tau(z, w))$$

with

$$I_{\underline{\alpha},\underline{\beta},\underline{\gamma}} := \frac{(4\pi)^d\, \Gamma_\Omega(\underline{\beta} - \frac{d}{r})\, \Gamma_\Omega(-\underline{\alpha}^* - \underline{\beta}^* - \underline{\gamma}^*)}{\Gamma_\Omega(-\underline{\alpha}^*)\, \Gamma_\Omega(-\underline{\gamma}^*)}.$$

Recall that $X \equiv \mathbb{R}^d$ is the Euclidean Jordan algebra whose positive cone is Ω.

Lemma 5.3. *Let* $(\underline{\alpha}, \underline{\beta}, \underline{\gamma}) \in \mathcal{D}$ *and let* $g \in NA$. *Then*

$$I_{\underline{\alpha},\underline{\beta},\underline{\gamma}}(g(z), g(w)) = \mathrm{N}_{\underline{\alpha}+\underline{\beta}+\underline{\gamma}}(\tau(g(ie)))\, I_{\underline{\alpha},\underline{\beta},\underline{\gamma}}(z, w)$$

for all $z, w \in T(\Omega)$.

Proof. Observe first that for all $\underline{s} \in \mathbb{C}^r$ and $g \in NA$

$$\mathrm{N}_{\underline{s}}(\tau(g(z), g(w))) = \mathrm{N}_{\underline{s}}(\tau(z, w))\, \mathrm{N}_{\underline{s}}(\tau(g(ie))), \quad z, w \in T(\Omega). \tag{5.6}$$

Indeed, both sides of (5.6) are holomorphic in z, anti-holomorphic in w, and coincide for $z = w$ (see (1.27)). Therefore they coincide everywhere. Using the invariance of μ_0, we obtain

$$\begin{aligned} I_{\underline{\alpha},\underline{\beta},\underline{\gamma}}(g(z), g(w)) &= \int_{T(\Omega)} \mathrm{N}_{\underline{\alpha}}(\tau(g(z), \xi))\, \mathrm{N}_{\underline{\beta}}(\tau(\xi, \xi))\, \mathrm{N}_{\underline{\gamma}}(\tau(\xi, g(w)))\, d\mu_0(\xi) \\ &= \int_{T(\Omega)} \mathrm{N}_{\underline{\alpha}}(\tau(g(z), g(\eta)))\, \mathrm{N}_{\underline{\beta}}(\tau(g(\eta), g(\eta)))\, \mathrm{N}_{\underline{\gamma}}(\tau(g(\eta), g(w)))\, d\mu_0(\eta) \\ &= \mathrm{N}_{\underline{\alpha}+\underline{\beta}+\underline{\gamma}}(\tau(g(ie)))\, I_{\underline{\alpha},\underline{\beta},\underline{\gamma}}(z, w). \end{aligned}$$

□

Corollary 5.4. *For $(\underline{\alpha}, \underline{\beta}, \underline{\gamma}) \in \mathcal{D}$ and all $z, w \in T(\Omega)$,*

$$I_{\underline{\alpha},\underline{\beta},\underline{\gamma}}(z, w) = I_{\underline{\alpha},\underline{\beta},\underline{\gamma}}(ie, ie)\ \mathrm{N}_{\underline{\alpha}+\underline{\beta}+\underline{\gamma}}(\tau(z, w)). \tag{5.7}$$

Proof. Lemma 5.3 implies (5.7) for $z = w$ (by using $g = g_z$). Since both sides of (5.7) are holomorphic in z and anti-holomorphic in w, they coincide for all $z, w \in T(\Omega)$.□

Lemma 5.5. *Fix $v \in \Omega$ and $\underline{\alpha} \in \mathbb{R}^r$. Then the function*

$$f_{\underline{\alpha},v}(x) := \mathrm{N}_{\underline{\alpha}}\left(\frac{x + iv}{2i}\right) \tag{5.8}$$

belongs to $L^2(X)$ if and only if

$$\alpha_j < (j-1)\frac{a}{4} - \frac{p-1}{2} \quad \textit{for } j = 1, 2, \ldots, r. \tag{5.9}$$

Thus, if $\underline{\alpha}, \underline{\gamma} \in \mathbb{R}^r$ satisfy

$$\alpha_j + \gamma_j < (j-1)\frac{a}{2} - p + 1 \quad \textit{for } j = 1, 2, \ldots, r \tag{5.10}$$

then $\int_X |\overline{f_{\underline{\alpha},v}(x)}\, f_{\underline{\gamma},v}(x)|\, dx < \infty$, and

$$\int_X \overline{f_{\underline{\alpha},v}(x)}\, f_{\underline{\gamma},v}(x)\, dx = (4\pi)^d \frac{\Gamma_\Omega(-\underline{\alpha}^* - \underline{\gamma}^* - \frac{d}{r})}{\Gamma_\Omega(-\underline{\alpha}^*)\,\Gamma_\Omega(-\underline{\gamma}^*)}\ \mathrm{N}_{\underline{\alpha}+\underline{\gamma}+\frac{d}{r}}(v).$$

Proof. Notice that if $-\alpha_{r+1-j} > (j-1)\frac{a}{2}$ for $1 \le j \le r$ then (1.29) implies that

$$f_{\underline{\alpha},v}(x) = \frac{1}{\Gamma_\Omega(-\underline{\alpha}^*)} \int_\Omega e^{-(\frac{x+iv}{2i}|t)}\, \mathrm{N}^*_{-\underline{\alpha}^*-\frac{d}{r}}(t)\, dt$$

and the integral converges absolutely. Thus the Fourier transform satisfies

$$\hat{f}_{\underline{\alpha},v}(t) = \frac{2^{-\sum_{j=1}^r \alpha_j}}{\Gamma_\Omega(-\underline{\alpha}^*)}\, \chi_\Omega(t)\, e^{-(v|t)}\, \mathrm{N}^*_{-\underline{\alpha}^*-\frac{d}{r}}(t)\,.$$

Thus by Parseval's formula

$$\begin{aligned}
\|f_{\underline{\alpha},v}\|^2_{L^2(X)} &= \frac{(2\pi)^d\, 2^{-2\sum_{j=1}^r \alpha_j}}{\Gamma_\Omega(-\underline{\alpha}^*)^2} \int_\Omega e^{-(v|2t)}\, \mathrm{N}^*_{-2\underline{\alpha}^*-2\frac{d}{r}}(t)\, dt \\
&= \frac{(4\pi)^d}{\Gamma_\Omega(-\underline{\alpha}^*)^2} \int_\Omega e^{-(v|s)}\, \mathrm{N}^*_{-2\underline{\alpha}^*-\frac{d}{r}}(s)\, d\mu_\Omega(s) \\
&= (4\pi)^d \frac{\Gamma_\Omega(-2\underline{\alpha}^* - \frac{d}{r})}{\Gamma_\Omega(-\underline{\alpha}^*)^2}\ \mathrm{N}_{2\underline{\alpha}+\frac{d}{r}}(v)
\end{aligned}$$

provided

$$-2\alpha_{r+1-j} > \frac{d}{r} + (j-1)\frac{a}{2} \quad \text{for } j = 1, 2, \ldots, r,$$

which is equivalent to (5.9). Suppose now that (5.10) holds. Then

$$\int_X \overline{f_{\alpha,v}(x)}\, f_{\gamma,v}(x)\, dx = \int_X |f_{\frac{\alpha+\gamma}{2},v}(x)|^2\, dx < \infty,$$

and Parseval's theorem yields

$$\begin{aligned}\int_X \overline{f_{\underline{\alpha},v}(x)}\, f_{\underline{\gamma},v}(x)\, dx &= \frac{2^{-\sum_{j=1}^{r}(\alpha_j+\gamma_j)}\,(2\pi)^d}{\Gamma_\Omega(-\underline{\alpha}^*)\,\Gamma_\Omega(-\underline{\gamma}^*)} \int_\Omega e^{-(v|2t)}\, \mathrm{N}^*_{-\underline{\alpha}^*-\underline{\gamma}^*-2\frac{d}{r}}(t)\, dt \\ &= (4\pi)^d \frac{\Gamma_\Omega(-\underline{\alpha}^* - \underline{\gamma}^* - \frac{d}{r})}{\Gamma_\Omega(-\underline{\alpha}^*)\,\Gamma_\Omega(-\underline{\gamma}^*)}\, \mathrm{N}_{\underline{\alpha}+\underline{\gamma}+\frac{d}{r}}(v).\end{aligned}$$ □

Proof of Theorem 5.2. In view of (5.7) we have only to find conditions for the finiteness of $I_{\underline{\alpha},\underline{\beta},\underline{\gamma}}(ie, ie)$, and to compute it. Notice first that, with the notation (5.8), we have

$$I_{\underline{\alpha},\underline{\beta},\underline{\gamma}}(ie, ie) = \int_\Omega \left(\int_X \overline{f_{\underline{\alpha},e+y}(x)}\, f_{\underline{\gamma},e+y}(x)\, dx \right) \mathrm{N}_{\underline{\beta}-p}(y)\, dy.$$

Therefore, if (5.10) holds, we obtain

$$\begin{aligned}I_{\underline{\alpha},\underline{\beta},\underline{\gamma}}(ie, ie) &= \frac{(4\pi)^d\,\Gamma_\Omega(-\underline{\alpha}^* - \underline{\gamma}^* - \frac{d}{r})}{\Gamma_\Omega(-\underline{\alpha}^*)\,\Gamma_\Omega(-\underline{\gamma}^*)} \int_\Omega \mathrm{N}_{\underline{\alpha}+\underline{\gamma}+\frac{d}{r}}(e+y)\, \mathrm{N}_{\underline{\beta}-p}(y)\, dy \\ &= \frac{(4\pi)^d}{\Gamma_\Omega(-\underline{\alpha}^*)\,\Gamma_\Omega(-\underline{\gamma}^*)} \int_\Omega \left(\int_\Omega e^{-(y|t)}\, \mathrm{N}_{\underline{\beta}-p}(y)\, dy \right) e^{-\operatorname{tr}(t)}\, \mathrm{N}^*_{-\underline{\alpha}^*-\underline{\gamma}^*-2\frac{d}{r}}(t)\, dt \\ &= \frac{(4\pi)^d\,\Gamma_\Omega(\underline{\beta} - \frac{d}{r})}{\Gamma_\Omega(-\underline{\alpha}^*)\,\Gamma_\Omega(-\underline{\gamma}^*)} \int_\Omega e^{-\operatorname{tr}(t)}\, \mathrm{N}^*_{-\underline{\alpha}^*-\underline{\beta}^*-\underline{\gamma}^*-\frac{d}{r}}(t)\, dt \\ &= \frac{(4\pi)^d\,\Gamma_\Omega(\underline{\beta} - \frac{d}{r})\,\Gamma_\Omega(-\underline{\alpha}^* - \underline{\beta}^* - \underline{\gamma}^*)}{\Gamma_\Omega(-\underline{\alpha}^*)\,\Gamma_\Omega(-\underline{\gamma}^*)},\end{aligned}$$

provided we have also

$$\beta_j > \frac{d}{r} + (j-1)\frac{a}{2}, \quad j = 1, 2, \ldots, r$$

and

$$\alpha_j + \beta_j + \gamma_j < (r-j)\frac{a}{2}, \quad j = 1, 2, \ldots, r.$$

This completes the proof of Theorem 5.2. □

For the proof of Theorem 5.1 we shall need also the following identity.

Lemma 5.6. *Let* $\underline{\alpha} \in \mathbb{C}^r$ *satisfy* $\mathrm{Re}(\alpha_j) > (j-1)\frac{a}{2}$ *for* $j = 1, 2, \ldots, r$. *Then*

$$\Gamma_\Omega(\underline{\alpha}) = \Gamma_\Omega\left(\underline{\alpha}^* + 2\underline{\rho} - \frac{d}{r}\right)$$

where $\underline{\rho}$ *is defined by* (5.3).

Proof. Notice first that for all $1 \le j \le r$

$$\left(\underline{\alpha}^* + 2\underline{\rho} - \frac{d}{r}\right)_j - (j-1)\frac{a}{2} = \alpha_{r+1-j} - (r-j)\frac{a}{2},$$

and so the real parts of both sides are positive simultaneously. Next, when $\mathrm{Re}\,\alpha_j > (j-1)\frac{a}{2}$ for all j,

$$\begin{aligned}
\Gamma_\Omega(\underline{\alpha}) &= (2\pi)^{\frac{d-r}{2}} \prod_{j=1}^{r} \Gamma\left(\alpha_j - (j-1)\frac{a}{2}\right) \\
&= (2\pi)^{\frac{d-r}{2}} \prod_{j=1}^{r} \Gamma\left(\alpha_{r+1-j} - (r-j)\frac{a}{2}\right) \\
&= (2\pi)^{\frac{d-r}{2}} \prod_{j=1}^{r} \Gamma\left(\left(\underline{\alpha}^* + 2\underline{\rho} - \frac{d}{r}\right)_j - (j-1)\frac{a}{2}\right) \\
&= \Gamma_\Omega\left(\underline{\alpha}^* + 2\underline{\rho} - \frac{d}{r}\right).
\end{aligned}$$

□

Proof of Theorem 5.1. In the case of the Toeplitz calculus in the context of the spaces $L^2_a(T(\Omega), \mu_\nu)$ the function $F(z, w)$ is given by

$$F(z, w) = \frac{K(z, w)\, K(w, ie)}{K(z, ie)\, K(w, w)}.$$

We claim that

$${}_z\widetilde{F}(\underline{\lambda}) = c(\nu, \underline{\lambda})\, E_{\underline{\lambda}}(z),$$

where

$$c(\nu, \underline{\lambda}) = \frac{\Gamma_\Omega(\underline{\lambda} + \underline{\rho} + \nu - \frac{d}{r})\, \Gamma_\Omega(-\underline{\lambda}^* - \underline{\rho}^* + \nu)}{\Gamma_\Omega(\nu)\, \Gamma_\Omega(\nu - \frac{d}{r})}$$

and

$$E_{\underline{\lambda}}(z) = E_{e_{\underline{\lambda}}}(z, ie) = \mathrm{N}_{\underline{\lambda}+\underline{\rho}}(\tau(z, ie)).$$

Indeed, using Theorem 5.2 and the definition (1.28)we find

$$
\begin{aligned}
{}_z\widetilde{F}(\underline{\lambda}) &= a(\nu) \int_{T(\Omega)} \frac{\mathrm{N}(\tau(z,w))^{-\nu}\, \mathrm{N}(\tau(w,ie))^{-\nu}}{\mathrm{N}(\tau(z,ie))^{-\nu}\, \mathrm{N}(\tau(w,w))^{-\nu}}\, \mathrm{N}_{\underline{\lambda}+\underline{\rho}}(\tau(w,w))\, d\mu_0(w) \\
&= a(\nu)\, \mathrm{N}(\tau(z,ie))^{\nu}\, I_{-\nu,\underline{\lambda}+\underline{\rho}+\nu,-\nu}(z,ie) \\
&= \frac{\Gamma_\Omega(\underline{\lambda}+\underline{\rho}+\nu-\frac{d}{r})\,\Gamma_\Omega(-\underline{\lambda}^*-\underline{\rho}^*+\nu)}{\Gamma_\Omega(\nu)\,\Gamma_\Omega(\nu-\frac{d}{r})}\, \mathrm{N}_{\underline{\lambda}+\underline{\rho}}(\tau(z,ie)).
\end{aligned}
$$

Using Theorem 5.2, we obtain

$$
\begin{aligned}
\tilde{\mathcal{B}}^{(\nu)}(\underline{\lambda}) &= \int_{T(\Omega)} {}_z\widetilde{F}(\underline{\lambda})\, \overline{{}_z\widetilde{F}(-\bar{\underline{\lambda}})}\, |K^{(\nu)}(z,ie)|^2\, d\mu_\nu(z) \\
&= c(\nu,\underline{\lambda})\, \overline{c(\nu,-\bar{\underline{\lambda}})} \int_{T(\Omega)} \mathrm{N}_{\underline{\lambda}+\underline{\rho}}(\tau(z,ie))\, \overline{\mathrm{N}_{-\underline{\lambda}+\underline{\rho}}(\tau(z,ie))}\, |K^{(\nu)}(z,ie)|^2\, d\mu_\nu(z) \\
&= a(\nu)\, c(\nu,\underline{\lambda})\, c(\nu,-\underline{\lambda})\, I_{-\underline{\lambda}+\underline{\rho}-\nu,\nu,\underline{\lambda}+\underline{\rho}-\nu}(ie,ie) \\
&= a(\nu)\, c(\nu,\underline{\lambda})\, c(\nu,-\underline{\lambda})(4\pi)^d\, \frac{\Gamma_\Omega(\nu-\frac{d}{r})\,\Gamma_\Omega(-2\underline{\rho}^*+\nu)}{\Gamma_\Omega(\underline{\lambda}^*-\underline{\rho}^*+\nu)\,\Gamma_\Omega(-\underline{\lambda}^*-\underline{\rho}^*+\nu)} \\
&= \frac{\Gamma_\Omega(\underline{\lambda}+\underline{\rho}+\nu-\frac{d}{r})\,\Gamma_\Omega(-\underline{\lambda}+\underline{\rho}+\nu-\frac{d}{r})}{\Gamma_\Omega(\nu)\,\Gamma_\Omega(\nu-\frac{d}{r})},
\end{aligned}
$$

where we used the definitions of $c(\nu,\underline{\lambda})$, $c(\nu,-\underline{\lambda})$, $a(\nu)$ and Lemma 5.6 to obtain

$$
\Gamma_\Omega(-2\underline{\rho}^*+\nu) = \Gamma_\Omega\left(-2\underline{\rho}+\nu+2\underline{\rho}-\frac{d}{r}\right) = \Gamma_\Omega\left(\nu-\frac{d}{r}\right).
$$

Notice also that by Lemma 5.6

$$
\begin{aligned}
\Gamma_\Omega\left(-\underline{\lambda}+\underline{\rho}+\nu-\frac{d}{r}\right) &= \Gamma_\Omega\left(-\underline{\lambda}^*+\underline{\rho}^*+\nu-\frac{d}{r}+2\underline{\rho}-\frac{d}{r}\right) \\
&= \Gamma_\Omega\left(-\underline{\lambda}^*-\underline{\rho}^*+\nu\right),
\end{aligned}
$$

since

$$
(2\underline{\rho}^*+2\underline{\rho})_j = \frac{2d}{r} \quad \text{for } j = 1,2,\dots,r.
$$

Therefore the eigenvalue of the Berezin transform $\mathcal{B}^{(\nu)}$ can be written in the form

$$
\widetilde{\mathcal{B}^{(\nu)}}(\underline{\lambda}) = \frac{\Gamma_\Omega(\underline{\lambda}+\underline{\rho}+\nu-\frac{d}{r})\,\Gamma_\Omega(-\underline{\lambda}^*-\underline{\rho}^*+\nu)}{\Gamma_\Omega(\nu)\,\Gamma_\Omega(\nu-\frac{d}{r})}.
$$

Finally, using (1.26) and (5.2) we obtain

$$\widetilde{\mathcal{B}^{(\nu)}}(\underline{\lambda}) = \prod_{j=1}^{r} \frac{\Gamma(\lambda_j + \rho_j + \nu - \frac{d}{r} - (j-1)\frac{a}{2})\,\Gamma(-\lambda_j + \rho_j + \nu - \frac{d}{r} - (j-1)\frac{a}{2})}{\Gamma(\nu - (j-1)\frac{a}{2})\,\Gamma(\nu - \frac{d}{r} - (j-1)\frac{a}{2})}$$
$$= \prod_{j=1}^{r} \frac{\Gamma(\lambda_j + t_\nu)\,\Gamma(-\lambda_j + t_\nu)}{\Gamma(\rho_j + t_\nu)\,\Gamma(-\rho_j + t_\nu)}$$

since $\rho_j - \frac{d}{r} - (j-1)\frac{a}{2} = -\frac{p-1}{2}$, $t_\nu = \nu - \frac{p-1}{2}$ and

$$\prod_{j=1}^{r} \Gamma\left(\nu - (j-1)\frac{a}{2}\right) = \prod_{j=1}^{r} \Gamma\left(\nu - (r-j)\frac{a}{2}\right) = \prod_{j=1}^{r} (\rho_j + t_\nu).$$

This completes the proof of Theorem 5.1. □

Remark 5.1. Quite generally, it is easy to see that the fundamental function $a_{\mathcal{T}}(\underline{\lambda})$ of the Toeplitz calculus $\mathcal{T}$ is equal to the eigenvalue of the Berezin transform $\mathcal{B} = \mathcal{T}'\,\mathcal{T}$:

$$a_{\mathcal{T}}(\underline{\lambda}) = \widetilde{\mathcal{B}}(\underline{\lambda}).$$

Therefore Theorem 5.1 yields (3.11).

Remark 5.2. The right hand side of (5.4) is an entire meromorphic function of $\underline{\lambda}$ which is analytic in the tube

$$Q_\nu := \{\underline{\lambda} \in \mathbb{C}^r;\ |\operatorname{Re}(\lambda_j)| < t_\nu\}. \tag{5.11}$$

Therefore the above proofs show that (5.4) holds for all $\underline{\lambda} \in Q_\nu$, $\nu > p-1$. We conjecture that it is possible to consider the Toeplitz quantization $\mathcal{T}^{(\nu)}$ and the associated link transform $\mathcal{B}^{(\nu)}$ in a canonical and explicit way for all $\nu > \frac{p-1}{2}$, and that (5.4) is valid for all $\underline{\lambda} \in Q_\nu$ in the extended range of ν. This may require the techniques of [AU97] and [AU99].

References

[A96] J. Arazy Boundedness and compactness of generalized Hankel operators on bounded symmetric domains, J. Funct. Anal. 137 (1996), 97–151.

[AU97] J. Arazy and H. Upmeier, Invariant Inner Products in Spaces of Holomorphic Functions on Bounded Symmetric Domains, Documenta Mathematica 2 (1997), 213–261.

[AU98] J. Arazy and H. Upmeier, Discrete Series Representations and Integration over Boundary Orbits of Symmetric Domains, in: Perspectives on Quantization (L. A. Coburn and M. A. Rieffel, eds.), Contemp. Math. 214 (1998), 1–22.

[AU99] J. Arazy and H. Upmeier, Boundary integration and the discrete Wallach points, ESI preprint No. 762 (1999), to be submitted to Mem. Amer. Math. Soc.

[AU01] J. Arazy and H. Upmeier, Invariant symbolic calculi and eigenvalues of invariant operators on real symmetric domains, in preparation.

[Be71] F. Berezin, Wick and anti-Wick operator symbols, Mat. Sb. 86 (4) (1971), 578–610; English translation: Math. USSR. Sb. 15 (1971), 577–606.

[Be72] F. Berezin, Covariant and contravariant symbols of operators, Math. USSR-Izv. 6 (5) (1972), 1117–1151.

[Be73] F. Berezin, Quantization in complex bounded domains, Soviet Math. Dokl. 14 (4) (1973), 1209–1213.

[Be74-1] F. Berezin, General concept of Quantization, Comm. Math. Phys. 40 (1975), 153–174.

[Be74-2] F. Berezin, Quantization, Math. USSR-Izv. 8 (5) (1974), 1109–1165.

[Be75] F. Berezin, Quantization in complex symmetric spaces, Math. USSR-Izv. 9 (2) (1975), 341–379.

[Be78] F. Berezin, On relation between covariant and contravariant symbols of operators for complex classical domains, Soviet Math. Dokl. 19 (1978), 786–789.

[E98] M. Englis, Invariant operators and the Berezin transform on Cartan domains, Math. Nachr. 195 (1998), 61–75.

[FK90] J. Faraut and A. Korányi, Function spaces and reproducing kernels on bounded symmetric domains, J. Funct. Anal. 88 (1990), 64–89.

[FK94] J. Faraut and A. Korányi, Analysis on Symmetric Cones, Clarendon Press, Oxford, 1994.

[Gi64] S. Gindikin, Analysis on homogeneous domains, Russian Math. Surveys 19 (1964), 1–89.

[Gi75] S. Gindikin, Invariant generalized functions in homogeneous domains, Funct. Anal. Appl. 9 (1975), 50–52.

[He78] S. Helgason, Differential geometry, Lie groups, and symmetric spaces, Grad. Stud. Math. 34, Amer. Math. Soc., Providence, RI, 2001.

[He84] S. Helgason, Groups and geometric analysis, Math. Surveys Monogr. 83, Amer. Math. Soc., Providence, RI, 2000.

[Hu63] L. K. Hua, Harmonic Analysis of Functions of Several Complex Variables in the Classical Domains, Transl. Math. Monogr. 6, Amer. Math. Soc., Providence, RI, 1963.

[La86] M. Lassalle, Noyau de Szegö, K-types et algèbres de Jordan, C. R. Acad. Sci. Paris 303 (1986), 1–4.

[La87] M. Lassalle, Algèbres de Jordan et ensemble de Wallach, Invent. Math. 89 (1987), 375–393.

[Lo77] O. Loos, Bounded Symmetric Domains and Jordan Pairs, Univ. California, Irvine, 1977.

[Sch69] W. Schmid, Die Randwerte holomorpher Funktionen auf hermitesch symmetrischen Räumen, Invent. Math. 8 (1969), 61–80.

[Un98] A. Unterberger, Quantization, Symmetries and Relativity, Contemp. Math. 214 (1998), 169–187.

[Up87] H. Upmeier, Jordan algebras in analysis, operator theory, and quantum mechanics, CBMS Ser. Math. 67, Amer. Math. Soc., Providence, RI, 1987.

[UU94] A. Unterberger and H. Upmeier, Berezin Transform and Invariant Differential Operators, Comm. Math. Phys. 164 (1994), 563–597.

[UU96] A. Unterberger and H. Upmeier, Pseudodifferential Analysis on symmetric cones, CRC Press, Boca Raton, 1996

[VR76] M. Vergne and H. Rossi, Analytic continuation of holomorphic discrete series of a semi-simple Lie group, Acta Math. 136 (1975), 1–59.

[W79] N. Wallach, The analytic continuation of the discrete series, I, II, Trans. Amer. Math. Soc. 251 (1979), 1–17, 19–37.

Jonathan Arazy, Department of Mathematics, University of Haifa, Haifa 31905, Israel
E-mail: `jarazy@math.haifa.ac.il`

Harald Upmeier, Fachbereich Mathematik, Universität Marburg, Hans-Meerwein Strasse, D-35032, Germany
E-mail: `upmeier@mathematik.uni-marburg.de`

Convex hulls of Coxeter groups*

Jeremy Brandman, James Fowler, Brian Lins,
Ilya Spitkovsky and Nahum Zobin

To Jaak Peetre at his 65th anniversary

Abstract. We survey known and new results concerning the geometric structure of the convex hulls of finite irreducible Coxeter groups. In particular we consider a conjecture concerning the normals to the faces of maximal dimension of these convex hulls. This conjecture is related to a theorem of Birkhoff and also to interpolation of operators. We describe various approaches to its proof as well as various computer calculations involved.

2000 Mathematics Subject Classification: Primary 51F15; Secondary 20F55, 52A20

1. Introduction

Let G be a finite irreducible Coxeter group naturally acting on a finite dimensional real Euclidean space V. So, G is a subset of the linear space End V of linear operators in V. Let $\mathbf{I}$ denote the identity operator.

We study the geometry of the convex hull of G, which we denote by conv G. This is a convex polytope in the linear space End V. What is its facial structure? In particular, what are its faces of maximal dimension? All these problems naturally arise in various disguises – we were mostly motivated by a duality approach to interpolation of operators discussed below.

Recently there has been substantial progress in this direction, see [10]. During the summer of 2000 we were able to move further, relying heavily on computer calculations. The goal of this article is to give a full account of the present state of the problem.

*A part of this research was conducted during the summer of 2000 at the College of William and Mary's Research Experiences for Undergraduates Program and was supported by NSF REU grant DMS-99-87803. J. Brandman and J. Fowler were student participants, N. Zobin was a mentor participant in the program. B. Lins was supported by a Monroe Scholar Summer Research Grant, I. Spitkovsky was partially supported by NSF Grant DMS-99-88579. I. Spitkovsky and N. Zobin were also partially supported by Summer Research Grants at the College of William and Mary.

1.1. Birkhoff's Theorem

To describe the results, we start with a formulation of a well known theorem due to G. Birkhoff [2].

Definition 1.1. Let $T = (t_{ij})$ be an $n \times n$ matrix. It is called *bistochastic* (or *doubly stochastic*) if its entries are non-negative and

$$\text{for every } j,\ 1 \leq j \leq n, \quad \sum_{i=1}^{n} t_{ij} = 1, \quad \sum_{i=1}^{n} t_{ji} = 1.$$

The set of all bistochastic $n \times n$ matrices is denoted by Ω_n.

Obviously, the set Ω_n is a convex polytope in the space of $n \times n$ matrices. Since there are $2n - 1$ independent linear equations involved in the definition, this polytope is actually in a subspace of dimension $n^2 - (2n - 1) = (n - 1)^2$.

Definition 1.2. Let Perm_n denote the group of $n \times n$ permutation matrices.

Theorem 1.3 (G. Birkhoff, [2]). *The set of vertices of the polytope Ω_n coincides with the set of permutation matrices* Perm_n.

So, the polytope of bistochastic matrices turns out to be nothing else but the convex hull of the group Perm_n. Let us reformulate this result. The group Perm_n acts reducibly on $\mathbb{R}^n$: it fixes the vector $e = (1, 1, \ldots, 1)$ and acts irreducibly on its orthogonal complement $e^\perp = \{(x_i) \in \mathbb{R}^n : \sum x_i = 0\}$. Consider the group $A_{n-1} = \mathrm{Perm}_n |_{e^\perp}$. This group is maybe the most important example of a finite irreducible Coxeter group. It spans an $(n-1)^2$-dimensional subspace in the space of $n \times n$ matrices. Note that all bistochastic matrices also fix the vector e and leave $e^\perp$ invariant – this is actually a part of the definition. So, Theorem 1.3 says that the polytope of bistochastic matrices is located in an $(n-1)^2$-dimensional subspace and its vertices are operators from A_{n-1}. Thus the definition of bistochastic matrices describes the convex hull of the Coxeter group A_{n-1} in terms of linear inequalities. One can rather easily see that this set of inequalities is the smallest possible: each inequality describes a half-space bounded by a face of the polytope. But what is the invariant meaning of these inequalities?

There exists a whole industry dealing with generalizations of results known for groups A_n to other Coxeter groups. Very often these generalizations turn out to be non-trivial and useful, providing deeper insights into the results. Sometimes they are simply exercises in the theory of Coxeter groups. It is usually very interesting if a result valid for A_n proves not to be valid for all Coxeter groups – such results are usually most challenging. This is exactly the case with generalizations of the Birkhoff Theorem: a natural generalization – see Conjecture 1.4 below – is not true for Coxeter groups whose graphs are branching, but it seems to be true for Coxeter groups whose graphs are non-branching (as of January 2001, Conjecture 1.4 has been verified for all finite irreducible Coxeter groups except H_4).

Actually, our main impetus came not from a simple (though natural) desire to generalize but from a rather unexpected source – the theory of Interpolation of Operators.

The problem becomes very natural in that setting and the main Conjecture arises from some deep results related to interpolation of operators in spaces with Coxeter-invariant norms. We describe this circle of ideas in Subsection 1.3 below.

1.2. Conjecture

Returning to the general setting of an irreducible Coxeter group G we may say that we are interested in calculation of the faces of conv G of maximal dimension $(\dim V)^2 - 1$. Each such face is a polytope of full dimension in an affine hyperplane $\{A \in \operatorname{End} V : f(A) = c\}$ where f is a linear functional on End V. These (properly scaled) functionals are naturally identified with elements of $\operatorname{Extr}(\operatorname{conv} G)^\circ$ – the set of extreme elements of the polar polytope $(\operatorname{conv} G)^\circ = \{h \in (\operatorname{End} V)^* : \forall A \in \operatorname{conv} G \; h(A) \leq 1\}$. We prefer to introduce a Euclidean structure into End V which allows to identify the spaces End V and $(\operatorname{End} V)^*$ and to treat the mentioned functionals as normals to the face. The needed scalar product on End V is given by the formula $(A, B) = \operatorname{trace}(AB^*)$, where B^* is the operator adjoint to B (to define B^* we need the Euclidean structure in V). Moreover, the Euclidean structure in V allows to identify V^* with V and therefore to identify $V \otimes V$ with $V \otimes V^*$. In turn, $V \otimes V^*$ is naturally identified with the space $(\operatorname{End} V)^*$, which is already identified with End V, so we may identify the spaces $V \otimes V$ and End V. In particular, for any $x, y \in V$ we identify $x \otimes y$ with the rank one operator $z \mapsto x\langle y, z\rangle$.

Let us describe the Conjecture.

Consider the set $\mathcal{W}(G)$ of all weights of the group G. Throughout this paper we consider "non-normalized weights", i.e., nonzero vectors directed along extreme rays of Weyl chambers – the term "weight" is often used to denote only specially normalized vectors of this type. Each weight ω is associated with a vertex $\pi(\omega)$ of the Coxeter graph $\Gamma(G)$.

Let E_G denote the set of *extremal weights* of the group G, i.e., those associated with the *end vertices* of the Coxeter graph $\Gamma(G)$.

Put $m_G(x, y) = \max\{\langle gx, y\rangle : g \in G\}$. One can show that in the case of an irreducible Coxeter group G the quantity $m_G(x, y)$ is strictly positive for any nonzero vectors $x, y \in V$.

Let

$$\mathcal{B}_G = \{\omega \otimes \tau / m_G(\omega, \tau) : \omega, \tau \in E_G, \pi(\omega) \neq \pi(\tau)\}.$$

We call the elements of $\mathcal{B}_G$ the *Birkhoff tensors*. Note that Birkhoff tensors all have rank one.

The importance of Birkhoff tensors for our problem is apparent because of the following result (see Theorem 4.4 below):

$$\mathcal{B}_G = (\operatorname{Extr}(\operatorname{conv} G)^\circ) \bigcap (\text{ rank 1 tensors }).$$

The following conjecture was first proposed in 1979 by Veronica Zobin [19] and later elaborated by the last author:

Conjecture 1.4. (a) *If the Coxeter graph* $\Gamma(G)$ *is non-branching then*

$$\mathcal{B}_G = \mathrm{Extr}(\mathrm{conv}\, G)^\circ.$$

(b) *If the Coxeter graph* $\Gamma(G)$ *is branching then*

$$\mathcal{B}_G \subsetneq \mathrm{Extr}(\mathrm{conv}\, G)^\circ.$$

Part (b) of the Conjecture was proved in [10]; we reproduce this proof in Section 8 below. As for Part (a), it was proved in [10] for all infinite families of Coxeter groups with non-branching graphs, and we have verified it for the groups F_4 and H_3 by rather nontrivial computer calculations. The only remaining group is H_4. The computer calculations that were successful for other groups could not be completed for H_4 on the available computers, mainly because of insufficient random access memory.

It should be noted that the success in proving Part (b) was achieved with a very strong computer component: a computer calculation found an essentially unique tensor of rank 3 belonging to the set $\mathrm{Extr}(\mathrm{conv}\, D_4)^\circ$, and then the general case of a Coxeter group with a branching graph was reduced to this one. We still do not quite understand the invariant meaning of this rank 3 tensor. However, after this tensor is found one can verify by hand that it really belongs to $\mathrm{Extr}(\mathrm{conv}\, D_4)^\circ$, so the proof does not formally depend upon the computer calculations.

Certainly, it would be very interesting to find a unified approach avoiding the case study of irreducible Coxeter groups and heavy use of computers. We believe that there must be a general simple reason for the validity of the Conjecture.

1.3. Interpolation of operators

The above conjecture naturally appeared in the theory of interpolation of operators in spaces with given symmetries – see [15, 16]. The main object studied in these papers was the convex set env G defined as the semigroup of all linear operators in V which transform every G-invariant convex closed set into itself:

$$\mathrm{env}\, G = \{T \in \mathrm{End}\, V : T(U) \subset U \text{ for every convex closed G-invariant } U \subset V\}.$$

Obviously,

$$\mathrm{conv}\, G \subset \mathrm{env}\, G.$$

If these two sets coincide then the only operators simultaneously contracting all G-invariant closed convex sets are those which have this property almost by definition. So this case is not interesting from the point of view of interpolation of operators. The opposite case is much more interesting – there are nontrivial operators that can be interpolated. In the case of an irreducible finite Coxeter group G we have a convenient dual description of the set env G – this description is one of the central results of [16]:

$$\mathrm{Extr}(\mathrm{env}\, G)^{\circ} = \mathcal{B}_G.$$

So the question is if conv G = env G, i.e., if $\mathrm{Extr}(G^{\circ}) = \mathcal{B}_G$.

Currently we know that the latter equality is not true for Coxeter groups with branching graphs. This leads to two difficult problems. First, what is $\mathrm{Extr}(G^{\circ})$ for such groups? Second, what are the extreme common contractions, i.e., the extreme elements of the semigroup env G? As of now, we have no viable conjectures.

1.4. Geometry of orbihedra

The problem of describing the facial structure of conv G is a particular case of a more general problem, which naturally arises in several areas of Operator Theory and Representation Theory. Consider a finite group G of linear operators acting on V. For every nonzero $x \in V$ consider the related G-orbihedron $\mathrm{Co}_G\, x$ – the convex hull of the G-orbit of x. The convex geometry of G-orbihedra is important in numerous problems. In the case when G is a Coxeter group, one can obtain very detailed information regarding the facial structure of $\mathrm{Co}_G\, x$ in convenient geometric terms – see [11] for the most comprehensive account. But as soon as we depart from Coxeter groups in their natural representations the situation becomes much more complicated. For example, the natural action of $G \times G$ on End V by pre- and post-multiplications is not generated by reflections across hyperplanes, and all of the powerful machinery developed in [11] is not applicable. Moreover, preliminary computer experiments (C. K. Li, I. Spitkovsky, N. Zobin) show that the geometry of the related orbihedra may be very complicated. Nevertheless, there are several cases when it is possible to understand this geometry pretty well. It is more natural to consider a larger group $S_2^{\otimes}(G)$ generated by $G \times G$ and the operator $T \mapsto T^*$, where T^* is the operator adjoint to T. First, conv G can be viewed as a $S_2^{\otimes}(G)$-orbihedron generated by the identity operator, and its facial structure does not seem too bad, at least in the case of a Coxeter group with a non-branching graph. The second example is the $S_2^{\otimes}(G)$-orbihedron generated by a Birkhoff tensor. One can easily see that the group $S_2^{\otimes}$ acts transitively on the set of Birkhoff tensors of a Coxeter group with a non-branching graph, so the set $\mathcal{B}_G$ is the set of extreme vectors of a $S_2^{\otimes}(G)$-orbihedron. Since $(\mathcal{B}_G)^{\circ} = \mathrm{conv}\, G$ in this case (not yet verified for $G = H_4$) then for $b \in \mathcal{B}_G$ we have $\mathrm{Extr}(\mathrm{Co}_{S_2^{\otimes}(G)}\, b)^{\circ} = G$. For what other elements $b \in \mathrm{End}\, V$ does the related $S_2^{\otimes}(G)$-orbihedron have a simple facial structure? This is a very interesting (but seemingly difficult) problem.

Let us remark that an analogous problem for infinite groups $O(V)$ and $U(V)$ of, respectively, orthogonal operators on a real Euclidean space V and unitary operators on a complex Hermitian space V, is closely related to the theory of Schatten–von Neumann ideals, which has been studied in great depth (though in different terms).

One can answer some of these questions by rethinking classical results in the geometric theory of Schatten–von Neumann ideals (see, e.g., [6]).

Another way to study the geometric structure of conv G is to explore the group of linear operators on End V preserving conv G. This is a sort of a linear preserver problem, rather popular in Linear Algebra, see [13]. There was considerable progress in this direction recently, see [8]. A general type of answer is as follows: the only operators preserving conv G are the so-called rigid embeddings, i.e., operators of the type $\phi(A) = gAh$ or $\phi(A) = gA^*h$, where g, h belong to the normalizer of G in $O(V)$, and $gh \in G$. Such results were known for groups $O(V)$ ([14]) and A_n ([9]). Rather unexpectedly, rigid embeddings are not the only operators preserving conv B_n, see [8].

Acknowledgments. We are greatly indebted to Chi-Kwong Li and Veronica Zobin for numerous valuable discussions of various aspects of the problem. We are also thankful to Val Spitkovsky whose computer expertise was so helpful to us.

2. A brief review of Coxeter groups

Let us now address several facts concerning the theory of Coxeter groups. For greater detail, consult [1], [3], or [7]. Let $G \subset \text{End } V$ be a group. Then G is a *Coxeter group* if it is finite, generated by orthogonal reflections across hyperplanes (containing the origin), and acts *effectively* (i.e., $gx = x$ for all $g \in G$ implies $x = 0$).

2.1. Roots and weights

Let $\mathcal{M}(G)$ denote the set of all *mirrors* – the hyperplanes in V such that the orthogonal reflections across them belong to the group G. Mirrors split the space V into connected components, whose closures are polyhedral cones. These cones are called *Weyl chambers*. It is known that Weyl chambers are actually *simplicial* cones, i.e., they have exactly dim V faces of codimension 1. These faces are called the *walls* of the chamber. The simpliciality of Weyl chambers implies that each Weyl chamber has exactly dim V *extreme rays*, and each extreme ray does not belong to exactly one wall of the chamber. Unit normals to the mirrors are called *roots*, the set of all roots is denoted by $\mathcal{R}_G$.

Fix a Weyl chamber C. Consider the roots $n_i(C)$, $1 \le i \le \dim V$, perpendicular to its walls, directed inwards with respect to the chamber. We call these *fundamental* or *simple roots*. It is known that the group G is generated by reflections across the walls (i.e., across the mirrors containing the walls) of a Weyl chamber, i.e., by the operators $R_i = \mathbf{I} - 2n_i \otimes n_i$, $1 \le i \le \dim V$. Consider the vectors $\omega_j(C)$, $1 \le j \le \dim V$, such that $\langle n_i, \omega_j \rangle = c_j \delta(i - j)$, $c_j > 0$. These vectors are called the *fundamental* *weights* associated with the chamber C. The exact values of c_j are not important for our purposes (for a standard normalization of fundamental weights see [3]). One can easily see that the fundamental weights are directed along the extreme rays of C. The set of all weights (i.e., those associated with any Weyl chamber) is denoted by $\mathcal{W}_G$. The sets $\mathcal{R}_G$ and $\mathcal{W}_G$ are fibered into G-orbits. It is known that the group G acts *simply transitively* on the set of Weyl chambers (i.e., for every two chambers there exists exactly one element of G transforming one of them onto the other). This immediately implies that the G-orbit of any vector x (we denote it by $\operatorname{Orb}_G x$) intersects a Weyl chamber at exactly one point, let $x^*(C)$ denote the only point of $C \cap \operatorname{Orb}_G x$. Consider $m_G(x, y) = \max_{g \in G} \langle gx, y \rangle$. It is known (see [16]) that for any Weyl chamber C

$$m_G(x, y) = \langle x^*(C), y^*(C) \rangle \geq 0.$$

In fact, one can easily show that $m_G(x, y) > 0$ for irreducible Coxeter groups, provided x, y are both nonzero.

2.2. Coxeter graphs

Since a Coxeter group G is generated by reflections across the walls of any Weyl chamber, all information about the group is encrypted in the geometry of a Weyl chamber. In its turn the whole geometry of a Weyl chamber is described by the angles between its walls. Since the group is finite, these angles must be π/k, $k \in \mathbb{Z}_+, k \geq 2$. There is a wonderful way to encode the information about the angles in a graph. Let $\Gamma(G)$ denote the *Coxeter graph* of G, constructed as follows: the set $\operatorname{vert}(G)$ of vertices of the graph is in a one-to-one correspondence with the set of walls of a fixed Weyl chamber C, and two vertices are joined by an edge if and only if the angle between the related walls is π/k, $k \geq 3$, and $k - 2$ is the multiplicity of the edge. Since the group acts transitively on the set of Weyl chambers, the Coxeter graph does not depend upon the choice of a Weyl chamber. Every fundamental root $n_i(C)$ is naturally associated with a wall of the Weyl chamber C, so it is associated with a vertex of $\Gamma(G)$. Every fundamental weight $\omega_j(C)$ is naturally associated with a unique wall of C (namely, with the one it does not belong to), so it is naturally associated with a vertex π of the graph $\Gamma(G)$. One can easily see that all weights from the same G-orbit are associated with the same vertex, so there is a one-to-one correspondence between the G-orbits of weights and the vertices of $\Gamma(G)$. Let $\pi(\omega)$ denote the vertex of $\Gamma(G)$ associated with the G-orbit of $\omega \in \mathcal{W}_G$.

An *end vertex* of the Coxeter graph $\Gamma(G)$ is any vertex connected to only one other vertex. A weight associated with an end vertex of $\Gamma(G)$ is called an *extremal weight*, the set of extremal weights is denoted by E_G.

A Coxeter graph is *branching* if it contains a vertex (called a *branching vertex*) connected to at least three other vertices. Otherwise, the graph is *non-branching*.

It is known that a Coxeter group G is irreducible if and only if its Coxeter graph $\Gamma(G)$ is connected. All connected Coxeter graphs are classified, so all irreducible

Coxeter groups are classified (see [3], [1], [7]). In particular, the classification shows that a connected Coxeter graph contains at most one branching vertex, the branching vertex is connected to exactly three other vertices, and all edges of a branching graph have multiplicity 1.

2.3. Supports and stabilizers

Fix a Weyl chamber C, let ω_i, $1 \leq i \leq \dim V$, denote the related fundamental weights, and for each i, $1 \leq i \leq \dim V$, let W_i denote the wall of C not containing ω_i. Since C is a simplicial cone then for every $a \in V$ there exists a unique decomposition

$$a^*(C) = \sum_i \lambda_i(a^*(C))\omega_i, \quad \lambda_i(a^*(C)) \geq 0.$$

Obviously,

$$\lambda_i(a^*(C)) = \frac{\langle a^*(C), n_i \rangle}{c_i}.$$

Introduce the *support* of a as follows:

$$\operatorname{supp}_G a = \{\pi_i \in \operatorname{vert}(G) : \lambda_i(a^*(C)) > 0\} = \{\pi_i \in \operatorname{vert}(G) : a^*(C) \notin W_i\}.$$

One can easily show that $\operatorname{supp}_G a$ does not depend upon the choice of the chamber C, and it actually depends only upon the G-orbit of a.

For $a \in C$ let

$$\operatorname{Stab}_G a = \{g \in G : ga = a\}$$

It is well known (see, e.g., [3]) that this subgroup is generated by reflections across the walls W_i of C, containing a. This subgroup is not a Coxeter group since the intersection of all mirrors containing a (we denote this intersection by V^a) is a nontrivial subspace of fixed vectors. Let us restrict the action of this subgroup to its invariant subspace $V_a = (V^a)^\perp$. Thus, the nontrivial fixed vectors are cut off, and we get a Coxeter group

$$G_a = \operatorname{Stab}_G a|_{V_a}$$

acting on this subspace V_a. Its Coxeter graph can be computed as follows (see [16]):

$$\Gamma(G_a) = \Gamma(G) \setminus \operatorname{supp}_G a.$$

The latter means that all the vertices from $\operatorname{supp}_G a$ are erased, as well as all adjacent edges. So, if ω is a fundamental weight then $V_\omega = \omega^\perp$ and $\Gamma(G_\omega) = \Gamma(G) \setminus \{\pi(\omega)\}$. If ω is an extremal fundamental weight then the group G_ω acts irreducibly on $\omega^\perp$, since the graph $\Gamma(G) \setminus \{\pi(\omega)\}$ is connected.

We recall the definitions of several Coxeter groups together with their extremal weights.

2.4. Classification of irreducible Coxeter groups.

The following are descriptions of all finite irreducible Coxeter groups. First let us examine the four infinite families A_n, B_n, D_n, and $I_2(n)$.

Define Perm_{n+1} to be the group of linear operators acting on $\mathbb{R}^{n+1}$ by permutations of the canonical basis $\{e_1, e_2, e_3, \ldots, e_{n+1}\}$. We see that $e = \sum_{i=1}^{n+1} e_i$ is a fixed vector; now restrict the action to the invariant subspace $e^{\perp}$.

Definition 2.1. $A_n = \{T|_{e^{\perp}} : T \in \mathrm{Perm}_{n+1}\}$.

The related Coxeter graphs are:

Vectors $\omega_1 = e_1 - e/(n+1)$, $\omega_n = e/(n+1) - e_{n+1}$ are extremal fundamental weights.

Note for future reference that ω_1 is the orthogonal projection of the vector e_1 onto the subspace $e^{\perp}$. Also, A_n does not contain $-\mathbf{I}$ (where $\mathbf{I}$ is the identity operator), and $\omega_n \in \mathrm{Orb}_{A_n}(-\omega_1)$.

Definition 2.2. B_n is the group of linear operators acting on $\mathbb{R}^n$ by taking e_i to $p(i)e_{\sigma(i)}$, where σ is a permutation of $\{1, \ldots, n\}$ and $p(i) = \pm 1$ for $1 \le i \le n$.

The related Coxeter graphs are:

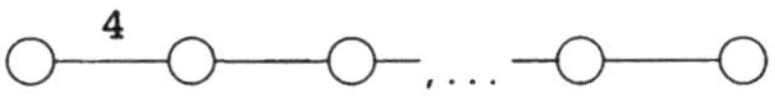

The vectors $\omega_1 = e_1$, $\omega_n = e = e_1 + \cdots + e_n$ are extremal fundamental weights.

Definition 2.3. $D_n = \{T \in B_n : T$ performs an *even* number of sign changes$\}$.

The related Coxeter graphs are:

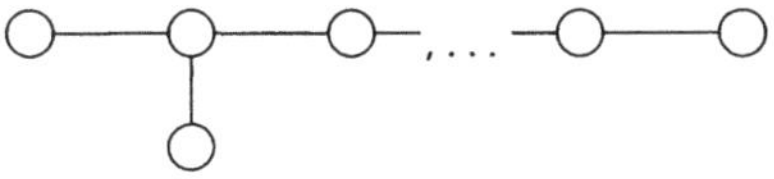

Definition 2.4. For $n \ge 3$, $I_2(n)$ is the dihedral group of order n, i.e., the group of symmetries of a regular n-gon. This is the group of operators acting on $\mathbb{R}^2$ generated by reflections across the lines $y = 0$ and $y = \tan(\pi/n)x$.

The related Coxeter graphs are:

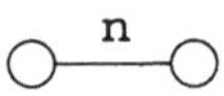

Now, let us list the exceptional groups F_4, H_3, H_4, E_6, E_7, E_8. For these groups we give their Coxeter graphs only (in the order they are listed):

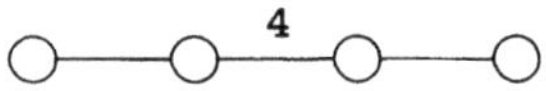

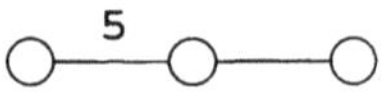

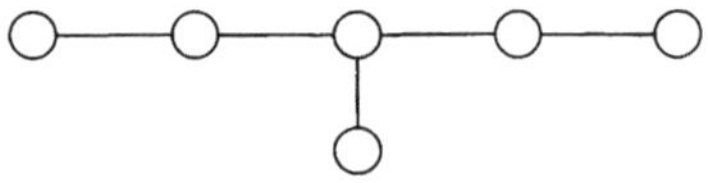

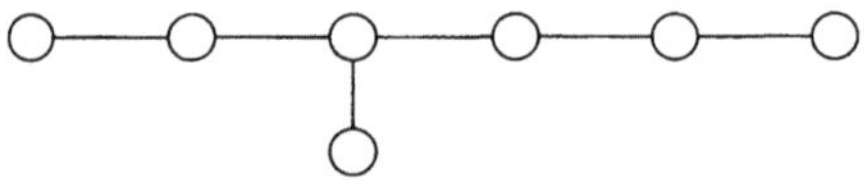

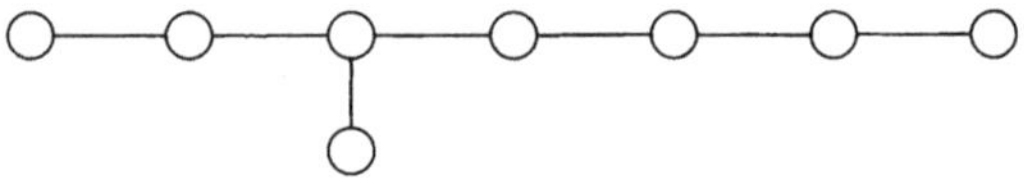

So, there exist four infinite families of irreducible Coxeter groups (A_n, B_n, D_n, $I_2(n)$) plus six exceptional groups (E_6, E_7, E_8, F_4, H_3, H_4). Each subscript indicates the dimension of the space V where the group naturally acts. Coxeter graphs of A_n, B_n, $I_2(n)$, F_4, H_3, H_4 are non-branching, Coxeter graphs of D_n, E_6, E_7, E_8 are branching.

3. The Main Theorem

Theorem 3.1. (a) *Let* $G = A_n, B_n, I_2(n), F_4$ *or* H_3. *Then*

$$\mathrm{Extr}(G^\circ) = \mathcal{B}_G.$$

(b) *Let* $G = D_n, E_6, E_7$ *or* E_8. *Then*

$$\mathrm{Extr}(G^\circ) \supsetneq \mathcal{B}_G.$$

We prove the first assertion of this theorem case by case. If $G = A_n$ or B_n, then the assertion is essentially the classical Birkhoff Theorem [2] – see details below. The case $G = I_2(n)$ was first proved in [10], but here we offer a much easier proof. The cases $G = F_4$ and $G = H_3$ are proved by computer calculations which we describe below. The second assertion – the case of the branching Coxeter graph – was proved in [10]; we reproduce the proof below.

4. Convex geometry and irreducible Coxeter groups

As it has been already mentioned in the Introduction, we equip the space End V with the scalar product $(T, S) =$ trace (TS^*), and we identify $x \otimes y$ with the rank 1 operator $z \to x\langle z, y\rangle$. One can easily check that $(x \otimes y)^* = y \otimes x$, trace $(x \otimes y) = \langle x, y\rangle$, $(x \otimes y)(w \otimes t) = (x \otimes t)\langle y, w\rangle$.

4.1. Polar sets

As usual, if we have a real Euclidean space W with a scalar product $(.,.)$ then for a subset $U \subset W$ we consider its *polar set* $U^\circ = \{z \in W : \forall\, x \in U,\ (x, z) \leq 1\}$. The set U° is a closed convex subset of W, containing 0. One can easily verify that $U^\circ = (\operatorname{conv} U)^\circ$. By the Bipolar Theorem, $(U^\circ)^\circ = \overline{\operatorname{conv}}\,(U \cup \{0\})$. So if $0 \in \operatorname{conv} G$ then $\operatorname{conv} G = ((\operatorname{conv} G)^\circ)^\circ$ (we may omit the closure since conv G is a closed polyhedron). We show that 0 is an *interior* point of conv G, see Lemma 4.2. This implies that the set $(\operatorname{conv} G)^\circ$ is compact and therefore, by the Krein–Milman Theorem, this set is the closed convex hull of its extreme points. So the set $\operatorname{Extr}(\operatorname{conv} G)^\circ$ provides a nice description of the set conv G:

$$\operatorname{conv} G = (\operatorname{Extr}(\operatorname{conv} G)^\circ)^\circ = \{T \in \operatorname{End} V : (T, b) \leq 1\ \forall\, b \in \operatorname{Extr}(\operatorname{conv} G)^\circ\}.$$

This formula and the definition of extreme points show that the elements of the set $\operatorname{Extr}(\operatorname{conv}(G)^\circ)$ are properly scaled normals to the faces of the polyhedron conv G.

4.2. Convex bodies associated with Coxeter groups

The following lemma can be deduced from the Burnside Theorem, but we prefer to give a simple direct proof, especially because the idea is also used in the proof of Theorem 4.4.

Lemma 4.1. *Let G be an irreducible Coxeter group. Then the set G spans the whole space* End V.

Proof. Fix a Weyl chamber C. As before let n_i, $i = 1, 2, \ldots, \dim V$, denote the related fundamental roots (i.e., the *unit* normals to the walls of C), associated with the vertices π_i of the Coxeter graph $\Gamma(G)$. These roots form a basis of V. The group G is generated by the reflections $R_i = \mathbf{I} - 2n_i \otimes n_i$, $i = 1, 2, \ldots, \dim V$. Since $\mathbf{I} \in G$, all operators $n_i \otimes n_i$ are in span G. Considering the products $R_i R_j = \mathbf{I} - 2n_i \otimes n_i - 2n_j \otimes n_j + 4\langle n_i, n_j \rangle n_i \otimes n_j$ such that the vertices π_i, π_j are connected by an edge (and therefore $\langle n_i, n_j \rangle \neq 0$), shows that all such operators $n_i \otimes n_j$ are in span G. Now choose three vertices π_i, π_j, π_k such that the second one is connected by edges to the first and the third ones. Considering the product $R_i R_j R_k \in G$ and using the previous remarks, we show that $n_i \otimes n_k \in$ span G. Repeating the same trick, we show that all operators $n_i \otimes n_j$ are in span G, provided the vertices π_i, π_j can be connected by a simple path in $\Gamma(G)$. Since the Coxeter graph of an irreducible group is connected, the lemma is proven. □

Lemma 4.2. *Let G be an irreducible Coxeter group. Then 0 is an interior point of the set* conv G.

Proof. Consider the arithmetic mean av_G of the elements of G. The group G obviously fixes every element in the range of av_G, but this irreducible group cannot have nonzero fixed vectors, therefore $av_G = 0$. So, $0 = av_G \in \operatorname{conv} G$. Assuming that 0 is *not* an interior point of conv G, we find a nonzero operator $b \in \operatorname{End} V$ such that $(g, b) \leq 0$ for all $g \in G$. Therefore either $G \subset \{a \in \operatorname{End} V : (a, b) = 0\}$, or $(av_G, b) < 0$. The first is impossible because G spans the space End V, and the second is impossible because $av_G = 0$. □

This result implies that the set $G^\circ = (\operatorname{conv} G)^\circ$ is compact and therefore $G^\circ = \operatorname{conv} \operatorname{Extr} G^\circ$.

Noticing that the set $\mathcal{B}_G$ consists of rank 1 operators and is invariant under pre- and post-multiplications by operators from G we arrive to the following result, which will be needed for Corollary 4.5:

Corollary 4.3. *Let G be an irreducible Coxeter group. Then* $0 \in \operatorname{conv}(\mathcal{B}_G)$.

Theorem 4.4. $\mathcal{B}_G = (\operatorname{Extr} G^\circ) \cap (\,$*rank* 1 *tensors*$\,)$

Proof. One of the main results of [16] asserts that

$$\mathcal{B}_G = \operatorname{Extr} \operatorname{conv}(G^\circ \cap (\text{ rank 1 tensors })).$$

This is a rather deep result closely connected with the approach to interpolation of operators outlined in the Introduction.

Let us prove that $\mathcal{B}_G \subset \operatorname{Extr} G^\circ$. This will obviously imply the assertion of the theorem.

Choose two extremal fundamental weights ω, τ belonging to a Weyl chamber C, such that $\pi(\omega) \neq \pi(\tau)$. It suffices to show that $(\omega \otimes \tau)/m_G(\omega, \tau) \in \operatorname{Extr} G^\circ$. Consider the set $\mathcal{M} = \{g \in G : (g, \omega \otimes \tau) = \langle g\tau, \omega \rangle = m_G(\tau, \omega) = m_G(\omega, \tau)\}$. Since for any $g \in G$ we have $(g, \omega \otimes \tau) = \langle g\tau, \omega \rangle \leq m_G(\omega, \tau)$, conv $\mathcal{M}$ is a face of conv G and all we need to show is that its dimension is maximal, i.e., to prove that $\mathcal{M}$ spans End V.

Define $\mathcal{P} = \{hg : h \in \mathrm{Stab}_G(\omega), g \in \mathrm{Stab}_G(\tau)\}$. Obviously, $\mathcal{P} \subset \mathcal{M}$.

Let n_i, $1 \le i \le N = \dim V$, denote the fundamental roots associated with the chamber C; we assume that all roots are of unit length. Let ω_i, $1 \le i \le N$, denote the related fundamental weights, we assume that $\tau = \omega_1$, $\omega = \omega_N$. Let $R_j = \mathbf{I} - 2n_j \otimes n_j$ be the corresponding reflections. Recall that $\mathrm{Stab}_G(\omega_i)$ is generated by $\{R_j : j \ne i\}$.

Obviously, $\mathbf{I} \in \mathrm{Stab}_G(\omega_1) \bigcap \mathrm{Stab}_G(\omega_N)$. Also, note $R_j \in \mathrm{Stab}_G(\omega_1)$ for all $1 < j \le N$, and $R_j \in \mathrm{Stab}_G(\omega_N)$ for all $1 \le j < N$. Thus, for all $1 < j \le N$, $n_j \otimes n_j \in \mathrm{span}\,\mathrm{Stab}_G(\omega_1)$. Similarly, for all $1 \le j < N$, $n_j \otimes n_j \in \mathrm{span}\,\mathrm{Stab}_G(\omega_N)$.

Choose any i, j such that $1 < i, j \le N$. Let $\pi_i = \pi_{k_1}, \pi_{k_2}, \dots, \pi_{k_r} = \pi_j$ be a simple path in $\Gamma(G)$, connecting π_i to π_j. Such a path exists since $\Gamma(G)$ is connected. For all $1 \le l \le r$, we see that $k_l \ne 1$, so $n_{k_l} \otimes n_{k_l} \in \mathrm{span}\,\mathrm{Stab}_G(\omega_1)$. Then the product $(n_{k_1} \otimes n_{k_1})(n_{k_2} \otimes n_{k_2}) \dots (n_{k_r} \otimes n_{k_r})$ is also in $\mathrm{span}\,\mathrm{Stab}_G(\omega_1)$. Since

$$\begin{aligned}(n_{k_1} \otimes n_{k_1})&(n_{k_2} \otimes n_{k_2}) \dots (n_{k_r} \otimes n_{k_r})\\ &= \langle n_{k_1}, n_{k_2}\rangle\langle n_{k_2}, n_{k_3}\rangle \dots \langle n_{k_{r-1}}, n_{k_r}\rangle (n_{k_1} \otimes n_{k_r})\end{aligned}$$

and for any $1 \le l < r$, $\langle n_{k_l}, n_{k_{l+1}}\rangle \ne 0$ (the vertices π_{k_l} and $\pi_{k_{l+1}}$ are joined in $\Gamma(G)$), $n_{k_1} \otimes n_{k_r} = n_i \otimes n_j \in \mathrm{span}\,\mathrm{Stab}_G(\omega_1)$ for all $1 < i, j \le N$. Repeating the same argument for $\mathrm{Stab}_G(\omega_N)$ yields $n_i \otimes n_j \in \mathrm{span}\,\mathrm{Stab}_G(\omega_N)$ for all $1 \le i, j < N$.

Choose π_m adjacent to π_1. Now $(n_1 \otimes n_1)(n_m \otimes n_N) \in \mathrm{span}(\mathcal{P})$. Since $\langle n_1, n_m\rangle \ne 0$, $n_1 \otimes n_N \in \mathrm{span}(\mathcal{P})$.

To show that $n_N \otimes n_1 \in \mathrm{span}(\mathcal{P})$ requires a slightly more refined argument. Since the system $\{n_i : 1 \le i \le N\}$ is a basis in the space V, and the system $\{\omega_i : 1 \le i \le N\}$ is biorthogonal to this basis, then one can easily show that

$$\mathbf{I} = \sum_{i,j=1}^{N} \langle \omega_i, \omega_j\rangle n_j \otimes n_i$$

Notice that $\langle \omega_i, \omega_j\rangle \ne 0$ (in fact, > 0) for any $1 \le i, j \le N$, because ω_i and ω_j are in the same Weyl chamber, and G is irreducible. For all $(i, j) \ne (N, 1)$, $n_i \otimes n_j \in \mathrm{span}(\mathcal{P})$, and $\mathbf{I} \in \mathrm{span}(\mathcal{P})$, so

$$\mathbf{I} - \sum_{\substack{1 \le i,j \le N\\ (i,j) \ne (N,1)}} \langle \omega_i, \omega_j\rangle n_i \otimes n_j = \langle \omega_n, \omega_1\rangle n_N \otimes n_1$$

is in $\mathrm{span}(\mathcal{P})$. Therefore, $n_N \otimes n_1 \in \mathrm{span}(\mathcal{P})$.

Thus, for all $1 \le i, j \le N$, $n_i \otimes n_j \in \mathrm{span}\,\mathcal{P}$. Since $\{n_i \otimes n_j : 1 \le i, j \le N\}$ form a basis for $\mathrm{End}\,V$ and $\mathcal{P} \subset \mathcal{M}$, we see that $\mathcal{M}$ spans $\mathrm{End}\,V$ as required. □

The next result easily follows from the previous ones and the Bipolar Theorem:

Corollary 4.5. *The following are equivalent:*

1. $\mathcal{B}_G{}^\circ \subset \mathrm{conv}\, G$.
2. $\mathcal{B}_G{}^\circ = \mathrm{conv}\, G$.

3. $\operatorname{Extr}(G^\circ) = \mathcal{B}_G$.

4. $\operatorname{Extr}(G^\circ) \subset \mathcal{B}_G$.

5. The proof of Theorem 3.1 for the groups A_n and B_n

Birkhoff's theorem can be reformulated as follows:

Theorem 5.1. $\operatorname{Extr} A_n^\circ = \mathcal{B}_{A_n}$.

Proof. Recall that $e = \sum_{i=1}^{n+1} e_i$. Definition 1.1 means that $T \in \Omega_{n+1}$ if and only if $Te = e$, $T^*e = e$ and T transforms the positive orthant of $\mathbb{R}^{n+1}$ into itself. So, $e^\perp$ is invariant under $T \in \Omega_{n+1}$. Therefore T transforms the intersection of the positive orthant with the affine hyperplane

$$\frac{1}{n+1} e + e^\perp$$

into itself. It is easy to see that this intersection is precisely $\operatorname{conv} \operatorname{Orb}_{\mathrm{Perm}_{n+1}} e_1$. Therefore T also transforms the set S – the orthogonal projection of this intersection onto the subspace $e^\perp$ – into itself. Since $\omega_1 = \operatorname{proj}_{d^\perp} e_1$ (see the description of A_n in Section 2) then

$$S = \operatorname{proj}_{e^\perp} \operatorname{conv} \operatorname{Orb}_{\mathrm{Perm}_{n+1}} e_1 = \operatorname{conv} \operatorname{Orb}_{A_n} \operatorname{proj}_{e^\perp} e_1 = \operatorname{conv} \operatorname{Orb}_{A_n} \omega_1.$$

It is known (see [16]) that

$$\operatorname{Extr}(\operatorname{Orb}_{A_n} \omega_1)^\circ = \frac{1}{m_G(\omega_1, \omega_n)} \operatorname{Orb}_{A_n} \omega_n.$$

So we conclude that $TS \subset S$ if and only if $(T, h\omega_n \otimes g\omega_1) = \langle Tg\omega_1, h\omega_n \rangle \le m_G(\omega_1, \omega_n)$ for all $g, h \in A_n$. Since $\omega_1 \in \operatorname{Orb}_{A_n}(-\omega_n)$ and $\omega_n \in \operatorname{Orb}_{A_n}(-\omega_1)$, the sets $\{h\omega_n \otimes g\omega_1 : g, h \in A_n\}$ and $\{g\omega_1 \otimes h\omega_n : g, h \in A_n\}$ coincide. Therefore $T \in \Omega_{n+1}$ if and only if $Te = e, Te^\perp \subset e^\perp$ and $T|_{e^\perp} \in (\mathcal{B}_{A_n})^\circ$. This means that $\operatorname{Extr}(\Omega_{n+1}|_{e^\perp})^\circ \subset \mathcal{B}_{A_n}$. By the Birkhoff Theorem, $\operatorname{Extr} \Omega_{n+1} = \mathrm{Perm}_{n+1}$, so $(\Omega_{n+1})^\circ = (\mathrm{Perm}_{n+1})^\circ$ and, since both Ω_{n+1} and Perm_{n+1} leave $e^\perp$ invariant, we get

$$\operatorname{Extr} A_n^\circ = \operatorname{Extr}(\mathrm{Perm}_{n+1}|_{e^\perp})^\circ = \operatorname{Extr}(\Omega_{n+1}|_{e^\perp})^\circ \subset \mathcal{B}_{A_n}.$$

According to Lemma 4.5, this proves the result. □

Definition 5.2. An $n \times n$ matrix (a_{ij}) is called *absolutely bistochastic* if

$$\text{for every } j,\ 1 \le j \le n, \quad \sum_{i=1}^{n} |a_{ij}| \le 1, \quad \sum_{i=1}^{n} |a_{ji}| \le 1.$$

Let $\mho_n$ be the set of all absolutely bistochastic $n \times n$ matrices.

The next lemma follows from the Birkhoff Theorem – see, for example, [12].

Lemma 5.3. $B_n = \mathrm{Extr}(\mho_n)$.

The desired description is now (almost) immediate.

Theorem 5.4. $\mathrm{Extr}\, B_n^\circ = \mathcal{B}_{B_n}$.

Proof. By Lemma 4.5, it suffices to prove $(\mathcal{B}_{B_n})^\circ \subset \mathrm{conv}(B_n)$. However, by Lemma 5.3, this statement is equivalent to $(\mathcal{B}_{B_n})^\circ \subset \mho_n$. Let $A = (a_{ij}) \in (\mathcal{B}_{B_n})^\circ$. Let $q = \sum_{j=1}^n \varepsilon_j e_j$, $\varepsilon_j = \pm 1$. All such q form the B_n-orbit of the extremal fundamental weight ω_n. Then $(A, q \otimes e_i) = \langle Ae_i, q\rangle \le 1$ for all $q \in Q$, $1 \le i \le n$. This is equivalent to $\sum_{j=1}^n \varepsilon_j a_{ij} \le 1$ for all $\varepsilon_j = \pm 1$, $1 \le i \le n$, or $\sum_{j=1}^n |a_{ij}| \le 1$. Similarly, using $(A, e_i \otimes q)$, deduce $\sum_{i=1}^n |a_{ij}| \le 1$. So $A \in \mho_n$. □

6. A proof of Theorem 3.1 for $I_2(n)$

Recall that the group $I_2(n)$ is a dihedral group acting on $\mathbb{R}^2$, i.e., the group of symmetries of a regular n-gon, with one vertex on the positive x-axis.

Let $\mathrm{Rot}(\theta)$ be the linear operator performing counter-clockwise rotation by the angle θ. Let

$$\mathrm{Refl}(\theta) = \mathrm{Rot}(\theta)\begin{pmatrix}1 & 0\\ 0 & -1\end{pmatrix}\mathrm{Rot}(-\theta) = \begin{pmatrix}\cos 2\theta & \sin 2\theta\\ \sin 2\theta & -\cos 2\theta\end{pmatrix}$$

be the linear operator performing reflection across the line at an angle θ from the x-axis in the counter-clockwise direction.

One can easily see that every orthogonal operator in $\mathbb{R}^2$ is either a rotation, or a reflection, depending upon whether its determinant is $+1$ or -1. (Indeed, its eigenvalues are either a pair of mutually conjugate complex numbers on the unit circle, or they are ± 1; in the first case the determinant is 1 and this operator is obviously a rotation, in the second case the determinant is -1 and the operator is obviously a reflection.)

Let

$$\mathrm{Rot}_n = \{\mathrm{Rot}(2\pi k/n) : 0 \le k < n\}, \quad \mathrm{Refl}_n = \{\mathrm{Refl}(\pi k/n) : 0 \le k < n\}.$$

Obviously, $I_2(n) \supset \mathrm{Rot}_n \cup \mathrm{Refl}_n$. Actually, these two sets coincide – a regular n-gon is not invariant under other rotations and reflections.

Lemma 6.1.

$$I_2(n) = \mathrm{Rot}_n \cup \mathrm{Refl}_n\,.$$

Each of the sets Rot_n, Refl_n spans a two-dimensional subspace in $\mathrm{End}(\mathbb{R}^2)$. There are exactly n elements in each of these two sets, and they are equidistributed on the unit circles in the related subspaces. So, conv Refl_n and conv Rot_n are regular n-gons in the related two-dimensional subspaces of the four-dimensional space End $\mathbb{R}^2$.

The following lemma is easily verified.

Lemma 6.2. *The two-dimensional subspaces spanned by* Rot_n *and* Refl_n *are mutually orthogonal.*

Corollary 6.3. *Every face* Φ *of* conv $I_2(n)$ *of maximal dimension is uniquely representable as*

$$\Phi = \mathrm{conv}(\phi \cup \psi)$$

where ϕ *is a side of the regular* n*-gon* conv Rot_n, *and* ψ *is a side of the regular* n*-gon* conv Refl_n.

Conversely, for any two sides $\phi \subset \mathrm{conv}\,\mathrm{Rot}_n$, $\psi \subset \mathrm{conv}\,\mathrm{Refl}_n$ *the set* $\mathrm{conv}(\phi \cup \psi)$ *is a face of* conv $I_2(n)$ *of maximal dimension.*

Proof. Let Φ be a face of conv $I_2(n)$ of maximal dimension. Then all operators from $I_2(n)$ are in a half-space defined by the hyperplane containing Φ. This hyperplane does not contain the origin, therefore it does not contain span Rot_n or span Refl_n. Therefore its intersections with these subspaces are hyperplanes in these subspaces. Φ must contain four linearly independent elements of $I_2(n)$. Since a hyperplane in a two-dimensional subspace can contain no more that two linearly independent elements, Φ must contain exactly two linearly independent rotations and exactly two linearly independent reflections. Since conv Rot_n and conv Refl_n are in a half-space defined by the hyperplane, Φ contains sides ϕ and ψ of the regular n-gons conv Rot_n and conv Refl_n, respectively. Thus $\Phi = \mathrm{conv}(\phi \cup \psi)$.

Now let ϕ and ψ be sides of the convex n-gons conv Rot_n and conv Refl_n, respectively. Since these convex sets are in mutually orthogonal subspaces, $\mathrm{conv}(\phi \cup \psi)$ contains four linearly independent elements of $I_2(n)$, so it defines a face of conv $I_2(n)$ of maximal dimension. □

Corollary 6.4. conv $I_2(n)$ *has exactly* n^2 *faces of maximal dimension.*

Lemma 6.5. $\mathrm{card}(\mathcal{B}_{I_2(n)}) = n^2$.

Proof. Obviously, the cone bounded by the lines $y = 0$ and $y = \tan(\pi/n)x$ is a Weyl chamber for $I_2(n)$. Therefore the vectors $\omega_1 = (1, 0)$ and $\omega_2 = (\cos(\pi/n), \sin(\pi/n))$ are the (extremal) fundamental weights. Therefore

$$\mathcal{B}_{I_2(n)} = \{g\omega_1 \otimes h\omega_2, h\omega_2 \otimes g\omega_1 : g, h \in I_2(n)\}.$$

Note that $\mathrm{card}(\mathrm{Orb}_{I_2(n)}\,\omega_i) = n$ for $i = 1, 2$. If n is even then $-\mathbf{I} \in I_2(n)$ and since $(-\omega) \otimes (-\tau) = \omega \otimes \tau$ we get $\mathrm{card}(\mathcal{B}_{I_2(n)}) = (2)(n)(n)/2 = n^2$. If n is odd then $-\mathbf{I} \notin I_2(n)$, and $-\omega_1 \in \mathrm{Orb}_{I_2(n)}\,\omega_2$. So again $\mathrm{card}(\mathcal{B}_{I_2(n)}) = (2)(n)(n)/2 = n^2$. □

So, the number of Birkhoff tensors $\mathrm{card}(\mathcal{B}_{I_2(n)})$ equals the overall number of faces of $\mathrm{conv}\, I_2(n)$. Using Theorem 4.4, we arrive at the following result, which proves Theorem 3.1 for $G = I_2(n)$.

Corollary 6.6. $\mathrm{Extr}(I_2(n))^\circ = \mathcal{B}_{I_2(n)}$.

7. Extreme elements of $(D_4)^\circ$

Theorem 7.1.

$$\mathrm{Extr}(D_4)^\circ = \mathcal{B}_{D_4} \bigcup \{gAh : g, h \in D_4\},$$

where

$$A = \frac{1}{4}\begin{pmatrix} -2 & 2 & 0 & -1 \\ 2 & -2 & 0 & -1 \\ -1 & -1 & -1 & 0 \\ 0 & 0 & 0 & 1 \end{pmatrix}.$$

This result was obtained by a computer calculation (in exact arithmetic). We used the cdd program, written by Komei Fukuda [4]. We discuss this program below in Section 10.

Obviously, the matrix A is of rank 3, so $\mathcal{B}_{D_4} \subsetneq \mathrm{Extr}(D_4)^\circ$. Actually what we need is the existence of a matrix of rank greater than 1 in $\mathrm{Extr}(D_4)^\circ$. It is not hard to check by hand that the matrix A belongs to $\mathrm{Extr}(D_4)^\circ$, i.e., to verify that the scalar product of A with every element of D_4 does not exceed 1 and to explicitly find 16 linearly independent elements of D_4 whose scalar products with this matrix are exactly 1. So, the proof of the Conjecture for the group D_4 does not formally depend upon the use of computer calculations.

8. Coxeter groups with branching graphs

In this section we prove Part (b) of the Conjecture.

Theorem 8.1. *Let G be a finite irreducible Coxeter group with a branching Coxeter graph $\Gamma(G)$. Then not all elements of* $\mathrm{Extr}(\mathrm{conv}\, G)^\circ$ *are of rank* 1, *i.e.*, $\mathcal{B}_G \subsetneq \mathrm{Extr}(\mathrm{conv}\, G)^\circ$.

Proof. It is known from the classification of connected Coxeter graphs (see, e.g., [3]) that every branching Coxeter graph contains a (branching connected) graph $\Gamma(D_4)$ as a subgraph. The statement of the theorem is valid for this group – see the previous Section. So we may assume that $\Gamma(G) \neq \Gamma(D_4)$. Therefore there exists an end vertex π such that the graph $\Gamma(G) \setminus \{\pi\}$ is a branching connected Coxeter graph.

Claim. If all elements of $\mathrm{Extr}(\mathrm{conv}\, G)^\circ$ are of rank 1, then the same is true for $\mathrm{Extr}(\mathrm{conv}\, H)^\circ$ where H is a Coxeter group such that $\Gamma(H) = \Gamma(G) \setminus \{\pi\}$.

This claim, together with the above considerations, easily leads to a proof of the theorem.

Let all elements of $\mathrm{Extr}(\mathrm{conv}\, G)^\circ$ be of rank 1. Let ω be an extremal fundamental weight associated with the vertex π. We may assume that its length is 1.

Consider the Coxeter group $G_\omega = \mathrm{Stab}_G\, \omega|_{\omega^\perp}$ and denote it by H. Then

$$\Gamma(H) = \Gamma(G) \setminus \{\pi\}.$$

Since π is an end vertex, this graph is connected and therefore H is an irreducible group. Consider the hyperplane $\Pi = \{T \in \mathrm{End}\, V : (T, \omega \otimes \omega) = 0\}$. Note that

$$\mathrm{Stab}_G\, \omega = G \cap (\Pi + \mathbf{I}).$$

This immediately follows from the fact that the elements of G are all orthogonal operators. Also, the affine hyperplane $(\Pi+\mathbf{I})$ is a support hyperplane of the polyhedron $\mathrm{conv}\, G$, i.e.,

$$G \subset \{T \in \mathrm{End}\, V : (T, \omega \otimes \omega) \leq \langle \omega, \omega \rangle\}.$$

Therefore the faces of maximal dimension of the polyhedron $\mathrm{conv}(\mathrm{Stab}_G\, \omega)$ are intersections of faces of $\mathrm{conv}\, G$ with the hyperplane $\Pi + \mathbf{I}$. We have assumed that the normals to all faces of $\mathrm{conv}\, G$ are of rank 1. We obtain the group H from the group $\mathrm{Stab}_G\, \omega$ by restricting the action of the latter to its invariant subspace $\omega^\perp$. We can view this as follows:

Let P denote the orthogonal projection of V onto the subspace $\omega^\perp$. Then the operator $T \mapsto PTP$ is an orthogonal projection in $\mathrm{End}\, V$. Then $H = P(\mathrm{Stab}_G\, \omega)P$. Therefore the faces of maximal dimension of $\mathrm{conv}\, H$ are projections of the faces of maximal dimension of $\mathrm{conv}(\mathrm{Stab}_G\, \omega)$. Thus the normals to faces of $\mathrm{conv}(H)$ are of the form PbP, where $b \in \mathrm{Extr}(\mathrm{conv}\, G)^\circ$. But all these tensors are of rank 1. So the Claim is proven, which completes the proof of the theorem. □

9. Computer tools

The results of the previous sections leave the following exceptional groups for which Conjecture 1.4 still needs verification: namely the groups F_4, H_3 and H_4. In this Section we discuss computer tools which have enabled us to verify the Conjecture for $G = F_4, H_3$, and we also discuss some approaches which hopefully will in the future allow us to verify the Conjecture for the remaining case $G = H_4$. All programs we have written are available at http://www.math.wm.edu/ zobin/. Our main tool is a cdd program which calculates the extreme elements of a polytope given by a system of linear inequalities, which is exactly what G° is. We must be able to write down the system of inequalities as an input file, so we need to obtain a list of matrices corresponding to the operators which are the elements of our group G. This is the first computational problem we address.

9.1. Matrix representation of Coxeter groups

All information about a Coxeter group is encoded in its graph, but going from the graph to a presentation of elements is not easy, and the computer can help here.

We wrote a program in C++ which, given a Coxeter graph, lists the matrices of all elements of the associated group in a natural orthonormal basis, together with Birkhoff tensors, fundamental roots and weights, generators, etc. To explain this program, we'll follow its logic and note its output for H_3. We assume that the input graph has n vertices, labeled with the associated fundamental weights $\omega_1, \omega_2, \ldots, \omega_n$.

Although the program takes a graph as its input, the computer is happier working with a matrix. Thus, the computer represents the given graph as a Cartan matrix:

Definition 9.1. Consider the Coxeter graph $\Gamma(G)$ on n vertices. Choose an ordering $\preccurlyeq$ of the vertices (for a non-branching graph there are two natural orderings of the vertices, going along the path). So, label the vertices $\pi_1, \pi_2, \ldots, \pi_n$ and assume that as i goes from 1 to n the vertices are arranged according to the chosen ordering. Then we have an ordering $W_1, W_2, \ldots, W_n$ of the walls in a Weyl chamber, as well as an ordering $r_1, r_2, \ldots, r_n$ of the fundamental roots. The *modified Cartan matrix* $\mathfrak{C}(G, \preccurlyeq)$ associated with this ordering of the fundamental roots is the $n \times n$ matrix (a_{ij}) with

$$a_{ij} = -\cos(\pi/(k(i,j)+2)) = \langle r_i, r_j \rangle$$

for $i \neq j$, and $a_{ii} = 1 = \langle r_i, r_i \rangle$. Here $k(i, j)$ is the multiplicity of the edge joining the vertices π_i and π_j of the graph, and r_i are the fundamental roots. So,

$$\mathfrak{C}(G, \preccurlyeq) = (\langle r_i, r_j \rangle)_{1 \leq i,j \leq n}.$$

For instance, choosing the ordering from the left to the right for the vertices of $\Gamma(H_3)$ (see 2.4) we get

$$\mathfrak{C}(H_3, \preccurlyeq) = \begin{pmatrix} 1 & -\cos(\pi/5) & 0 \\ -\cos(\pi/5) & 1 & -\cos(\pi/3) \\ 0 & -\cos(\pi/3) & 1 \end{pmatrix}$$

$$= \begin{pmatrix} 1 & -(1+\sqrt{5})/4 & 0 \\ -(1+\sqrt{5})/4 & 1 & -1/2 \\ 0 & -1/2 & 1 \end{pmatrix}.$$

The natural orthonormal basis $e_1, e_2, \ldots, e_n$ which we are going to use is obtained from the basis of fundamental roots $r_1, r_2, \ldots, r_n$ by the Gram–Schmidt orthogonalization procedure.

Clearly then

$$\begin{aligned} r_1 &= \lambda_{11}e_1, \\ r_2 &= \lambda_{21}e_1 + \lambda_{22}e_2, \\ &\vdots \\ r_n &= \lambda_{n,1}e_1 + \cdots + \lambda_{n,n}e_n, \end{aligned}$$

for some λ_{ij}. It is, in fact, these scalars λ_{ij} that we are after. These scalars form a lower-triangular matrix Λ. Obviously, $\lambda_{11} = 1$.

One can immediately see that

$$\Lambda\Lambda^t = (\langle r_i, r_j\rangle)_{1\le i,j\le n} = \mathfrak{C}(G, \preccurlyeq).$$

So, the matrix Λ is nothing else but the Cholesky factor of $\mathfrak{C}(G, \preccurlyeq)$. There are numerous programs for efficient Cholesky factorization.

For H_3, this gives us approximately

$$\Lambda(H_3) \approx \begin{pmatrix} 1 & 0 & 0 \\ 0.809\ldots & 0.588\ldots & 0 \\ 0 & 0.851\ldots & 0.526\ldots \end{pmatrix}.$$

Next, the program computes the fundamental weights. The fundamental weights form a basis dual to the basis of the fundamental roots. So if $W(G, \preccurlyeq)$ is the "matrix of fundamental weights" of G in which each column gives the coordinates of a fundamental weight in the basis $e_1, e_2, \ldots, e_n$, then $W(G, \preccurlyeq) = \Lambda^{-1}$. This is an easy calculation for the program to perform. The program gives

$$(W(H_3, \preccurlyeq))^t \approx \begin{pmatrix} 1 & -1.376\ldots & 2.227\ldots \\ 0 & 1.701\ldots & -2.753\ldots \\ 0 & 0 & 1.902\ldots \end{pmatrix}.$$

Now we want the matrices for the elements of G. Getting a representation requires generators; consider the generators $R_i = \mathbf{I} - 2r_i \otimes r_i$. Their matrices in our basis are $g_i = \mathbf{I} - 2\Lambda_i^t\Lambda_i$ where Λ_i is the i-th row of the matrix Λ.

Generating all the elements requires iterating. We initialize a list with the n generators. At each iteration, multiply each element in the list with every other element in the list; if the product is not in the list, add the new matrix to the list. If at any iteration, no new matrices are created, then stop.

This naïve algorithm generates the elements of H_3 quite quickly, but it takes too long to enumerate the elements of the much larger group H_4. A more intelligent approach is required; we can keep the previous method, but we have to reduce the number of spurious matrix multiplications. With each element in the list, store the *length* of the element, i.e., the minimal number of generators whose product is the element. Clearly, for the generators themselves, this number is 1. On the k-th iteration, then, we simply take all elements of length k, and pre- and post-multiply them by the generators to get all elements of length $k + 1$. This is a pretty simple optimization, but it pays off quite well.

We intend to use this program to compile the input file for the cdd program, which actually computes the extreme elements of G°.

With a list of elements in hand, writing down the Birkhoff tensors is easy. Since we've generated the fundamental weights, we can find a particular Birkhoff tensor $\omega \otimes \tau$. To find the others, we iterate through all $g, h \in G$ and add $g\omega \otimes h\tau$ to the list of tensors. Each time we add a tensor to the list, however, the program must check that the new tensor is in fact different from all the previously generated tensors.

Now that we can explore H_3 with the computer, we are ready to tackle the conjecture.

10. Computer proofs of the Conjecture for F_4 and H_3

We want to show that $\operatorname{Extr} G^\circ = \mathcal{B}_G$ for $G = F_4, H_3$. We'll verify this by finding $\operatorname{Extr} G^\circ$ with the aid of a computer. The algorithm we employ is the Double Description Method, otherwise known as Chernikova's algorithm. A (non-computer) calculation based on this algorithm was used in [18] for a solution of a problem regarding the geometry of some orbihedra, see also [11] for another approach to this problem.

Our exposition of this algorithm follows the one given in [5].

10.1. The double description method

Given a finite set S in $\mathbb{R}^n$, we want to find $\operatorname{Extr} S^\circ$. First we homogenize the problem by switching from S to the cone $S' = \{\lambda(\{1\} \times S) \subset \mathbb{R}^{n+1} : \lambda \geq 0\}$, and from the polar set S° to the *dual cone*

$$D(S') = \{v \in \mathbb{R}^{n+1} : \forall s \in S',\ \langle s, v\rangle \geq 0\}.$$

Then the projections of extreme rays of the cone $P(S')$ to $\mathbb{R}^n$ are directed along the vectors from $\operatorname{Extr} S^\circ$, so the problem of finding the convex hull is merely a disguise for the problem of enumerating the extreme rays of a polyhedral cone with vertex at the origin.

The algorithm we use takes a polyhedral cone given by a system of homogeneous linear inequalities and finds its extreme rays. As input, we have a matrix A, whose rows are the coefficients of the linear inequalities defining the cone, in other words, the normals to the faces of the cone. The matrix A describes a cone

$$P(A) = \{v \in \mathbb{R}^{n+1} : Av \geq 0\}.$$

As output, we want an $(n + 1) \times m$ matrix R whose columns are vectors whose linear combinations with non-negative coefficients give the whole cone $P(A)$. So, R

provides an alternative description of this cone

$$P(A) = \{v \in \mathbb{R}^{n+1} : \exists \lambda \in \mathbb{R}^m, \lambda > 0,\ v = R\lambda\}.$$

The ordered pair (A, R) is called a *double description* pair; such a pair provides two descriptions of the same cone.

Here of course there is a slight abuse of terminology, since we are identifying matrices with the (non ordered) sets of their column or row vectors.

The columns of R include vectors directed along all of the extreme rays. But R may also have other redundant columns. Clearly there exist many different matrices R which form a double description pair with A.

The redundancy may be eliminated rather simply – we must omit all vectors belonging to less than n independent boundary hyperplanes, i.e., turning less than n independent inequalities (given by the rows of A) into equalities. The obtained matrix is already unique – up to permutations of columns and scaling of each column.

Simply put, the algorithm takes A and finds a double description pair (A, R). The method is iterative. Let A_k be the matrix of the first k rows of A. Since A_{n+1} is a simplicial cone, finding a matrix R_{n+1} is easy: just solve $A_{n+1}R_{n+1} = I$ to get a double description pair (A_{n+1}, R_{n+1}). Now iterate: suppose that for some A_k we have a double description pair (A_k, R_k); we'd like to find an R_{k+1} for A_{k+1}. Let a_{k+1} be the $(k+1)$-st row vector of A; this vector determines a hyperplane cutting $\mathbb{R}^{n+1}$ into three pieces (two half-spaces and a hyperplane) as follows:

$$\begin{aligned} H_{k+1}^{+} &= \{v \in \mathbb{R}^{n+1} : \langle v, a_{k+1}\rangle > 0\}, \\ H_{k+1}^{0} &= \{v \in \mathbb{R}^{n+1} : \langle v, a_{k+1}\rangle = 0\}, \text{ and} \\ H_{k+1}^{-} &= \{v \in \mathbb{R}^{n+1} : \langle v, a_{k+1}\rangle < 0\}. \end{aligned}$$

Let $r_1, \ldots, r_j$ be the columns of R_k. The vector a_{k+1} will also partition these rays into three sets

$$\begin{aligned} J_{k+1}^{+} &= \{r_i : r_i \in H_{k+1}^{+}\}, \\ J_{k+1}^{0} &= \{r_i : r_i \in H_{k+1}^{0}\}, \text{ and} \\ J_{k+1}^{-} &= \{r_i : r_i \in H_{k+1}^{-}\}. \end{aligned}$$

What is the relationship between R_{k+1} and R_k? Clearly, $J_{k+1}^{+} \subset R_{k+1}$ and $J_{k+1}^{0} \subset R_{k+1}$, but there might be something we are missing. Indeed it is not hard to prove the following lemma (see [5]):

Lemma 10.1. *Let (A_k, R_k) be a double description pair. Then so is (A_{k+1}, R_{k+1}) where*

$$R_{k+1} = J_{k+1}^{+} \cup J_{k+1}^{0} \cup M,$$

and

$$M = \{\operatorname{span}\{u, v\} \cap H_{k+1}^{0} : u \in J_{k+1}^{-}, v \in J_{k+1}^{+}\}.$$

By applying this lemma iteratively, we eventually find a matrix R forming a double description pair (A, R). As it was mentioned earlier, this R might include some extraneous rays; to ensure minimality of R_{k+1} at each step with this naïve approach, we assume that R_k is not redundant, and then we remove rays in R_{k+1} that lie on fewer than n hyperplanes of A_{k+1}.

It is always better if such a program works incrementally, outputting the extreme rays it has already calculated. This can be done by verifying feasibility of the vectors from the non-redundant version of R_k, i.e., checking if they satisfy *all* inequalities defining the cone. Then the feasible vectors should be listed and outputted as a partial result. Note that non-feasible vectors from R_k should not be included in this partial list, but they are needed to perform the next iteration.

To actually perform this iterative process by computer, we use the cdd program, an implementation in C++ of the double description algorithm [4]. Although cdd is essentially the algorithm presented above, it is packed with many optimizations. These optimizations are not enough to make cdd the best ray enumeration algorithm for all problems – there are more efficient algorithms for simplicial polyhedra – but cdd is excellent for the degenerate (i.e., non-simplicial) case. Since the convex hulls of Coxeter groups are degenerate polyhedra in operator space, cdd is particularly well suited for finding Extr G° for G a Coxeter group.

10.2. Computing the convex hulls of F_4 and H_3

To compute Extr G° with the help of cdd we first need to prepare an input file, where we write down the system of linear inequalities describing G°:

$$\sum_{1\le i,j\le n} s_{ij} g_{ji} \le 1, \quad g \in G.$$

Preparing such an input file is not too hard for relatively small groups like F_4 and H_3, but it is not at all easy for H_4, consisting of 14, 400 elements. So, we have compiled the input files for F_4 and H_3 by hand, and for the case of H_4 we have used the input file compiled by Val Spitkovsky, who has applied quite sophisticated programming tools to do this.

In the future we plan to generate these files with the help of our program listing the matrices (g_{ij}) of all operators from the group G. This could be important for verification of the Conjecture for the group H_4, since we hope to introduce some additional optimization into the cdd program (exploiting symmetries, a clever choice of the ordering, etc) and we shall need a significant flexibility in preparing the input file.

The double description method can be performed exactly in arithmetic over $\mathbb{Q}$. Since the matrices representing operators from the Coxeter group F_4 in our basis e_1, e_2, e_3, e_4 have rational entries, cdd can find $\mathrm{Extr}(F_4)^\circ$ exactly.

Theorem 10.2. $\mathrm{Extr}(F_4)^\circ = \mathcal{B}_{F_4}$.

The situation is more complicated for H_3: there is no basis in which the matrices for the elements of H_3 have rational entries (this is equivalent to the fact that H_3 and H_4 are not *crystallographic* groups, see [3]). Nonetheless, the matrices of the elements of H_3 in our basis e_1, e_2, e_3 are over the field $\mathbb{Q}(\sqrt{5})$, the algebraic extension of $\mathbb{Q}$ by $\sqrt{5}$.

To capitalize on this fact, we extended cdd to perform exact arithmetic over $\mathbb{Q}(\sqrt{5})$. Such arithmetic is easy to work with: an element of $\mathbb{Q}(\sqrt{5})$ is identified with an ordered pair (p, q) where $p, q \in \mathbb{Q}$, i.e., $(p, q) \cong p + q\sqrt{5}$. Elementary algebra quickly verifies the following:

$$\begin{aligned}(p, q) + (p', q') &= (p + p', q + q'),\\ (p, q)(p', q') &= (pp' + 5qq', pq' + p'q),\\ (p', q')^{-1} &= \left(\frac{p}{p^2 - 5q^2}, \frac{-q}{p^2 - 5q^2}\right).\end{aligned}$$

Our modified version of cdd computes $\mathrm{Extr}(H_3)^\circ$ exactly over $\mathbb{Q}(\sqrt{5})$. We did this in under a half hour on a Pentium III. By Theorem 4.4,

$$\mathcal{B}_G = (\mathrm{Extr}\, G\circ) \cap (\text{rank 1 tensors}).$$

So it suffices to verify that for all $v \in \mathrm{Extr}(H_3)^\circ$, $\mathrm{rank}(v) = 1$. The cdd output verified this, thereby proving

Theorem 10.3. $\mathrm{Extr}(H_3)^\circ = \mathcal{B}_{H_3}$.

11. Computer attacks on H_4

The matrices of operators from H_4 also belong to the field $\mathbb{Q}(\sqrt{5})$. Although in theory cdd can compute $\mathrm{Extr}(H_4)^\circ$, we can't even come close in practice. The group H_3 has only 120 elements; H_4 has 14,400. The initial problem is memory: as cdd iterates, it generates a plethora of extraneous rays. After a couple of hundred iterations, the number of extraneous rays easily fills up all of memory.

However, our problem has a lot of symmetries and it is natural to try to use these symmetries to reduce the volume of computations. Here we discuss some steps which we have already taken in this direction and others which we hope to carry out in the future.

Since the group acts by multiplications on itself and this action is transitive, we need only consider the faces of conv H_4 containing $\mathbf{I}$. This means that we are in fact interested in whether the extreme rays of the cone

$$\{S \in \mathrm{End}\, V : \forall g \in H_4 \ \ (S, g) \le (S, \mathbf{I})\}$$

are of rank 1. So we stay in dimension 16 (instead of going to dimension 17 while homogenizing). The memory requirements are reduced substantially in this way. Nevertheless, the computation still takes too long.

A very important tool in reducing the amount of calculations in cdd is the choice of ordering of the inequalities describing G°, i.e., the choice of ordering in the Coxeter group G. It seems that a clever choice of ordering could produce a sharp drop in the amount of extraneous rays.

The cdd program calculates elements of $\mathrm{Extr}(H_4)^\circ$. But we already know a lot of these elements, since $\mathcal{B}_{H_4}$ is a subset of $\mathrm{Extr}(H_4)^\circ$. So what we actually need is not the calculation of all extreme vectors but rather a verification that the convex hull of known extreme vectors is already the set we are studying.

11.1. Birkhoff faces

Here we outline another approach to our problem. Let us say that a face of conv G is a *Birkhoff face* if some Birkhoff tensor is orthogonal to this face. Because H_4 is non-branching, the group generated by the operator $T \mapsto T^*$ and pre- and post-multiplications by elements of H_4 acts transitively on the set of Birkhoff tensors, so each Birkhoff face can be transformed to any other Birkhoff face by this group. So we may consider only Birkhoff faces containing $\mathbf{I}$. It is easy to list all operators from G belonging to a Birkhoff face containing $\mathbf{I}$ orthogonal to $\omega \otimes \tau \in \mathcal{B}_G$:

$$\langle g\omega, \tau\rangle = (g, \omega \otimes \tau) = (\mathbf{I}, \omega \otimes \tau) = \langle \omega, \tau\rangle = 1 \quad \text{if and only if}$$

$$g = hk, \quad h \in \mathrm{Stab}_G\, \tau,\ k \in \mathrm{Stab}_G\, \omega.$$

To confirm the Conjecture we need to show that every Birkhoff face is adjacent to only Birkhoff faces, i.e., every subface of a Birkhoff face comes from the intersection with another Birkhoff face. So we need to study the adjacencies of Birkhoff faces. We have written a computer program computing the graph of adjacencies of Birkhoff faces. It still takes too long to run it for H_4. Description of Birkhoff subfaces (i.e., intersections of Birkhoff faces) is a challenging problem closely related to many interesting topics, including a Word Problem for Coxeter groups, Bruhat orderings, etc.

11.2. An application of Poincaré's Theorem

Let us describe another possible approach. Consider the following differential 15-form on the 16-dimensional space End $\mathbb{R}^4$:

$$\Omega(x) = \sum_{i=1}^{16} \frac{x_i}{\|x\|^2} dx_1 \wedge \cdots \wedge^i \cdots \wedge dx_{16}.$$

As usual, $\wedge^i$ means that dx_i is omitted. This form is orthogonally invariant, closed outside of the origin and its integral over any 15-dimensional surface, which is starlike with respect to the origin, is non-negative. By the Poincaré Theorem the integrals of this form over the boundaries of all 16-dimensional bodies containing the origin are the same. In particular they are equal to the easily computable integral over a sphere centered at the origin. Let's suppose that we could calculate $\int_\Phi \Omega$ where Φ is a Birkhoff face. Since the form is orthogonally invariant the integral over the union of all the Birkhoff faces of conv H_4 equals this integral multiplied by the number of Birkhoff tensors. Since the integral over any face of conv H_4 is non-negative, we conclude that all faces are Birkhoff faces if and only if the integral over the union of Birkhoff faces equals the integral over a sphere.

This approach, while certainly interesting, is still not efficient enough. To perform the needed numerical calculations would require integrating over a Birkhoff face of conv H_4. But to numerically integrate, we have to be able to test whether a point is in a face of conv H_4. Knowing the vertices of a face isn't enough to perform this test; we need to know the faces of a face, and this is again a problem for cdd. But there are 480 points in the 15-dimensional Birkhoff face of conv H_4. And sadly, 480 vertices is still too many points for cdd to handle.

12. Open problems

Problem 12.1. *Does Conjecture* 1.4 *hold for* H_4*?*

Of course this problem could be solved by brute force, by simply using more powerful computers. But it would be much more interesting and useful to optimize the cdd program, to find clever orderings, etc.

It would be very interesting to develop a version of the cdd program taking symmetries into account.

Problem 12.2. *Find a "classification-free" proof of Conjecture* 1.4.

This is definitely the heart of the matter. We believe that a promising approach is further study of Birkhoff faces and their subfaces. We have a reason to believe that the structure of Birkhoff subfaces of lower dimensions is simpler.

Problem 12.3. *Calculate* Extr G° *for irreducible branching Coxeter groups.*

We believe that real progress in this problem will depend upon progress in the previous problem.

Problem 12.4. *Calculate* Extr env G *for irreducible branching Coxeter groups.*

References

[1] Benson, C., and Grove, L., Finite Reflection Groups, 2nd ed., Springer-Verlag, New York, 1985.

[2] Birkhoff, G., Tres observaciones sobre el algebra lineal, University Nac. Tucuman Rev. Ser. A5 (1946),147–150.

[3] Bourbaki, N., Groupes et Algebres de Lie, Ch. IV–VI, Hermann, Paris, 1968.

[4] Fukuda, K., cdd+: C++ implementation of the double description method for computing all vertices and extremal rays of a convex polyhedron given by a system of linear inequalities, version 0.76, March 1999, available from ftp://ftp.ifor.math.ethz.ch/pub/fukuda/cdd.

[5] Fukuda, K., and Prodon, A., Double description method revisited, Lecture Notes in Comput. Sci. 1120, Springer-Verlag, 1996, 91–111.

[6] Gohberg, I. C., and Krein, M. G., Introduction to the Theory of Linear Nonselfadjoint operators, Transl. Math. Monogr. 18, Amer. Math. Soc., Providence, RI, 1969.

[7] Humphreys, J., Reflection Groups and Coxeter Groups, Cambridge University Press, Cambridge, 1990.

[8] Li, C. K., Spitkovsky, I., and Zobin, N., Finite reflection groups and linear preserver problems, Rocky Mountain J. Math., to appear, preprint available.

[9] Li, C. K., Tam, B. S., and Tsing, N. K., Linear maps preserving permutation and stochastic matrices, Linear Algebra Appl. 341 (2002), 5–22.

[10] McCarthy, N., Ogilvie, D., Spitkovsky, I., and Zobin, N., Birkhoff's Theorem and convex hulls of Coxeter groups, Linear Algebra Appl., to appear, preprint available.

[11] McCarthy, N., Ogilvie, D., Zobin, N., and Zobin, V., Convex geometry of Coxeter-invariant polyhedra, submitted, preprint available.

[12] Mityagin, B.S., An interpolation theorem for modular spaces, Mat. Sb 66 (4) (1965), 473–482 (in Russian); English translation in: Interpolation spaces and allied topics in analysis, Lecture Notes in Math. 1070, Springer-Verlag, New York, 1983, 10–23.

[13] Pierce, S., et. al., A Survey of Linear Preserver Problems, Linear and Multilinear Algebra 33 (1992), 1–130.

[14] Wei, A., Linear transformations preserving the real orthogonal group, Canad. J. Math. 27 (1975), 561–572.

[15] Zobin, N., and Zobina, V., Duality in operator spaces and problems of interpolation of operators, Pitman Research Notes in Math. 257 (1992), 123–144.

[16] Zobin, N., and Zobina, V., Coxeter groups and interpolation of operators, Integral Equations Operator Theory 18 (1994), 335–367.

[17] Zobin, N., and Zobina, V., A general theory of sufficient collections of norms with a prescribed semigroup of contractions, Oper. Theory Adv. Appl. 73 (1994), 397–416.

[18] Zobin, N., and Zobin, V., Geometric structure of $B_{2,2}$-orbihedra and interpolation of operators, Linear and Multilinear Algebra 48 (2000), 67–91.

[19] Zobina, V., Operator interpolation in spaces with prescribed symmetries, Ph.D. thesis (in Russian), Kazan State University, Russia, 1979.

Jeremy Brandman, P.O. Box 200631, New Haven, CT 06520, U.S.A.
E-mail: jeremy.brandman@yale.edu

Jim Fowler, 132 Lowell House Mail Center, Cambridge, MA 02138, U.S.A.
E-mail: jafowler@fas.harvard.edu

Brian Lins, Department of Mathematics, Rutgers University, 110 Frelinghuysen Road, Piscataway, NJ 08854-8019, U.S.A.
E-mail: bclins@math.rutgers.edu

Ilya Spitkovsky, Department of Mathematics, College of William and Mary, Williamsburg, VA 23187, U.S.A.
E-mail: ilya@math.wm.edu

Nahum Zobin, Department of Mathematics, College of William and Mary, Williamsburg, VA 23187, U.S.A.
E-mail: zobin@math.wm.edu

On embedding properties of some extrapolation spaces *

María J. Carro and Joaquim Martín

Dedicated to Jaak Peetre

Abstract. Given a sublinear operator T satisfying that $\|Tf\|_{L^p(\nu)} \leq \frac{C}{p-1}\|f\|_{L^p(\mu)}$, for every $1 < p \leq p_0$, with C independent of f and p, it has been recently proved that $T : L\log L \to M(\varphi)$, where $M(\varphi)$ is the maximal Lorentz space with $\varphi(t) = t(1+\log^+ t)^{-1}$. Also, if T satisfies that $\|Tf\|_{L^p(\nu)} \leq Cp\|f\|_{L^p(\mu)}$, for every $p \geq p_0$, then $T : \Lambda^1\big(\min(t^{-1}, 1)\big) \cap L^\infty \to M(\phi)$, where $\phi(t) = \big(1 + \log^+(1/t)\big)^{-1}$.

The purpose of this note, is to study embedding properties of the extrapolation spaces $L\log L$ and $M(\varphi)$ with respect to L^1, and also embedding properties of $\Lambda^1\big(\min(t^{-1}, 1)\big) \cap L^\infty$ and $M(\phi)$ with respect to L^∞. We shall also extend these type of results to more general extrapolation theorems.

2000 Mathematics Subject Classification: 46M35.

1. Introduction

In 1951, Yano (see [9]) proved that for every sublinear operator satisfying that

$$T : L^p(\mu) \longrightarrow L^p(\nu)$$

is bounded, for every $1 < p \leq p_0$, with constant less than or equal to $\frac{C}{p-1}$, where μ and ν are two finite measure, it holds that $T : L\log L(\mu) \longrightarrow L^1(\nu)$ is bounded. If the measures involved are not finite, then an easy modification in the proof of this result shows that $T : L\log L(\mu) \longrightarrow L^1(\nu) + L^\infty(\nu)$ is bounded.

This theorem has recently been improved in [3] and [4], showing that, if μ and ν are σ-finite measures and T satisfies that

$$T : L^{p,1}(\mu) \longrightarrow L^{p,\infty}(\nu),$$

*This work has been partially supported by CICYT BFM2001-3395 and by CURE 2001SGR 00069.
Key words and phrases: Real interpolation, extrapolation, maximal and minimal Lorentz spaces

is bounded with constant less than or equal to $\frac{C}{p-1}$, where $L^{p,\infty}(\nu)$ is endowed with the norm $\|f\|_{L^{p,\infty}} = \sup_t \big(t^{1/p} f_\nu^{**}(t)\big)$, then

$$T : L\log L(\mu) \longrightarrow M(\varphi; \nu)$$

where $\varphi(t) = t(1+\log^+ t)^{-1}$ and, the maximal Lorentz space $M(\varphi) = M(\varphi; \nu)$ is defined (see [1], p. 69]) as the set of measurable functions such that

$$\|f\|_{M(\varphi)} = \sup_{t>0} \big(\varphi(t) f_\nu^{**}(t)\big) < \infty,$$

where $f_\nu^{**}(t) = \frac{1}{t}\int_0^t f_\nu^*(s)\,ds$ and f_ν^* is the decreasing rearrangement of f with respect to the measure ν, (in what follows, we shall omit the subindices ν or μ whenever it is clear the measure we are working with). In particular, if $\varphi(t) = t^{1/p}$, $M(\varphi) = L^{p,\infty}$.

Also, in the setting of Lorentz spaces, it holds that $L\log L$ is the minimal Lorentz space $\Lambda(\varphi)$, where $\varphi(t) = t\big(1+\log^+(1/t)\big)$ and

$$\|f\|_{\Lambda(\varphi)} = \int_0^\infty f^*(t)\,d\varphi(t).$$

If $\varphi(t) = t^{1/p}$, $\Lambda(\varphi)$ is the Lorentz space $L^{p,1}$, where

$$\|f\|_{L^{p,1}} = \frac{1}{p}\int_0^\infty f^*(t) t^{1/p-1}\,dt. \tag{1}$$

Therefore, in this context of minimal-maximal Lorentz spaces, the new version of Yano's theorem can be stated as follows:

Theorem 1.1 (Yano). *Let $\varphi_\theta(t) = t^{1-\theta}$ and let T be a sublinear operator such that*

$$T : \Lambda(\varphi_\theta; \mu) \to M(\varphi_\theta; \nu)$$

is bounded with $\|T\| \le C/\theta$, $(0 < \theta < \theta_0 \le 1)$. Then

$$T : \Lambda(\varphi_{D_+}; \mu) \to M(\varphi_{R_+}; \nu),$$

where $\varphi_{D_+}(t) = t(1+\log^+\frac{1}{t})$ and $\varphi_{R_+}(t) = t(1+\log^+ t)^{-1}$.

We also have a dual version. That is, if

$$\|Tf\|_{L^p(\nu)} \le Cp\|f\|_{L^p(\mu)},$$

for every $p \ge p_0$, then it was proved in [4] that

$$T : \Lambda^1\big(\min(t^{-1}, 1); \mu\big) \cap L^\infty(\mu) \to M(\phi; \nu),$$

where $\phi(t) = 1/\big(1+\log^+(1/t)\big)$, improving a previous result due to Zygmund (see [10], p. 119). The formulation of this result in the above terminology is the following:

Theorem 1.2 (Zygmund). *Let* $\varphi_\theta(t) = t^{1-\theta}$ *and let* T *be a sublinear operator such that*

$$T : \Lambda(\varphi_\theta; \mu) \to M(\varphi_\theta; \nu)$$

is bounded with $\|T\| \le C/(1-\theta)$, $(\theta_0 < \theta < 1)$. *Then*

$$T : \Lambda(\varphi_{D_-}; \mu) \to M(\varphi_{R_-}; \nu),$$

where $\varphi_{D_-}(t) = (1 + \log^+ t)$ *and* $\varphi_{R_-}(t) = (1 + \log^+(1/t))^{-1}$.

Now, let us consider *compatible pairs* of Banach spaces $\bar{A} = \big(A_0, A_1\big)$. That is, we assume that there is a large topological vector space $\mathcal{V}$ such that $A_i \subset \mathcal{V}$, $i = 0, 1$, continuously. Usually we drop the terms "compatible" and "Banach" and refer to a compatible Banach pair simply as a "pair".

Let us recall that given a pair $\bar{A} = (A_0, A_1)$, the Peetre *K-functional* is defined, for $a \in A_0 + A_1$ and $t > 0$, by

$$K(a, t) = K(a, t; \bar{A}) = \inf\{\|a_0\|_{A_0} + t\,\|a_1\|_{A_1} : a = a_0 + a_1,\ a_i \in A_i\}.$$

It is easy to see that $K(t, a)$ is a nonnegative and concave function of $t > 0$, (and thus also continuous). Therefore

$$K(a, t; \bar{A}) = K(a, 0^+; \bar{A}) + \int_0^t k(a, s; \bar{A})\, ds,$$

where the k–functional, $k(a, s; \bar{A}) = k(a, s)$, is a uniquely defined, nonnegative, decreasing and right-continuous function of $s > 0$.

In particular, if $\bar{A} = \big(L^1(\nu), L^\infty(\nu)\big)$, we have that $k(a, s; \bar{A}) = f^*(s)$ and

$$K(f, t) = \int_0^t f^*(s)\, ds.$$

The new point of view of Yano's and Zygmund's theorems presented above gives us the idea of defining, for a pair $\bar{A}$, the corresponding minimal and maximal spaces as follows:

Definition 1.3. The minimal Lorentz space, $\Lambda(\varphi; \bar{A})$, is the set of elements $a \in A_0 + A_1$ such that $K(a, 0^+; \bar{A}) = 0$ and

$$\|a\|_{\Lambda(\varphi; \bar{A})} = \int_0^\infty k(a, s; \bar{A})\, d\varphi(s) < \infty,$$

and the maximal Lorentz space, $M(\varphi; \bar{A})$, is the set of elements $a \in A_0 + A_1$ such that

$$\|a\|_{M(\varphi; \bar{A})} = \sup_{t>0} \left(\frac{K(a, t; \bar{A})}{t} \varphi(t) \right) < \infty.$$

Then, the two following extrapolation results have been obtained in [5] (see also [6]).

Theorem 1.4. *Let $\bar{A} = (A_0, A_1)$ and $\bar{B} = (B_0, B_1)$ be two pairs and let T be a linear operator such that*

$$T : \Lambda(\varphi_\theta; \bar{A}) \to M(\varphi_\theta; \bar{B})$$

is bounded with $\|T\| \leq \frac{C}{\theta}$, $(0 < \theta < \theta_0)$. Then

$$T : \Lambda(\varphi_{D_+}; \bar{A}) \to M(\varphi_{R_+}; \bar{B}).$$

Theorem 1.5. *Let $\bar{A} = (A_0, A_1)$ and $\bar{B} = (B_0, B_1)$ be two pairs and let T be a linear operator such that*

$$T : \Lambda(\varphi_\theta; \bar{A}) \to M(\varphi_\theta; \bar{B})$$

is bounded with $\|T\| \leq \frac{C}{1-\theta}$, $(\theta_0 < \theta < 1)$. Then

$$T : \Lambda(\varphi_{D_-}; \bar{A}) \to M(\varphi_{R_-}; \bar{B}).$$

The purpose of this note is to study embedding properties of the extrapolation spaces $\Lambda(\varphi_{D_+}; \bar{A})$, $M(\varphi_{R_+}; \bar{B})$, $\Lambda(\varphi_{D_-}; \bar{A})$ and $M(\varphi_{R_-}; \bar{B})$ with respect to the corresponding end-point spaces A_0, A_1, B_0 and B_1 respectively. For example: it is clear that the domain space $\Lambda(\varphi_{D_+}; \bar{A}) \subset A_0$, while the opposite embedding does not clearly hold. However, if we consider the Lions–Peetre real interpolation spaces $\bar{A}_{\theta,p}$ defined by (see [1])

$$\|a\|_{\bar{A}_{\theta,p}} = \left(\theta(1-\theta) \int_0^\infty \left(\frac{K(t,a;\bar{A})}{t^\theta}\right)^p \frac{dt}{t}\right)^{1/p},$$

then, for every $0 < \theta < 1$ and $p \geq 1$,

$$\bar{A}_{\theta,1} \subset \cdots \subset \bar{A}_{\theta,p} \subset \cdots \subset \bar{A}_{\theta,\infty},$$

and, we obtain (see Theorem 2.3 below) that if we intersect A_0 with the biggest space of the above chain then

$$A_0 \cap \bar{A}_{\theta,\infty} \subset \Lambda(\varphi_{D_+}; \bar{A}).$$

Similar results will be proved for the other three extrapolation spaces.

Constants such as C will denote universal constants (independent of the parameters involved) and may change from one occurrence to the next. As usual, the symbol $f \approx g$ will indicate the existence of a universal positive constant C so that $f/C \leq g \leq Cf$, while the symbol $f \preceq g$ means that $f \leq Cg$.

Remark 1.6. It is important to mention, and we want to thank the referee for this comment, that the space $\Lambda(\varphi; \bar{A})$ can be expressed in term of the K-functional which is usually more convenient since it is subadditive. For particular φ's, this is a simple consequence of an integration by parts and, for example, it can be trivially proved that

$$\|a\|_{\Lambda(\varphi_{D_+};\bar{A})} = \sup_{t>0} K(t,a;\bar{A}) + \int_0^1 \frac{K(t,a;\bar{A})}{t}\,dt,$$

and

$$\|a\|_{\Lambda(\varphi_{D_-};\bar{A})} = \sup_{t>0} \frac{K(t,a;\bar{A})}{t} + \int_1^\infty \frac{K(t,a;\bar{A})}{t^2}\,dt.$$

For general $\varphi's$, we have to use Theorem 3.1 in [7].

Acknowledgement. The authors want to express their deepest gratitude to Eugene Pustylnik for his numerous and useful comments on the subject. In particular, he has informed us that our embeddings of Theorems 2.1 and 2.2 can also be obtained (after some easy lemmas) using Theorem 3 in [8].

2. Relationship between the extrapolation and the end-point spaces

Let us start by analyzing the case $\bar{L} = (L^1, L^\infty)$.

Proposition 2.1. *For every* $p > 1$,

$$L^1 \cap L^{p,\infty} \subset L\log L \subset L^1,$$

and

$$L^1 \subset M(\varphi_{R_+}) \subset L^1 + L^{p,1},$$

where the constant of the first and last embeddings are less than or equal to $Cp/(p-1)$.

Proof. To show the first embedding, we observe that

$$\|f\|_{L\log L} \approx \|f\|_1 + \int_0^1 f^{**}(t)\,dt,$$

and therefore

$$\begin{aligned}\|f\|_{L\log L} &\preceq \|f\|_{L^1} + \int_0^1 \frac{t^{1/p} f^{**}(t)}{t^{1/p}}\,dt \le \|f\|_{L^1} + \|f\|_{L^{p,\infty}} \int_0^1 t^{-1/p}\,dt \\ &= \|f\|_{L^1} + \frac{p}{p-1}\|f\|_{L^{p,\infty}}.\end{aligned}$$

The second and third embeddings are trivial. To prove the last embedding, let $f \in M(\varphi_{R_+})$. Then, for every $t > 0$,

$$\int_0^t f^* \le \|f\|_{M(\varphi_{R_+})}\big(1 + \log^+ t\big).$$

Hence, if we define $\overline{f} = f\chi_{\{|f|>f^*(1)\}}$, we have that

$$\|\overline{f}\|_{L^1} = \int_0^1 f^*(t)\,dt \le \|f\|_{M(\varphi_{R_+})}.$$

Now, set $\underline{f} = f - \overline{f}$ and recall that the norm in $L^{p,1}$ is given by (1). Then, if $p > 1$, an integration by parts shows that

$$\begin{aligned}
\|\underline{f}\|_{L^{p,1}} &= \frac{1}{p}\int_0^\infty \underline{f}^*(t)t^{1/p-1}\,dt \leq f^*(1) + \frac{1}{p}\int_1^\infty f^*(t)t^{1/p-1}\,dt \\
&\preceq \|f\|_{M(\varphi_{R_+})} + \frac{p-1}{p}\|f\|_{M(\varphi_{R_+})} + \frac{p-1}{p}\int_1^\infty \Big(\int_0^t f^*\Big)t^{1/p-2}\,dt \\
&\leq \frac{p-1}{p}\|f\|_{M(\varphi_{R_+})} + \|f\|_{M(\varphi_{R_+})}\frac{p-1}{p}\int_1^\infty (1+\log^+ t)t^{1/p-2}\,dt \\
&\approx \frac{p}{p-1}\|f\|_{M(\varphi_{R_+})}
\end{aligned}$$

from which the result follows. □

Proposition 2.2. *For every $p > 1$, it holds that*

$$L^\infty \cap L^{p,\infty} \subset \Lambda(\varphi_{D_-}) \subset L^\infty,$$

and

$$L^\infty \subset M(\varphi_{R_-}) \subset L^\infty + L^{p,1},$$

where the constants of the first and last embedding are less than or equal to Cp.

Proof. The proof follows the same pattern than Proposition 2.1. Also, it can be deduced using duality in Proposition 2.1, since the associated space of $L^{p,1}$ is equal to $L^{p',\infty}$ and it was proved in [4] that $\Lambda(\varphi_{D_-})$ is the associated space of $M(\varphi_{R_+})$ and $M(\varphi_{R_-})$ is the associated space of $L \log L$. □

Let us consider now, the general case $\bar{A} = (A_0, A_1)$.

Theorem 2.3. *Let $\bar{A} = (A_0, A_1)$ be a pair. Then, for every $0 < \theta < 1$,*

$$A_0 \cap \bar{A}_{\theta,\infty} \subset \Lambda(\varphi_{D_+}; \bar{A}) \subset A_0,$$

and

$$A_0 \subset M(\varphi_{R_+}; \bar{A}) \subset A_0 + \bar{A}_{\theta,1},$$

where the constants of the first and last embedding are less than or equal to C/θ.

Proof. The second and the third embeddings are trivial. The first embedding follows from the fact that if $a \in A_0 \cap \bar{A}_{\theta,\infty}$, then $k(.; a) \in L^1 \cap L^{1/(1-\theta),\infty}$, and, by Proposition 2.1, we have that $k(.; a) \in L \log L$, which is equivalent to

$$\|a\|_{\Lambda(\varphi_{D_+}; \bar{A})} \approx \int_0^\infty k(t; a)\Big(1 + \log^+ \frac{1}{t}\Big)\,dt < \infty.$$

To prove the last embedding, we have to proceed as in Proposition 2.1. Let $a \in M(\varphi_{R_+}; \bar{A})$. Then,

$$\int_0^t k(s,a)\,ds \le \|a\|_{M(\varphi_{R_+};\bar{A})}\big(1+\log^+ t\big).$$

Hence, if we define $\bar{k}(s,a) = k(s,a)\chi_{(0,1)}$ and $\underline{k}(s,a) = k(s,a)\chi_{(1,\infty)}$ we have that

$$\begin{aligned} K(t,a) &\le \int_0^t \bar{k}(s,a)\,ds + \int_0^t \underline{k}(s,a)\,ds \\ &\le \int_0^t \bar{k}(s,a)ds + \int_0^t \big(\underline{k}(s,a) + k(1,a)\chi_{(0,1)}\big)ds, \end{aligned}$$

and since the last two functions are concave, we can use the K-divisibility theorem (see [2], Theorem 3.2.7) to have that there exist a_0 and a_1 such that $a = a_0 + a_1$,

$$K(t,a_0) \le \int_0^t \bar{k}(s,a)\,ds,$$

and

$$K(t,a_1) \le \int_0^t \big(\underline{k}(s,a) + k(1,a)\chi_{(0,1)}\big)\,ds.$$

Now, if we define $\tilde{A}_0$ as the set of elements in $A_0 + A_1$ such that $\sup_t K(t,a_0) < \infty$ then, using Holmstedt's formula and Theorem 1.5 of [1] (p. 297), we have that $A_0 + A_{\theta,1} = \tilde{A}_0 + A_{\theta,1}$ with equivalent norms and, hence,

$$\begin{aligned} \|a\|_{A_0+A_{\theta,1}} &\approx \|a\|_{\tilde{A}_0+A_{\theta,1}} \le \|a_0\|_{\tilde{A}_0} + \|a_1\|_{A_{\theta,1}} \\ &= \sup_t K(t,a_0) + \theta(1-\theta)\int_0^\infty \frac{K(t,a_1)}{t^{1+\theta}}\,dt \\ &\le \int_0^\infty \bar{k}(s,a)\,ds + \theta(1-\theta)K(1,a)\int_0^1 t^{-\theta}\,dt \\ &\quad + \theta(1-\theta)\int_1^\infty \frac{K(t,a)}{t^{1+\theta}}\,dt \\ &\preceq \int_0^1 k(s,a)\,ds + K(1,a) \\ &\quad + \|a\|_{M(\varphi_{R_+};\bar{A})}\theta(1-\theta)\int_1^\infty \frac{\big(1+\log^+ t\big)}{t^{1+\theta}}\,dt \\ &\preceq \frac{1}{\theta}\|a\|_{M(\varphi_{R_+};\bar{A})}, \end{aligned}$$

from which the result follows. □

And, similarly:

Theorem 2.4. *Let $\bar{A}$ be a pair. Then, for every* $0 < \theta < 1$,

$$A_1 \cap \bar{A}_{\theta,\infty} \subset \Lambda(\varphi_{D_-}; \bar{A}) \subset A_1,$$

and

$$A_1 \subset M(\varphi_{R_-}; \bar{A}) \subset A_1 + \bar{A}_{\theta,1},$$

where the constants of the first and last embeddings are less than or equal to $C/(1-\theta)$.

References

[1] C. Bennett and R. Sharpley, Interpolation of Operators, Academic Press, Boston, 1988.

[2] Yu. A. Brudnyi and N. Ya. Krugljak, Interpolation Functors and Interpolation Spaces, North-Holland, Amsterdam, 1991.

[3] M. J. Carro, New extrapolation estimates, J. Funct. Anal. 174 (2000), 155–166.

[4] M. J. Carro, On the range space of Yano's extrapolation theorem and new extrapolation estimates at infinity, Preprint, 2000.

[5] M. J. Carro and J. Martín, Extrapolation theory for the real interpolation method, Collect. Math. (2002), to appear.

[6] B. Jawerth and M. Milman, Extrapolation Theory with Applications, Mem. Amer. Math. Soc. 89, Amer. Math. Soc., Providence, RI, 1991.

[7] J. Martín and J. Soria, New Lorentz spaces for the restricted weak type Hardy's inequalities, Preprint, 2001.

[8] E. Pustylnik, Minimal and maximal intermediate Banach spaces, Ukranian Math. J. 29 (1997), 102–107.

[9] S. Yano, An extrapolation theorem, J. Math. Soc. Japan 3 (1951), 296–305.

[10] A. Zygmund, Trigonometric Series, vol. I, Cambridge University Press, Cambridge–New York, 1959.

María J. Carro, Departament de Matemàtica Aplicada i Anàlisi, Universitat de Barcelona, 08071 Barcelona, Spain

E-mail: `carro@mat.ub.es`

Joaquim Martín, Departament de Matemàtica Aplicada i Anàlisi, Universitat de Barcelona, 08071 Barcelona, Spain

E-mail: `jmartin@mat.ub.es`

Revisiting almost orthogonality and eigenexpansions

*Mischa Cotlar and Cora Sadosky**

Dedicated to Jaak Peetre

Abstract. In this paper we consider almost orthogonal series and expansions in generalized eigenvectors to give norm estimate for self-adjoint operators.

2000 Mathematics Subject Classification: Primary: 47A55; Secondary: 47A60

1. Introduction

Let H be a Hilbert space and $\mathcal{L}(H)$ the algebra of all bounded linear operators acting in H. If (e_n) is an orthonormal (o.n.) basis of H, and A a symmetric linear operator whose domain contains (e_n), then (e_n) is called an *A-eigenbasis* if $Ae_n = a_n e_n, \forall n$, where a (a_n) is a scalar sequence. An operator sequence (A_n) is called *orthogonal* if $A_n^* A_m = A_n A_m^* = 0$ for $n \neq m$. The following elementary properties show that the notions of orthogonality of vector sequences and of operator sequences are related to those of A-eigenvectors and A-eigenforms, providing estimates for $\|A\|$.

1. If $A_n f = \langle f, e_n \rangle a_n e_n$, $\sup_n |a_n| = c < \infty$, for (e_n) an o.n. basis, then (A_n) is an orthogonal sequence, and $\sum_n A_n f$ converges to Af, where $A \in \mathcal{L}(H)$, and $\|A\| \leq c$. Moreover, (e_n) is an o.n. A-eigenbasis.

2. If the linear operator A has an o.n. A-eigenbasis (e_n), with $\sup_n |a_n| = c < \infty$, for the corresponding (a_n), then $\|A\| \leq c$, and $Af = \sum_n A_n f$, where $A_n f = \langle f, e_n \rangle a_n e_n$.

3. If $A = A^* \in \mathcal{L}(H)$ and has an o.n. A-eigenbasis (e_n), then the sesquilinear form $B(f,g) = \langle f, g \rangle$ is A-symmetric, and $B = \sum_n B_n$, where $B_n(f,g) = \langle f, e_n \rangle \langle e_n, g \rangle$ are A-eigenforms (see Section 2 for notation), for all n.

*First author partially supported by a grant from CONICYT, Venezuela; second author partially supported by a grant from the U.S. Department of Energy.

Unfortunately, the class of operators A with an o.n. A-eigenbasis is very small. Yet, the assertion on the series $\sum_n A_n$ in (1) holds for the class of all orthogonal sequences satisfying $\sup \|A_n\| = c < \infty$, as well as for the much larger class of *almost orthogonal* (a.o.) series (as defined in Section 2). However the sum of an a.o. series need not have eigenvectors, and it may be difficult to verify if a given series is a.o.. Thus it is not clear how "large" is the class of a.o. series which can be explicitly described (so that an explicit estimate can be given for the sum of the series).

Therefore, it is natural to ask whether there is a generalized notion of A-eigenbasis for which analogues of the properties (1), (2) and (3) still hold, producing a "large" class of explicitly given a.o. series, and a "large" class of operators A with such generalized A-eigenbases. Furthermore this consideration should include also the bases with generalized orthogonality properties, such as the frames and the systems of coherent states.

Here we seek to recover the connection with eigenvectors and sesquilinear eigenforms for a significantly large class of operators and a.o. series by introducing two different yet related notions of generalized A-eigenbases. The definition of A-eigenvector involves both the operator A and the vector, and to relax it, we can, for instance, replace A by its compression to a subspace varying with the vector, or replace the vector by a sum of vectors obtained through a partition of the unit operator. The approach through compressions of A leads to a notion that we call here *compression* A-eigenbases, while the other approach leads to what we call here *expanded* A-eigenbases.

The paper is organized as follows.

In Section 2 we emphasize the relation between *almost orthogonal* (a.o.) and *weakly orthogonal* (WO) sequences and functions, state some of the basic results on generalized orthogonality, and describe some of their possibilities for norm estimation.

In Section 3 we introduce the notion of compression A-eigenbases such that the class of operators for which properties (1) and (2) hold for them is the largest possible, i.e., the whole of $\{A \in \mathcal{L}(H) : A = A^*\}$, and the corresponding operators A_n are of a very simple type, given through WO (weak orthogonal) systems, or their continuous analogues in the case of coherent states. Here the sequence (A_n) is not necessarily orthogonal, but it is almost orthogonal. There is also an algorithm, given by infinite recursion, assigning to each $A = A^* \in \mathcal{L}(H)$ a compression A-eigenbasis, and a representation of the form $A = A_1 + A_2 + A_3$ where A_1, A_2, A_3 are simple a.o. operators given by a tensor product of two WO systems with explicitly estimated norms. This result gives a large subclass of operators for which the norm estimates are explicitly controlled.

In Section 4 we give the second notion of A-eigenbases, where the condition $Ae_n = a_n e_n$ is replaced by $Ae_n = \sum_{k \le N_n} a_{nk} e_{nk}$, i.e., Ae_n belongs to a predetermined N_n-dimensional subspace, with a predetermined o.n. basis $(e_{n1}, \dots, e_{nN_n})$ instead to a 1-dimensional space, as in the ordinary eigenvector condition. The advantage of these expanded A-eigenbases is that the underlying o.n. basis (e_n) can be chosen arbitrarily, and that there is a simple (theoretical) algorithm which assigns to each self-adjoint $A \in \mathcal{L}(\mathcal{H})$ a well-determined expanded A-eigensystem, which simplifies its use, and provides functional properties for the application $A \to A$-eigensystem. The development of a corresponding functional calculus, and its applications to singular integrals and to the Krein trace formula will be given elsewhere.

A related way to a functional calculus is given in Section 5, for the special case when the summands A_n in an a.o. series A belong to a commutative C^*-algebra. In such case, norm estimates for a large class of operator sequences $(f(A_n))$ are derived from that of A.

2. Almost orthogonality

Here we recall a few facts concerning the notion of *almost orthogonal* operator series, first studied in [1, 2, 3, 4, 5, 6, 7] (see also the books by E. M. Stein [8], Folland [9] and Coifman–Meyer [10]), and emphasize its relation to *weakly orthogonal* (WO) sequences, developed by Rochberg and Semmes into a *nearly weakly orthogonal* (NWO) theory that provides good estimates for operator norms and singular values [11].

Each vector sequence (e_n) of H produces an associated scalar kernel or matrix $k(m,n) =: |\langle e_m, e_n\rangle|$ and the sequence is orthogonal in the usual sense if $k(m,n) = 0$ whenever $m \neq n$. Relaxing this condition on k leads to generalizations of the notion of orthogonal vector sequences, in particular to the refinements of the notion of Riesz bases defined below.

Similarly, each operator-valued sequence (A_n), $A_n \in \mathcal{L}(H)$, produces a pair of kernels

$$k_1(m,n) := \|A_m^* A_n\|^{1/2} \quad \text{and} \quad k_2(m,n) = \|A_m A_n^*\|^{1/2}, \tag{2.1}$$

and (A_n) is called orthogonal if, for $i = 1, 2$, $k_i(m,n) = 0$ whenever $m \neq n$. Again relaxing this condition on k_i leads to generalizations of the notion of orthogonal operator sequences, in particular to the notion of almost orthogonal operator sequences.

Orthogonality appears in the properties (1) and (2) of the Introduction in two ways: both the vector sequence (e_n) and the operator sequence (A_n) are orthogonal. The vector orthogonality of (e_n) implies that the associated kernel $k(m,n) = |\langle e_m, e_n\rangle|$ defines a bounded operator in ℓ^2 , i.e.

$$\Big|\sum_{m,n} k(m,n) a_m \overline{b_n}\Big| \le c\Big(\sum |a_m|^2\Big)^{1/2}\Big(\sum |b_n|^2\Big)^{1/2} \tag{2.2}$$

whenever $a = (a_n) \in \ell^2, b = (b_n) \in \ell^2$. Similarly the operator orthogonality of (A_n) implies that each of the two associates kernels $k_1(m,n)$ and $k_2(m,n)$ defined in (2.1) produces a bounded operator in ℓ^2. An operator sequence (A_n) is called *almost orthogonal* (a.o.) *with constants* c_1, c_2 if (2.2) holds for $k = k_i$ and $c = c_i$, when $i = 1, 2$.

More generally, if (E, dy) is a measure space whose measure dy is positive and σ–finite, and if $y \to A_y : E \to \mathcal{L}(H)$ is a weakly measurable $\mathcal{L}(H)$-valued function in E, then the function A_y is called almost orthogonal with constants c_1 and c_2 if the associated kernels

$$k_1(y, z) = \|A_y^* A_z\|^{1/2} \quad \text{and} \quad k_2(y, z) = \|A_y A_z^*\|^{1/2}$$

satisfy, for $i = 1, 2$,

$$\left| \iint_{E\times E} k_i(y, z) f(y) g(z) dy dz \right| \leq c_i \Big(\int_E |f(y)|^2 dy\Big)^{1/2} \Big(\int_E |g(y)|^2 dy\Big)^{1/2} \quad (2.3)$$

whenever $f(y)$ and $g(y)$ are measurable functions in E.

The following proposition is the basic property of almost orthogonal functions (see [9]).

Proposition 2.1 (The Almost Orthogonal Lemma for operator-valued functions). *If $y \to A_y : E \to \mathcal{L}(H)$ is an a.o. function with constants c_1, c_2, then the integral $\int A_y dy$ converges in the weak sense to an operator $A \in \mathcal{L}(H)$, i.e. $\forall f, g \in H$*

$$\langle Af, g\rangle = \int \langle A_y f, g\rangle dy \quad \textit{and} \quad \|A\| \leq (c_1 c_2)^{1/2}.$$

Following Rochberg and Semmes we say that a vector sequence $(e_n) \subset H$ is *weakly orthogonal*, WO, if its associated kernel $k(m, n)$ satisfies (2.2), and for such sequence the following discrete analogue of Proposition 2.1 holds.

Proposition 2.2 (Operators defined by a tensor product of two WO sequences). *If (ϕ_n) and (ψ_n) are two WO sequences with associated constants c_1, c_2, then the operator series $\sum_n \phi_n \otimes \psi_n$, which assigns to each $f \in H$ the vector series $\sum_n \langle f, \phi_n\rangle \psi_n$ converges in the weak sense to an operator A, i.e. $\forall$, $f, g \in H$,*

$$\langle Af, g\rangle = \sum_n \langle f, \phi_n\rangle\langle \psi_n, g\rangle = \sum_n \langle (\phi_n \otimes \psi_n) f, g\rangle, \quad (2.4)$$

and $\|A\| \leq (c_1 c_2)^{1/2}$.

More generally, if $y \to A_y : E \to H$ is a weakly measurable vector-valued function in E, then ϕ_y will be called weakly orthogonal, WO, with constant c, if its associated kernel $k(y, z) := |\langle \phi_y, \phi_z\rangle|$ satisfies

$$\iint k(y, z) |f(y)| |g(z)| dy dz \leq c \|f\|_2 \|g\|_2 \quad (2.5)$$

for all $f, g \in L^2(E, dy)$.

The following proposition is due to A. Unterberger [7].

Proposition 2.3 (Operators defined by a tensor product of two WO functions). *If ϕ_y and ψ_y are two WO functions in (E, dy), with constants c_1, c_2, then the integral $\int \langle f, \phi_y \rangle \psi_y dy$ converges for each f to a limit Af, and defines an operator $A \in \mathcal{L}(H)$, in the sense that $\langle Af, g \rangle = \int \langle f, \phi_y \rangle \langle \psi_y, g \rangle dy$, $\forall f, g$, and $\|A\| \le (c_1 c_2)^{1/2}$.*

Moreover, Unterberger showed that both Proposition 2.1 and Proposition 2.3, are corollaries of the following general result [12].

Proposition 2.4 (Lemma on μ-almost orthogonal operator functions). *Let $y \to A_y$ be a weakly measurable H-valued function in (E, dy), let $\mu : E \to \mathbb{R}$ be a strictly positive scalar function, and set*

$$k_{\mu,1}(y, z) := (\mu(y)\mu(z))^{1/2} \| \, |A_y|^{1/2} |A_z|^{1/2} \|,$$

$$k_{\mu,2}(y, z) := (\mu(y)\mu(z))^{-1/2} \| \, |A_y^*|^{1/2} |A_z^*|^{1/2} \|.$$

If, for each $i = 1, 2$, $k_{\mu,i}$ defines a bounded operator in $L^2(E, y)$ with norm $\le c_i$, then there exists $A \in \mathcal{L}(H)$ such that $A = \int A_y dy$ in the weak sense, and $\|A\| \le (c_1 c_2)^{1/2}$.

The notion of WO sequences has been shown as closely related to orthogonality of operators, and it is interesting to observe that it is also closely related to orthogonality of vectors, since it is a refinement of the classical notion of Riesz bases. In fact, a sequence of vectors $(e_n) \in H$ is called a *Riesz sequence* if there exist constants $0 < c_1 \le c_2 < \infty$ such that

$$c_1\Big(\sum |a_n|^2\Big) \le \Big\| \sum a_n e_n \Big\|^2 \le c_2\Big(\sum |a_n|^2\Big), \tag{2.6}$$

or, equivalently, if (e_n) is the image of an o.n. sequence under an isometric operator, while (e_n) is WO if one of the following equivalent conditions is satisfied

$$\Big\| \sum a_n e_n \Big\|^2 \le c\Big(\sum |a_n|^2\Big), \quad \forall \text{ scalar sequences } (a_n), \tag{2.7}$$

$$\sum |\langle f, e_n \rangle|^2 \le c\|f\|^2, \quad \forall f \in H, \tag{2.8}$$

or, equivalently, if (e_n) is the image of an o.n. sequence under a bounded linear operator. Moreover, if a Riesz sequence (e_n) is also a basis, then it satisfies

$$\forall f \in H, \quad c_1\|f\|^2 \le \sum |\langle f, e_n \rangle|^2 \le c_2\|f\|^2 \tag{2.9}$$

A sequence (e_n) satisfying (2.9) is called a *frame*, and when $c_1 = c_2$ it is called a *tight frame*. A similar notion for a function is that of a system of coherent states: If (E, dy) is a measure space as above, and $y \mapsto e_y : E \to H$ is a weakly measurable function, then, following Unterberger, we say that this function is a *system of coherent states* in H if,

$$\int_E |\langle f, e_y \rangle|^2 dy = \|f\|^2, \quad \forall f \in H. \tag{2.10}$$

In further developments of the theory, Unterberger combined Proposition 2.3 and Proposition 2.4, with the theory of coherent states to obtain a very deep approach to pseudodifferential operators [12], while Rochberg and Semmes combined Proposition 2.2 with Carleson techniques in a theory of *nearly weakly orthogonal* (NWO) sequences [11].

Rather than going into similar developments of Propositions 2.1 – 2.4 we will be concerned here with the fact that the operators in those Propositions need not have associated eigenbases, so that the property (3) in the Introduction has no sense for them.

Let us remark that there are generalizations of property (3) for a large class of self-adjoint operators, Every o.n. basis (e_n) produces an expansion $B(f, g) = \sum_n B_n(f, g)$ of the positive form $B(f, g) = \langle f, g\rangle$ into elementary positive forms $B_n(f, g) = \langle f, e_n\rangle\langle e_n, g\rangle$. If $A \in \mathcal{L}(\mathcal{H})$ is self-adjoint, as in (3), then B is A-symmetric, i.e. $\langle Af, g\rangle = \langle f, Ag\rangle$, but the same does not hold for the elementary components B_n. But if (e_n) is an A-eigenbasis, $Ae_n = a_n e_n$, then each B_n is not only A-symmetric but also an A-eigenform: $B_n(Af, g) = a_n B_n(f, g)$. Thus the o.n. A-eigenbases produce expansions of the metric $B(f, g) = \langle f, g\rangle$ of H into elementary A-eigenforms B_n. M.G. Krein developed a general theory of Fourier expansions of A-invariant forms into elementary A-eigenforms, systematized by Iu. Berezansky in [13], and later the authors extended this theory for the so-called generalized Toeplitz (GT) forms (see [14]).

While in [1] the boundedness of symmetric operators is studied through expansions into almost orthogonal operators, in [14] GT forms are considered through expansions into A-eigenforms, and this second investigation developed away from almost orthogonality. To solve the limitations in dealing with a.o. expansions mentioned in the Introduction, in the next two sections we combine both types of expansions through two different notions of generalized A-eigenbases.

3. Compression A-eigenbases

Let $A = A^* \in \mathcal{L}(H)$ be a self-adjoint operator. We want to generalize the notion of o.n. A-eigenbasis without losing contact with a.o. series and the explicit control on the operator norms, and to produce, in addition, the largest possible class of operators A having such generalized A-eigenbases.

A natural generalization of $Ae_n = a_n e_n$ is $Ae_n = a_n e_n + v_n$ where $\|v_n\| = o(1/n)$. Another variant is to require the matrix $\langle v_n, e_k\rangle$ to be almost diagonal, i.e. $\langle v_n, e_k\rangle$ decays rapidly when $|n-k| \to \infty$; a similar condition is used in wavelet theory, where the (non-orthogonal) basis is written as a double sequence (e_{nk}). The first variant is simpler, and provides easily an estimate for $\|A\|$, but it still leads to a rather restricted class of operators. To obtain a larger class, observe that if Q_n is an orthogonal projection satisfying $Q_n e_n = e_n$ then the condition

$$Ae_n = a_n\, e_n + v_n \tag{3.1}$$

implies the weaker condition

$$Q_n A e_n = a_n e_n + w_n, \quad \text{whith } \|w_n\| \le \|v_n\|. \tag{3.2}$$

Furthermore, by subdividing the sequence (e_n) into disjoint intervals $(e_1, \dots, e_{N_1})$, $(e_{N_1+1}, \dots, e_{N_1+N_2}), \dots$, we may assume that (e_n) is a truncated double sequence defined as follows.

Fix an increasing integer sequence $N = (N_k)$ such that $\sum_n N_k^{-1} = \|N^{-1}\| < \infty$, set

$$\mathcal{D}(N) = \{(n,k) : k < N_n,\ n \in \mathbb{N}\}, \tag{3.3}$$

and consider an o.n. truncated double sequence (e_{nk}), $(n,k) \in \mathcal{D}(N)$, that is $(e_{11}, \dots, e_{1N_1}, e_{21}, \dots, e_{2N_2}, \dots)$. For each such double sequence, and for all $n > 1$, we call H_n the subspace spanned by $\{e_{sk} : (s,k) \in \mathcal{D}(N),\ s < n\}$, and P_n the orthoprojector onto H_n. Now set $Q_n = I - P_n$, so that $P_n e_{sk} = e_{sk}$ for $s < n$, and $Q_n e_{nk} = e_{nk}$ whenever $n > 1$.

Then the condition $Ae_{nk} = a_{nk} e_{nk} + v_{nk}$ implies the weaker condition $Q_n A\, e_{nk} = a_{nk}\, e_{nk} + Q_n\, v_{nk} = a_{nk}\, e_{nk} + w_{nk}$, $\|w_{nk}\| \le \|v_{nk}\|$, where $Q_n A$ can be replaced by the compression of A to the subspace $Q_n(H)$, and we introduce the following

Definition 3.1. Let $A \in \mathcal{L}(H)$, let $\mathcal{D}(N)$ be as in (3.3), and let $(e_{nk} : (n,k) \in \mathcal{D}(N))$ be an o.n. basis. This basis is called a *compression* A-eigenbasis, if the following two conditions are satisfied:

1. $\forall j < N_1,\ Ae_{1j} = a_{1j} e_{1_j} + w_{1j}$ with $|a_{1j}| \le c,\ \|w_{1j}\| \le cN_1^{-1}$,
2. $\forall n > 1,\ (n,k) \in \mathcal{D}(N),\ Q_n A e_{nk} = a_{nk} e_{nk} + w_{nk}$,

   with $|a_{nk}| \le c,\ \|w_{nk}\| \le cN_k^{-1}$.

Since our basis is orthogonal, it follows from the definition of P_n that $n \le i$ implies $\langle P_n A\, e_{nk}, e_{ij}\rangle = 0$, and similarly, $i \le n$ implies $\langle A P_n\, e_{nk}, e_{ij}\rangle = 0$, for all k, so that for every constant β the following conditions hold

$$\sum_{i \ge n,\, j} |\langle P_n A e_{nk}, e_{ij}\rangle|^2 \le \frac{1}{4}\beta N_n^{-1} \|N^{-1}\| \sum_{i,j} |\langle A e_{nk}, e_{ij}\rangle|^2, \tag{3.4}$$

$$\sum_{i < n,\, j} |\langle A P_n\, e_{nk}, e_{ij}\rangle|^2 \le \frac{1}{4}\beta N_n^{-1} \|N^{-1}\| \sum_{i,j} |\langle A e_{nk}, e_{ij}\rangle|^2, \tag{3.5}$$

$$\sum_{i < n,\, j} |\langle e_{nk}, a_{ij} e_{ij} + w_{ij}\rangle|^2 \le (1 + \sup_{n,k} |a_{nk}|) \sum_{ij} |\langle e_{nk}, w_{ij}\rangle|^2. \tag{3.6}$$

Definition 3.1 imposes conditions *1*, and *2*, which together with the orthogonality of the basis imply conditions (3.4)–(3.6). Therefore an equivalent definition of a compression A-eigenbasis, would be that it satisfies *1*, and *2* as well as (3.4)–(3.6) for all $n > 1$. This equivalent definition has the advantage that it can be used for non-orthogonal bases as well.

Remark 3.2. A similar further relaxation can be combined with the second variant where $\langle v_n, e_k\rangle$ decays as $|n-k| \to \infty$, but we shall not consider this wavelet type variant here.

Since we want to have the largest possible set of compression A-eigensystems we replace orthonormal bases (e_n) by systems of coherent states (e_y) as defined in Section 2.

Recall that a weakly measurable function $y \mapsto e_y : E \to H$ is a system of coherent states over (E, dy) if (2.10) is satisfied. If $E = \mathbb{Z}^+$, dy is the discrete Lebesgue measure in $\mathbb{Z}^+$ and $L^2(E, dy) = \ell^2_+$, then every ordinary o.n. basis (e_n) is a system of coherent states over the measure space $(\mathbb{Z}^+, dy)$. By the polarization identity for the scalar product, condition (2.10) is equivalent to

$$\int_E \langle f, e_y\rangle\langle e_y, g\rangle\, dy = \langle f, g\rangle. \tag{3.7}$$

Here, and in what follows, we assume $\|e_y\| \le 1$, $\forall y$.

Setting $\widehat{f}(y) = \langle f, e_y\rangle$ then the operator $\mathcal{F}$ which assigns to each $f \in H$ the function $\widehat{f} \in L^2(E, dy)$ is an isometry of H onto a subspace of $L^2(E, dy)$. A weakly measurable function $y \mapsto \phi(y) : E \to H$ is said to be integrable in the weak sense if, for all $g \in H$, the scalar function $\langle\phi(y), g\rangle$ is (absolutely) integrable, so that $\int_E |\langle\phi(y), g\rangle|\, dy < \infty$, and $\int_E \phi(y)\, dy = \Phi \in H$ is defined by $\langle\Phi, g\rangle = \int_E \langle\phi(y), g\rangle\, dy$. Therefore from (2.10) and (3.7) it follows that for all $u \in H$ the integral $\int_E \langle f, e_y\rangle\, e_y\, dy$ exists in the weak sense, and $f = \int_E \langle f, e_y\rangle\, e_y\, d_y = \int_E \widehat{f}(y) e_y\, dy$.

Similarly, if $T \in \mathcal{L}(H)$, then

$$Tf = \int_E \langle f, e_y\rangle\, Te_y\, dy = \int_E \langle f, e_y\rangle\, \phi_y\, dy, \quad \phi_y = Te_y, \tag{3.8}$$

in the weak sense, since $\langle Tf, g\rangle = \langle f, T^*g\rangle = \int_E \langle f, e_y\rangle\langle e_y, T^*g\rangle\, dy$ $= \int_E \langle f, e_y\rangle\langle Te_y, g\rangle\, dy$. Moreover we have the following result.

Proposition 3.3. *Let $y \mapsto s(y)$ be a bounded weakly measurable function from E to H. Then there exists an $S \in \mathcal{L}(H)$, $\|S\| \le M$ satisfying, $Se_y = s(y)$ for all y, if and only if for each $g \in H$ there is a $g' \in H$ such that, for all y,*

$$\langle s(y), g\rangle = \langle e_y, g'\rangle \quad \textit{and} \quad \|g'\| \le M\|g\|. \tag{3.9}$$

Proof. If for each g there is g' satisfying (3.9) then, $\forall f \in H$, the integral $B(f, g) := \int_E \langle f, e_y\rangle\langle s(y), g\rangle dy$ exists and defines a bounded sesquilinear form

$B(f, g) = \langle f, g' \rangle$, $|B(f, g)| \leq M\|g\| \, \|f\|$, so that the operator S associated to B satisfies $\langle Sf, g \rangle = \int \langle f, e_y \rangle \langle s(y), g \rangle dy$, and by (3.9),

$$\int_E \langle f, e_y \rangle \langle s(y), g \rangle dy = \int \langle f, e_y \rangle \langle Se_y, g \rangle dy. \tag{3.10}$$

Since $\overline{\langle Se_y, g \rangle} = \langle S^* g, e_y \rangle = \widehat{S^* g}(y)$ belongs to $\mathcal{F}(H)$, and the same is true for $\overline{\langle S(y), g \rangle}$, (3.10) implies that $\langle S(y), g \rangle = \langle Se_y, g \rangle$ for almost all y. Choosing an o.n. basis (g_n) in H, for almost all y, $\langle S(y), g_n \rangle = \langle Se_y, g_n \rangle$ holds for all g_n, and thus $S(y) = Se_y$. The converse is immediate. □

A bounded weakly measurable function $\lambda : E \to \mathbb{C}$ is called a *multiplier function* for the given system (e_y) of coherent states, if for all $g \in H$, there is a $g' \in H$ such that for all y $\langle \lambda(y) e_y, g \rangle = \langle e_y, g' \rangle$. In this case $|\langle e_y, g' \rangle| = |\lambda(y)| |\langle e_y, g \rangle|$ implies, by (3.7), that $\|g'\| \leq M\|g\|$ with $M = \sup |\lambda(y)| < \infty$. Hence by Proposition 3.3, if the bounded function $\lambda = \lambda(y) = (\lambda_y)$ is a multiplier, then there exists an operator $S \in \mathcal{L}(H)$ such that $\lambda(y) e_y = Se_y$ holds for all y. Thus every bounded multiplier $\lambda = (\lambda_y)$ defines a bounded operator S, in the weak sense, by

$$Sf = \int_E \langle f, e_y \rangle \lambda_y e_y \, dy, \tag{3.11}$$

which satisfies $\lambda_y = Se_y$, for almost all y. Such an operator S, defined through a bounded multiplier function, is called a *multiplier operator* with respect to the system of coherent states (e_y).

Observe that when $E = Z^+$, $L^2(E, dy) = \ell^2_+$, and (e_n) is an o.n. basis, then every bounded sequence $(\lambda_y) = (\lambda_n)$ is a multiplier. Proposition 3.3 says that (S_y) is a WO of special type if for each g there is a g' satisfying $\langle S_y, g \rangle = \langle e_y, g' \rangle$ with $\|g'\| \leq M\|g\|$. However, if (e_y) is a general coherent system, then a bounded function (λ_y) need not be a multiplier, since in general $\mathcal{F}(H) \neq L^2(E, dy)$.

If the system of coherent states (e_y) is WO (in the case $L^2(E, dy) = \ell^2_+$ and (e_y) an o.n. basis this is always the case), then every multiplier operator, given by (3.11), can be written as $Sf = \int_E \langle f, e_y \rangle Se_y dy = \int_E \langle f, e_y \rangle \phi_y dy$ where $\phi_y = Se_y = \lambda_y e_y$ is also WO, so that S is a tensor product of two WO functions, i.e. the integral of a special μ-almost orthogonal operator-valued function $(S_y = e_y \otimes \varphi_y)$, in the sense of Unterburger (see the lines preceding Proposition 2.4).

Still more generally, we replace the system of coherent states over (E, dy) by a weakly measurable vector function $y \mapsto e_y : E \to H$, called a *frame over* E (see (2.9)), if there exist two constants, $\beta_1, \beta_2 > 0$, such that the following two conditions are satisfied

$$\int_E \|\langle f, e_y \rangle\|^2 dy \leq \beta_1 \|f\|^2, \quad \forall f \in H, \tag{3.12}$$

$$|\langle f, g \rangle| \leq \beta_2 \int_E |\langle f, e_y \rangle| \, |\langle e_y, g \rangle| \, dy, \quad \forall f, g \in H. \tag{3.13}$$

A system of coherent states is the same as a frame with $\beta_1 = \beta_2 = 1$.

Let $d\mu(x)$ and $d\nu(y)$ be two positive σ-finite measures in $[1, \infty)$, where $\mu([1, N]) \leq N$, $\nu([1, N]) \leq N$ for every $N \in [1, \infty)$, and consider the measure space (E, dz) where $E = \{z = (x, y) \in [1, \infty) \times [1, \infty] : y < N_x\}$ and $dz = d(x, y)$ is the restriction of $d\mu(x) \times d\nu(y)$ to E, where $x \mapsto N_x$ is a given function in $[1, \infty)$, such that N_x^{-1} is a positive μ-integrable function in the variable x, $\int_1^\infty N_x^{-1} d\mu(x) =: \|N^{-1}\| < \infty$, and let $z = (x, y) \mapsto e_z = e_{xy}$ be a frame over (E, dz) with constants $\beta_1, \beta_2 > 0$. With Definition 3.1 and inequalities (3.4)–(3.6) as reference, and the notation above, we give the following

Definition 3.4. If $A = A^* \in \mathcal{L}(H)$ is a self-adjoint operator, and $(e_z) := (e_{xy})$ is a frame over the measure space (E, dz), we call this frame an *A-eigenframe* if for all $z = (x, y) \in E$ the following conditions are satisfied

$$Q_x A e_{xy} = a_{xy} e_{xy} + w_{xy} \tag{3.14}$$

where $Q_x = I - P_x$, P_x is the orthoprojector over the subspace spanned by the vectors $\{e_{st} : s < x\}$, a_{xy}, w_{xy} are bounded weakly measurable scalar functions, such that

$$\sup_{(x,y)} |a_{xy}| = \|a\| < \infty, \quad \text{and} \quad \|w_{xy}\| \leq c N_x^{-1},$$

and, for

$$c_{xy} = \frac{1}{4\beta_1\beta_2} N_x^{-1} \|N^{-1}\| \int_E |\langle A e_{xy}, e_{st}\rangle|^2 \, d(s, t), \tag{3.15}$$

$$\int_{s \geq x} |\langle P_x A e_{xy}, e_{st}\rangle|^2 \, d(s, t) \leq c_{xy}, \tag{3.16}$$

$$\int_{s<x} |\langle A P_s e_{xy}, e_{st}\rangle|^2 d(s, t) \leq c_{xy}, \tag{3.17}$$

and

$$\int_{s<x} |\langle e_{xy}, a_{st} e_{st} + w_{st}\rangle|^2 \leq (1 + \|a\|) \int |\langle e_{xy}, w_{st}\rangle|^2 \, d(s, t), \tag{3.18}$$

where the function given by $(z, \tau) \mapsto \langle e_z, w_\tau \rangle$, $z = (x, y)$, $\tau = (s, t)$, is a measurable function in the two variables (z, τ).

Remark 3.5. Basic examples of systems of coherent states or frames over (E, dz) are obtained when H is a reproducing kernel Hilbert space of analytic functions in a space $L^2(E, dz)$, and $z \mapsto e_z$ is its reproducing kernel. In this case it is not difficult to define functions $z \to e_z$ satisfying the last measurability condition.

Theorem 3.6. (a) *If $A = A^* \in L(H)$ is a self-adjoint operator and (e_{xy}) is a compression A-eigensystem of coherent states over (E, dz), then there exist five operators*

$A_1, \ldots, A_5$ *of a simple WO type, such that*

$$A = A_1 + \cdots + A_5,$$

$$\|A_j\| < c \text{ for } j = 1, 2, 5, \quad \|A_j\| < \frac{1}{4}\|A\| \text{ for } j = 3, 4, \quad \text{and} \quad \|A\| \le 6c.$$

If $(e_{xy}) = (e_{nk})$ *is a discrete o.n. basis, then* $A_3 = A_4 = 0$ *and* $\|A\| \le 3c$.

(b) *Conversely, for every* $A = A^* \in L(H)$, *and for every fixed* N_x *with* $\|N^{-1}\| < \infty$, *there exits a discrete compression o.n. A-eigenbasis with constant* $c \le 2\|A\|$. *Moreover, there is a theoretical algorithm which assigns to each* $A = A^*$ *such an A-eigenbasis.*

Proof of (a). Let (e_{xy}) be a compression A-eigensystem of coherent states over (E, dz), and for every pair $f, g \in H$ set

$$\begin{aligned}
I_1 &= I_1(f, g) = \int_E a_{xy}\langle f, e_{xy}\rangle\langle e_{xy}, g\rangle\, d(x, y)\\
I_2 &= \int_E \langle f, e_{xy}\rangle\langle w_{xy}, g\rangle\, d(x, y)\\
I_3 &= \int_E \langle f, e_{xy}\rangle \left[\int_{s\ge x} \langle P_x A e_{xy}, e_{st}\rangle\langle e_{st}, g\rangle\, d(s, t)\right] d(x, y)\\
I_4 &= \int_E \langle f, e_{xy}\rangle \left[\int_{s<x} \langle A P_s e_{st}\rangle\langle e_{st}, g\rangle\, d(s, t)\right] d(x, y),\\
I_5 &= \int_E \langle f, e_{xy}\rangle \left[\int_{s<x} \langle e_{xy} a_{st} e_{st} + w_{st}\rangle\langle e_{st}, g\rangle\, d(s, t)\right] d(x, y)
\end{aligned}$$

Then

$$|I_1| \le \left[\int_E |\langle f, e_{xy}\rangle|^2\, d(x, y)\right]^{1/2} \left[\int_E |a_{xy}\langle g, e_{xy}\rangle|^2\, d(x, y)\right]^{1/2}. \tag{3.19}$$

Similarly, since $\int_E |\langle w_{xy}, g\rangle|^2 \le \left(\int |N_x^{-1}|^2\right) \|g\|^2$,

$$|I_2| \le \int |\langle f, e_{xy}\rangle|\, |\langle w_{xy}, g\rangle|\, d(x, y) \le \|N^{-1}\|\, \|f\|\, \|g\|. \tag{3.20}$$

Since now $\beta_1 = \beta_2 = 1$ and $\int_E |\langle f, e_{xy}\rangle| N_x^{-1}\, dxdy \le \|f\|\, \|N^{-1}\|$, we have

$$\begin{aligned}
|I_3| &\le \int_E |\langle f, e_{xy}\rangle| \left[\int_{s\ge x} |\langle P_x A e_{xy}, e_{st}\rangle|\, |\langle e_{st}, g\rangle|\, d(s, t)\right] dxy\\
&\le \frac{1}{4}\|A\|\, \|f\|\, \|g\|.
\end{aligned} \tag{3.21}$$

Similarly

$$|I_4| \le \frac{1}{4}\|A\|\, \|f\|\, \|g\| \tag{3.22}$$

Since by (3.18),

$$\int_E |\langle e_{st}, g\rangle| \left[\int_{x>s} |\langle f, e_{xy}\rangle|\,|\langle e_{xy}, a_{st}e_{st} + w_{st}\rangle|\, d(x,y)\right] d(s,t)$$

$$\leq (1 + \sup |a_{st}|) \int_E |\langle e_{st}, g\rangle|\, \|f\|\, \|w_{st}\|\, d(s,t),$$

$$\begin{aligned} |I_5| &\leq \int_E |\langle f, e_{xy}\rangle| \left[\int_{s<x} |\langle e_{xy}, a_{st}e_{st} + w_{st}\rangle|\,|\langle e_{st}, g\rangle|\, d(s,t)\right] d(x,y) \\ &\leq (1+\sup |a_{st}|)\|f\| \left[\int |\langle g, e_{st}\rangle|^2 d(s,t)\right]^{1/2} \left(\int w_{st}^2 d(s,t)\right)^{1/2}. \end{aligned} \tag{3.23}$$

From (3.20) and (3.21) it follows that $I_1(f,g) = \langle A_1 f, g\rangle$, $I_2(f,g) = \langle A_2 f, g\rangle$ where

$$A_1, A_2 \in L(H) \quad \text{with norms } \leq c\, \|N^{-1}\|. \tag{3.24}$$

From (3.22) it follows that for fixed (x,y), $\int_{s\geq x} \langle P_x A e_{xy}, e_{st}\rangle \langle e_{st}, g\rangle d(s,t)$ is an antilinear functional in g, of norm $\leq \frac{1}{4}\|A\|\, N_x^{-1}\, \|N^{-1}\|$, so that there exists $u_{xy} \in H$ such that $\|u_{xy}\| \leq \frac{1}{4}\|A\|\, N_x^{-1}\, \|N^{-1}\|$ and such that

$$I_3(f,g) = \int_E \langle f, e_{xy}\rangle\, \langle u_{xy}, g\rangle\, d(x,y) = \left\langle \int_E \langle f, e_{xy}\rangle u_{xy}\, d(x,y) g\right\rangle,$$

so that $I_3(f, g = \langle A_3 f, g\rangle$ where $A_3 f = \int_E \langle f, e_{xy}\rangle u_{xy}\, d(x,y)$, with

$$\|A_3 f\| \leq \left(\int |\langle f, e_{xy}\rangle|^2 d(x,y)\right)^{1/2} \left(\int u_{xy}^2\right)^{1/2} \leq 1/4\|A\|,$$

and thus A_3 is a simple WO type operator with norm $\leq 1/4\|A\|$, and the same is true for A_4. A similar argument applies to I_5, so that $I_5(f,g) = \langle A_5 f, g\rangle$ with $\|A_5\| \leq c\|N^{-1}\|$, and of WO type, and the same holds for I_1, I_2, from (3.20), (3.21).

It remains to see that $A = A_1 + \cdots + A_5$. We have that

$$\begin{aligned} Af &= \int_E \langle f, e_{xy}\rangle A e_{xy} = \int \langle f, e_{xy}\rangle Q_x A e_{xy} + \int \langle f, e_{xy}\rangle P_x A e_{xy} \\ &= \int_E \langle f, e_{xy}\rangle (a_{xy}e_{xy} + w_{xy})\, d(x,y) + \int_E \langle f, e_{xy}\rangle P_x A e_{xy}\, d(x,y). \end{aligned}$$

Thus

$$\begin{aligned}\langle Af, g\rangle &= \int \langle f, e_{xy}\rangle a_{xy}\langle e_{xy}, g\rangle + \int \langle f, e_{xy}\rangle\langle w_{xy}, g\rangle \\ &\quad + \int \langle f, e_{xy}\rangle\langle P_x A e_{xy}, g\rangle \\ &= I_1(f,g) + I_2(f,g) + \int \langle f, e_{xy}\rangle\langle P_x A e_{xy}, g\rangle.\end{aligned} \tag{3.25}$$

Now

$$\begin{aligned}\langle P_x A e_{xy}, g\rangle &= \int \langle P_x A e_{xy}, e_{st}\rangle\langle e_{st}, g\rangle d(s,t) \\ &= \left(\int_{s\geq x} + \int_{s<x}\right)\langle P_x A e_{xy}, e_{st}\rangle\langle e_{st}, g\rangle d(s,t),\end{aligned}$$

so that

$$\langle Af, f\rangle = I_1 + I_2 + I_3 + II, \tag{3.26}$$

where

$$\begin{aligned}II &= \int \langle f, e_{xy}\rangle \left[\int_{s<x} \langle P_x A e_{xy}, e_{st}\rangle \cdot \langle e_{st}, g\rangle d(s,t)\right] d(x,y) \\ &= \int \langle f, e_{xy}\rangle \left[\int_{s<x} \langle e_{xy}, A e_{st}\rangle \langle e_{st}, g\rangle d(s,t)\right] d(x,y),\end{aligned}$$

since $P_x e_{st} = e_{st}$ for $s < x$ by definition of P_x. Writing $\langle e_{xy}, Ae_{st}\rangle = \langle e_{xy}, P_s A e_{st}\rangle + \langle e_{xy}, Q_s A e_{st}\rangle$ and using $Q_s A e_{st} = a_{st} e_{st} + w_{st}$ it follows that $II = I_4 + I_5$.

Thus $\langle Af, g\rangle = I_1(f,g) + \cdots + I_5(f,g)$, and hence $A = A_1 + \cdots + A_5$. □

Remark 3.7. As mentioned in Section 2, condition (2.7) is equivalent to condition (2.8) and each of these two conditions can be taken as definition of a WO sequence. But in the case of functions, $(x, y) \mapsto e_{xy}$, the analogues of these conditions give rise to two different versions of WO: the one in Proposition 2.3, and the one in Theorem 3.6, and for both Proposition 2.4 holds.

Proof of (b). This part of the proof follows from a well-known argument. Let us first see that given an integer $N > 0$, an element $e^0 \in H$, and an operator $A = A^* \in L(H)$, we can construct and o.n. system $(e_1, \dots, e_n) \subset H$, and two systems $(a_1, \dots, a_n) \subset \mathbb{C}$, $\{g_1, \dots, g_n\} \subset H$, such that $n \leq N$, for every $k = 1, \dots, n$ the relation $Ae_k = a_k e_k + v_k$ holds with $|a_k| \leq \|A\|$, $\|v_k\| \leq \|A\| N^{-1}$, and $e^0 \in L$, where L is the span of $(e_1, \dots, e_n)$. Let $\Delta \mapsto E(\Delta)$ be the spectral measure of A, acting in the interval $I = [-\|A\|, \|A\|]$, let $I = \Delta_1 \cup \cdots \cup \Delta_N$ be a partition of I into N equal intervals of length $2\|A\|N^{-1}$, let a_k be the center of Δ_k, and $H_k = E(\Delta_k)H$, so that $H = H_1 \oplus \cdots \oplus H_N$. Let e'_k be the normalized projection of e^0 onto H_k and $e_k = e'_k$ if $e'_k \neq 0$, $e_k = 0$ otherwise. Then, since $e_k \in H_k$, $Ae_k = \int a\, d_a E e_k = a_k e_k + v_k$

where

$$\|v_k\| = \left\| \int_{\Delta_k} (a - a_k) d_a E e_k \right\| \leq \sup\{|a - a_k| : a \in \Delta_k\} \leq \|A\| \, N^{-1}\},$$

where $|a_k| \leq \|A\|$, and where e^0 is clearly in the span of $(e_1, \ldots, e_N)$.

Now fix a dense sequence (e_j^0) in H and use the preceding property to construct the desired o.n. basis (e_n) by induction, as follows. Given (N_j), $\sum N_j^{-1} < \infty$, we construct as above an o.n. system $(e_{11}, \ldots, e_{1n_1})$ and two sequences $(a_{1k} : k = 1, \ldots, n_1) \subset \mathbb{C}$, $(v_{1k} : k = 1, \ldots, n_1) \subset H$ such that $n_1 \leq N_1$, $Ae_{1k} = a_{1k} e_{1k} + v_{1k}$ with $|a_{1k}| \leq \|A\|$, $\|v_{1k}\| \leq \|A\| N_1^{-1}$ and $e_1^0 \in L_1 = \operatorname{span}\{e_{11}, \ldots, e_{1n_1}\}$.

Let P_1 be the orthoprojector on L_1, $Q_1 = I - P_1$ be the orthoprojector on $L_1^{\perp}$, and apply the preceding construction to the operator $A_i = P_{L_1^{\perp}} A|_{L_1^{\perp}} = Q_1 A|_{L_1^{\perp}}$ to obtain an o.n. system $(e_{21}, \ldots, e_{2n_2})$ and two sequences $(a_{2k} : k = 1, \ldots, n_2)$, $(v_{2k} : k = 1, \ldots, n_2)$ such that $n_2 \leq N_2$, $e_2^0 \in L_2 = \operatorname{span}(\{e_{11}, \ldots, e_{1n_1}\} \cup \{e_{21}, \ldots, e_{2n_2}\})$ and $Q_1 A e_{2k} = a_{2k} e_{2k} + v_{2k}$, $|a_{2k}| \leq \|A\|$, $\|v_{2k}\| \leq \|A\| N_2^{-1}$ for $k = 1, \ldots, n_2$. Continuing by induction this construction we obtain the desired o.n. compression A-eigenbasis. □

Remark 3.8. Taking $N_j = (2c/\varepsilon)2^j$, one obtains from part (b) of Theorem 3.6 an o.n. compression A-eigenbasis for any given operator $A = A^*$. Then, by part (a), one has $A = A_1 + A_2 + A_5$ where A_1 is a diagonal operator and A_2, A_5 are operators defined as $A_2 f = \sum \langle f, e_{n_k} \rangle v_{n_k}$ and $A_5 f = \sum \langle f, v_{n_k} \rangle e_{nk}$, respectively, where $\|v_{nk}\| < cN_n^{-1}$. It is easy to see from the estimates in the proof that A_2 and A_5 are not only bounded but also Hilbert–Schmidt operators of norm $< \varepsilon$. Thus as a corollary of Theorem 3.6 follows the Weyl–von Neumann Theorem asserting that each operator $A = A^* \in \mathcal{L}(H)$ can be given as the sum of one diagonal operator and one Hilbert–Schmidt operator of norm less a prescribed $\varepsilon > 0$.

The proof of part (a) of Theorem 3.6 also gives the following

Corollary 3.9. *If $A = A^* \in L(H)$ and (e_{xy}) is a compression A-eigenframe over (E, dz), then there exist five a. o. operators $A_1, \ldots, A_5$ of simple type as in 3.6, such that for all $f, g \in H$,*

$$|\langle Af, g \rangle| \leq \sum_{i=1}^{5} |\langle A_i f, g \rangle|, \quad \|A_3\| \leq \frac{1}{4} \|A\|,$$

$$\|A_4\| \leq \|A\|, \quad \text{and} \quad \|A\| \leq 6c \|N^{-1}\|.$$

Remark 3.10. The algorithm in part (b) of 3.6 assigns to each $A = A^*$ an compression o.n. A-eigenbasis, but it does not assign to each $A = A^*$ an compression A-eigenframe or system of coherent states in a predetermined measures space (E, dz). This question will be considered elsewhere.

4. An A-eigenbasis given by an explicit simple algorithm

In Theorem 3.6 a theoretical algorithm was given, assigning to each operator $A = A^*$ a compression A-eigenbasis providing essential information on A. However there are few explicitly known o.n. bases, and the algorithm to obtain the A-eigenbasis from A and the given o.n. basis requires an infinite number of steps. Thus we seek a different type of generalized A-eigenbases, still providing information on A, but given by a simple algorithm assigning to each operator $A = A^*$ a generalized A-eigensystem.

For a fixed sequence (N_j), of strictly increasing positive integers, let $\mathcal{D}(N_j) = \{(j,k) \in \mathbb{Z}^+ \times \mathbb{Z}^+ : k \leq N_j\}$ and consider double sequences (e_{jk}) of triangular type-(N_j), where the pair (j,k) varies in the set $\mathcal{D}(N_j)$, so that $(e_{jk}) = (e_{11}, \ldots, e_{1N_1}, e_{21}, \ldots, e_{2N_2}, \ldots, e_{j1}, \ldots, e_{jN_j}, \ldots)$.

An *expanded basis-*(N_j) is an o. n. basis (e_j) of H with an associated sequence (e_{jk}) of triangular type-(N_j), such that for each $j = 1, 2, \ldots, e_j = e_{j1} + \cdots + e_{jN_j}$ and $(e_{j1}, \ldots, e_{jN_j})$ is an orthogonal(not necessarily normalized) sequence, $\langle e_{jk}, e_{j\ell}\rangle = 0$ for $k \neq \ell$.

A natural way to produce an expanded basis is to take an ordinary o.n. basis (e_j) and, for every $j = 1, 2, \ldots$, fix a partition of unity $I = P_{j1} + \cdots + P_{jN_j}$ where the $P_{jk} \in \mathcal{L}(H)$ are the orthoprojectors satisfying $P_{jk}P_{j\ell} = 0$ for $k \neq \ell$, and then set $e_{jk} = P_{jk}e_j$.

Here we shall concentrate on dyadic expanded bases that we now describe. Take $N_1 = 2, \ldots, N_j = 2^j, \ldots$, and identify each $(j,k) \in \mathcal{D}(2^j) =: \mathcal{D}$ with a dyadic interval $\Delta_{jk} \subset [0,1]$,such that, for each j, $[0,1] = \Delta_{j1} + \cdots + \Delta_{j2^j}$ is a dyadic partition, and chose the P_{jk} so that $P_{jk}P_{\ell m} = P_{\ell m}$ whenever $\Delta_{\ell m} \subset \Delta_{jk}$, that is, whenever the projection $P_{j+1,\ell}$ of generation $j+1$ is obtained by partitioning each $P_{j\ell}$ into a sum of two projections $P_{j\ell} = P_{j+1,\ell} + P_{j+1,\ell+1}$. We write also $P_{jk} = P(\Delta_{jk})$, so that $\Delta_{jk} \supset \Delta_{m\ell}$ implies $P(\Delta_{jk}) \geq P(\Delta_{m\ell})$. These expanded bases are called of *dyadic type* (2^j), and they are given by fixing an ordinary o.n. basis (e_j) and a dyadic family of partitions of unity $P(\Delta_{jk})$, and setting $e_{jk} = P(\Delta_{jk})e_j$, for all $(j,k) \in \mathcal{D}$, $(j,k) \sim$ dyadic Δ_{jk}. The dyadic expanded bases are *locally* orthogonal, in the sense that, for each $j = 1, 2, \ldots$, $\langle e_{jk}, e_{j\ell}\rangle = 0$ for $k \neq \ell$, and they are *globally* almost orthogonal, in the sense that for each e_{jk} and each $m < j$ there is only one $e_{m\ell}$ not orthogonal to e_{jk} , namely the only ascendant $\Delta_{m\ell}$ of generation m such that $\Delta_{m\ell} \supset \Delta_{jk}$. An expanded basis (e_{jk}) is called an *expanded* A-eigenbasis if $Ae_{jk} = a_{jk}e_{jk} + w_{jk}$, where $\|w_{jk}\| \leq cN_j^-1$ and there is a fixed $\delta > 0$ such that $\delta a_{jk} \in \Delta_{jk}, \forall (j,k) \in \mathcal{D}$.

To each self-adjoint operator $A \in \mathcal{L}(H)$, $\|A\| \leq 1$, is associated a well-determined dyadic expanded basis $(e(\Delta_{jk}))$ as follows. By the spectral theorem, for each $A = A^*$ there exists a measure space (X, μ) with a finite measure $\mu \geq 0$, a unitary isomorphism U of H onto $L^2(X,\mu)$, and a bounded measurable function $f = f(A)$, such that $UAU^{-1} = A_f$, where A_f is the operator in $L^2(X,\mu)$ defined by $(A_f g)(x) = f(x)g(x)$, for each $g \in L^2(X,\mu)$. For each interval $\Delta \subset [0,1]$, let $\widetilde{\Delta} = f^{-1}(\Delta) = \{x \in X : f(x) \in \Delta\}$, and let $\widetilde{P}(\Delta)$ be the orthoprojector in $L^2(X,\mu)$

defined by $\widetilde{P}(\Delta)g = 1_{\widetilde{\Delta}}g$, $1_{\widetilde{\Delta}}$ the characteristic function of $\widetilde{\Delta}$, $1_{\widetilde{\Delta}} = 0$ if $\widetilde{\Delta} = \emptyset$. If (Δ_{jk}) is the family of dyadic intervals in $[0, 1]$ then, for each $j = 1, 2, \ldots$, the $(\widetilde{P}(\Delta_{jk}))$ form a dyadic family of projector partitions of unity, which through U^{-1} give a corresponding dyadic family of orthoprojectors $(P(\Delta_{jk}))$ of H. Choosing a fixed o.n. basis (e_j) of H and setting $e_{jk} = P(\Delta_{jk})e_j$ (if $\widetilde{\Delta}_{jk} = \emptyset$ then $e_{jk} = 0$), a well-determined dyadic expanded basis in H is obtained, and it is called *the expanded basis associated to* A. If A has a simple spectrum, then X can be taken as $X = [0, 1]$ and $f(x) = x$, and then $\widetilde{P}(\Delta_{jk})$ will be given by $\widetilde{P}(\Delta_{jk})g = 1_{\Delta_{jk}}g$, for all $g \in L^2(X, \mu)$.

Let $A \in \mathcal{L}(H)$ be a self-adjoint operator, and let $(e_{jk}) = (e(\Delta_{jk}))$ be the dyadic expanded basis associated with A. Since $e_{j1} + \cdots + e_{j2^j} = e_j$ where (e_j) is an ordinary o.n. basis, each $f \in H$ is written as $f = \Sigma_j \langle f, e_j\rangle e_j$, and if an operator $D \in \mathcal{L}(H)$ is *diagonal* with respect to (e_{jk}), in the sense that

$$De_{jk} = a_{jk}(D)e_{jk} \quad \text{for each } \Delta_{jk} \sim (j, k) \in \mathcal{D}, \tag{4.1}$$

then, for every $(j, k) \in \mathcal{D}$, the inequality

$$|a_{jk}|\,\|e_{jk}\| = \|a_{jk}e_{jk}\| = \|De_{jk}\| \leq \|D\|\,\|e_{jk}\| \tag{4.2}$$

implies

$$|a_{jk}| \leq \|D\|, \quad a_{jk} = a_{jk}(D), \forall (j, k) \in \mathcal{D}, \tag{4.3}$$

and D is given by

$$Df = \sum_j \langle f, e_j\rangle De_j = \sum_j \sum_{k=1}^{2^j} \langle f, e_j\rangle a_{jk}(D)e_{jk}. \tag{4.4}$$

Observe that in the last sum the factor $\langle f, e_j\rangle$ appears instead of $\langle f, e_{jk}\rangle$. The expression of D in (4.4) is not equivalent to (4.1) but to

$$De_j = \sum_{k=1}^{2^j} De_{jk} = \sum_{k=1}^{2^j} a_{jk}e_{jk}, \quad \forall j, \tag{4.5}$$

and an operator $D \in \mathcal{L}(H)$ will be called an *expanded diagonal* operator with respect to A, if D satisfies (4.5) for each $j = 1, 2, \ldots$.

Though here we consider mainly expanded bases where $e_{jk} = P_{jk}e_j$, with P_{jk} orthoprojectors, the basic property of expanded diagonal operators will be stated for the more general expanded bases where, for each j, the system of orthoprojectors $P_{j1}, \ldots, P_{j2^j}$ is replaced by a general orthogonal sequence of commuting self-adjoint operators (P_{jk}), with $\|P_{jk}\| \leq c_P$ for all j, k, for fixed c_P. Still more generally, we replace the sequence (P_{jk}) of projections by an almost orthogonal sequence (P_{jk}). Thus we formulate the following general definition: Let (e_j) be an ordinary o.n. basis, and for each $j = 1, 2, \ldots$, let $P_{j1}, \ldots, P_{j2^j}$ be a finite almost orthogonal sequence of self-adjoint commuting operators, satisfying (2.9) with constants c_1 and c_2, such that $(c_1c_2)^{1/2} = c_P$. $\|P_{jk}\| \le c_P$, fixed, for all j, k. Assume that (P_{jk}) satisfies the dyadic partition condition

$$P_{jk} = P_{j+1,2k-1} + P_{j+1,2k} \quad \text{for } k = 1, 2, \dots, 2^j \text{ and all } j. \tag{4.6}$$

As above, we identify (j, k) with Δ_{jk}, write $P(\Delta_{jk})$ instead of P_{jk} and set $e_{jk} = P(\Delta_{jk})e_j$. Setting $P_j = \sum_{k=1}^{2^j} P(\Delta_{jk})$ we have $\Sigma_k e_{jk} = P_j e_j$. From (5.4) follows that

$$P_{jk} = P_{m,2^{m-j}k-2^{m-j}+1} + \cdots + P_{m,2^{m-j}k} \quad \text{if } j < m. \tag{4.7}$$

Proposition 4.1. *Let $(e_{jk}) = (P(\Delta_{jk})e_j)$, $(j, k) \in \mathfrak{D}$, be a dyadic expanded basis, where, for each $j = 1, 2, \dots$, $(P(\Delta_{jk}) : k = 1, \dots, 2^j)$ is a finite a.o.. sequence of self-adjoint commuting operators satisfying (2.9) with $(c_1c_2)^{1/2} = c_P$, and (4.6), and let D be a expanded diagonal operator with respect to this expanded basis. i.e., $De_{jk} = a_{jk}e_{jk}$, $a_{jk} = a_{jk}(D)$. If there is a fixed $\delta > 0$, $\delta = \delta_P$, such that*

$$\delta a_{jk} \in \Delta_{jk}, \quad \forall (j, k) \in \mathfrak{D}, \tag{4.8}$$

then every partial sum δD_m, for $D_m f = \sum_{j=1}^m \sum_{k=1}^{2^j} \langle f, e_j\rangle a_{jk} e_{jk}$, can be written as $\delta D_m = T_m + V_m$, where $\|T_m\| \le c_P$, and V_m is an a.o. Hilbert–Schmidt operator of WO type, satisfying $\|V_m\| \le \|V_m\|_2 \le c_P$. Thus, for all $m = 1, 2, \dots$,

$$\|D_m\| \le 2c_P\delta_P^{-1}. \tag{4.9}$$

Proof. By (4.8), $\delta|a_{jk}| \le 1$, so that by the property of a.o. series, the operator $P_{m\delta} := \sum_{n=1}^{2^m} \delta a_{mn} P(\Delta_{mn})$ is bounded in norm by the fixed constant c_P. Setting $T_m f = \sum_{j=1}^m \sum_{n=1}^{2^m} \langle f, e_j\rangle \delta a_{mn} P(\Delta_{mn}) e_j = \sum_{j=1}^m \langle f, e_j\rangle P_{m\delta} e_j = P_{m\delta}\big(\sum_{j=1}^m \langle f, e_j\rangle e_j\big)$ we have that $\|T_m f\| \le \|P_{m\delta}\| \big(\sum_{j=1}^m |\langle f, e_j\rangle|^2\big)^{1/2} \le c_P\|f\|$, wich implies $\|T_m\| \le c_P$.

From the definition of T_m and from (4.7) it follows easily that

$$\begin{aligned} V_m &= \delta D_m f - T_m f \\ &= \sum_{j=1}^m \sum_{k=1}^{2^j} \langle f, e_j\rangle \delta a_{jk} P(\Delta_{jk}) e_j - \sum_{j=1}^m \sum_{n=1}^{2^m} \langle f, e_j\rangle \delta a_{mn} P(\Delta_{mn}) e_j \\ &= \sum_{j=1}^m \sum_{k=1}^{2^j} \langle f, e_j\rangle \big[(\delta a_{jk} - \delta a_{m,2^{m-j}k-1}) P_{m,2^{m-j}k-2^{m-j}+1} \\ &\qquad + \cdots + (\delta a_{jk} - \delta a_{m,2^{m-j}k}) P_{m,2^{m-j}k}\big] e_j = \sum_{j=1}^m \langle f, e_j\rangle w_j, \end{aligned} \tag{4.10}$$

where δa_{jk} and $\delta a_{m,2^{m-j}k-1}$ both belong to the interval Δ_{jk} of length 2^{-j}, and their difference, as well as the other differences in parentheses in the last expression are all less than or equal 2^{-j}, and (w_j) is an almost orthogonal sequence with constant less than or equal $c_P 2^{-j}$, so

$$V_m f = \delta D_m f - T_m f = \sum_{j=1}^{m} \langle f, e_j \rangle w_j \quad \text{with } \|w_j\| \leq c_P 2^{-j}. \tag{4.11}$$

By Propositions 2.2 and 2.4, (4.11) implies that $\sum_j A_j f = \sum_{j=1}^{m} \langle f, e_j \rangle w_j$ is almost orthogonal. In fact, this is easily verified in the present case, since for $i = 1, 2$, $k_i(j,m) = |\langle w_j, w_m \rangle| \leq c_P 2^{-j} c_P 2^{-m} < \infty$. Thus V is an almost orthogonal operator and, from (4.11) follows that $V_m e_j = w_j$ and $\sum_j \|V_m e_j\|^2 \leq c_P^2$, so that V_m is a Hilbert–Schmidt operator with $\|V_m\|_2 \leq c_P$. □

Remark 4.2. From the proof of Proposition 4.1 follows that every expanded diagonal operator D satisfying condition (4.8) is the sum of two almost orthogonal operators.

Proposition 4.3. *For every $A = A^* \in L(H)$, the associated expanded basis of A is an expanded A-eigenbasis, and A can be written as $A = D + V$, where D is an expanded diagonal operator with respect to the expanded basis (e_{jk}), $(j,k) \in \mathcal{D}_P$, associated with A, satisfying* (4.8) *for $\delta = 1$, and V is a Hilbert–Schmidt operator, $\|V\|_2 < \varepsilon$, for a prescribed $\varepsilon > 0$. Moreover, here the associated o.n. basis (e_n) can be arbitrarily given.*

Remark 4.4. While in Section 3 the o.n. basis varied with A, in Proposition 4.3 it can be chosen arbitrarily, and requires only a simple expansion, and this construction allows to extend the decomposition to all operators $F(A)$, for "good" Lipschitz functions F, even not real ones, thus providing new variants that cannot be obtained by the method in Section 3, which varies with F, even for real F. The interest of this variant lies on the possibility of a functional calculus, as well as in its computability.

Proof. Without loss of generality we may consider A such that $\|A\| \leq 1$. By the definition of the expanded basis associated with A, we may take $H = L^2(X, \mu)$, $A = A_f$ defined by $A_f g(x) = f(x)g(x)$ for a fixed bounded $f \in L^2(X, \mu)$ and all $g \in L^2(X, \mu)$, so that the expanded basis $(e_j))$ is given by $e_{jk}(x) = 1_{\widetilde{\Delta}_{jk}}(x) e_j(x)$, where $\widetilde{\Delta}_{jk} = \{x \in X : f(x) \in \Delta_{jk}\}$ and (e_j) is a fixed o.n. basis of $L^2(X, \mu)$ and for all $j = 1, 2, \ldots$, $A_f e_j(x) = \sum_{k=1}^{2^j} 1_{\widetilde{\Delta}_{jk}}(x) f(x) e_j(x) = \sum_{k=1}^{2^j} 1_{\widetilde{\Delta}_{jk}}(x) f(x) e_{jk}(x)$. $A_f g(x) = \sum_{k=1}^{2^j} 1_{\widetilde{\Delta}_{jk}}(x) f(x) g(x)$. Take the center of the dyadic interval $\Delta_{jk} \in \mathcal{D}$ as a_{jk}, and define a formal series $\sum_{j=1}^{\infty} \sum_{k=1}^{2^j} D_{jk}$ with $D_{jk} f = \langle f, e_j \rangle a_{jk}\, e_{jk}$, with partial sums $D_m f = \sum_{j=1}^{m} \sum_{k=1}^{2^j} \langle f, e_j \rangle a_{jk} e_{jk}$, where the a_{jk} satisfy (4.8) for $\delta = 1$.

In this expanded basis A is expressed as

$$Af = \sum_{j=1}^{\infty} \langle f, e_j \rangle A e_j = \sum_j \langle f, e_j \rangle \sum_{k=1}^{2^j} f 1_{\widetilde{\Delta}_{jk}} e_j = \sum_j \langle f, e_j \rangle \sum_{k=1}^{2^j} f 1_{\widetilde{\Delta}_{jk}} e_{jk},$$

with partial sums

$$A_m f = \sum_{j=1}^{m} \langle f, e_j \rangle \sum_{k=1}^{2^j} f 1_{\widetilde{\Delta}_{jk}} e_{jk}.$$

Since $f(x) \in \Delta_{jk}$ for $x \in \widetilde{\Delta}_{jk}$, and since Δ_{jk} is an interval of length 2^{-j}, and both numbers a_{jk} and $1_{\widetilde{\Delta}_{jk}}(x) f(x)$ are in Δ_{jk} if $1_{\widetilde{\Delta}_{jk}}(x) f(x) \neq 0$ and $e_{jk}(x) = 1_{\widetilde{\Delta}_{jk}}(x) e_j(x) = 0$ if $x \notin \widetilde{\Delta} jk$, we have that $\|1_{\widetilde{\Delta}_{jk}}(x) f(x) e_{jk}(x) - a_{jk} e_{jk}(x)\| = \|(1_{\widetilde{\Delta}_{jk}}(x) f(x) - a_{jk}) e_{jk}(x)\| \leq 2^{-j} \|e_{jk}\|$.

Thus $A_m f - D_m f = \sum_{j=1}^{m} \langle f, e_j \rangle w_j$ where, since for each j, (e_{jk}) is an orthogonal system, for each j, $\|w_j\| \leq \left(\sum_{k=1}^{2^j} (2^{-j} \|e_{jk}\|)^2 \right)^{1/2} \leq 2^{-j}$. Hence, $\|D_m f - D_{m+p} f\| \leq \|A_m f - A_{m+p} f\| + \|\sum_{j=m+1}^{m+p} \langle f, e_j \rangle w_j\| \leq \|A_m f - A_{m+p} f\| + \left(\sum_{j=m+1}^{m+p} 2^{-2j} \right)^{1/2} \|f\|$. Since $\|A_m f - A_{m+p} f\|$ tends to 0 when m tends to infinity, it follows that $D_m f$ converges strongly to a limit Df where D is expanded diagonal with respect to the expanded basis (e_{jk}), with $a_{jk} = a_{jk}(D)$ satisfying (4.8) for $\delta = 1$, and that $V := A - D$ is the limit of $V_m = A_m - D_m$ with $V_m f = \sum_{j=1}^{m} \langle f, e_j \rangle w_j$ and $\|w_j\| \leq 2^{-j}$. Thus, each V_m is a Hilbert–Schmidt operator, the operator sequence (V_m) converges to the Hilbert–Schmidt operator V, and $A = D + V$.

Observe that if instead of choosing an expanded basis of type (2^j) we choose one of type $(2^j 2^p)$, we get $\|w_j\|^2 \leq 2^{-p} 2^{-j}$, and then $\|V\|_2 \leq 2^{-p} < \varepsilon$ for any prescribed $\varepsilon > 0$, taking p sufficiently large. □

Remark 4.5. Proposition 4.3 produces a very large set of expanded A-eigenbases that can be explicitly described, and which provide representations $A = A_1 + A_2 + A_3$ with an explicit control on the norms of the WO type summands A_1, A_2, A_3. We shall not go here into generalizations to frames and systems of coherent states over a given measure space which provide still large sets of representations of A of the above type.

5. Almost orthogonal series in commutative C^*-algebras

The Almost Orthogonal Lemma and its variants (see Proposition 2.1) provide norm estimates for a large class of operators in classical analysis, and some of those operators are given by special a.o. sequences in a commutative C*-algebra. This suggests to study in more detail the a.o. sequences in particular C*-algebras. Here we show that each a.o. sequence in a commutative C*-algebra gives rise to a large set of explicitly given strongly convergent operator series, with an explicit estimate for the norm of their sums. Moreover, in such case, the series not only converge in the strong topology of $\mathcal{L}(H)$, but also in a natural pointwise sense.

The proof of Theorem 5.1 below is inspired on an approach to the original Almost Orthogonal Lemma [1] due to B. Sz.-Nagy [2] (see also [15]).

Let $\mathcal{A} \subset L(H)$ be a commutative C^*-algebra, W its associated Gelfand compact space, $\{A_n\} \subset \mathcal{A}$ a fixed sequence of operators such that, $\|A_n\| \leq 1$, $\forall n$, and let $k(m,n) = \|(A_m A_n\|^{1/2}$ be its associated kernel. By the spectral theorem for C^*-algebras there is a sequence of orthogonal subspaces $(H_\alpha) \subset H$, for each α a measure μ_α in W, and an isometry J_α mapping H_α on to $L^2(W, \mu_\alpha)$, such that

1. $\forall f \in H$, $f = \sum f_\alpha$ with $f_\alpha \in H_\alpha$ and $\|f\|^2 = \sum \|f_\alpha\|^2$;
2. $\forall\, n$ and $\forall\, \alpha$, $A_n(H_\alpha) \subset H_\alpha$, and
3. $\forall\, A_n$ corresponds a continuous function $\widehat{A_n} \in C(W)$ such that, for each $f_\alpha \in H_\alpha$, $J_\alpha(A_n f_\alpha) = \widehat{A_n}(J_\alpha f_\alpha)$, i.e, under the isomorphism J_α, the restriction of A_n to H_α transforms into the operator $\widehat{f_\alpha}(w) \to \widehat{A_n}(w)\widehat{f_\alpha}(w)$, where $\widehat{f_\alpha} = J_\alpha f_\alpha$.

Every continuous scalar function F in $\mathbb{R}$ assigns to each self-adjoint operator $A \in \mathcal{A}$ an operator $F(A)$, and gives rise to the sequence $(F(A_n))$, so that $\widehat{F(A)} = F(\widehat{A})$, $(\widehat{F(A_n)}) = F(\widehat{A_n})$, and, if $|A|$ is defined by $|A| = |\widehat{A}|$, then $|A| \geq 0$, $F(|A|) = F(|\widehat{A}|)$.

If F is continuously differentiable, and if $F_+(r) := F(r)$ when $F'(r) > 0$ and zero otherwise, $F_-(r) := F(r)$ when $F'(r) < 0$ and zero otherwise, we may write

$$F = G_1 + G_2 = G_1 - |G_2| \tag{5.1}$$

where $G_1 \geq 0$ and $G_2 \leq 0$ are continuously differentiable functions such that

$$G_1 = F_+, \quad G_2 = F_-. \tag{5.2}$$

Theorem 5.1. *With the above notation, let F be a continuously differentiable function in $\mathbb{R}$ such that the functions G_1, G_2 in (5.2) satisfy*

$$\sup_m \Big(\sum_n G_1(k(m,n))\Big) = c_1 < \infty,$$

$$\sup_m \Big(\sum_n |G_2(k(m,n))|\Big) = c_2 < \infty.$$

Then the scalar series $\sum_n |G_1(\widehat{A_n})(w)|$ and $\sum_n |G_2(\widehat{A_n}(w))|$ converge for every $w \in W$, $\sum_n G_1(A_n)$ to an operator A', of norm $\|A'\| \leq c_1$, $\sum_n G_2(A_n)$ to an operator A'', of norm $\|A''\| \leq c_2$, and $\sum_n F(A_n)$ converges to an operator A, of norm $\|A\| \leq c_1 + c_2$.

Remark 5.2. In the particular case where $F(r) = r$, Theorem 5.1 reduces to a known version of Lemma 2.1 (see [9]). Of course, $\sum_n F(A_n)$ should not be confused here with $F(\sum_n A_n)$.

Proof. In view of above properties 1, 2, 3, we may assume that $H = L^2(W, \mu)$, μ a non-negative measure on the compact space W, and that the given sequence of operators (A_n) is defined through the sequence $(\widehat{A}_n)$, $\widehat{A}_n \in C(W)$, as follows.

$$\forall f \in H = L^2(W, \mu), \quad A_n f = \widehat{A}_n f, \tag{5.3}$$

so that, $\forall w \in W$, $(A_n f)(w) = \widehat{A_n}(w) f(w)$, and by hypothesis, $\|A_n\| = \|\widehat{A}_n\|_\infty = \sup\{|\widehat{A_n}(w)| : w \in W\} \leq 1$ holds for all n. For a fixed $w \in W$ and a fixed $\varepsilon > 0$, set $M = \sup\{|\widehat{A : n(x)}| : n \in \mathbb{N}\}$, let $m(\varepsilon) = m(\varepsilon, w)$ be such that

$$|\widehat{A}_{m(\varepsilon)}(w)| \geq M - \varepsilon, \tag{5.4}$$

and set $k(n) = k(m(\varepsilon), n)$, so that $\sup\{k(n) : n \in \mathbb{N}\} \leq 1$, since $\|A_n\| \leq 1$ for all n. Since w is fixed, we can define the distribution functions $a_*(a) = \operatorname{card}\{n : |\widehat{A_n}(w)| > a\}$ and $k_*(a) = \operatorname{card}\{n : k(n) > a\}$, of the scalar functions $n \to |\widehat{A_n}(w)|$ and $n \to k(n)$, so that $a_*(a) = 0$ if $a \geq M$, and $k_*(a) = 0$ if $a \geq 1$. Let us first show that (assuming w fixed)

$$\forall a \in (0, M - \varepsilon), \quad a_*(a) \leq k_*(a). \tag{5.5}$$

In fact, if $|\widehat{A_n}(w)| > a$ then, by (5.4), and since $a < M - \varepsilon < |\widehat{A}_{m(\varepsilon)}(w)|$, it follows that $a^2 < |\widehat{A_n}(w)||\widehat{A}_{m(\varepsilon)}| \leq \|A_n A_{m(\varepsilon)}\| = k(m(\varepsilon), n)^2 = k(n)^2$, so $k(n) > a$. Thus $|\widehat{A_n}(w)| > a$ implies $k(n) > a$, which proves (5.5).Since $G_1(a) \geq 0$ it follows from (5.5) that

$$\begin{aligned}\int_0^{M_\varepsilon} G_1(a) da_*(a) &= \int_0^{M-\varepsilon} G_1^2(a) a_*(a) da \leq \int_0^{1-\varepsilon} G_1^1(a) a_*(a) da \\ &\leq \int_0^{1-\varepsilon} G_1^1(a) k_*(a) da \leq \int_0^1 G_1(a) da_*(a).\end{aligned}$$

and, since by a known property of distribution functions,

$$\sum_n G_1(|\widehat{A_n}(w)|) = \int_0^{M-\varepsilon} G_1(a) da_*(a) \quad \text{and}$$

$$\sum_n G_1(k(m(\varepsilon)), n) = \sum G_1(k(n)) = \int_0^1 G_1(a) dk_*(a),$$

letting $\varepsilon \to 0$, we get $\sum_n G(|\widehat{A_n}(w)| \leq \sum_n G_1(k(m(\varepsilon), n)))$, for all $w \in W$. Then the assertions on G_1 follow easily, and similarly those on G_2, and, as stated, $F = G_1 + G_2$.

In particular if $F(\lambda) = \lambda$, $F'(\lambda) = 1$, so that $F = G_1$, $G_2 = 0$ and (5.5) gives in this case

$$\sum_n |\widehat{A_n}(w)| \leq \sup_m \Big(\sum_n k(m, n)\Big),$$

which implies the strong convergence of $\sum_n \widehat{A}_n(w)f(w)$ in the $L^2(\mu)$-norm for all $f \in L^2(W, \mu)$, and therefore, the strong convergence of $\sum A_n$ in $L(H)$, which is a known version of Proposition 2.1 (see [9], [7]). □

References

[1] Cotlar, M. A Combinatorial Inequaliy and its Applications to L^2 space, Rev. Mat. Cuyana 1 (1955), 41–55.

[2] Sz.Nagy, B., Note on sums of almost orthogonal operators, Acta Sci. Math. Szeged 18 (1959), 189–191.

[3] Peetre, J. Une caractérisation abstraite des opérateurs différentiels, Math. Scand. 7 (1959), 211–218.

[4] Knapp, A. W. and E. M. Stein, Intertwining operators for semi-simple groups, Ann. of Math. 93 (1971), 489–578.

[5] Calderón, A.P. and R. Vaillancourt, A class of bounded pseudo-differential operators, Proc. Nat. Acad. Sci. U.S.A. 69 (1972), 1185–1187.

[6] Fefferman, C., Pointwise convergence of Fourier series, Ann. of Math.(2) 98(1973), 551–571.

[7] Unterberger, A., Extensions du lemme de Cotlar et applications, C. R. Acad. Sci. Paris Ser.A 288 (1979), 249–252.

[8] Stein, E. M., Harmonic Analysis, Princeton Math. Ser. 43, Princeton University Press, Princeton, 1993.

[9] Folland, G. B., Harmonic Analysis in Phase Space, Ann. of Math. Stud. 122, Princeton University Press, Princeton, 1989.

[10] Coifman, R. R. and Y. Meyer, Au delà des opérateurs pseudo-différentiels, Astérisque 57, Société Mathématique de France, 1978.

[11] Rochberg, R. and S. Semmes, Nearly weakly orthonormal sequences, singular values estimates and Calderón–Zygmund operators, J. Funct. Anal. 86 (1989), 237–306.

[12] Unterberger, A., États Cohérents, Chapitre II, Cours de D.E.A., Université de Reims, 1991.

[13] Berezansky, Iu., Eigenexpasions of Self-Adjoint Operators, Transl. Math. Monographs, Amer. Math. Soc., Providence, RI, 1968.

[14] Cotlar, M. and C. Sadosky, On the Helson-Szegő theorem and a related class of modified Toeplitz kernels, Proc. Symp. Pure Math., Amer. Math. Soc. 25 (1979), 383–407.

[15] Cotlar, M., Convolution operators and factorization, McGill Analysis Seminar Lecture Notes, McGill University, Montréal, 1972.

Mischa Cotlar, Central University of Venezuela, UCV, Residencias Paris ap.82, Av. El Mirador, La Campiña, Caracas 1050, Venezuela

E-mail: `mcotlar@euler.ciens.ucv.ve`

Cora Sadosky, Department of Mathematics, Howard University, Washington, DC 20059-0001, U.S.A.
E-mail: cs@scs.howard.edu

Localization and extrapolation in Orlicz–Lorentz spaces

David Cruz-Uribe, SFO and Miroslav Krbec

Dedicated to Jaak Peetre on occasion of his 65th birthday

Abstract. We apply a decomposition technique introduced in [5] to give a variety of characterizations of the Orlicz space $L^{\exp t^r}$ and the Orlicz–Lorentz spaces $L^{\exp t^{r'},t^r}$. Our results extend and unify classical results and those found in [5]. We apply the decomposition method to inequalities for convolutions with the Riesz kernel, corresponding to Sobolev imbeddings in the critical case.

2000 Mathematics Subject Classification: 46E30, 46E35

1. Introduction

In this paper we prove extrapolation theorems for the Orlicz and Orlicz–Lorentz spaces which occur naturally in the limiting cases of Sobolev imbeddings. We first consider the exponential Orlicz spaces $L^{\exp t^a}$ on sets of bounded measure in $\mathbb{R}^n$. (For simplicity we will assume throughout the paper that all spaces are defined on sets of Lebesgue measure one.) Recall that for $\Omega \subset \mathbb{R}^n$, $|\Omega| = 1$, the Orlicz space $L^{\exp t^a} = L^{\exp t^a}(\Omega)$ is the Banach function space of all measurable f on Ω such that

$$\|f\|_{L^{\exp t^a}} = \inf\left\{\lambda > 0 : \int_\Omega \exp([f(x)/\lambda]^a) < 2\right\} < \infty.$$

A now classical result due to Trudinger and others, proved by different techniques and in various forms in [13, 10, 8, 11] (and in a more general setting in a number of recent papers), is that when $kr = n$, the Sobolev space $W^{k,r}$ is contained in $L^{\exp t^{n/(n-k)}}$.

It is well known that $f \in L^{\exp t^a}$ is equivalent to the extrapolation inequality $\sup_k k^{-1/a}\|f\|_k < \infty$, where $\|\cdot\|_k$ is the norm on the Lebesgue space L^k. We use the concept of a "decomposition" developed in [5] to generalize this characterization of $L^{\exp t^a}$. To state our result, we need two definitions which we will also use below. First, for $k \geq 1$, define the intervals $I_k = (e^{-k}, e^{-k+1})$. Note that up to a set of measure zero, $\bigcup_k I_k = [0, 1]$.

Definition 1.1. Given an interval $I \subset [0,1]$ and the non-increasing rearrangement f^* of a function f on $[0,1]$, define the *localized Lorentz norm* by

$$\|f\|_{p,q,I} = \left(\frac{q}{p}\int_I t^{q/p} f^*(t)^q \frac{dt}{t}\right)^{1/q}$$

if $1 \le p, q < \infty$, and

$$\|f\|_{p,\infty,I} = \sup_{t\in I} t^{1/p} f^*(t)$$

if $1 \le p \le \infty$.

Theorem 1.2. *Given a function f and $a > 0$, the following are equivalent:*

(i) $f \in L^{\exp t^a}$;

(ii) $\displaystyle\sup_k \frac{\|f\|_{k,\infty}}{k^{1/a}} < \infty$;

(iii) $\displaystyle\sup_k \frac{\|f\|_{k,\infty,I_k}}{k^{1/a}} < \infty$.

Remark 1.3. The equivalence of (i) and (ii) is known in the setting of the abstract Σ-method (cf. [7]). Here we give a direct proof.

In [4] it was shown that the classical limiting imbedding could be refined by passing to a richer scale of spaces – the Orlicz–Lorentz spaces $L^{\exp t^{r'}, t^r}$, $1 < r < \infty$. More precisely, it was shown that if $f \in W^{k,r}$ then

$$\int_0^1 \left(\frac{f^*(t)}{\log(e/t)}\right)^r \frac{dt}{t} < \infty. \tag{1.1}$$

In [5] it was observed that if $r > 1$ then (1.1) is equivalent to $f \in L^{\exp t^{r'}, t^r}$. We note in passing that this refinement is analogous to the refinement possible when $kr < n$ by passing from the Lebesgue space $L^{n/(n-kr)}$ to the Lorentz space $L^{n/(n-kr),r}$. (See, for example, [1] for a detailed treatment of refined Sobolev imbeddings into the Lorentz scale. Also see [14, Chapter 2].)

In this paper we will take (1.1) as the definition of $L^{\exp t^{r'}, t^r}$; we will not discuss the Orlicz–Lorentz scale in general and we refer the reader to [9], [5] and [6]. We also note in passing that these spaces are equivalent to the Lorentz–Zygmund spaces with norm $\|\cdot\|_{\infty,r,-r}$. See [2] for details.

Our next result is analogous to Theorem 1.2 and gives global and local extrapolation characterizations of $L^{\exp t^{r'}, t^r}$ in terms of both Lebesgue and Lorentz norms. Here and below, given an interval $I \subset [0,1]$, $\|f\|_{k,I}$ denotes the norm of f^* in $L^k(I)$.

Theorem 1.4. *Given a function f and r, $1 < r < \infty$, the following are equivalent:*

(i) $f \in L^{\exp t^{r'}, t^r}$;

(ii) $\displaystyle\sum_{k=1}^{\infty}\left(\frac{\|f\|_{k,I_k}}{k}\right)^r < \infty;$

(iii) $\displaystyle\sum_{k=1}^{\infty}\left(\frac{\|f\|_{k,r,I_k}}{k^{1/r'}}\right)^r < \infty;$

(iv) $\displaystyle\sum_{k=1}^{\infty}\left(\frac{\|f\|_{k}}{k}\right)^r < \infty;$

(v) $\displaystyle\sum_{k=1}^{\infty}\left(\frac{\|f\|_{k,r}}{k}\right)^r < \infty.$

Remark 1.5. The lack of symmetry in Condition (iii) can be removed (or, more precisely, hidden) if we omit the constant q/p in the definition of the local Lorentz norm.

Remark 1.6. Theorem 1.4 is still true when $r = 1$, but the space $L^{\exp t^{r'},t^r}$ is undefined and we must replace the first condition with (1.1), which still makes sense.

Theorems 1.2 and 1.4 can be combined (along with results which are either classical or in [5]) into a single theorem. Here, when $r = \infty$ we identify $L^{\exp t^{r'},t^r}$ with $L^{\exp t}$.

Corollary 1.7. *Given a function f and r, $1 \le r \le \infty$, the following are equivalent:*

(i) $f \in L^{\exp t^{r'},t^r}$ (*if* $r > 1$);

(ii) $\displaystyle\frac{f^*(t)}{\log(e/t)} \in L^r([0,1], dt/t);$

(iii) $\displaystyle\left\{\frac{\|f\|_{k,I_k}}{k}\right\} \in \ell^r;$

(iv) $\displaystyle\left\{\frac{\|f\|_{k,r,I_k}}{k^{1/r'}}\right\} \in \ell^r;$

(v) $\displaystyle\left\{\frac{\|f\|_{k}}{k}\right\} \in \ell^r;$

(vi) $\displaystyle\left\{\frac{\|f\|_{k,r}}{k}\right\} \in \ell^r.$

Proof. We first consider the case $r = \infty$. Then the following implications hold:

(a) (i) $\Leftrightarrow$ (ii). Recall that $L^\infty([0,1], dt/t) = L^\infty([0,1])$. Then this is classical; see, for example, [3, Chapter 4].

(b) (ii) $\Leftrightarrow$ (iii) $\Leftrightarrow$ (v). [5, Theorems 2.1, 2.2].

(c) (i) $\Leftrightarrow$ (iv) $\Leftrightarrow$ (vi). This is Theorem 1.2.

Now suppose that $1 \le r < \infty$. If $r > 1$ then, as we noted above, the equivalence (i) $\Leftrightarrow$ (ii) is shown in [5, Section 1]. The rest of the implications are in Theorem 1.4. □

The remainder of this paper is organized as follows: in Section 2 we prove Theorem 1.2 and in Section 3 we prove Theorem 1.4. Finally, in Section 4, as an application of our results, we use Theorem 1.4 to give a simple proof of the fact that if $f \in W^{k,r}$ then f satisfies (1.1).

Throughout this paper we will make the following assumptions and use the following notation. As we noted above, all function spaces are assumed to be defined on sets with Lebesgue measure 1. Given a function f in such a space, f^* will denote its non-increasing rearrangement on $[0, 1]$.

We decompose the interval $[0, 1]$ (up to a set of measure zero) into the union of disjoint intervals $I_k = (e^{-k}, e^{-k+1})$, $k \ge 1$.

The notation $A \sim B$ means that there exist positive constants c and C, independent of any functions involved, such that $cA \le B \le CA$. Furthermore, we explicitly point out the key fact that in all such relations in the following, c and C are independent of the parameter $k \in \mathbb{N}$. On the other hand, the parameter r in our results is a fixed real number, $1 \le r \le \infty$, and the constants c and C may depend on it.

2. Proof of Theorem 1.2

We first note that it suffices to consider the case $a = 1$, since given arbitrary $a > 0$ we have that $f \in L^{\exp t^a}$ if and only if $f^a \in L^{\exp t}$, and $(f^*)^a = (f^a)^*$.

Second, in [5] it was shown that $f \in L^{\exp t}$ if and only if

$$\sup_k \frac{\|f\|_{k,I_k}}{k} < \infty. \tag{2.1}$$

(ii) $\Rightarrow$ (i). We estimate (2.1):

$$\begin{aligned}
\sup_k \frac{1}{k}\left(\int_{I_k} f^*(t)^k\,dt\right)^{1/k} &\le \sup_k \frac{1}{k} f^*(e^{-k})|I_k|^{1/k} \\
&\le \sup_k \frac{1}{k} f^*(e^{-k})\left(e^{-k}\right)^{1/k}\left(e^{k}\right)^{1/k} \le \sup_k \frac{e}{k}\left(\sup_{t\in I_k} t^{1/k} f^*(t)\right) \\
&\le e \sup_k \frac{1}{k} \sup_{t\in(0,1)} t^{1/k} f^*(t) = e \sup_k \frac{\|f\|_{k,\infty}}{k}.
\end{aligned}$$

(i) $\Rightarrow$ (ii). By the definition of the I_k's, we have that $|I_{j+1}|\left(\frac{e^2}{e-1}\right) = e^{-j+1}$.

We now prove the converse of the previous inequality:

$$\begin{aligned}
\sup_k \frac{\|f\|_{k,\infty}}{k} &= \sup_k \frac{1}{k} \sup_{t\in(0,1)} t^{1/k} f^*(t) = \sup_k \frac{1}{k} \sup_j \sup_{t\in I_j} t^{1/k} f^*(t) \\
&= \sup_k \frac{1}{k} \sup_j e^{(-j+1)/k} f^*(e^{-j}) \\
&\le \sup_k \frac{1}{k} \sup_j \left(\frac{e^2}{e-1}\right)^{1/k} f^*(e^{-j})|I_{j+1}|^{1/k} \\
&\le \left(\frac{e^2}{e-1}\right) \sup_k \frac{1}{k} \sup_j \left(\int_{I_{j+1}} f^*(t)^k\,dt\right)^{1/k} \\
&\le \left(\frac{e^2}{e-1}\right) \sup_k \frac{1}{k} \left(\int_0^1 f^*(t)^k\,dt\right)^{1/k} \\
&= \left(\frac{e^2}{e-1}\right) \sup_k \frac{\|f\|_k}{k}.
\end{aligned}$$

(ii) $\Rightarrow$ (iii). This is immediate.
(iii) $\Rightarrow$ (i). Again we estimate (2.1). For each $k \ge 1$:

$$\begin{aligned}
\frac{\|f\|_{k,I_k}}{k} &\le \frac{1}{k}|I_k|^{1/k} \sup_{t\in I_k} f^*(t) = \frac{1}{k} e(1-e^{-1})^{1/k} \sup_{t\in I_k} f^*(t) \\
&\le \frac{1}{k} \sup_{t\in I_k} t^{1/k} f^*(t) = \frac{\|f\|_{k,\infty,I_k}}{k}.
\end{aligned}$$

This completes the proof. □

3. Proof of Theorem 1.4

(ii) $\Rightarrow$ (i). We estimate (1.1). On I_k, $\log(e/t) \ge \log(e/e^{-k+1}) = k$, so

$$\begin{aligned}
\int_0^1 \left(\frac{f^*(t)}{\log(e/t)}\right)^r \frac{dt}{t} &= \sum_{k=1}^{\infty} \int_{I_k} \left(\frac{f^*(t)}{\log(e/t)}\right)^r \frac{dt}{t} \le \sum_{k=1}^{\infty} \frac{e^k}{k^r} f^*(e^{-k})^r |I_k| \\
&= (e-1) \sum_{k=1}^{\infty} \left(\frac{f^*(e^{-k})}{k}\right)^r \quad (\text{since } e^k|I_k| = (e-1)) \\
&\le (e-1) \sum_{k=1}^{\infty} k^{-r} \left(\frac{1}{|I_{k+1}|} \int_{I_{k+1}} f^*(t)^{k+1}\,dt\right)^{r/(k+1)} \\
&\le (e-1)e^r(1-e^{-1})^{-r} \sum_{k=1}^{\infty} \left(\frac{k+1}{k}\right)^r \left(\frac{\|f^*\|_{k+1,I_{k+1}}}{k+1}\right)^r
\end{aligned}$$

$$(\text{here we used the estimate } |I_{k+1}|^{1/(k+1)} \geq e^{-1}(1-e^{-1}))$$
$$\leq 2^r(e-1)e^r(1-e^{-1})^{-r} \sum_{k=1}^{\infty} \left(\frac{\|f^*\|_{k,I_k}}{k} \right)^r .$$

(i) ⇒ (ii). By an almost identical argument we prove the reverse inequality:

$$\begin{aligned}
\sum_{k=1}^{\infty} \left(\frac{\|f^*\|_{k,I_k}}{k} \right)^r &\leq \sum_{k=1}^{\infty} k^{-r} f^*(e^{-k})^r |I_k|^{r/k} \\
&\leq e^{-r}(e-1)^r \sum_{k=1}^{\infty} k^{-r} f^*(e^{-k})^r \\
&\leq e^{-r}(e-1)^r(1-e^{-1})^{-1} \sum_{k=1}^{\infty} \frac{e^k}{k^r} \int_{I_{k+1}} f^*(t)^r\, dt \\
&\quad (\text{here we used equality } e^k|I_{k+1}| = 1-e^{-1}) \\
&\leq e^{-r}(e-1)^r(1-e^{-1})^{-1} \sum_{k=1}^{\infty} \frac{(k+2)^r}{k^r} \int_{I_{k+1}} \left(\frac{f^*(t)}{\log(e/t)} \right)^r \frac{dt}{t} \\
&\leq 3^r e^{-r}(e-1)^r(1-e^{-1})^{-1} \sum_{k=1}^{\infty} \int_{I_{k+1}} \left(\frac{f^*(t)}{\log(e/t)} \right)^r \frac{dt}{t} \\
&\leq 3^r e^{-r}(e-1)^r(1-e^{-1})^{-1} \int_0^1 \left(\frac{f^*(t)}{\log(e/t)} \right)^r \frac{dt}{t}.
\end{aligned}$$

(i) ⇔ (iii). The proof is similar to the proof of Theorem 3.1 in [5]:

$$\begin{aligned}
\sum_{k=1}^{\infty} \int_{I_k} \left(\frac{f^*(t)}{\log(e/t)} \right)^r \frac{dt}{t} &\sim \sum_{k=1}^{\infty} \frac{1}{k^r} \int_{I_k} [f^*(t)]^r \frac{dt}{t} \\
&\sim \sum_{k=1}^{\infty} \frac{1}{k^r} \int_{I_k} t^{r/k} [f^*(t)]^r \frac{dt}{t} \\
&\sim c \sum_{k=1}^{\infty} \frac{1}{k^{r-1}} \left[\frac{r}{k} \int_{I_k} t^{r/k} [f^*(t)]^r \frac{dt}{t} \right] \\
&\sim \sum_{k=1}^{\infty} \left(\frac{\|f\|_{k,r,I_k}}{k^{1/r'}} \right)^r .
\end{aligned}$$

(ii) ⇔ (v). We estimate as follows:

$$\sum_{k=1}^{\infty} \frac{\|f^*\|_{k,r}^r}{k^r} = r \sum_{k=1}^{\infty} \frac{1}{k^{r+1}} \int_0^1 t^{r/k} f^*(t)^r \frac{dt}{t}$$

$$= r \sum_{k=1}^{\infty} \frac{1}{k^{r+1}} \sum_{j=1}^{\infty} \int_{I_j} t^{r/k} f^*(t)^r \frac{dt}{t}$$

$$\sim r(e-1) \sum_{k=1}^{\infty} \frac{1}{k^{r+1}} \sum_{j=1}^{\infty} f^*(e^{-j})^r e^{-rj/k}$$

$$= r(e-1) \sum_{j=1}^{\infty} f^*(e^{-j})^r \sum_{k=1}^{\infty} k^{-r-1} e^{-rj/k}$$

$$\sim re^{-r}(e-1) \sum_{j=1}^{\infty} \|f^*\|_{j,I_j}^r \sum_{k=1}^{\infty} k^{-r-1} e^{-rj/k}$$

$$\sim re^{-r}(e-1) \sum_{j=1}^{\infty} \|f^*\|_{j,I_j}^r \int_0^{\infty} x^{-r-1} e^{-rj/x}\, dx,$$

$$= re^{-r}(e-1) \sum_{j=1}^{\infty} \|f^*\|_{j,I_j}^r \int_0^{\infty} y^{r-1} e^{-rjy}\, dy$$

(after the change of variables $y = 1/x$)

$$= r^{1-r}\Gamma(r) e^{-r}(e-1) \sum_{j=1}^{\infty} \frac{\|f^*\|_{j,I_j}^r}{j^r},$$

(here Γ is the Gamma function).

(v) $\Rightarrow$ (iv). By the sharp form of Calderón's lemma (see [12, Chapter 5]), if $k \geq r$, $\|f\|_k \leq \|f\|_{k,r}$. (We remark in passing that the usual statement of this result (e.g. in [14, Chapter 1]) is weaker and has a constant on the right-hand side which depends on k and r.) Given this inequality, this implication is immediate.

(iv) $\Rightarrow$ (ii). This is immediate.

This completes the proof. □

4. An application to convolution inequalities

As we mentioned in the Introduction, one of the major reasons for studying the extrapolation properties of $L^{\exp t^{r'}, t^r}$, $1 < r < \infty$, is their use in the imbedding theory for Sobolev-type spaces. Here we will consider the integrability properties of the convolution $I_{n/p} * g$, where $I_{n/p}$ is (a multiple of) the Riesz kernel $x \mapsto |x|^{n/p-n}$ in $\mathbb{R}^n$ and $g \in L_p(\Omega)$, where Ω is a domain in $\mathbb{R}^n$, $|\Omega| = 1$. This convolution corresponds to the critical case of the Sobolev imbedding. It is well known that functions in the Sobolev space $W^{k,p}$, where $kp = n$, are in every L^p space, $p < \infty$, but, in general, they are not bounded.

A careful analysis of properties of this convolution due to Brézis and Wainger [4] lead to (1.1) (with $r = p$) as the characterization of the target space. Here we will use Theorem 1.4 to give a simple proof of the integrability characterization in this critical case. To avoid unnecessary technical details we shall consider the convolution

$$(I_{n/p} * g)(x) = \int_{\mathbb{R}^n} \frac{g(y)\,dy}{|x-y|^{n-n/p}}$$

for $g \in L_p(\Omega)$, $1 < p < \infty$, where $|\Omega| = 1$ and g is assumed to be identically zero on $\mathbb{R}^n \setminus \Omega$.

Theorem 4.1. *Fix p, $1 < p < \infty$, and let Ω and $g \in L^p(\Omega)$ be as above. Then*

$$\|I_{n/p} * g\|_{L^{\exp t^{p'},t^p}} \le c\|g\|_p,$$

with $c > 0$ independent of g.

Remark 4.2. Our assumption about the support of g is just technical and it is not necessary. Alternatively, one can consider functions $g \in L_p(\mathbb{R}^n)$ taken by convolutions with Bessel kernels to functions in the potential Sobolev space $H^{n/p,p}$, supported in a given set of finite Lebesgue measure.

Naturally, one can tune more finely the target space in by making additional assumptions about the space where g lives. This will be done elsewhere.

Proof. First note that we can assume that $g \in L^\infty$ since bounded functions are dense in $L^p(\Omega)$. Second, by homogeneity we may assume that $\|g\|_p = 1$.

Now let $f = I_{n/p} * g$ and for $t > 0$ let

$$f^{**}(t) = \frac{1}{t}\int_0^t f^*(\tau)\,d\tau.$$

By O'Neil's convolution inequality (cf. e.g. [14, Lemma 1.8.8]),

$$f^*(t) \le f^{**}(t) \le pt^{-1/p'}\int_0^t g^*(s)\,ds + \int_t^1 g^*(\tau)\tau^{-1/p'}\,d\tau.$$

If $t \in (0,1)$ then

$$f^*(t) \le p\|g\|_p + \int_t^1 g^*(\tau)\tau^{-1/p'}\,d\tau.$$

Now define

$$B(t) = \int_t^1 g^*(\tau)\tau^{-1/p'}\,d\tau, \qquad B_k = B(e^{-k}).$$

Then for $t \in I_k$,

$$f^*(t) \le p + B(t) \le p + B_k. \tag{4.1}$$

By Condition (ii) of Theorem 1.4 (with $r = p$) it will suffice to show that

$$\sum_{k=1}^{\infty} k^{-p} \|f^*\|_{k,I_k}^p \leq c. \tag{4.2}$$

By inequality (4.1),

$$\begin{aligned}\|f^*\|_{k,I_k}^p &\leq 2^p \|p\|_{k,I_k}^p + 2^p \|B_k\|_{k,I_k}^p \\ &\leq 2^p p^p |I_k|^{p/k} + 2^p B_k^p |I_k|^{p/k} \\ &\leq c + cB_k^p.\end{aligned}$$

Therefore, to complete the proof we need to show that

$$\sum_{k=1}^{\infty} \frac{B_k^p}{k^p} \leq c. \tag{4.3}$$

Note that since g is bounded, $B_k \leq p\|g\|_\infty$, and so this sum is finite. By the mean value theorem,

$$\begin{aligned}\sum_{k=1}^{\infty} \frac{B_k^p}{k^p} &\leq c \sum_{k=1}^{\infty} \left(\frac{1}{k^{p-1}} - \frac{1}{(k+1)^{p-1}} \right) B_k^p \\ &= c \Big(\sum_{k=1}^{\infty} \frac{B_k^p}{k^{p-1}} - \sum_{k=1}^{\infty} \frac{B_k^p}{(k+1)^{p-1}} \Big) \\ &= cB_1^p + c \sum_{k=2}^{\infty} \frac{B_k^p - B_{k-1}^p}{k^{p-1}}.\end{aligned}$$

Clearly, $B_1^p \leq c\|g\|_p^p = c$. To bound the sum we again apply the mean value theorem to get

$$\begin{aligned}B_k^p - B_{k-1}^p &\leq pB_k^{p-1} \frac{g^*(e^{-k})}{e^{-k/p'}} e^{-k+1} \\ &\leq cpB_k^{p-1} g^*(e^{-k}) e^{-k/p+1} \\ &\leq cpB_k^{p-1} g^*(e^{-k}) e^{-k/p} (e-1)^{1/p}.\end{aligned}$$

Thus

$$\begin{aligned}\sum_{k=2}^{\infty} \frac{B_k^p - B_{k-1}^p}{k^{p-1}} &\leq cp \sum_{k=2}^{\infty} \frac{B_k^{p-1}}{k^{p-1}} g^*(e^{-k}) e^{-k/p} (e-1)^{1/p} \\ &\leq cp \Big(\sum_{k=1}^{\infty} \frac{B_k^p}{k^p} \Big)^{1/p'} \Big(\sum_{k=2}^{\infty} (g^*(e^{-k})^p e^{-k} (e-1) \Big)^{1/p}\end{aligned}$$

$$\leq cp\|g\|_p \Big(\sum_{k=2}^{\infty} \frac{B_k^p}{k^p}\Big)^{1/p'}.$$

Hence,

$$\sum_{k=1}^{\infty} \frac{B_k^p}{k^p} \leq c + c\Big(\sum_{k=1}^{\infty} \frac{B_k^p}{k^p}\Big)^{1/p'}. \tag{4.4}$$

If $\sum_{k=1}^{\infty}(B_k/k)^p \leq 1$ then (4.3) holds. If the sum is greater than one then (4.4) gives us that

$$\Big(\sum_{k=1}^{\infty} \frac{B_k^p}{k^p}\Big)^{1/p} \leq c$$

and (4.3) again holds. This completes the proof. □

Acknowledgements. The research of the first author was partly supported by Ford Foundation postdoctoral fellowship. The second author wishes to acknowledge partial support by Grant No. 201/97/0744 of GA ČR.

References

[1] Alvino, A., Lions, P.-L. and Trombetti, G., On optimization problems with prescribed rearrangements, Nonlinear Anal. 13 (1989), 185–220.

[2] Bennett, C. and Rudnick, K., On Lorentz-Zygmund spaces, Dissertationes Math. (Rozprawy Mat.) CLXXV (1980), 1–72.

[3] Bennett, C. and Sharpley, R., Interpolation of operators, Academic Press, New York, 1988.

[4] Brézis, H. and Wainger, S., A Note on limiting cases of Sobolev embeddings and Convolution Inequalities, Comm. Partial Differential Equations 5 (1980), 773–789.

[5] Edmunds, D. E. and Krbec, M., On decomposition in exponential Orlicz spaces, Math. Nachr. 213 (2000), 77–88.

[6] Krbec, M. and Lang, J., On imbeddings between weighted Orlicz–Lorentz spaces, Georgian Math. J. 4 (1997), 117–128.

[7] Milman, M., Extrapolation and optimal decompositions: with applications to analysis, Lecture Notes in Math. 1580, Springer-Verlag, Berlin, 1994.

[8] Moser, J., A sharp form of an inequality by N. Trudinger, Indiana Univ. Math. J. 20 (1971), 1077–1092.

[9] Montgomery-Smith, S. J., Comparison of Orlicz–Lorentz spaces, Studia Math. 103 (1992), 161–189.

[10] Peetre, J., Espaces d'interpolation et théorème de Soboleff, Ann. Inst. Fourier 16 (1966), 279–317.

[11] Strichartz, R. S., A Note on Trudinger's extension of Sobolev's inequality, Indiana Univ. Math. J. 21 (1972), 841–842.

[12] Stein, E. M. and Weiss, G., Introduction to Fourier analysis in Euclidean spaces, Princeton University Press, Princeton, 1971.

[13] Trudinger, N., On imbeddings into Orlicz spaces and some applications, J. Math. Mech. 17 (1967), 473–483.

[14] Ziemer, W. P., Weakly differentiable functions, Grad. Texts in Math. 120, Springer-Verlag, Berlin, 1989.

David Cruz-Uribe, Department of Mathematics, Trinity College, Hartford, CT 06106-3100, U.S.A.

E-mail: `david.cruzuribe@mail.trincoll.edu`

Miroslav Krbec, Institute of Mathematics, Academy of Sciences of the Czech Republic, Žitná 25, 115 67 Prague 1, Czech Republic

E-mail: `krbecm@matsrv.math.cas.cz`

Green functions for powers of the Laplace–Beltrami operator

Miroslav Engliš *

Dedicated to Jaak Peetre on the occasion of his 65th birthday

Abstract. We show that for $\mathbf{\Delta}$ the Laplace–Beltrami operator on a complete Riemannian manifold Ω satisfying a certain condition, the powers $\mathbf{\Delta}^m$ are essentially self-adjoint operators on L^2 whose Green functions are of constant sign $(-1)^m$ on $\Omega \times \Omega$. This applies, in particular, to smoothly bounded domains in the complex plane $\mathbb{C}$ with the Poincaré metric, and to the unit ball $\mathbb{B}^d$ of $\mathbb{C}^d$ with the invariant metric; for the former, this contrasts markedly with the situation for the ordinary Laplace operator Δ. For the plane domains, we also obtain, using uniformization, a Myrberg-like formula for the corresponding Green functions.

2000 Mathematics Subject Classification: Primary 58G20 ; Secondary 30F35

0. Introduction

It is well known that for any smoothly bounded domain Ω in $\mathbb{R}^n$, the Green function for the Laplace operator Δ on Ω, with Dirichlet boundary conditions, is negative. The similar assertion for the biharmonic operator Δ^2 already turns out to be false: there exist domains in $\mathbb{R}^2$ – even very nice ones, such as sufficiently eccentric ellipses [Gr] – for which the Green function for Δ^2, with Dirichlet boundary conditions, changes sign. (See [Du], [Lo], [Sz], [Os], [CD], [KKM], [ST] for further examples; on the other hand, examples of domains for which the sign does not change are the unit balls [Bo], their small perturbations [St] [GS] and quadrature domains on negatively curved Riemann surfaces [HJS].) The same phenomenon occurs for the corresponding Green functions for Δ^m when $m \geq 3$: there exist smoothly bounded convex domains for which the Green functions change sign [KKM], as well as others on which they do not [GS]. Boggio [Bo] proved that for Ω the unit ball in $\mathbb{R}^n$, the Green function Γ_m for Δ^m with Dirichlet boundary conditions satisfies $(-1)^m \Gamma_m(x, y) > 0 \, \forall x, y$, for all

*The author was supported by GA AV ČR grant no. A1019005.

$m \geq 1$. (For another approach see [EP1]). Hayman and Korenblum [HK] conjectured that balls are the only domains with this property.

In this paper, we wish to consider the similar problem for the Green functions of powers of the "invariant" Laplacian, that is, the Laplace–Beltrami operator $\mathbf{\Delta}_\Omega$ with respect to the Poincaré metric, on smoothly bounded domains $\Omega \subset \mathbb{C}$. (For the unit disc $\mathbb{D}$, $\mathbf{\Delta}_\mathbb{D} \equiv \mathbf{\Delta}$ is indeed invariant under the action of the group of all holomorphic automorphisms of $\mathbb{D}$, and is, in fact, the unique differential operator on $\mathbb{D}$ with this property satisfying $\mathbf{\Delta} f(0) = \Delta f(0)$ for all f; for other domains $\Omega \subset \mathbb{C}$, $\mathbf{\Delta}_\Omega$ is invariant under all biholomorphic automorphisms of Ω if there are any. For simplicity, we will often use the term "invariant" also in the general situation when there is no group action present.) Our first main result is the following theorem, which shows that for $\mathbf{\Delta}$, the question of positivity of the Green functions has a much nicer answer than for Δ.

Theorem 0.1. *Let Ω be a smoothly bounded domain in $\mathbb{C}$ and $\mathbf{\Delta}_\Omega$ the invariant Laplacian on Ω. Then the Green function G_m^Ω for $\mathbf{\Delta}_\Omega^m$ satisfies*

$$(-1)^m G_m^\Omega(x, y) > 0 \quad \forall x, y \in \Omega.$$

Further, there is a Myrberg-type formula expressing the Green function G_m^Ω in terms of the Green function $\mathbf{G}_m^\mathbb{D}$ for the disc. (Cf. Theorem 2.8.)

We pause to explain what precisely is meant by the Green function. First of all, we show that $\mathbf{\Delta}_\Omega$ is essentially self-adjoint, so there is no need to specify any boundary conditions. Second, $\mathbf{\Delta}_\Omega^m$ has a bounded right inverse $\mathbf{G}_\Omega^m$, and for each m, $\mathbf{G}_\Omega^m$ is an integral operator with kernel G_m^Ω:

$$\mathbf{G}_\Omega^m f(x) = \int_\Omega f(y) G_m^\Omega(x, y)\, d\mu_\Omega(y),$$

where $d\mu_\Omega$ is the Poincaré measure on Ω.

The proof builds on two earlier papers on this subject by J. Peetre and the present author, [EP2] and [EP3]. In the former, the functions G_m^Ω were studied for Ω the unit disc (see also [EP1]). In particular, explicit formulas for $G_m^\mathbb{D}$ were given there for m up to 4, and a description of the analytic continuation and boundary behaviour for general m was obtained which will prove useful below. In the latter, among other things, a method based on uniformization was employed to prove Theorem 0.1 for $m = 2$; we will use a similar method here to derive from the result for the disc a Myrberg-type formula for G_m^Ω for any smoothly bounded Ω (and arbitrary m).

Our second main result is a generalization of Theorem 0.1 to an arbitrary complete Riemannian manifold. Here the point is that the Laplace–Beltrami operator $\mathbf{\Delta}$ on Ω is then again essentially self-adjoint, so the positivity of G_m^Ω follows from that of $G := G_1^\Omega$ by simple iteration; the latter, in turn, is immediate from the familiar maximum principle once we can identify G with the ordinary Green function g of Ω used in potential theory. This we can do rigorously only under some additional assumptions on $\mathbf{\Delta}$ or g.

Theorem 0.2. *Let Ω be a complete Riemannian manifold, $\mathbf{\Delta}$ the Laplace–Beltrami operator on Ω, and assume that $\mathbf{\Delta}$ has a bounded right inverse* **G** *on $L^2(\Omega)$ or that the potential-theoretic Green function $g(x, y)$ defines a bounded integral operator* **g** *on $L^2(\Omega)$. Then $\mathbf{\Delta}^m$ has a bounded right inverse $\mathbf{G}^m = \mathbf{g}^m$, and for each m, $\mathbf{G}^m$ is an integral operator whose kernel G_m satisfies $(-1)^m G_m(x, y) > 0\ \forall x, y \in \Omega$.*

We remark that $\mathbf{\Delta}$ has a bounded right inverse (or, mutatis mutandis, is bounded below on L^2) if and only if a certain Poincaré inequality holds on Ω; see below for details.

The paper is organized as follows. In Section 1, we prove Theorem 0.1 for Ω the unit disc. Then we use uniformization to extend the result to all smoothly bounded domains Ω, and to obtain the Myrberg-type formula; in fact, most proofs even work for any Riemann surface whose covering group is finitely generated, of convergence type, and does not contain parabolic elements. (All smoothly bounded plane domains belong to this class.) This is done in Section 2. The simplest higher-dimensional case, namely the invariant Laplacian $\mathbf{\Delta}_{\mathbb{B}^d}$ on the unit ball $\mathbb{B}^d$ of $\mathbb{C}^d$, $d > 1$, is briefly treated in Section 3 in order to demonstrate the differences against the one-dimensional situation which can arise (the most notable being that the Green functions $G(\cdot, y)$ no longer belong to L^2). Finally, in the final Section 4 we discuss the general case of the Laplace–Beltrami operator $\mathbf{\Delta}$ on an arbitrary complete Riemannian manifold satisfying a certain Poincaré inequality, and give a proof of Theorem 0.2.

The author would like to thank the referee for remarks that improved the presentation and for pointing out additional references.

1. The disc

Dropping for simplicity the subscript $\mathbb{D}$ from the notation, let

$$\mathbf{\Delta} = (1 - |z|^2)^2 \Delta, \quad d\mu(z) = \frac{dm(z)}{(1 - |z|^2)^2}$$

be the invariant Laplacian and the invariant (Poincaré) measure on the unit disc $\mathbb{D}$. Here dm stands for the Lebesgue area measure. The invariance refers to the fact that

$$\mathbf{\Delta}(f \circ \phi) = (\mathbf{\Delta} f) \circ \phi, \tag{1.1}$$

$$d\mu(\phi(z)) = d\mu(z) \tag{1.2}$$

for any biholomorphic automorphism ϕ of $\mathbb{D}$.

Let A_0 be the restriction of $\mathbf{\Delta}$ to the subspace $C_0^\infty(\mathbb{D})$ of compactly supported C^∞ functions in $L^2 \equiv L^2(\mathbb{D}, d\mu)$. The operator A_0 is symmetric:

$$\int_{\mathbb{D}} A_0 f\, \overline{g}\, d\mu = \int_{\mathbb{D}} \Delta f\, \overline{g}\, dm = \int_{\mathbb{D}} f\, \overline{\Delta g}\, dm = \int_{\mathbb{D}} f\, \overline{A_0 g}\, d\mu \quad \forall f, g \in C_0^\infty(\mathbb{D}).$$

Denote by $A = A_0^{**}$ the closure of A_0 and by $A^* = A_0^*$ their adjoint. Explicitly, A^* is the restriction of $\mathbf{\Delta}$ to the subspace of all functions f in L^2 for which $\mathbf{\Delta} f$, taken in the sense of distributions, belongs to L^2.

It follows from the result of Roelcke [R1] and Gaffney [Gf] (which even applies to a general complete Riemannian manifold, see Section 4 below) that $A = A^*$, i.e. the operator A is self-adjoint. Moreover, the same is true for any power A_0^m ($m = 1, 2, \dots$); that is, $(A_0^m)^* = (A_0^m)^{**}$. See Cordes [Co].

Our first goal will be to show that the operators $(A_0^m)^*$ are all injective.

Proposition 1.1. $\ker(A_0^m)^* = \{0\}$; *that is, any* $f \in L^2$ *satisfying* $\mathbf{\Delta}^m f = 0$ *in the sense of distributions is identically zero.*

Proof. Recall that for any two densely defined operators X and Y, $X \subset Y$ implies $Y^* \subset X^*$ and $X^m \subset Y^m$, and, if XY is also densely defined, $Y^*X^* \subset (XY)^*$. Further, if X is selfadjoint, then so is X^m for any m. From $A_0 \subset A = A^*$ we thus get

$$A_0^m \subset A^m = (A^m)^* \subset (A_0^m)^*.$$

Taking adjoints gives

$$(A_0^m)^{**} \subset A^m \subset (A_0^m)^*,$$

so by Cordes' result $(A_0^m)^* = A^m = (A_0^*)^m$. Thus it suffices to show that $\ker A_0^* = \{0\}$.

So let $h \in \ker A_0^*$, that is, $h \in L^2$ and $\mathbf{\Delta} h = 0$ in the sense of distributions; since $\mathbf{\Delta} = (1 - |z|^2)^2 \Delta$, this means, by Weyl's lemma, that h is (coincides a.e. with) a harmonic function. Thus $|h|^2$ is subharmonic, so by the sub-mean-value property

$$\int_{|z|<R} |h(z)|^2 \, d\mu(z) \geq |h(0)|^2 \int_{|z|<R} d\mu(z), \qquad \forall R < 1.$$

However, as $R \nearrow 1$, the second integral tends to infinity, while the first one is bounded by $\|h\|^2$. Consequently, $h(0) = 0$. Replacing h by $h \circ \omega$ where ω is a fractional linear self-map of $\mathbb{D}$ interchanging 0 and z, we see that $h(z) = 0$ for all $z \in \mathbb{D}$. □

We digress briefly to mention an alternative characterization of the operator A. Let $X : L^2 \to L^2$ be the restriction of the operator

$$f \mapsto (1 - |z|^2) \frac{\partial f}{\partial \bar{z}}$$

to the subspace of all functions in L^2 for which the right-hand side, taken in the sense of distributions, belongs to L^2. Then X is densely defined and closed, hence X^*X is selfadjoint (see [RN], §118).

Proposition 1.2. $X^*X = -\frac{1}{4}A$.

Proof. Let temporarily Y_0 stand for the restriction of the formal adjoint $f \mapsto -(1 - |z|^2)^2 \partial[(1 - |z|^2)^{-1} f]$ of X to C_0^∞, and let $Y = Y_0^{**}$ be the closure of Y_0. Thus

$Y_0^* = Y^* = X$ and $X^* = Y$. Since both X and Y are defined on C_0^∞ and map it into itself, the same is true for $YX = X^*X$. As $4X^*X\phi = -\boldsymbol{\Delta}\phi\ \forall\phi \in C_0^\infty$ by a simple computation, we thus have $-A_0 \subset 4X^*X$. Since X^*X is closed, it follows that also $-A \subset 4X^*X$. But A and X^*X are both selfadjoint; thus $A = -4X^*X$. □

Returning from the digression, we now proceed to construct a right inverse for A. Let

$$l(x, y) = \frac{1}{4\pi} \log \left| \frac{x - y}{1 - \bar{x}y} \right|^2. \tag{1.3}$$

It is well known that

$$l(\phi x, \phi y) = l(x, y) \tag{1.4}$$

for any holomorphic automorphism ϕ of $\mathbb{D}$.

Theorem 1.3. *Consider the integral operator*

$$\mathbf{G}f(x) = \int_{\mathbb{D}} f(y)\, l(x, y)\, d\mu(y). \tag{1.5}$$

Then the following holds.

(a) $\mathbf{G}f$ *is defined for any* $f \in L^2$ *and belongs to* L^∞.

(b) $\mathbf{G}$ *is a bounded selfadjoint operator on* L^2.

(c) $\operatorname{Ran}\mathbf{G} = \operatorname{dom} A$ *and* $A\mathbf{G} = I$. *(That is,* $\mathbf{G} = A^{-1}$ *in the sense of the functional calculus of self-adjoint operators.)*

Proof. (a) By Cauchy–Schwarz, $|\mathbf{G}f(x)| \le \|f\| \cdot \|l(x, \cdot)\|$, so it is enough to show that $l(x, \cdot) \in L^2\ \forall x \in \mathbb{D}$. Owing to (1.2) and (1.4), it suffices to show this for $x = 0$. But upon making the change of variables $z = \sqrt{t}e^{i\theta}$, we have

$$\int_{\mathbb{D}} (\log |z|^2)^2\, d\mu(z) = 2\pi \int_0^1 \left(\frac{\log t}{1 - t}\right)^2 dt < \infty.$$

(b) By the well-known Schur test ([HS], Theorem 5.2) and the symmetry and negativity of $l(x, y)$, it suffices to exhibit a positive function u such that $-\mathbf{G}u \le cu$ on $\mathbb{D}$ with some finite c. We claim that any of the functions

$$u(z) = (1 - |z|^2)^s, \quad 0 < s < 1,$$

does this job. Indeed, one has

$$\begin{aligned} -\mathbf{G}u(y) &= \int_{\mathbb{D}} (1 - |x|^2)^s\, |l(x, y)|\, d\mu(x) \\ &= \int_{\mathbb{D}} \left(1 - \left|\frac{x - y}{1 - \bar{x}y}\right|^2\right)^s |l(x, 0)|\, d\mu(x) \quad \text{by invariance} \end{aligned}$$

$$= \frac{1}{4\pi} \int_{\mathbb{D}} \frac{(1-|x|^2)^s (1-|y|^2)^s}{|1-\overline{x}y|^{2s}} \left|\log |x|^2\right| d\mu(x).$$

Expanding $(1-z)^{-s}$ into Taylor series and integrating term by term, one checks that

$$\frac{1}{2\pi} \int_0^{2\pi} |1-\rho e^{i\theta}|^{-2s} \, d\theta = F(s,s,1,\rho^2), \tag{1.6}$$

where F denotes the hypergeometric function. Consequently, switching to polar coordinates and setting $|x|^2 = t$, we can continue with

$$\begin{aligned} -\mathbf{G}u(y) &= \tfrac{1}{4}(1-|y|^2)^s \int_0^1 (1-t)^{s-2} |\log t| \, F(s,s,1,|y|^2 t) \, dt \\ &\le \tfrac{1}{4} u(y) \int_0^1 (1-t)^{s-2} |\log t| \, F(s,s,1,t) \, dt \end{aligned}$$

as the function $F(s,s,1,\cdot)$ is increasing on $(0,1)$. However, it is known (cf. [BE], §2.10) that as $t \nearrow 1$,

$$|F(\nu, \nu+|m|, |m|+1, t)| \sim \begin{cases} -\log(1-t) & \text{if } \nu = \frac{1}{2}, \\ (1-t)^{\min(0,1-2\operatorname{Re}\nu)} & \text{otherwise.} \end{cases} \tag{1.7}$$

Thus the last integral is finite for $0 < s < 1$.

(c) For any $\phi \in C_0^\infty$, we have

$$\mathbf{G}A_0\phi(y) = \int_{\mathbb{D}} l(x,y) \mathbf{\Delta}\phi(x) \, d\mu(x) = \int_{\mathbb{D}} l(x,y) \Delta\phi(x) \, dm(x) = \phi(y),$$

since $l(x,y)$ is the Green function for the ordinary Laplacian Δ. Consequently, by the self-adjointness of $\mathbf{G}$, for any $g \in L^2$

$$\langle \mathbf{G}g, A_0\phi \rangle = \langle g, \mathbf{G}A_0\phi \rangle = \langle g, \phi \rangle \quad \forall \phi \in C_0^\infty.$$

Thus $\mathbf{G}g$ lies in the domain of $A_0^* = A$ and $A\mathbf{G}g = g$; that is, $\operatorname{Ran}\mathbf{G} \subset \operatorname{dom} A$ and $A\mathbf{G} = I$. As A is injective (Proposition 1.1), it follows from $A\mathbf{G}A = A$ that $\mathbf{G}A = I$ on $\operatorname{dom} A$, so $\operatorname{Ran}\mathbf{G} = \operatorname{dom} A$ and $\mathbf{G} = A^{-1}$. □

For $m = 1, 2, \ldots$, define the functions $G_m(x,y)$ by

$$G_m(x,\cdot) = \mathbf{G}^{m-1}(l(x,\cdot)). \tag{1.8}$$

As we have just seen in the proof of part (a) of the last theorem, $l(x,\cdot) \in L^2 \ \forall x$, so part (b) ensures that this definition makes sense. Note that in view of (1.4) and (1.2), the functions G_m are also invariant (i.e. (1.4) holds for G_m in place of l). As $l(x,y) = l(y,x)$ and $l(x,y) < 0 \ \forall x, y \in \mathbb{D}$, it also follows from (1.5) that $G_m(x,y) = G_m(y,x)$ and

$$(-1)^m G_m > 0 \quad \text{on } \mathbb{D} \times \mathbb{D}. \tag{1.9}$$

We can now state the main result of this section.

Theorem 1.4. *The function G_m is the Green function for the operator A^m; that is, for any $f \in L^2$,*

$$\mathbf{G}^m f(x) = \int_{\mathbb{D}} f(y) G_m(x, y)\, d\mu(y) \tag{1.10}$$

is the unique solution in L^2 of the equation $\mathbf{\Delta}^m u = f$.

Proof. We only need to show that $\mathbf{G}^m$ is given by the formula (1.10), the rest is just a restatement of part (c) of the preceding theorem. However, by (1.8),

$$\begin{aligned}\int_{\mathbb{D}} f(y) G_m(x, y)\, d\mu(y) &= \langle f, G_m(x, \cdot)\rangle = \langle f, \mathbf{G}^{m-1}(l(x, \cdot))\rangle \\ &= \langle \mathbf{G}^{m-1} f, l(x, \cdot)\rangle = \mathbf{G}(\mathbf{G}^{m-1} f)(x) = \mathbf{G}^m f(x),\end{aligned}$$

owing to the self-adjointness of $\mathbf{G}$. □

We conclude this section by making contact with the results in [EP2] and giving estimates on the boundary behaviour of G_m, which will be used in the next section.

Proposition 1.5. *As $x \to \partial\mathbb{D}$, we have*

$$G_m(0, x) = O((1 - |x|^2) \log^{m-1}(1 - |x|^2)), \tag{1.11}$$

and as $x \to 0$,

$$G_m(0, x) = O(|x|^{2(m-1)} \log |x|^2). \tag{1.12}$$

Proof. It was shown in [EP2] (Corollaries 2.5 and 2.10) that there are unique functions (denoted there also by G_m) on $\mathbb{D}$ such that (1.11) and (1.12) hold and

$$\mathbf{\Delta}^m G_m(0, x) = \delta(x) \tag{1.13}$$

(the delta function at the origin). On the other hand, it follows from the last theorem that the functions $G_m(0, x)$ of the present paper are the unique functions satisfying

$$G_m(0, \cdot) \in L^2 \tag{1.14}$$

and (1.13). As (1.11) and (1.12) implies (1.14), it follows that the two functions $G_m(0, x)$ coincide, and the assertion follows. □

2. Plane domains and Fuchsian groups

Let now Ω be a plane domain of hyperbolic type (i.e. $\mathbb{C} \setminus \Omega$ contains at least two points) and $\phi : \mathbb{D} \to \Omega$ the uniformization map. Recall that the *Poincaré metric* on Ω is given by $ds^2 = w(z)^{-2}|dz|^2$, where w is given by

$$w(\phi(x)) = (1 - |x|^2)|\phi'(x)|. \tag{2.1}$$

(It is easy to see that this definition is consistent). It is known that w vanishes on $\partial\Omega$ (in fact, $w(z) \leq \operatorname{dist}(z, \partial\Omega)$, where dist stands for the Euclidean distance; cf. [Kr], p. 45). The Laplace–Beltrami operator ("invariant Laplacian") $\mathbf{\Delta}_\Omega$ on Ω is given by

$$\mathbf{\Delta}_\Omega f(z) = w(z)^2 \Delta f(z). \tag{2.2}$$

One has

$$(\mathbf{\Delta}_\Omega f) \circ \phi = \mathbf{\Delta}(f \circ \phi). \tag{2.3}$$

The operator $\mathbf{\Delta}_\Omega$ is formally selfadjoint with respect to the *Poincaré measure*

$$d\mu_\Omega(z) = w(z)^{-2}\, dm(z).$$

The *covering group* of ϕ is $G = \{\omega \in \operatorname{Aut}(\mathbb{D}) : \phi \circ \omega = \phi\}$. We will often write ωx instead of $\omega(x)$, and similarly for $\phi(x)$.

A subgroup G of the group of all holomorphic automorphisms of $\mathbb{D}$ is called *discontinuous* if for any point $x \in \mathbb{D}$, there exists a neighbourhood U of x in $\mathbb{D}$ such that $\omega \in G$ and $\omega x \in U$ implies $\omega x = x$. For any discontinuous subgroup G, the quotient space $\mathbb{D}/G$ inherits the holomorphic structure from $\mathbb{D}$ and thus is a Riemann surface, with the canonical projection $\phi : \mathbb{D} \to \mathbb{D}/G$ as a holomorphic covering map (possibly with branch points). Conversely, all Riemann surfaces not biholomorphic to the Gauss sphere $\mathbb{C} \cup \{\infty\}$, the complex plane $\mathbb{C}$, or the punctured plane $\mathbb{C} \setminus \{0\}$, arise in this way (Koebe's Uniformization Theorem).

Recall that a holomorphic automorphism ϕ of $\mathbb{D}$ is called hyperbolic, parabolic or elliptic, respectively, depending on whether it has two fixed points on $\partial\mathbb{D}$, one double fixed point on $\partial\mathbb{D}$, or one fixed point in $\mathbb{D}$ and one in $\mathbb{C} \setminus \mathbb{D}$ (no other possibilities can occur). If G does not contain elliptic transformations, the open set $\mathcal{O} = \{x \in \mathbb{D} : |x| < |\omega x| \,\forall \omega \in G, \omega \neq \mathrm{id}\}$ is a *fundamental domain* for G. A discontinuous subgroup containing only hyperbolic elements is called *purely hyperbolic*. The *limit set* Λ of G is the set of accumulation points of the orbit $\{\omega(0);\ \omega \in G\}$ (equivalently, of $\{\omega x;\ \omega \in G\}$ for any fixed $x \in \mathbb{D}$); if G is discontinuous, Λ is contained in $\partial\mathbb{D}$, and G is said to be of the *first* or the *second kind*, respectively, depending on whether $\Lambda = \partial\mathbb{D}$ or Λ is a proper subset of $\partial\mathbb{D}$. Finally, a discontinuous subgroup G is said to be of *convergence type* if $\sum_{\omega \in G}(1 - |\omega(0)|) < \infty$; otherwise it is of *divergence type*. For finitely generated groups without parabolic elements, the notions of second kind and convergence type coincide (though for arbitrary discontinuous groups they do not).

It is known that the covering group of a plane domain Ω of hyperbolic type is discontinuous and contains no elliptic elements; further, it is finitely generated if and only if Ω has finite connectivity, contains no parabolic elements if and only if $\partial\Omega$ has no isolated points, and is of convergence type if and only if $\mathbb{C} \setminus \Omega$ has positive capacity (see Tsuji [Ts], Theorems XI.13 and III.35).

In particular, if Ω is *smoothly bounded* (i.e. bounded and with C^∞ boundary), then its covering group G is finitely generated, purely hyperbolic, of convergence type and of second kind. It can further be shown that in that case the boundary of the fundamental domain $\mathcal{O}$ consists of finitely many circular arcs, each of which lies either on $\partial\mathbb{D}$ or on a circle orthogonal to $\partial\mathbb{D}$, that ϕ extends C^∞-smoothly to the closure $\overline{\mathcal{O}}$ of $\mathcal{O}$ and ϕ' does not vanish there (see Goluzin [Go], Chapter VI, §2), the limit set Λ is a closed, perfect, nowhere dense subset of $\partial\mathbb{D}$ which is disjoint from $\overline{\mathcal{O}}$, and the function $w(z)/\operatorname{dist}(z, \partial\Omega)$, where dist stands for the Euclidean distance, is bounded and bounded away from zero on Ω.

We introduce the notation

$$\log_1 x := \max\{1, 1 - \log x\} \quad (x > 0).$$

It is easily verified that $\log_1$ has the submultiplicativity property

$$\frac{\log_1 a}{\log_1(1/b)} \le \log_1 ab \le \log_1 a \cdot \log_1 b \quad \text{for any } a, b > 0. \tag{2.4}$$

Throughout the rest of this section, G will be a discontinuous group of holomorphic automorphisms of $\mathbb{D}$, and we enumerate by $\omega_0 = \text{id}, \omega_1, \omega_2, \dots$ the elements of G and set $a_n = \omega_n^{-1}(0)$.

Lemma 2.1. *Assume that G is finitely generated and contains no parabolic elements. Then there exists $\delta > 0$ such that*

$$\delta \le |1 - \bar{a}_n x| \le 2 \quad \forall n\ \forall x \in \overline{\mathcal{O}}, \tag{2.5}$$

and, consequently,

$$\frac{1}{4}(1 - |x|^2)(1 - |a_n|^2) \le 1 - |\omega_n x|^2 \le \frac{1}{\delta^2}(1 - |x|^2)(1 - |a_n|^2) \tag{2.6}$$

for all n and all $x \in \overline{\mathcal{O}}$.

Proof. As we have already recalled above, for such G the cluster points of the sequence $\{a_n\}$ form a subset of the unit circle (the limit set Λ) lying at a positive distance from $\overline{\mathcal{O}}$ (see [B2], Theorems 10.2.3 and 10.2.5). As the continuous function $(x, a) \mapsto |1 - \bar{a}x|$ on the compact set $(\{a_n\} \cup \Lambda) \times \overline{\mathcal{O}}$ attains its minimum, (2.5) follows. The second assertion is immediate from the equality

$$1 - |\omega_n x|^2 = \frac{(1 - |x|^2)(1 - |a_n|^2)}{|1 - \bar{a}_n x|^2}. \tag{2.7}$$

□

A beautiful result of Dalzell [Da] says that if a discontinuous subgroup G is of convergence type, i.e. $\sum_n (1 - |a_n|) < \infty$, then one also has

$$\sum_n (1 - |a_n|) \log^k \frac{1}{1 - |a_n|} < \infty \quad \forall k \ge 1. \tag{2.8}$$

From (2.7) it therefore follows that the series

$$\gamma_k(x) := \sum_n (1 - |\omega_n x|^2) \log^k \frac{1}{1 - |\omega_n x|^2} \tag{2.9}$$

converges for any $k \geq 0$ and $x \in \mathbb{D}$, uniformly for x in compact subsets of $\mathbb{D}$. Obviously, $\gamma_k(\omega x) = \gamma_k(x)$ for any $\omega \in G$; hence one can define a function Γ_k on the Riemann surface $\mathbb{D}/G =: \Omega$ by the recipe

$$\Gamma_k(\phi x) := \gamma_k(x), \quad x \in \mathbb{D}.$$

Instead of γ_0 and Γ_0 we will write simply γ and Γ, respectively.

Lemma 2.2. *Let G be finitely generated, of convergence type and without parabolic elements. Then for each $k \geq 0$ there exists $C < \infty$ such that*

$$\frac{1}{C} \leq \frac{\gamma_k(x)}{(1 - |x|^2) \log_1^k (1 - |x|^2)} \leq C \quad \forall x \in \overline{\mathcal{O}} \cap \mathbb{D}. \tag{2.10}$$

Proof. Consider the terms $n = 0$ and $n = 1$ in (2.9),

$$(1 - |x|^2) \log^k \frac{1}{1 - |x|^2} + (1 - |\omega_1 x|^2) \log^k \frac{1}{1 - |\omega_1 x|^2}.$$

Since $\omega_1 \overline{\mathcal{O}} \cap \mathcal{O} = \emptyset$ and $0 \in \mathcal{O}$, we have $\log(1/(1 - |\omega_1 x|^2)) \geq \delta_1 > 0 \; \forall x \in \overline{\mathcal{O}}$. Also by (2.7), $(1 - |x|^2)/(1 - |\omega_1 x|^2) = |1 - \bar{a}_1 x|^2/(1 - |a_1|^2)$ is bounded on $\overline{\mathcal{O}}$ (even on $\overline{\mathbb{D}}$). Thus the second summand is $\geq \delta_2(1 - |x|^2) \; \forall x \in \overline{\mathcal{O}}$, for some $\delta_2 > 0$. Consequently, the sum of the two terms is

$$\begin{aligned} \geq (1 - |x|^2)\Big[\log^k \frac{1}{1 - |x|^2} + \delta_2\Big] &\geq \min(\delta_2, 1)(1 - |x|^2)2^{1-k}\Big[\log \frac{1}{1 - |x|^2} + 1\Big]^k \\ &= \min(\delta_2, 1)2^{1-k} \cdot (1 - |x|^2) \log_1^k (1 - |x|^2), \end{aligned}$$

so the left inequality in (2.10) follows.

On the other hand, by (2.6) and the submultiplicativity of $\log_1$, for any $x \in \overline{\mathcal{O}} \cap \mathbb{D}$ and any n,

$$(1 - |\omega_n x|^2) \log_1^k (1 - |\omega_n x|^2) \leq \frac{1}{\delta^2}(1 - |x|^2)(1 - |a_n|^2) \log_1^k (1 - |x|^2) \log_1^k (1 - |a_n|^2),$$

so

$$\gamma_k(x) \leq \frac{1}{\delta^2} \gamma_k(0)\,(1 - |x|^2) \log_1^k (1 - |x|^2).$$

Thus the second inequality to be proved follows from (2.8). □

Corollary 2.3. *Let G be finitely generated, of convergence type and without parabolic elements. Then for each $k \geq 0$ there exists $C < \infty$ such that*

$$\frac{1}{C}\Gamma(z) \log_1^k \Gamma(z) \leq \Gamma_k(z) \leq C\Gamma(z) \log_1^k \Gamma(z) \quad \forall z \in \Omega.$$

Proposition 2.4. *Let Ω be a smoothly bounded domain in $\mathbb{C}$ and G the covering group of its uniformization map ϕ. Then there exists $C < \infty$ such that*

$$\frac{1}{C} w(z) \leq \Gamma(z) \leq C\, w(z) \quad \forall z \in \Omega. \tag{2.11}$$

Consequently, in the last corollary we may replace Γ by w.

Proof. As noted above, the hypotheses imply that

$$\phi' \text{ is bounded and bounded away from zero on } \overline{\mathcal{O}} \tag{2.12}$$

(cf. Goluzin [Go], Chapter VI, §2), so by (2.10) with $k = 0$ and (2.1)

$$\frac{1}{C} w(\phi(x)) \leq \gamma(x) \leq C\, w(\phi(x)) \quad \forall x \in \overline{\mathcal{O}},$$

and the assertion follows. □

For a discontinuous group G of convergence type, consider the series

$$s(x, y) := \sum_{\omega \in G} l(\omega x, y),$$

with the function l given by (1.3). It follows from (2.7) that the series on the right-hand side converges uniformly for x, y in compact subsets of $\mathbb{D}$ (in fact, it is majorized by $\dfrac{c_\delta \gamma(x)}{1 - |y|}$ if $|x - y| > \delta > 0$). In view of (1.4), the function $s(x, y)$ satisfies $s(x, y) = s(y, x)$ and $s(x, y) = s(\omega x, \omega y)\ \forall \omega \in G$. Consequently, the recipe

$$l_\Omega(\phi x, \phi y) := s(x, y) \tag{2.13}$$

defines consistently a function l_Ω on $\Omega \times \Omega$, $\Omega := \mathbb{D}/G$, which satisfies

$$l_\Omega(x, y) = l_\Omega(y, x) < 0, \tag{2.14}$$

$$l_\Omega(x, y) \leq c_{\delta, y} \Gamma(x) \quad \text{if } |x - y| > \delta, \tag{2.15}$$

and

$$l_\Omega(x, y) \leq C_{\delta, y} \log |x - y|^2 \quad \text{if } |x - y| < \delta. \tag{2.16}$$

Observe that, in particular, (2.15), (2.16) and Lemma 2.2 imply that

$$l_\Omega(x, \cdot) \in L^2(\Omega, d\mu_\Omega) \quad \forall x \in \Omega. \tag{2.17}$$

We will again abbreviate $L^2(\Omega, d\mu_\Omega)$ to L^2 if there is no danger of confusion.

Proposition 2.5. *Let G be finitely generated, of convergence type, and without parabolic elements. Then*

$$\int_\Omega l_\Omega(z, t)\, \mathbf{\Delta}_\Omega f(z)\, d\mu_\Omega(z) = f(t)$$

for any $f \in C_0^\infty(\Omega)$ and $t \in \Omega$.

Proof. Fix $y \in \mathbb{D}$ with $\phi(y) = t$. There exists $R \in (0, 1)$ such that $\mathcal{O} \cap \operatorname{supp}(f \circ \phi) \subset R\mathbb{D}$. The image of $R\mathbb{D}$ under a holomorphic automorphism σ of $\mathbb{D}$ is a disc (in the Euclidean metric) centered at $\frac{1-R^2}{1-R^2|\sigma(0)|^2}\sigma(0)$ and of radius $\frac{R(1-|\sigma(0)|^2)}{1-R^2|\sigma(0)|^2}$. A simple calculation then reveals that

$$\int_{\sigma R\mathbb{D}} \big|\log|x|^2\big|\, d\mu(x) \le C_R(1 - |\sigma(0)|^2).$$

Applying this to $\sigma(x) = \dfrac{\omega(x) - y}{1 - \overline{y}\omega(x)}$, $\omega \in G$, it transpires that

$$\int_{R\mathbb{D}} |l(\omega x, y)|\, d\mu(x) \le C_{R,y}(1 - |\omega(0)|^2).$$

Thus by (1.2),

$$\begin{aligned}
\int_{\mathbb{D}} |(\mathbf{\Delta}_\Omega f)(\phi x)\, l(x, y)|\, d\mu(x) &= \sum_{\omega\in G} \int_{\mathcal{O}} |(\mathbf{\Delta}_\Omega f)(\phi(\omega x))\, l(\omega x, y)|\, d\mu(x) \\
&\le \sum_{\omega\in G} \sup_{\Omega} |\mathbf{\Delta}_\Omega f| \int_{R\mathbb{D}} |l(\omega x, y)|\, d\mu(x) \\
&\le C_{f,y,R} \sum_{\omega\in G} (1 - |\omega(0)|^2) \\
&= C_{f,y,R} \cdot \gamma(0) < \infty.
\end{aligned}$$

Consequently, the integral

$$\int_{\mathbb{D}} (\mathbf{\Delta}_\Omega f)(\phi x)\, l(x, y)\, d\mu(x)$$

exists and is equal to

$$\begin{aligned}
\sum_{\omega\in G} \int_{\omega\mathcal{O}} (\mathbf{\Delta}_\Omega f)(\phi x)\, l(x, y)\, d\mu(x) &= \sum_{\omega\in G} \int_{\mathcal{O}} (\mathbf{\Delta}_\Omega f)(\underbrace{\phi(\omega x)}_{=\phi(x)})\, l(\omega x, y)\, d\mu(x) \\
&= \int_{\mathcal{O}} (\mathbf{\Delta}_\Omega f)(\phi x) \underbrace{\sum_{\omega\in G} l(\omega x, y)}_{=l_\Omega(\phi x, t)}\, d\mu(x) \\
&= \int_{\Omega} \mathbf{\Delta}_\Omega f(z)\, l_\Omega(z, t)\, d\mu_\Omega(z).
\end{aligned}$$

On the other hand, denoting by χ_E the characteristic function of a set E, the functions $g_\omega := \chi_{\omega\mathcal{O}} \cdot (f \circ \phi)$ belong to $C_0^\infty(\mathbb{D})$ (since $f \circ \phi \in C_0^\infty(\mathcal{O})$) and $\mathbf{\Delta} g_\omega = \chi_{\omega\mathcal{O}} \mathbf{\Delta}(f \circ \phi)$. Hence by (2.3),

$$\sum_{\omega\in G} \int_{\omega\mathcal{O}} (\mathbf{\Delta}_\Omega f)(\phi x)\, l(x, y)\, d\mu(x) = \sum_{\omega\in G} \int_{\omega\mathcal{O}} \mathbf{\Delta} g_\omega(x)\, l(x, y)\, d\mu(x)$$

$$= \sum_{\omega \in G} \int_{\mathbb{D}} \boldsymbol{\Delta} g_\omega(x)\, l(x, y)\, d\mu(x)$$

$$= \sum_{\omega \in G} g_\omega(y) = f(\phi y) = f(t),$$

by an application of Proposition 1.4 to g_ω. This completes the proof. □

As in Section 1, denote by A_0 the restriction of $\boldsymbol{\Delta}_\Omega$ to $C_0^\infty(\Omega)$, by $A = A_0^{**}$ its closure, and by $A_0^* = A^*$ their Hilbert space adjoint in L^2; explicitly, A_0^* is the restriction of $\boldsymbol{\Delta}_\Omega$ to the subspace of all functions f in L^2 for which $\boldsymbol{\Delta}_\Omega f$, taken in the distributional sense, belongs to L^2. By the results of Gaffney and Cordes, the operator A is again selfadjoint, and so are all $(A_0^m)^*$, $m = 1, 2, \ldots$.

Theorem 2.6. *Let G be finitely generated, of convergence type, and without parabolic elements. Consider the integral operator*

$$\mathbf{G}_\Omega f(x) := \int_\Omega f(y)\, l_\Omega(x, y)\, d\mu_\Omega(y).$$

Then the following holds.

(a) $\mathbf{G}_\Omega$ *is a bounded selfadjoint operator on* L^2.

(b) $\operatorname{Ran} \mathbf{G}_\Omega \subset \operatorname{dom} A$ *and* $A\mathbf{G}_\Omega = I$.

(c) *If in addition A is injective, then* $\operatorname{Ran} \mathbf{G}_\Omega = \operatorname{dom} A$, *and, consequently,* $\mathbf{G}_\Omega = A^{-1}$ *in the sense of the functional calculus of selfadjoint operators.*

Proof. We again use the Schur test to prove boundedness; the self-adjointness then follows from (2.14) (cf. [HS], Corollary 10.6).

For $s > 0$, consider the series

$$\zeta(s) := \sum_n (1 - |a_n|^2)^s.$$

It is a result of Beardon [B1] that for any finitely generated G of second kind, there exists a δ, $0 < \delta < 1$ (called the *exponent of convergence*) such that $\zeta(s)$ converges for all $s > \delta$. Now for any $\delta < c < 1$ and $u = \Gamma^c$,

$$\begin{aligned} -(\mathbf{G}_\Omega u)(\phi y) &= \int_\Omega \Gamma(z)^c\, |l_\Omega(z, \phi y)|\, d\mu_\Omega(z) \\ &= \int_{\mathcal{O}} \Gamma(\phi x)^c \sum_{\omega \in G} |l(\omega x, y)|\, d\mu(x) \\ &\le C \int_{\mathcal{O}} (1 - |x|^2)^c \sum_{\omega \in G} |l(\omega x, y)|\, d\mu(x) \quad \text{by Lemma 2.2} \\ &\le C \int_{\mathbb{D}} (1 - |x|^2)^c \sum_{\omega \in G} |l(\omega x, y)|\, d\mu(x) \end{aligned}$$

$$= C \sum_{\omega \in G} \big(- \mathbf{G}[(1 - |x|^2)^c]\big)(\omega y)$$

$$\le C \sum_{\omega \in G} (1 - |\omega y|^2)^c \quad \text{by the proof of part (b) of Theorem 1.3}$$

$$\le C(1 - |y|^2)^c \sum_n (1 - |a_n|^2)^c \quad \text{by Lemma 2.1}$$

$$= C\,\zeta(c)\,(1 - |y|^2)^c$$

$$\le C\zeta(c)\gamma(y)^c \quad \text{by Lemma 2.2}$$

$$= C\zeta(c)u(\phi y),$$

for any $y \in \overline{\mathcal{O}}$. This proves (a). Using the last proposition, parts (b) and (c) follow in the same way as part (c) in Theorem 1.3. □

From now on, we will assume that the hypotheses of the last theorem are fulfilled, i.e. that G is finitely generated, of convergence type, and without parabolic elements.

As in Section 1, let us define functions $G_m^\Omega(x, y)$ on $\Omega \times \Omega$ by the recipe

$$G_m^\Omega(x, \cdot) = \mathbf{G}_\Omega^{m-1}(l_\Omega(x, \cdot)), \quad y \in \Omega,\ m = 1, 2, \dots. \tag{2.18}$$

Owing to (2.17) the definition makes sense, and from (2.14) we see that $G_m^\Omega(x, y) = G_m^\Omega(y, x)$ and

$$(-1)^m G_m^\Omega > 0 \quad \text{on } \Omega \times \Omega. \tag{2.19}$$

In the same way as Theorem 1.4, one proves the following.

Theorem 2.7. *The operators $\mathbf{G}_\Omega^m$ are integral operators with kernels G_m^Ω.*

We now give a formula for G_m^Ω in terms of the functions $G_m^{\mathbb{D}} \equiv G_m$ from the preceding section. For $x, y \in \mathbb{D}$ and $m = 1, 2, \dots$, denote

$$s_m(x, y) := \sum_{\omega \in G} G_m(\omega x, y).$$

Owing to (1.11) and (2.8), the series on the right-hand side converges for any discontinuous group of convergence type, uniformly on compact subsets, and satisfies

$$s_m(x, y) = s_m(y, x) = s_m(\omega x, \omega y) \quad \forall x, y \in \mathbb{D}, \omega \in G, \tag{2.20}$$

$$(-1)^m s_m > 0 \quad \text{on } \mathbb{D} \times \mathbb{D}, \tag{2.21}$$

and

$$s_m(x, y) \le c(\delta, y)\,\gamma_{m-1}(x) \quad \text{if } |x - y| > \delta. \tag{2.22}$$

Theorem 2.8. $s_m(x, y) = G_m^\Omega(\phi x, \phi y)$ *for all* $x, y \in \mathbb{D}$ *and* $m = 1, 2, \dots$.

Proof. For $m = 1$, this is true by definition. Proceeding inductively, assume that the assertion holds for $m - 1$ in place of m. Then

$$\begin{aligned}
G_m^\Omega(\phi x, \phi y) &= \mathbf{G}_\Omega(G_{m-1}^\Omega(\cdot, \phi y))(\phi x) \\
&= \int_\Omega l_\Omega(t, \phi x)\, G_{m-1}^\Omega(t, \phi y)\, d\mu_\Omega(t) \\
&= \int_\mathcal{O} l_\Omega(\phi z, \phi x)\, G_{m-1}^\Omega(\phi z, \phi y)\, d\mu(z) \\
&= \int_\mathcal{O} s_1(z, x)\, s_{m-1}(z, y)\, d\mu(z) \quad \text{by the induction hypothesis} \\
&= \int_\mathcal{O} \sum_{\omega,\sigma \in G} l(z, \omega x)\, G_{m-1}(z, \sigma y)\, d\mu(z) \\
&= \int_\mathcal{O} \sum_{\omega,\sigma \in G} l(z, \omega x)\, G_{m-1}(z, \omega\sigma y)\, d\mu(z) \\
&= \sum_{\omega,\sigma \in G} \int_{\omega^{-1}\mathcal{O}} l(z, x)\, G_{m-1}(z, \sigma y)\, d\mu(z) \quad \text{by invariance} \\
&= \sum_{\sigma \in G} \int_\mathbb{D} l(z, x)\, G_{m-1}(z, \sigma y)\, d\mu(z) \\
&= \sum_{\sigma \in G} \mathbf{G}[G_{m-1}(\cdot, \sigma y)](x) \\
&= \sum_{\sigma \in G} G_m(x, \sigma y) = s_m(x, y),
\end{aligned}$$

the various interchanges of integration and summation signs being justified by the fact that all the integrands are of the same sign $(-1)^m$. □

We now show that for smoothly bounded domains we can also fulfill the hypothesis required in part (c) of Theorem 2.6 above. Namely, it is another result of Roelcke ([R2], Satz 5.1) that $\ker A \neq \{0\}$ if and only if $\mu(\Omega) < \infty$, and this can happen only for groups of the first kind. (See also Proposition 4.1 below.) For smoothly bounded domains, however, a much more elementary proof can be given.

Theorem 2.9. *Let Ω be a smoothly bounded domain in $\mathbb{C}$ and $h \in L^2(\Omega, d\mu_\Omega)$. Then $\mathbf{\Delta}_\Omega^m h = 0$ in the sense of distributions only if $h = 0$ a.e. In other words,* $\ker(A_0^m)^* = \{0\}$.

Proof. As we saw in the proof of Proposition 1.1, we need only to show that $\ker A_0^* = 0$, that is, by (2.2) and the Weyl Lemma, that the only harmonic function in L^2 is the constant zero. Thus let $h \in L^2$ be harmonic. Applying a suitable conformal mapping, we may assume that the outer boundary of Ω (i.e. the boundary of the unbounded connected component of $\mathbb{C} \setminus \Omega$) is the unit circle; thus h is a function which is harmonic in some annulus $\mathcal{A} = \{z : R < |z| < 1\}$ and, by (2.12) and (2.1), square-integrable there with respect to the measure $d\mu_{\mathbb{D}}$. The former fact implies that

$$h(z) = \sum_{m \in \mathbf{Z} \setminus \{0\}} (a_m z^m + b_m \bar{z}^{-m}) + a_0 + b_0 \log |z|^2 \quad \text{in } \mathcal{A} \tag{2.23}$$

with the series uniformly convergent on compact subsets of $\mathcal{A}$, i.e.

$$\begin{aligned} &\limsup_{m\to+\infty} |a_m|^{1/m} < 1, \qquad &&\limsup_{m\to+\infty} |b_{-m}|^{1/m} < 1, \\ &\limsup_{m\to+\infty} |a_{-m}|^{1/m} < R, \qquad &&\limsup_{m\to+\infty} |b_m|^{1/m} < R. \end{aligned} \tag{2.24}$$

The functions

$$z \mapsto \frac{1}{2\pi} \int_0^{2\pi} h(ze^{i\theta}) e^{-mi\theta}\, d\theta \quad (m = 0, \pm 1, \pm 2, \dots)$$

clearly belong to $L^2(\mathcal{A}, d\mu_{\mathbb{D}})$ whenever h does. It transpires that each of the functions

$$a_m z^m + b_m \bar{z}^{-m} \quad (m \neq 0), \quad a_0 + b_0 \log |z|^2,$$

belongs to $L^2(\mathcal{A}, d\mu_{\mathbb{D}})$. A small computation reveals that this is only possible if

$$a_0 = 0 \quad \text{and} \quad a_m + b_m = 0 \quad \forall m \neq 0. \tag{2.25}$$

However, in view of (2.24), this implies that the series (2.23) converges even for all $R < |z| < 1/R$, so, in particular, h extends continuously to the unit circle and, again by (2.25), vanishes there. Performing the same argument for the other boundary components, we see that $h \in C(\overline{\Omega})$ and $h = 0$ on $\partial\Omega$. As h is harmonic, h must vanish identically. □

In the general case, the assertion of Theorem 2.9 may fail (for instance, if Ω is $\mathbb{C}$ minus two points).

Returning to the smoothly bounded case, we can neatly summarize our findings in the following theorem.

Theorem 2.10. *Let Ω be a smoothly bounded domain in $\mathbb{C}$. Then the following assertions hold.*

(a) *The closure $A_0^{**} =: A$ of the operator A_0 is selfadjoint and injective on L^2.*

(b) *The inverse $\mathbf{G}_\Omega^m$ of the operator A^m is a bounded self-adjoint operator on L^2 $(m = 1, 2, \dots)$.*

(c) *$\mathbf{G}_\Omega^m$ are integral operators with kernels $G_m^\Omega(z, t)$ given by*

$$G_m^\Omega(\phi x, \phi y) = \sum_{\omega \in G} G_m^{\mathbb{D}}(\omega x, y) \quad (x, y \in \mathbb{D}).$$

(d) *The Green functions* G_m^Ω *are of constant sign* $(-1)^m$ *on* $\Omega \times \Omega$ *and satisfy the boundary estimates*

$$G_m^\Omega(x,y) \le c(\delta,y)\,\Gamma_{m-1}(x) \le c(\delta,y)\,w(x)\log_1^{m-1} w(x) \quad \textit{if}\ |x-y| > \delta.$$

(e) $G_m^\Omega(\cdot,y)$ *is the unique function in* L^2 *satisfying* $\mathbf{\Delta}_\Omega^m G_m^\Omega(\cdot,y) = \delta_y$ *(the point mass at* y*) in the sense of distributions, i.e.*

$$\int_\Omega G_m^\Omega(x,y)\,\mathbf{\Delta}_\Omega^m f(x)\,d\mu_\Omega(x) = f(y) \quad \forall f \in C_0^\infty(\Omega). \tag{2.26}$$

Proof. (a) and (b) are immediate from Theorem 2.9 and the parts (a) and (c) of Theorem 2.6. (c) follows from Theorem 2.7 and Theorem 2.8, and (d) from (2.19), (2.22) and Proposition 2.4. Finally (2.26) is just a restatement of $\mathbf{G}_\Omega^m \mathbf{\Delta}_\Omega^m f = f$ $\forall f \in C_0^\infty(\Omega)$, which is immediate from parts (a) and (b), and the uniqueness assertion in (e) again follows from Theorem 2.9. □

We conclude by giving briefly an analogue of Proposition 1.2, since this alternative definition of the Laplacian is common in other contexts.

Let $X : L^2 \to L^2$ be the restriction of the operator

$$f \mapsto w\,\frac{\partial f}{\partial \bar z}$$

to the subspace of all functions in L^2 for which the right-hand side, taken in the sense of distributions, belongs to L^2. In other words, $X = Y_0^*$ where Y_0 is the restriction to $C_0^\infty(\Omega)$ of the operator

$$f \mapsto -w^2\,\frac{\partial}{\partial z}(w^{-1}f). \tag{2.27}$$

Then X is densely defined and closed, X^* is the restriction of the operator (2.27) to dom X^*, X^*X is selfadjoint and $4X^*Xf = -\mathbf{\Delta}_\Omega f$ $\forall f \in C_0^\infty(\Omega)$. Arguing as in the proof of Proposition 1.2, we thus arrive at the following proposition.

Proposition 2.11. *Let* Ω *be of hyperbolic type. Then* $X^*X = -\frac{1}{4}A$.

3. The ball

We now show how all the results of Section 1 can be extended to the case of $\Omega = \mathbb{B}^d$, the unit ball in $\mathbb{C}^d$. The proofs are essentially the same, with the following modifications. The invariant Laplacian and the invariant measure on $\mathbb{B}^d$ are given, respectively, by

$$\mathbf{\Delta} = (1-|z|^2)\Big[\Delta - \sum_{i,j} z_i \bar z_j \frac{\partial^2}{\partial z_i \partial \bar z_j}\Big],$$

$$d\mu(z) = \frac{dm(z)}{(1-|z|^2)^{d+1}}.$$

By the results of Gaffney and Cordes, the operator A, defined as the closure of the restriction A_0 of $\mathbf{\Delta}$ to $C_0^\infty(\mathbb{B}^d)$, is self-adjoint, and so are all $(A_0^m)^*$, $m = 1, 2, \dots$. As in Proposition 1.1, one verifies that functions annihilated by $\mathbf{\Delta}$ (which are known to possess the mean-value property with respect to spheres centered at the origin) again never belong to $L^2 = L^2(\mathbb{B}^d, d\mu)$ unless they are identically zero, and it follows in the same way as before that A, as well as all its powers $A^m = (A_0^m)^*$, have trivial kernel. (It is also possible to give an alternative characterization of A along the lines of Proposition 1.2, i.e. as $-4X^*X$ for a certain weak maximal first-order differential operator X from L^2 into the Cartesian product of d copies of L^2 – essentially, X is a kind of invariant complex gradient on $\mathbb{B}^d$.) One then defines the operator $\mathbf{G}$ on C_0^∞ as in (1.5), with (1.3) replaced by the function

$$l(x, y) = -\frac{1}{2\,\mathrm{area}(\partial\mathbb{B}^d)} \int_{1-\frac{(1-|x|^2)(1-|y|^2)}{|1-\langle x,y\rangle|^2}}^{1} \frac{(1-s)^{d-1}}{s^d}\,ds, \tag{3.1}$$

since this corresponds to the fundamental solution of the radial part

$$f(t) \mapsto 4\frac{(1-t)^{d+1}}{t^{d-1}}\Big[\frac{t^d}{(1-t)^{d-1}}f'\Big]'$$

of $\mathbf{\Delta}$ (see [EP2], Section 1). Indeed, for any $\phi \in C_0^\infty(\mathbb{B}^d)$, denoting by $f \in C_0^\infty[0, 1)$ the radialization of ϕ (i.e. $f(t)$ is the mean value of ϕ over the sphere $|z|^2 = t$), we have by the rotation-invariance of $\mathbf{\Delta}$

$$\int_{\mathbb{B}^d} l(0, y)\,\mathbf{\Delta}\phi(y)\,d\mu(y) = \int_{\mathbb{B}^d} l(0, y)\,\mathbf{\Delta} f(|y|^2)\,d\mu(y),$$

which further equals to

$$\begin{aligned}
&\int_0^1 l(0, \sqrt{t})\,\frac{4(1-t)^{d+1}}{t^{d-1}}\Big[\frac{t^d}{(1-t)^{d-1}}f'\Big]'\,\frac{t^{d-1}\,\mathrm{area}(\partial\mathbb{B}^d)\,dt}{2(1-t)^{d+1}} \\
&\quad = \int_0^1 \Big(\int_1^t \frac{(1-s)^{d-1}}{s^d}\,ds\Big)\Big[\frac{t^d}{(1-t)^{d-1}}f'\Big]'\,dt \\
&\quad = \Big[\frac{t^d}{(1-t)^{d-1}}f' \int_1^t \frac{(1-s)^{d-1}}{s^d}\,ds\Big]_0^1 - \int_0^1 f'\,dt \\
&\quad = f(0) = \phi(0),
\end{aligned}$$

since $f(1) = 0$ as ϕ has compact support, and the boundary term in the partial integration vanishes because the $\int_1^t$ there is $O(t^{1-d})$ as $t \to 0$ ($O(\log t)$ if $d = 1$). The statements and the proofs of Theorems 1.3 and 1.4 are the only places where some real changes occur, since $l(0, \cdot) \notin L^2$ if $d > 1$; one therefore defines $\mathbf{G}f$ by (1.5) only for $f \in C_0^\infty$ and then shows instead that $\mathbf{G}$ is bounded on L^2, and that $A\mathbf{G}f = f$ $\forall f \in C_0^\infty$. The latter is basically the calculation we made a few lines above when justifying the definition of l. The former is again proved by applying the Schur test, with test functions $(1-|z|^2)^s$, $0<s<d$, the equality (1.6) replaced by

$$\frac{1}{\operatorname{area}(\partial\mathbb{B}^d)}\int_{\partial\mathbb{B}^d}|1-\langle x,y\rangle|^{-2s}\,d\sigma(x)=F(s,s,d,|y|^2)$$

($d\sigma$ being the surface element on $\partial\mathbb{B}^d$), and using the growth estimates

$$\begin{aligned} l(0,x) &\sim |x|^{2(1-d)} \quad \text{as } x\to 0 \quad (\sim \log|x|^2 \text{ for } d=1),\\ l(0,x) &\sim (1-|x|^2)^d \quad \text{as } x\to\partial\mathbb{B}^d, \end{aligned}$$

which are immediate from (3.1). This settles the analogue of Theorem 1.3. Finally, for $m=1,2,\dots$ define the functions $G_m(x,y)$ by

$$G_m(x,y)=\int_{\mathbb{B}^d}\dots\int_{\mathbb{B}^d} l(x,z_1)l(z_1,z_2)\dots l(z_{m-1},y)\,d\mu(z_1)\,\dots\,d\mu(z_{m-1}).$$

Though a priori it is not even clear that this integral converges, repeated application of the computation made in the above Schur test, in conjunction with the Fubini theorem, shows that for $u(z)=(1-|z|^2)^s$, $0<s<d$, one has

$$\int_{\mathbb{B}^d}(-1)^m G_m(x,y)u(y)\,d\mu(y)\le C\,u(x)\qquad \forall x\in\mathbb{B}^d.$$

It follows that G_m is finite almost everywhere and, in fact, defines a bounded selfadjoint integral operator on L^2. One more application of Fubini shows that this operator coincides with $\mathbf{G}^m$ on positive C_0^∞ functions; by continuity, they therefore coincide on all of L^2, and the analogue of Theorem 1.4 follows. (Alternatively, we could instead refer directly to Theorem 10.7 in [HS] on composition of bounded integral operators with nonnegative kernels.)

(The analogue of Proposition 1.5 for the ball is unknown to the present author.)

Consequently, we can summarize the situation in the following theorem.

Theorem 3.1. *For Ω the unit ball $\mathbb{B}^d$ in $\mathbb{C}^d$, there exists a bounded selfadjoint operator $\mathbf{G}$ on $L^2(\mathbb{B}^d,d\mu)$ such that $u=\mathbf{G}f$ is the unique L^2 solution to the equation $\mathbf{\Delta}u=f$, $\forall f\in L^2$; further, for each $m=1,2,\dots$, $\mathbf{G}^m$ is an integral operator whose kernel $G_m(x,y)$ – the Green function for $\mathbf{\Delta}^m$, by definition – has the constant sign $(-1)^m$ on $\mathbb{B}^d\times\mathbb{B}^d$. Finally, $\mathbf{G}^{-1}=A$, the closure of the symmetric operator $(\mathbf{\Delta}|C_0^\infty)$ on L^2.*

4. Complete Riemannian manifolds

For a Riemannian manifold Ω with the metric $ds^2=g_{ij}\,dx^i\,dx^j$, the Laplace–Beltrami operator (on functions) is defined in local coordinates as

$$\mathbf{\Delta}f=\frac{1}{\sqrt{g}}\frac{\partial}{\partial x^i}\Big(\sqrt{g}\,g^{ji}\frac{\partial f}{\partial x^j}\Big)=\operatorname{div}\operatorname{grad}f.$$

Here g^{ji} is the inverse matrix to g_{ij}; $g = \det(g_{ij})$, so that $\sqrt{g}$ is the density of the volume form $d\mu(x) = \sqrt{g(x)}\, dx^1 \dots dx^n$ induced by the metric; grad is the operator

$$(\operatorname{grad} f)^j = g^{ji} \frac{\partial f}{\partial x^i}$$

from functions into vector fields on Ω, and $-\operatorname{div}$ is its formal adjoint with respect to the $L^2 = L^2(\Omega, d\mu)$ inner products. (Here we have adopted the usual differential-geometric notation for co- and contravariant indices, and used the summation convention.)

Functions annihilated by $\mathbf{\Delta}$ are, by definition, the *harmonic functions* on Ω. They share most properties of the ordinary harmonic functions on $\mathbb{R}^n$ – they are always C^∞, are characterized by a certain mean-value property, obey the Harnack inequalities, the maximum principle, the Harnack theorem, etc.; see, for instance, Keller [Ke], §§ 2.3–2.6.

Let A_0 and X_0 be the restrictions to C_0^∞ of $\mathbf{\Delta}$ and grad, respectively, and $A = A_0^{**}$ and $X = X_0^{**}$ their closures. It is a fundamental result of Gaffney [Gf] and Cordes [Co] (for a simpler proof see Chernoff [Ch]), already alluded to above, that if Ω is complete then the operator A is selfadjoint, and so are the operators $(A_0^m)^*$ (which then must coincide with A^m), for all $m = 1, 2, \dots$. Arguing as in the proofs of Propositions 1.2 and 2.11, one then sees that $A = -X^*X$.

In contrast to the previous sections, however, it is not in general true that A is injective. (For a simple example, take $\Omega = \mathbb{R}^2 \simeq \mathbb{C}$ with a radial metric $ds^2 = \gamma(|z|^2)\, dz\, d\bar{z}$. It is easily seen that Ω is complete if and only if $\sqrt{\gamma} \notin L^1(0, \infty)$, while the (harmonic) function constant one belongs to $L^2(\Omega)$ if and only if $\gamma \in L^1(0, \infty)$. The function $\gamma(t) = (1 + t^2)^{-3/4}$ meets both requirements. Another example is $\Omega = \mathbb{C} \setminus \{0, 1\}$ with the Poincaré metric, see the remarks after Theorem 2.9 above.) The following criterion tells us when this happens.

Proposition 4.1. *Let Ω be a complete Riemannian manifold. Then* $\ker A \neq \{0\}$ *if and only if $\mu(\Omega) < \infty$, and in that case* $\ker A$ *consists of the constant functions.*

Proof. By Satz 5 on p. 145 in Roelcke [R1], we have

$$\langle A\phi, \phi \rangle = -\int_\Omega \|\operatorname{grad} \phi\|^2\, d\mu \quad \forall \phi \in C_0^\infty = \operatorname{dom} A_0. \tag{4.1}$$

(Here, of course, $\|\operatorname{grad}\phi\|^2 = g_{ij} (\operatorname{grad}\phi)^i \overline{(\operatorname{grad}\phi)^j} = g^{ji} (\partial\phi/\partial x^i) \overline{(\partial\phi/\partial x^j)}$.) Since A is the closure of A_0, this equality in fact remains in force for all $\phi \in \operatorname{dom} A$ (in particular, $\operatorname{dom} A$ is a subset of the Sobolev space $H^1(\Omega)$). It follows that any $\phi \in \ker A$ must be constant. Since the function constant one belongs to L^2 if and only if Ω has finite volume, the conclusion follows. □

For the rest of this section, we confine ourselves to the situation when

$$\ker A = \{0\}. \tag{4.2}$$

Then the operator A has a self-adjoint (possibly unbounded) inverse $A^{-1} =: \mathbf{G}$. Now if $\mathbf{G}$, or, more generally, some power $\mathbf{G}^m$ is an integral operator, it is natural to call its kernel the Green function for A^m and, prompted by the results of the preceding sections, ask about its sign.

There is another definition of a Green function for A, coming from potential theory (see, for instance, Nakai and Sario [NS]). Namely, let $\Omega_1 \subset \Omega_2 \subset \cdots$ be a sequence of regular domains (i.e. smoothly bounded and with compact closures) exhausting Ω. For each y and k so large that $y \in \Omega_k$, let $g_k(\cdot, y)$ be the Green function, with pole at y, of the operator $\mathbf{\Delta}$ on Ω_k with the Dirichlet boundary conditions. (Thus $\mathbf{\Delta} g_k(\cdot, y) = \delta_y$ in the sense of distributions, and $g_k(x, y) = 0\ \forall x \in \partial\Omega_k$.) As k increases, the functions $g_k(\cdot, y)$ decrease, so by the Harnack theorem (cf. [Ke], p. 242), either g_k diverge to $-\infty$, or $g_k(\cdot, y) \searrow g(\cdot, y)$, locally uniformly on Ω, for some harmonic function $g(\cdot, y)$. In the former case Ω is called *parabolic,* in the latter it is called *hyperbolic* and $g(x, y)$ is called the Green function of Ω. (The hyperbolicity is independent of the choice of $y \in \Omega$, and both the hyperbolicity and $g(x, y)$ are independent of the choice of the exhausting sequence Ω_k.)

It is known ([Ke], p. 237) that the function $g(x, y)$ is symmetric in x and y, is C^∞ outside the diagonal (by elliptic regularity), and satisfies $g(x, y) = O(d(x, y)^{2-n})$ as $x \to y$ ($O(\log d(x, y))$ if $n = 2$), where $n \geq 2$ is the dimension of Ω and $d(x, y)$ stands for the geodesic distance of x and y. In particular, both $g(\cdot, y)$, for any y, and g itself are locally integrable. Consequently, the integral operator

$$\mathbf{g} f(x) := \int_\Omega g(x, y) f(y)\, d\mu(y)$$

is well-defined on C_0^∞, maps it into L^1_{loc}, and

$$\langle \mathbf{g}\phi, \psi \rangle = \langle \phi, \mathbf{g}\psi \rangle \qquad \forall \phi, \psi \in C_0^\infty. \tag{4.3}$$

Finally, the functions g_k are negative on $\Omega_k \times \Omega_k$, by the maximum principle ([Ke], p. 242), thus g is negative on $\Omega \times \Omega$.

Denote now by $\mathbf{g}_0$ the restriction of $\mathbf{g}$ to

$$\operatorname{dom} \mathbf{g}_0 = \{\mathbf{\Delta}\phi : \phi \in C_0^\infty\} \quad (= \operatorname{Ran} A_0),$$

and by $\mathbf{g}_{\max}$ the restriction of $\mathbf{g}$ to

$$\operatorname{dom} \mathbf{g}_{\max} = \{f \in L^2 : \mathbf{g}|f| < \infty \text{ a.e. and } \mathbf{g} f \in L^2\}.$$

We will say that $\mathbf{g}$ is bounded on L^2 if $\operatorname{dom} \mathbf{g}_{\max} = L^2$. (This already implies that $\mathbf{g}_{\max}$ is a bounded linear operator on L^2, see [HS], Theorem 3.10.) Under the hypothesis (4.2), which implies that $\mathbf{g}_0$ is densely defined, this clearly happens if and only if $\mathbf{g}_0$ extends to a bounded linear operator on L^2.

Theorem 4.2. *Let Ω be a complete Riemannian manifold satisfying* (4.2). *Then $\mathbf{g}_0$ takes values in L^2 and its closure coincides with A^{-1}.*

Proof. Let $\mathbf{g}_k$ be the integral operator on Ω_k with kernel $g_k(x, y)$. For any $\phi \in C_0^\infty$, we then have $\mathbf{g}_k \mathbf{\Delta} \phi = \phi$ as soon as $\operatorname{supp} \phi \in \Omega_k$; as g_k tend monotonely to g, it follows that $\mathbf{g\Delta}\phi = \phi$. In other words,

$$\mathbf{g}_0 A_0 \phi = \phi \qquad \forall \phi \in \operatorname{dom} A_0. \tag{4.4}$$

Since $\operatorname{dom} \mathbf{g}_0 = \operatorname{Ran} A_0$ by definition, this means that $\mathbf{g}_0 = A_0^{-1}$ (in particular, $\operatorname{Ran} \mathbf{g}_0 \subset L^2$). As, by (4.2), $\operatorname{dom} \mathbf{g}_0$ is dense in L^2 and A_0^* is injective, it follows that $\mathbf{g}_0^* = (A_0^*)^{-1} = A^{-1}$. Taking adjoints we thus get

$$\mathbf{g}_0^{**} = A^{-1}, \tag{4.5}$$

which is the desired assertion. □

Corollary 4.3. *Under the hypotheses of the last theorem, assume that either A has a bounded inverse, or that $\mathbf{g}$ is bounded on L^2. Then*

$$A^{-1} = \mathbf{g}_{\max}.$$

In particular, $\mathbf{G} := A^{-1}$ is an integral operator whose kernel $G(x, y) = g(x, y)$ – the Green function for A, by definition – satisfies $G(x, y) < 0$ on $\Omega \times \Omega$.

Since integral operators are not closed in general (cf. [HS], Examples 3.6 and 3.7), it is not clear whether the equality $A^{-1} = \mathbf{g}_{\max}$ must prevail for unbounded A^{-1} and $\mathbf{g}$. The point is that, for instance, for any harmonic function h on Ω (not assumed to be square-integrable), the operators $\mathbf{g}_0$ and $\mathbf{g}_0'$ determined by the functions $g(x, y)$ and $g'(x, y) := g(x, y) + h(x) + h(y)$, respectively, are the same, but the corresponding maximal integral operators $\mathbf{g}_{\max}$ and $\mathbf{g}'_{\max}$ are different; hence $\mathbf{g}_{\max}$ and $\mathbf{g}'_{\max}$ cannot both be equal to A^{-1}, even though $\mathbf{g}_0'^{**} = A^{-1} = \mathbf{g}_0^{**}$. Thus even if we know a priori that A^{-1} is an integral operator, it is in general impossible to read off its kernel from its restriction to the (dense) subspace $\operatorname{dom} \mathbf{g}_0$.

The condition that A^{-1} be bounded is equivalent to the validity of the Poincaré (or Hardy-type) inequality

$$\int_\Omega |u|^2 \, d\mu \le C \int_\Omega \| \operatorname{grad} u \|^2 \, d\mu \qquad \forall u \in C_0^\infty(\Omega), \tag{4.6}$$

by (4.1). There is an extensive literature on this topic, though more often in the context of incomplete manifolds, i.e. in the presence of a boundary; see e.g. [CrD], [KO], [Dv], and the references therein. For instance, one sees from Theorem 1.14 in [KO] that on the unit disc $\mathbb{D}$ with the radial metric $ds^2 = (1 - |z|^2)^\alpha dz \, d\bar{z}$, $\alpha \le -2$, A^{-1} is bounded if and only if $\alpha = -2$, i.e. for the Poincaré metric. (For $\alpha > -2$ the metric ds^2 is not complete.) – We should also remark that there exists a characterization of hyperbolicity similar to (4.6): namely, (Ω, g_{ij}) is hyperbolic if and only if there exists a positive function $\beta(x)$ on Ω such that $\int_\Omega \beta |u|^2 \, d\mu \le \int_\Omega \| \operatorname{grad} u \|^2 \, d\mu \; \forall u \in C_0^\infty(\Omega)$. See Ancona [An].

Finally, for $m = 1, 2, \dots$, define

$$G_m(x, y) = \int_\Omega \dots \int_\Omega g(x, z_1) g(z_1, z_2) \dots g(z_{m-1}, y)\, d\mu(z_1) \dots d\mu(z_{m-1}). \quad (4.7)$$

Combining the negativity of g with Theorem 10.7 in [HS], we arrive at the following generalization of Corollary 4.3.

Corollary 4.4. *Let Ω be a complete Riemannian manifold satisfying* (4.2), *and assume that either A has a bounded inverse or that $\mathbf{g}$ is bounded on L^2. Then G_m is finite a.e. on $\Omega \times \Omega$, and $A^{-m} = \mathbf{G}^m$ is a bounded integral operator with kernel G_m. In particular, this kernel – the Green function for A^m, by definition – is of constant sign $(-1)^m$ on $\Omega \times \Omega$.*

Remark. Observe that the boundedness of $\mathbf{G}$ thus implies that G_m are finite a.e. on $\Omega \times \Omega$ for any m. For $m = 2$, in particular, the latter is easily seen to be equivalent to the square-integrability of $g(\cdot, y)$ over Ω minus a small neighbourhood of y (for any y). Thus, for instance, the operator $\mathbf{G} = A^{-1}$ is unbounded on $\Omega = \mathbb{D}$ with a metric $ds^2 = \gamma(z)\, dz\, d\bar{z}$ as soon as $(1 - |z|)^2 \gamma(z) \notin L^1(\mathbb{D}, dm)$.

References

[An] A. Ancona, Théorie du potentiel sur les graphes et les variétés, in: École d'été de Probabilités de Saint-Flour XVIII (1988), Lecture Notes in Math. 1427, Springer-Verlag, Berlin, 1990, 1–112.

[BE] H. Bateman, A. Erdélyi, Higher transcendental functions I, McGraw-Hill, New York–Toronto–London, 1953.

[B1] A. F. Beardon, Inequalities for certain Fuchsian groups, Acta Math. 127 (1971), 221–258.

[B2] A. F. Beardon, The geometry of discrete groups, Grad. Texts in Math. 91, Springer-Verlag, Berlin–Heidelberg–New York, 1983.

[Bo] T. Boggio, Sulle funzioni di Green d'ordine m, Rend. Circ. Mat. Palermo 20 (1905), 97–135.

[Ch] P. R. Chernoff, Essential self-adjointness of powers of generators of hyperbolic equations, J. Funct. Anal. 12 (1973), 401–414.

[CD] C. V. Coffman, R. J. Duffin, On the structure of biharmonic functions satisfying the clamped plate conditions on a right angle, Adv. Appl. Math. 1 (1980), 373–389.

[Co] H. O. Cordes, Selfadjointness of powers of elliptic operators on noncompact manifolds, Math. Ann. 195 (1972), 257–272.

[CrD] C. B. Croke, A. Derdziński, A lower bound for λ_1 on manifolds with boundary, Comm. Math. Helv. 62 (1987), 106–121.

[Da] D. P. Dalzell, Convergence of certain series associated with Fuchsian groups, J. London Math. Soc. 23 (1948), 19–22.

[Dv] E. B. Davies, A review of Hardy inequalities, The Maz'ya anniversary collection, Vol. 2 (Rostock, 1998), Oper. Theory Adv. Appl. 110, Birkhäuser, Basel, 1999, 55–67.

[Du] R. J. Duffin, On a question of Hadamard concerning super-biharmonic functions, J. Math. Phys. 27 (1949), 253–258.

[EP1] M. Engliš, J. Peetre, Covariant differential operators and Green functions, Ann. Polon. Math. 66 (1997), 77–103.

[EP2] M. Engliš, J. Peetre, Green's functions for powers of the invariant Laplacian, Canadian J. Math. 50 (1998), 40-73.

[EP3] M. Engliš, J. Peetre, Green functions and eigenfunction expansions for the square of the Laplace-Beltrami operator on plane domains, Ann. Mat. Pura Appl., to appear.

[Gf] M. P. Gaffney, The harmonic operator for exterior differential forms, Proc. Nat. Acad. Sci. USA 37 (1951), 48–50.

[Gr] P. R. Garabedian, A partial differential equation arising in conformal mapping, Pacific J. Math. 1 (1951), 485–524.

[Go] G. M. Goluzin, Geometric theory of functions of a complex variable, Nauka, Moscow, 1966.

[GS] H.-C. Grunau, G. Sweers, Positivity for perturbations of polyharmonic operators with Dirichlet boundary conditions in two dimensions, Math. Nachr. 179 (1996), 89–102.

[HS] P. R. Halmos, V. S. Sunder, Bounded integral operators on L^2 spaces, Springer-Verlag, Berlin–Heidelberg–New York, 1978.

[HK] W. K. Hayman, B. Korenblum, Representation and uniqueness of polyharmonic functions, J. Anal. Math. 60 (1993), 113-133.

[HJS] H. Hedenmalm, S. Jakobsson, S. Shimorin, A maximum principle à la Hadamard for biharmonic operators with applications to the Bergman space, C. R. Acad. Sci. Paris Sér. Math. 328 (1999), 973–978.

[Ke] S. Keller, Die Existenz einer Greenschen Funktion auf Riemannschen Mannigfaltigkeiten, Comment. Math. Helv. 48 (1973), 234–253.

[KKM] V. A. Kozlov, V. A. Kondrat'ev, V. G. Maz'ya, On sign variability and the absence of "strong" zeros of solutions of elliptic equations, Izv. Akad. Nauk SSSR Ser. Mat. 53 (1989), 328–344 (Russian); English translation: Math. USSR-Izv. 34 (1990), 337–353.

[Kr] I. Kra, Automorphic functions and Kleinian groups, Benjamin, Reading, 1972.

[KO] A. Kufner, B. Opic, Hardy-type inequalities, Pitman Research Notes in Math. 219, Longman, Harlow, 1990.

[Lo] C. Loewner, On generation of solutions of the biharmonic equation in the plane by conformal mappings, Pacific J. Math. 3 (1953), 417–436.

[NS] L. Sario, M. Nakai, C. Wang, L.O. Chung, Classification theory of Riemannian manifolds, Lecture Notes in Math. 605, Springer-Verlag, Berlin, 1977.

[Os] S. Osher, On Green's function for the biharmonic equation in a right angle wedge, J. Math. Anal. Appl. 43 (1973), 705–716.

[RN] F. Riesz, B. Sz.-Nagy, Leçons d'analyse fonctionnelle, Akademiai Kiado, Budapest, 1972.

[R1] W. Roelcke, Über den Laplace-Operator auf Riemannschen Mannigfaltigkeiten mit diskontinuierlichen Gruppen, Math. Nachr. 21 (1960), 131–149.

[R2] W. Roelcke, Das Eigenwertproblem der automorphen Formen in der hyperbolischern Ebene. I, Math. Ann. 167 (1966), 292–337.

[ST] H. S. Shapiro, M. Tegmark, An elementary proof that the biharmonic Green function of an eccentric ellipse changes sign, SIAM Rev. 36 (1994), 99–101.

[St] M. Stessin, An extension of a theorem of Hadamard and domination in the Bergman space, J. Funct. Anal. 115 (1993), 212–226.

[Sz] G. Szegö, Remark on the preceding paper of Charles Loewner, Pacific J. Math. 3 (1953), 437–446.

[Ts] M. Tsuji, Potential theory in modern function theory, Maruzen, Tokyo, 1959.

Miroslav Engliš, MÚ AV ČR, Žitná 25, 11567 Prague 1, Czech Republic
E-mail: `englis@math.cas.cz`

Symmetric linear functionals on function spaces

Tadeusz Figiel and Nigel Kalton *

To Professor Jaak Peetre on his 65th birthday

Abstract. We discuss the properties of linear functionals defined on spaces of measurable functions which are invariant under rearrangements.

2000 Mathematics Subject Classification: 42A50, 46E30

1. Introduction

Suppose (Ω, Σ, μ) is any nonatomic σ-finite measure space. We let $\mathcal{M}(\Omega)$ be the space of all Σ-measurable complex-valued functions f such that

$$d_f(x) := \mu\{|f| > x\} < \infty, \qquad x > 0.$$

We define $\mathcal{M}_{\mathbb{R}}(\Omega)$ and $\mathcal{M}_+(\Omega)$ to be the subsets of all real-valued or all non-negative real-valued functions respectively. If $f \in \mathcal{M}$, then the decreasing rearrangement of $|f|$ is denoted by f^*; this is defined by

$$f^*(t) = \inf_{\mu E = t} \sup_{\omega \notin E} |f(\omega)|.$$

From now we assume that either $\mu(\Omega) = \infty$ or $\mu(\Omega) = 1$. If Ω is a standard measure space we may assume that either $\Omega = (0, \infty)$ or $\Omega = (0, 1]$ with Lebesgue measure. We define a *symmetric ideal* $\mathcal{X}$ to be a linear subspace of $\mathcal{M}$ such that:

- If $f \in \mathcal{X}$ and $|g| \le |f|$ then $g \in \mathcal{X}$,
- If $f \in \mathcal{X}$ and $d_g \le d_f$ then $g \in \mathcal{X}$,
- $\mathcal{X} \neq \{0\}$.

*Research supported by NSF grant DMS-9870027

Of course if $\mathcal{X}$ is any symmetric ideal on Ω where $\mu(\Omega) = 1$ (respectively, $\mu(\Omega) = \infty$) then we can define the associated symmetric ideal $\mathcal{X}(0, 1]$ (respectively $\mathcal{X}(0, \infty)$ by $f \in \mathcal{X}(0, 1]$ (respectively $\mathcal{X}(0, \infty)$) if and only there exists $g \in \mathcal{X}$ with $d_g = d_f$.

Note that $\mathcal{X} \neq \{0\}$ implies that $\mathcal{X}$ contains all bounded measurable functions with support of finite measure. It follows from this remark that if $f \in \mathcal{X}$ and $g \in \mathcal{M}$ satisfies $g^*(t) \leq Cf^*(t/C)$ for some constant C then $g \in \mathcal{X}$.

By a symmetric quasi-Banach ideal we mean a symmetric ideal $\mathcal{X}$ equipped with a quasi-norm $f \to \|f\|_{\mathcal{X}}$ such that:

- $\|f\|_{\mathcal{X}} \leq \|g\|_{\mathcal{X}}$ whenever $f^* \leq g^*$.
- $\mathcal{X}$ is complete for the quasi-norm $\|\cdot\|_{\mathcal{X}}$.

In this case we define the dilation operators $D_s : \mathcal{X}(0, 1] \to \mathcal{X}(0, 1]$ by

$$D_s f(t) = \begin{cases} f(t/s) & \text{if } t \leq s \\ 0 & \text{if } t > s, \end{cases}$$

or $D_s : \mathcal{X}(0, \infty) \to \mathcal{X}(0, \infty)$ by

$$D_s f(t) = f(t/s), \quad 0 < t < \infty.$$

The Boyd indices $p_{\mathcal{X}}$ and $q_{\mathcal{X}}$ are defined by

$$\frac{1}{p_{\mathcal{X}}} = \lim_{s\to\infty} \frac{\log \|D_s\|}{\log s}.$$

and

$$\frac{1}{q_{\mathcal{X}}} = \lim_{s\to 0} \frac{\log \|D_s\|}{\log s}.$$

Suppose $\mathcal{X}$ is a symmetric ideal on (Ω, Σ, μ). Let us say that $f \sim g$ if f, g have the same distributions i.e. $\mu \circ f^{-1} = \mu \circ g^{-1}$ coincide as Borel measures on $\mathbb{C} \setminus \{0\}$. We shall say a linear functional $\varphi : \mathcal{X} \to \mathbb{C}$ is *symmetric* if $\varphi(f) = \varphi(g)$ whenever $f \sim g$. We say φ is a *regular symmetric functional* if additionally $\varphi(\chi_A) = 1$ whenever $\mu(A) = 1$. We emphasize here that we have no topological assumptions in general: $\mathcal{X}$ need not be a quasi-Banach ideal, and φ need not have any continuity properties.

The study of such symmetric functionals goes back around twenty years. In fact the corresponding notion for sequence spaces is closely related to the problem of defining unusual traces on certain ideals of operators on a Hilbert space; this subject can be traced back to the work of Dixmier [5] in the 1960's. In the 1980's Pietsch, motivated by the study of traces on ideals of operators on Banach spaces again drew attention to the problem of characterizing symmetric functionals on sequence spaces (see for example [14]). For more recent work in this direction see [11], [6] and [12].

In this paper, we want to draw a neat and perhaps unexpected connection between analytic functions and symmetric functionals. Before doing this in Section 2 we develop the basic theory of symmetric functionals on an arbitrary symmetric ideal of functions. Our main result Theorem 2.8 here characterizes the set $Z_{\mathcal{X}}$ (which we call the center) of all functions f so that $\varphi(f) = 0$ for all symmetric functionals φ. These results were essentially known to the first author in the 1980's; see for example [13] and [14]. It is interesting that $Z_{\mathcal{X}}$ can be characterized purely in terms of a generalized K-functional.

In Section 3 we prove our main results. For the case of the unit circle $\mathbb{T}$, this says that for a symmetric ideal $\mathcal{X}$ which satisfies a very mild condition, (in particular for any quasi-Banach symmetric ideal), that if f is analytic function on $\mathbb{D}$ with $f(0) = 0$, belonging to the Smirnov class such that its boundary values, also denoted by f, belong to $\mathcal{X}(\mathbb{T})$ then f actually belongs to the center $Z_{\mathcal{X}}$. For the special case of $\mathcal{X} = L_1$ this result is proved in two different ways in [10]; the center of L_1 is there denoted by $H^{\text{sym}}_{1,0}$. It is also equivalent (again for the case $\mathcal{X} = L_1$) to an older result of Ceretelli [3], later independently found by Davis [4]. This result is also discussed in [10] and [9]; in [10] it is connected to the theory of commutators in interpolation. Theorems 3.3 and 3.4 are inspired by the analogous results for traces in [12].

Finally in Section 4 we show that for quasi-Banach ideals a real function belongs to $\mathcal{X}(\mathbb{R})$ if and only if it is the real part of a function f in the corresponding Hardy class $H_{\mathcal{X}}$.

2. Symmetric functionals

In order to treat the case $\mu(\Omega) = 1$ let us introduce the notion of a symmetric ideal *of finite type*. If $\mu(\Omega) = \infty$ we say that $\mathcal{X}$ is a symmetric ideal of finite type if $f \in \mathcal{X}$ implies that $\mu(\operatorname{supp} f) < \infty$. If $\mathcal{Y}$ is a symmetric ideal on Ω where $\mu(\Omega) = 1$ we can embed (Ω, μ) in a larger nonatomic measure space (Ω', μ'), by taking a countable disjoint union of copies of (Ω, μ). We can then define $\mathcal{X}(\Omega')$ of finite type by the property that $f \in \mathcal{X}$ if and only if $\mu'(\operatorname{supp} f) < \infty$ and $f\chi_E \in \mathcal{Y}(E)$ whenever $\mu'(E) = 1$. We call $\mathcal{X}$ the *infinite extension* of $\mathcal{Y}$. It is then easy to see that a symmetric linear functional φ on $\mathcal{Y}$ extends to a symmetric linear functional on $\mathcal{X}$. Using this idea of an infinite extension, we shall restrict our attention unless otherwise stated to the case when $\mu(\Omega) = \infty$.

For $f \in \mathcal{M}_+$ we define the *generalized K-functional* of f for $0 < u < v < \infty$ by

$$K(u, v; f) = \int_u^v f^*(t)dt. \tag{2.1}$$

We extend the definitions of K to general complex functions by linearity. To be precise if $f = g + ih \in \mathcal{M}$ where g and h are real we define

$$K(u, v; f) = K(u, v; g_+) - K(u, v; g_-) + iK(u, v; h_+) - iK(u, v; h_-)$$

and where, as usual $g_+ = \max(g, 0)$ and $g_- = \max(-g, 0)$ etc.

We will also need a dual notion which we term the K^**-functional* which we define for all $f \in \mathcal{M}$:

$$K^*(r, s; f) = \int_{r \le |f| < s} f \, d\mu. \tag{2.2}$$

If $f \in \mathcal{M}$ we can define measurable sets $E_u(f)$ for $u > 0$ so that $\mu E_u(f) = u$, $E_u(f) \subset E_v(f)$ if $u < v$ and $E_u \subset \{|f| \ge f^*(u)\}$. Let $E_{u,v}(f) = E_v(f) \setminus E_u(f)$ when $u < v$.

Lemma 2.1. *Suppose* $f, g \in \mathcal{M}$ *and* $|f| \le |g|$.

(1) *Suppose* $0 < u < v < \infty$. *Then*

$$\left| \int_{E_{u,v}(f)} f \, d\mu - \int_{E_{u,v}(g)} f \, d\mu \right| \le 2(ug^*(u) + vg^*(v)). \tag{2.3}$$

(2) *Suppose* $0 < r < s < \infty$. *Then*

$$\left| \int_{r \le |f| < s} f \, d\mu - \int_{r \le |g| < s} f d\mu \right| \le 2(rd_g(r) + sd_g(s)). \tag{2.4}$$

Proof. (1) Let $h = \chi_{E_v(f)} - \chi_{E_u(f)} + \chi_{E_u(g)} - \chi_{E_v(g)}$. If $\omega \in E_{u,v}(f)$ and $h(\omega) = 1$ then either $\omega \in E_u(g)$ or $|g(\omega)| \le g^*(v)$. Thus

$$\left| \int_{E_{u,v}(f)} hf \, d\mu \right| \le uf^*(u) + (v - u)g^*(v).$$

If $\omega \in E_u(f)$ then $h(\omega) = -1$ if $\omega \in E_{u,v}(g)$;

$$\left| \int_{E_u(f)} hf \, d\mu \right| \le ug^*(u).$$

If $\omega \in \Omega \setminus E_v(f)$ then $|f(\omega)| \le f^*(v)$ and $h(\omega) = 1$ if $\omega \in E_{u,v}(g)$.

$$\left| \int_{\Omega \setminus E_v(f)} hf \, d\mu \right| \le (v - u) f^*(v).$$

Combining gives (1).

For (2) note that

$$\left| \int_{\substack{r \le |f| < s \\ |g| < r}} f \, d\mu \right| \le rd_f(r), \quad \left| \int_{\substack{r \le |f| < s \\ |g| \ge s}} f \, d\mu \right| \le sd_g(s),$$

$$\left| \int_{\substack{r \le |g| < s \\ |f| < r}} f \, d\mu \right| \le rd_g(r), \quad \text{and} \left| \int_{\substack{r \le |g| < s \\ |f| \ge s}} f \, d\mu \right| \le sd_f(s).$$

This concludes the proof. □

The following is an immediate consequence of Lemma 2.1:

Lemma 2.2. *Suppose $f, g \in \mathcal{M}$ and $|f| \leq |g|$. Then:*

$$\left|K(u, v; f) - \int_{E_{u,v}(g)} f\,d\mu\right| \leq 8(ug^*(u) + vg^*(v)) \tag{2.5}$$

and

$$\left|K^*(r, s; f) - \int_{r \leq |g| < s} f\,d\mu\right| \leq 2(rd_g(r) + sd_g(s)). \tag{2.6}$$

Proof. Note that (2.6) simply restates (2.4). To prove (2.5) simply note that (2.3) gives this with constant 2 if f is positive. We can apply it in turn to $(\Re f)_+$, $(\Re f)_-$, $(\Im f)_+$ and $(\Im f)_-$. □

Our next lemma shows that the K-functional has approximate linearity properties:

Lemma 2.3. *Suppose $f, g \in \mathcal{M}$ and suppose $h = f + g$ and $\psi = |f| + |g|$. Then if $0 < u < v < \infty, 0 < r < s < \infty$,*

$$|K(u, v; h) - K(u, v; f) - K(u, v; g)| \leq 24(u\psi^*(u) + v\psi^*(v)) \tag{2.7}$$

and

$$|K^*(r, s; h) - K^*(r, s; f) - K^*(r, s; g)| \leq 6(rd_\psi(r) + sd_\psi(s)). \tag{2.8}$$

Proof. For (2.7) it is enough to note that

$$\left|K(u, v; \phi) - \int_{E_{u,v}(\psi)} \phi\,d\mu\right| \leq 8(u\psi^*(u) + v\psi^*(v))$$

when $\phi = f, g$ and h. (2.8) is similar. □

Next we define the notion of the *center* $Z_{\mathcal{X}}$ of a symmetric ideal $\mathcal{X}$ supported on a measure space Ω with $\mu(\Omega) = 1$ or $\mu(\Omega) = \infty$. We shall say that $f \in Z_{\mathcal{X}}$ if there exists $h \in \mathcal{X}_+$ such that whenever $0 < u < v < \infty$ we have:

$$|K(u, v; f)| \leq uh^*(u) + vh^*(v). \tag{2.9}$$

Let us make the remark that if $\mathcal{X}$ is of finite type and $\mu(\operatorname{supp} f) = 1$ then by taking v large enough (2.9) will imply that

$$|K(u, 1; f)| \leq uh^*(u).$$

Hence it suffices to consider in this case h with $\mu(\operatorname{supp} h) = 1$. This implies that if $\mathcal{Y}$ is a symmetric ideal on a probability space and $\mathcal{X}$ is the infinite extension of $\mathcal{Y}$ then $Z_{\mathcal{Y}}$ coincides with the restriction of $Z_{\mathcal{X}}$ to Ω.

Lemma 2.4. *$f \in Z_{\mathcal{X}}$ if and only if there exists $h \in \mathcal{X}_+$ and $C > 0$ with*

$$|K^*(r, s; f)| \leq C(rd_h(r) + sd_h(s)), \quad 0 < r < s < \infty. \tag{2.10}$$

Proof. Assume first $f \in Z_{\mathcal{X}}$. Choose $u = \mu(|f| > r)$ and $v = \mu(|f| > s)$. Then we can assume $\{r \leq |f| < s\} = E_{u,v}(f)$ and so

$$K^*(r, s; f) = \int_{E_{u,v}(f)} f\, d\mu$$

and (2.10) follows from (2.5) with $f = g$.

Conversely suppose (2.10) holds and $0 < u < v < \infty$; we can assume $h^* \geq f^*$. Let $r = 2h^*(u)$ and $s = 2h^*(v)$. Then $d_h(r) \leq u$ and $d_h(s) \leq v$ and so

$$\int_{r \leq |f| < s} f\, d\mu \leq 2C(uh^*(u) + vh^*(v)).$$

Now

$$\int_{\substack{E_{u,v}(f)\\ |h| \leq s}} f\, d\mu \leq 2vh^*(v)$$

and

$$\int_{\substack{E_{u,v}(f)\\ |h| > r}} d\mu \leq 2uh^*(u).$$

Combining again with (2.5) gives that

$$|K(u, v; f)| \leq (2C + 10)(uh^*(u) + vh^*(v)).$$

□

Proposition 2.5. *The center $Z_{\mathcal{X}}$ is a linear subspace of $\mathcal{X}$ with the property that $f \in Z_{\mathcal{X}}$ if and only if $\Re f, \Im f \in \mathcal{X}$.*

Proof. The only statement that requires proof is linearity; this follows directly from (2.7). □

We will need a result of Kwapień from 1983:

Theorem 2.6 ([13]). *Let $f \in \mathcal{M}$ be a bounded real function with support of finite measure. If $\int f d\mu = 0$ then there exist $g_1, g_2 \in \mathcal{M}$ with $\operatorname{supp} g_j \subset \operatorname{supp} f$ for $j = 1, 2$, $\|g_j\|_\infty \leq 6\|f\|_\infty$, $g_1 \sim g_2$ and $f = g_1 - g_2$ (μ-a.e).*

Proof. In effect Kwapień proves this for a standard measure space with $g_2 = g_1 \circ \sigma$ where σ is a measure preserving transformation. For the general case we apply Kwapień's theorem to $(\mathbb{R}, \nu)$ where $\nu(B) = \mu(f^{-1}B \cap \operatorname{supp} f)$. We can then write $\phi(x) = x$ in the form $\phi = \psi_1 - \psi_2$ where $\psi_1 \sim \psi_2$ and $\|\psi_j\|_\infty \leq 6\|f\|_\infty$. Let $g_j = \psi_j \circ f$. □

If $\mathcal{X}$ is any symmetric ideal let us introduce a corresponding sequence space $\mathcal{S}\mathcal{X}$. Let $(A_n)_{n\in\mathbb{Z}}$ be a disjoint sequence of measurable sets in Ω such that $\mu A_n = 2^{-n}$. We define $\mathcal{S}\mathcal{X}$ as the space of sequences $\xi = (\xi_n)$ such that $L\xi := \sum_{n\in\mathbb{Z}} 2^n \xi_n \chi_{A_n} \in \mathcal{X}$.

It is clear that this definition is independent of the choice of (A_n); indeed we may take $\Omega = (0, \infty)$ and $A_n = (2^{-n}, 2^{1-n})$.

It is easy to see from the symmetry condition on $\mathcal{X}$ that if $\xi \in \mathcal{S}\mathcal{X}$ then we have $\sum_{n\in\mathbb{Z}} \xi_{n-1}\chi_{A_n} \in \mathcal{X}$. On $\mathcal{S}\mathcal{X}$ we may therefore define the translation operator $\mathcal{T}((\xi_n)_{n\in\mathbb{Z}}) = (\xi_{n-1})_{n\in\mathbb{Z}}$. We say that a linear functional ψ on $\mathcal{S}\mathcal{X}$ is *translation-invariant* if $\psi(\mathcal{T}(\xi)) = \psi(\xi)$ for every $\xi \in \mathcal{S}\mathcal{X}$. It is trivial that $\psi(\xi) = 0$ for every translation-invariant ψ if and only if ξ is in the range $\mathcal{R}$ of $I - \mathcal{T}$ where I is the identity on $\mathcal{S}\mathcal{X}$.

Lemma 2.7. (1) *If φ is a symmetric linear functional on $\mathcal{X}$ then $\varphi \circ L$ is translation-invariant functional on $\mathcal{S}\mathcal{X}$.*

(2) *$\xi \in \mathcal{R}$ if and only if there exists $\eta \in \mathcal{S}\mathcal{X}_+$ such that*

$$\Big|\sum_{k=m+1}^{n} \xi_k\Big| \le \eta_m + \eta_n, \qquad -\infty < m < n < \infty. \tag{2.11}$$

Proof. (1) Let $A_n = B_n \cup C_n$ be a partition of A_n into two measurable sets such that $\mu B_n = \mu A_n = 2^{-(n+1)}$. If $\xi \in \mathcal{S}$ then

$$\begin{aligned}\varphi(L\xi) &= \varphi\Big(\sum_{n\in\mathbb{Z}} 2^n \xi_n \chi_{B_n} + \sum_{n\in\mathbb{Z}} 2^n \xi_n \chi_{C_n}\Big) = 2\varphi\Big(\sum_{n\in\mathbb{Z}} 2^n \xi_n \chi_{B_n}\Big)\\ &= 2\varphi\Big(\sum_{n\in\mathbb{Z}} 2^{n-1}\xi_{n-1}\chi_{A_n}\Big)\\ &= \varphi(L\mathcal{T}\xi).\end{aligned}$$

(2) If $\xi = \eta - \mathcal{T}\eta$ then

$$\Big|\sum_{k=m+1}^{n} \xi_k\Big| = |\eta_n - \eta_m| \le |\eta_m| + |\eta_n|.$$

Conversely assume (2.11). It suffices, by splitting into real and imaginary parts, to treat the case when ξ is real. Let

$$\zeta_n = \begin{cases} \sum_{k=0}^{n} \xi_k & n \ge 0 \\ -\sum_{k=1}^{-n} \xi_{-k} & n < 0. \end{cases}$$

Then by (2.11) we have

$$\sup_n(\zeta_n - \eta_n) \le \inf_m(\zeta_m + \eta_m).$$

Pick $\lambda \in \mathbb{R}$ so that

$$\zeta_n - \eta_n \le \lambda \le \zeta_m + \eta_m, \qquad m, n \in \mathbb{Z}.$$

Then $(\zeta_n - \lambda)_{n\in\mathbb{Z}} \in \mathcal{S}\mathcal{X}$ and $(I - \mathcal{T})(\zeta_n - \lambda)_{n\in\mathbb{Z}} = \xi$. □

We now come to the main theorem of the section. As mentioned in the Introduction, results of this type were known to the first author in the mid 1980's. See Pietsch [14] where sequence space analogues are considered. A non-commutative analogue of this result appears in Dykema, Figiel, Weiss and Wodzicki [6]; see also [11] for the earlier special case of the trace-class, analogous to the case $\mathcal{X} = L_1$.

Theorem 2.8. *Let $\mathcal{X}$ be a symmetric ideal on (Ω, Σ, μ). Then $f \in Z_{\mathcal{X}}$ if and only if $\varphi(f) = 0$ for every symmetric linear functional on $\mathcal{X}$.*

Proof. Let us introduce the space $\mathcal{N}$ of $\mathcal{X}$ defined by $f \in \mathcal{N}$ if and only if $\varphi(f) = 0$ for every symmetric linear functional. We will prove that $\mathcal{N} = Z_{\mathcal{X}}$. First suppose $f \in \mathcal{N}$; we will show that $f \in Z_{\mathcal{X}}$. Indeed f can be written in the form $f = \sum_{j=1}^{m}(g_j - h_j)$ where $g_j, h_j \in \mathcal{X}$ and $g_j \sim h_j$. It thus suffices to show that $g - h \in Z_{\mathcal{X}}$ whenever $g \in \mathcal{X}$ and $g \sim h$. Note that for any u, v we have $K(u, v; g) = K(u, v; h) = -K(u, v; -h)$. Hence by Lemma 2.3 (2.7) we have

$$|K(u, v; g - h)| \leq 8(u\psi^*(u) + v\psi^*(v))$$

where $\psi = |g| + |h|$. We conclude that $\mathcal{N} \subset Z_{\mathcal{X}}$.

For the converse we shall prove the following statements:

$$\mathcal{X} = \mathcal{N} + L(\mathcal{S}\mathcal{X}). \tag{2.12}$$

$$Z_{\mathcal{X}} = \mathcal{N} + L(\mathcal{R}). \tag{2.13}$$

Let us suppose $f \in \mathcal{X}$. Then we may find $g \in \mathcal{X}$ so that $g \sim f$ and for every n, $g\chi_{A_n} \sim f\chi_{E_{2^{-n},2^{1-n}}(f)}$. Let

$$\xi_n = \int_{A_n} g \, d\mu.$$

By Theorem 2.6 applied to $(g - \xi_n)\chi_{A_n}$ for each n we can write $g - L\xi = h_1 - h_2$ where $h_1 \sim h_2$ and $|h_j\chi_{A_n}| \leq 12 f^*(2^{-n})$ for every n and $j = 1, 2$. Hence $h_1, h_2 \in \mathcal{X}$. Now

$$f - L\xi = (f - g) + (h_1 - h_2) \in \mathcal{N}.$$

This proves (2.12). If $f \in Z_{\mathcal{X}}$ we note that

$$\sum_{k=m+1}^{n} \xi_k = \int_{\bigcup_{k=m+1}^{n} A_k} g \, d\mu = \int_{E_{2^{-n},2^{-m}}(f)} f \, d\mu.$$

Hence by Lemma 2.2 (2.5)

$$\Big|\sum_{k=m+1}^{n} \xi_k\Big| \leq |K(2^{-n}, 2^{-m}; f)| + 8(2^{-m} f^*(2^{-m}) + 2^{-n} f^*(2^{-n})).$$

It follows that there exists $\psi \in \mathcal{X}_+$ so that

$$\Big|\sum_{k=m+1}^{n} \xi_k\Big| \leq 2^{-m}\psi^*(2^{-m}) + 2^{-n}\psi^*(2^{-n}), \quad m < n.$$

Now by (2) of Lemma 2.7 this implies $\xi \in \mathcal{R}$. This proves 2.13.

By (1) of Lemma 2.7 we have $L(\mathcal{R}) \subset \mathcal{N}$ and so $Z_{\mathcal{X}} \subset \mathcal{N}$. This completes the proof. □

In fact we have also proved the following statement:

Corollary 2.9. *There is a natural isomorphism between the space of symmetric functionals on $\mathcal{X}$ and the space of translation-invariant functionals on $\mathcal{S}\mathcal{X}$ implemented by $\varphi \to \varphi \circ L$.*

Proof. We first show that $L(\mathcal{R}) = \mathcal{N} \cap L(\mathcal{X})$. Indeed by Lemma 2.7 (1) we know $L(\mathcal{R}) \subset \mathcal{N}$. Now suppose $L\xi = f \in \mathcal{N}$. Then $|\xi_n| \leq f^*(2^{-n})$. If $g = \sum_{n\in\mathbb{Z}} f^*(2^{-n})\chi_{A_n}$ then $g \in \mathcal{X}$ and $|f| \leq g$. Hence by Lemma 2.2, (2.5),

$$\Big|\sum_{k=m+1}^{n} \xi_k\Big| \leq |K(2^{-n}, 2^{-m}; f)| + 8(2^{-m}g^*(2^{-m}) + 2^{-n}g^*(2^{-n})).$$

Now since $f \in Z_{\mathcal{X}}$ this inequality implies that $\xi \in \mathcal{R}$ by Lemma 2.7 (2). Now by (2.12) and the fact that $Z_{\mathcal{X}} = \mathcal{N}$ it follows that L induces an isomorphism between $\mathcal{S}\mathcal{X}/\mathcal{R}$ and $\mathcal{X}/\mathcal{N}$. This implies the corollary. □

We have stated all these results for infinite measure spaces. As noted at the beginning of the section we can quickly derive from this the corresponding results for probability spaces. Indeed it is not difficult to show that for ideals on probability spaces one may consider the sequence space $\mathcal{S}\mathcal{X}(\mathbb{N})$ modelled on the natural numbers, and, in this case, we define translation to be the right shift, $\mathcal{T}(\xi) = (\xi_{n-1})_{n\in\mathbb{N}}$ with the understanding that $\xi_0 = 0$. The result corresponding to Corollary 2.9 then holds.

Let us consider a few examples.

Suppose $\int_0^1 f^*(s)ds < \infty$ for all $f \in \mathcal{X}$. Then we may define $K(0, v; f) = \lim_{u\to 0} K(u, v; f)$. Conversely if $\int_1^\infty f^*(s)ds < \infty$ for all $f \in \mathcal{X}$, then we may define $K(u, \infty; f) = \lim_{v\to\infty} K(u, v; f)$. The latter situation arises if $\mathcal{X}$ is of finite type, or if $\mu(\Omega) = 1$.

Proposition 2.10. (1) *If $\int_0^1 f^*(s)ds < \infty$ for all $f \in \mathcal{X}$ then $f \in Z_{\mathcal{X}}$ and only if*

$$|K(0, u; f)| \leq uh^*(u), \quad 0 < u < \infty, \tag{2.14}$$

for some $h \in \mathcal{X}$.

(2) *If $\mathcal{X}$ has the property that $\int_1^\infty f^*(s)ds < \infty$ for every $f \in \mathcal{X}$ then*

$$|K(u, \infty; f)| \le uh^*(u), \quad 0 < u < \infty, \tag{2.15}$$

for some $h \in \mathcal{X}$.

Proof. In case (1) note that the hypothesis implies $\lim_{u\to 0} uh^*(u) = 0$ for all $h \in \mathcal{X}$; it is then easy to see that (2.9) reduces to (2.14). Similarly if (2) holds then $\lim_{u\to\infty} uh^*(u) = 0$ for all $h \in \mathcal{X}$ and the argument is similar. □

Let us use these remarks to relate the property $Z_{\mathcal{X}} = \mathcal{X}$ to the Boyd indices. We first summarize some properties of the Boyd indices in the next lemma.

Lemma 2.11. *Suppose $p < p_{\mathcal{X}} \le q_{\mathcal{X}} < q$. Suppose $h \in \mathcal{X}(\mathbb{R})$. Then there exists a constant C such that*

$$\Big\| \sum_{k\in\mathbb{Z}} 2^{\min(k/p,k/q)} D_{2^{-k}} h \Big\|_{\mathcal{X}} \le C\|h\|_{\mathcal{X}} \tag{2.16}$$

$$C^{-1}\|h\|_{L_p+L_q} \le \|h\|_{\mathcal{X}} \le C\|h\|_{L_p\cap L_q}. \tag{2.17}$$

It is easy then to see that the following proposition holds.

Proposition 2.12. *Suppose $\mathcal{X}$ is a quasi-Banach ideal.*
(1) *If $q_{\mathcal{X}} < 1$ then $Z_{\mathcal{X}} = \mathcal{X}$.*
(2) *If $p_{\mathcal{X}} > 1$ and $\mu(\Omega) = \infty$ then $Z_{\mathcal{X}} = \mathcal{X}$.*
(3) *If $p_{\mathcal{X}} > 1$ and $\mu(\Omega) = 1$ then $Z_{\mathcal{X}} = \{f \in \mathcal{X} : \int f\, d\mu = 0\}$.*

Proof. Just note that if $f \in \mathcal{X}_+$ then, where applicable,

$$K(0, u; f) = \int_0^1 f^*(su)\, ds \le \sum_{k\le 0} 2^k D_{2^{-k}} f(u)$$

and

$$K(u, \infty; f) = \int_1^\infty f^*(su) ds \le \sum_{k\ge 1} 2^k D_{2^{-k}} f(u).$$

Then (1) and (2) follow from Lemma 2.11 (2.16) and the remarks preceding it. (3) follows from the fact that if $f \in L_1$, is real-valued and $\int f d\mu = 0$ then $K(0, u; f) = -K(u, \infty; f)$ This implies that $f \in Z_{\mathcal{X}}$. □

Thus for quasi-Banach ideals the only case where one gets a non-trivial symmetric linear functional is when $p_{\mathcal{X}} \le 1 \le q_{\mathcal{X}}$. We now discuss the cases of L_1.

If f is a real function in $\mathcal{M}$ we introduce the function $f_d : (-\infty, \infty) \to \mathbb{R}$ defined by the conditions $f_d \sim f$, f_d is decreasing and non-positive on $(-\infty, 0)$ and f_d is decreasing and non-negative on $(0, \infty)$. Then it is clear that

$$K(u, v; f) = \int_{u \leq |t| \leq v} f_d(t)dt.$$

In the case of L_1 the center was introduced as the *symmetric Hardy class* H_1^{sym} in [10]. A real function $f \in H_1^{\text{sym}}(\Omega)$ if and only if

$$\int_0^\infty |M(t)| \frac{dt}{t} < \infty$$

where

$$M(t) = \int_{-t}^{t} f_d(s)ds.$$

It follows that there are many discontinuous symmetric functionals on L_1 in both the probability and infinite measure cases.

An interesting case is to take $\mathcal{X}$ to be the space of all f so that $\lim_{t\to 0} tf^*(t) = \lim_{t\to\infty} tf^*(t)dt = 0$. Then $f \in Z_{\mathcal{X}}$ if and only if

$$\lim_{\substack{u\to 0 \\ v\to\infty}} \int_{|u|<t<|v|} f_d(s)ds = 0$$

or equivalently

$$\lim_{\substack{r\to 0 \\ s\to\infty}} \int_{r<|f|<s} f(s)ds = 0.$$

Symmetric continuous linear functionals on weak L_1 were used in [8] (and are implicit in the work of Cwikel and Fefferman [2] on the dual of weak L_1).

3. Analytic functions

Let $\mathbb{D}$ be the unit disk and $\mathbb{T}$ be the unit circle. We recall that the Nevanlinna class $N(\mathbb{D})$ consists of all functions f which are analytic in $\mathbb{D}$ and satisfy

$$\sup_{r<1} \int_0^{2\pi} \log_+ |f(re^{i\theta})| \frac{d\theta}{2\pi} < \infty.$$

If $f \in N$ then the boundary values $f(e^{i\theta}) = \lim_{r\to 1} f(re^{i\theta})$ exist almost everywhere. The Smirnov class $N^+(\mathbb{D})$ consists of all functions $f \in N$ such that

$$\lim_{r\to 1} \int_0^{2\pi} \log_+ |f(re^{i\theta}| \frac{d\theta}{2\pi} = \int_0^{2\pi} \log_+ |f(e^{i\theta}| \frac{d\theta}{2\pi}.$$

We can regard N^+ as both a space of analytic functions on $\mathbb{D}$ and as a space of measurable functions on $\mathbb{T}$. We shall use the notation $f_r(e^{i\theta}) = f(re^{i\theta})$ so that $f_r : \mathbb{T} \to \mathbb{C}$. We will use μ to denote the normalized measure $d\theta/2\pi$.

We also need analogous spaces in the upper half-plane $\mathbb{U} = \{z : \Im z > 0\}$. We define $N^+(\mathbb{U})$ to be the class of all functions f analytic in $\mathbb{U}$ so that $f \circ \varphi \in N^+$ where $\varphi : \mathbb{D} \to \mathbb{U}$ is given by $\varphi(z) = i(1+z)/(1-z)$. If $f \in N^+(\mathbb{U})$ then

$$\lim_{y\to 0} f(x+iy) = f(x)$$

exists a.e. and

$$\log|f(x+iy)| \le \frac{y}{\pi}\int_{-\infty}^{\infty} \frac{\log|f(x)|}{(t-x)^2+y^2}dt. \tag{3.1}$$

We will write $f_y(x) = f(x+iy)$ and in this case μ is Lebesgue measure. As before we can consider $N^+(\mathbb{U})$ as a space of analytic functions in the upper half-plane or as a space of measurable functions on the real line.

In the next technical theorem we prove a result on the distribution of boundary values of functions in $N^+(\mathbb{D})$ and $N^+(\mathbb{U})$.

Theorem 3.1. *There exists a constant C so that:*

(1) *If $f \in N^+(\mathbb{D})$ and $f(0) = 0$ then, whenever $0 < r < s < \infty$,*

$$\begin{aligned}\left|\int_{r\le|f|<s} f(e^{i\theta})\frac{d\theta}{2\pi}\right| &\le C\Bigg(r\mu(|f|>r) + s\mu(|f|>s) \\ &\quad + \int_0^{2\pi} r\log_+\frac{|f(e^{i\theta})|}{r} + s\log_+\frac{|f(e^{i\theta})|}{s}\frac{d\theta}{2\pi}\Bigg)\end{aligned} \tag{3.2}$$

(2) *If $f \in N^+(\mathbb{U})$ then*

$$\begin{aligned}\left|\int_{r\le|f|<s} f(x)dx\right| &\le C\Bigg(r\mu(|f|>r) + s\mu(|f|>s) \\ &\quad + \int_{-\infty}^{\infty} r\log_+\frac{|f(x)|}{r} + s\log_+\frac{|f(x)|}{s}dx\Bigg)\end{aligned} \tag{3.3}$$

Proof. We begin by fixing a smooth bump function $b : \mathbb{R} \to \mathbb{R}$ such that $\operatorname{supp} b \subset (0, 1/2)$, $b \ge 0$, $\int b(x)dx = 1$. Let $\beta(t) = 2|b(t)| + |b'(t)|$.

Note first that the estimates are trivial when $s \le 2r$.

Now suppose $0 < r < s < \infty$, with $s \ge 2r$. We define

$$\varphi(\tau) = \varphi_{r,s}(\tau) = \int_{-\infty}^{\tau} b(t-\log r) - b(t-\log s)dt.$$

Notice that the two terms in the integrand are never simultaneously positive (since $\log 2 > \frac{1}{2}$), and φ is a bump function which satisfies $\varphi(\tau) = 0$ if $\tau < \log r$ or $\tau > \frac{1}{2} + \log s$ while $\varphi(\tau) = 1$ if $\log r + \frac{1}{2} \le \tau \le \log s$. Then let $\psi = \psi_{r,s}$ be defined to be the function such that $\psi(\tau) = 0$ if $\tau < \log r$ and

$$\psi''(\tau) = e^{\tau}(2|\varphi''(\tau)| + |\varphi'(\tau)|).$$

In fact, this implies that

$$\psi''(\tau) = e^{\tau}(\beta(\tau - \log r) + \beta(\tau - \log s))$$

and then

$$\psi(\tau) = \psi_{r,s}(\tau) = \int_{-\infty}^{\tau} (\tau - t)e^{t}(\beta(t - \log r) + \beta(t - \log s))\, dt,$$

and

$$\psi'(\tau) = \int_{-\infty}^{\tau} e^{t}(\beta(t - \log r) + \beta(t - \log s))dt.$$

Thus, if we set

$$C_0 = \int_{-\infty}^{\infty} e^{t}\beta(t)dt$$

then

$$\psi'(\tau) \le C_0(r\chi_{(\tau > \log r)} + s\chi_{(\tau > \log s)})$$

and so

$$\psi_{r,s}(\tau) \le C_0(r(\tau - \log r)_+ + s(\tau - \log s)_+). \tag{3.4}$$

Now we use the argument of Lemma 2.6 of [12]. If we define

$$h(z) = h_{r,s}(z) = \psi(\log|z|) - x\varphi(\log|z|), \quad z = x + iy \ne 0,$$

and $h(0) = 0$ then h is a C^2-subharmonic function which vanishes on a neighborhood of 0. We note the estimates (from (3.4))

$$\psi(\log|z|) \le C_0\left(r\log_+\frac{|z|}{r} + s\log_+\frac{|z|}{r}\right) \tag{3.5}$$

and

$$0 \le h(z) \le C_0\left(r\log_+\frac{|z|}{r} + s\log_+\frac{|z|}{r}\right), \quad |z| \ge 2s. \tag{3.6}$$

Note of course that C_0 is independent of r, s.

Let us first treat the case when $f \in N^+(\mathbb{D})$ and $f(0) = 0$. Then $h \circ f$ is subharmonic on $\mathbb{D}$. Hence

$$0 \le \int_0^{2\pi} h \circ f_r \frac{d\theta}{2\pi}$$

for $0 \le r < 1$. Now $h_+(z) := \max(h(z), 0) \le M(\log_+ |z| + 1)$ for a suitable constant M and $h_+ - h$ is bounded. From the definition of the Smirnov class we have:

$$\lim_{r\to 1} \int_{|f_r|>2|f|} \log_+ |f_r| \frac{d\theta}{2\pi} = 0.$$

Hence

$$\lim_{r\to 1} \int_{|f_r|>2|f|} h_+ \circ f_r \frac{d\theta}{2\pi} = 0.$$

Now by the Dominated Convergence Theorem we have

$$\lim_{r\to 1} \int_{|f_r|\le 2|f|} h_+ \circ f_r \frac{d\theta}{2\pi} = \int_0^{2\pi} h_+ \circ f \frac{d\theta}{2\pi}.$$

Thus

$$\lim_{r\to 1} \int_0^{2\pi} h_+ \circ f_r \frac{d\theta}{2\pi} = \int_0^{2\pi} h_+ \circ f \frac{d\theta}{2\pi}.$$

By the Bounded Convergence Theorem,

$$\lim_{r\to 1} \int_0^{2\pi} (h - h_+) \circ f_r \frac{d\theta}{2\pi} = \int_0^{2\pi} (h - h_+) \circ f \frac{d\theta}{2\pi}.$$

We thus conclude that

$$\int_0^{2\pi} h \circ f \frac{d\theta}{2\pi} \ge 0$$

and thus

$$\Re \int_0^{2\pi} f\varphi(\log|f|) \frac{d\theta}{2\pi} \le \int_0^{2\pi} \psi(\log|f|) \frac{d\theta}{2\pi}.$$

Applying this inequality to αf for every α with $|\alpha| = 1$ gives

$$\left| \int_0^{2\pi} f\varphi(\log|f|) \frac{d\theta}{2\pi} \right| \le \int_0^{2\pi} \psi(\log|f|) \frac{d\theta}{2\pi}. \tag{3.7}$$

Now note that

$$\begin{aligned} \left| \int_0^{2\pi} f(\varphi(\log|f|) - \chi_{(r\le|f|<s)}) \frac{d\theta}{2\pi} \right| &\le \int_0^{2\pi} |f|(\chi_{(r<f<2r)} + \chi_{(s<|f|<2s)}) \frac{d\theta}{2\pi} \\ &\le 2r\mu(|f| > r) + 2s\mu(|f| > s). \end{aligned}$$

Now combining with (3.4) and (3.7) gives (3.2).

The proof of (3.3) is somewhat similar. We only sketch the details. It is clear that we may assume that

$$\int_{-\infty}^{\infty} \log_+ |f| dx < \infty.$$

It then follows from (3.1) that

$$\int_{-\infty}^{\infty} \log_+ |f_y| dx \le \int_{-\infty}^{\infty} \log_+ |f| dx \tag{3.8}$$

whenever $y > 0$ and also that $\lim_{y\to\infty} \sup |f(x+iy)| = 0$. As above we note that $h \circ f$ is subharmonic on $\mathbb{U}$; in this case $h \circ f$ vanishes for y large enough. It follows using (3.6) that $\int_{-\infty}^{\infty} h \circ f_y dx$ is a convex function of $y > 0$ and so that

$$\lim_{y\to 0} \int_{-\infty}^{\infty} h \circ f_y dx \ge 0.$$

In this case we use (3.8) in place of the Smirnov condition to deduce

$$\int_{-\infty}^{\infty} h \circ f \, dx \ge 0.$$

The remaining details are then similar. □

We now define (cf. [12]) a symmetric ideal $\mathcal{X}$ to be *geometrically stable* if given $f \in \mathcal{X}$ there exists $h \in \mathcal{X}$ with

$$\exp\left(\frac{1}{t}\int_0^t \log f^*(s)\, ds\right) \le h^*(t), \quad 0 < t < \infty. \tag{3.9}$$

Note of course that we require $\log_+ |f| \in L_1$ for all $f \in \mathcal{X}$.

Proposition 3.2. *Every symmetric quasi-Banach ideal is geometrically stable.*

Proof. For $0 < p \le 1$ we note that for an appropriate constant C_1,

$$\begin{aligned}\exp\left(\frac{1}{t}\int_0^t \log f^*(s)\, ds\right) &\le \left(\frac{1}{t}\int_0^t f^*(s)^p\, ds\right)^{\frac{1}{p}} \\ &\le \left(\sum_{n=1}^{\infty} 2^{-n} D_{2^n} f^*(t)^p\right)^{\frac{1}{p}} \\ &\le C_1 \sum_{n=1}^{\infty} 2^{-\frac{n}{2}-\frac{n}{2p}} D_{2^n} f^*(t).\end{aligned}$$

Now if $\frac{1}{2} + \frac{1}{2p} > \frac{1}{p_{\mathcal{X}}}$ the series

$$C_1 \sum_{n=1}^{\infty} 2^{-\frac{n}{2}-\frac{n}{2p}} D_{2^n} f^*$$

converges in $\mathcal{X}(0, 1)$ or $\mathcal{X}(0, \infty)$ to some $h = h^*$ satisfying (3.9). □

If $\mathcal{X}$ is a symmetric ideal on $\mathbb{T}$ let us define $H_{\mathcal{X}}(\mathbb{T})$ to be the space of all $f \in \mathcal{X}(\mathbb{T})$ so that $f \in N^+(\mathbb{D})$. Similarly $H_{\mathcal{X}}(\mathbb{R})$ consists of all $f \in \mathcal{X}(\mathbb{R})$ so that $f \in N^+(\mathbb{U})$.

Theorem 3.3. *Suppose $\mathfrak{X}$ is a geometrically stable symmetric ideal on $\mathbb{T}$ (in particular this applies when $\mathfrak{X}$ is a quasi-Banach ideal). Suppose $f \in H_{\mathfrak{X}}$ with $f(0) = 0$. Then $f \in Z_{\mathfrak{X}}(\mathbb{T})$ and so for every symmetric linear functional φ on $\mathfrak{X}$ we have $\varphi(f) = 0$.*

Theorem 3.4. *Suppose $\mathfrak{X}$ is a geometrically stable symmetric ideal on $\mathbb{R}$ (in particular this applies when $\mathfrak{X}$ is a quasi-Banach ideal). Suppose $f \in H_{\mathfrak{X}}(\mathbb{R})$. Then $f \in Z_{\mathfrak{X}}$ and so for every symmetric linear functional φ on $\mathfrak{X}$ we have $\varphi(f) = 0$.*

Proof. The proofs of both theorems are essentially the same. We prove only Theorem 3.3. Assume first f satisfies the condition that $\mu(|f| = r) = 0$ for all $r > 0$. Then if $0 < u < v < \infty$ we have $E_{u,v}(|f|) = \{f^*(u) \le |f| < f^*(v)\}$ and so

$$\left| K(u, v; f) - \int_{f^*(u) \le |f| < f^*(v)} f \frac{d\theta}{2\pi} \right| \le 2(uf^*(u) + vf^*(v))$$

by Lemma 2.1, (2.3). Thus using Theorem 3.1 (3.2),

$$|K(u, v; f)|$$
$$\le (C + 2)(uf^*(u) + vf^*(v)) + \int_0^{2\pi} f^*(u) \log_+ \frac{|f|}{f^*(u)} + f^*(v) \log_+ \frac{|f|}{f^*(v)} \frac{d\theta}{2\pi}.$$

Now by (3.9) we can find h with

$$\frac{1}{t} \int_0^t \log f^*(s)\, ds \le \log h^*(t).$$

Now

$$\begin{aligned}\int_0^{2\pi} f^*(u) \log_+ \frac{|f|}{f^*(u)} \frac{d\theta}{2\pi} &= f^*(u) \int_0^u (\log f^*(t) - \log f^*(u)) dt \\ &\le uf^*(u) \log \frac{h^*(u)}{f^*(u)} \\ &\le uh^*(u).\end{aligned}$$

This together with a similar calculation for v gives the estimate

$$K(u, v; f) \le ug^*(u) + vg^*(v)$$

where $g^* = (C + 2) f^* + h^*$.

Now for the general case we may clearly define some $g \in N^+(\mathbb{D})$ so that $g \in \mathfrak{X}(\mathbb{T})$ and $\mu(|g| = r) = 0$ for every choice of r. It then follows that for almost every choice of $(e^{i\theta}, e^{i\phi}) \in \mathbb{T}^2$ we have $|g(e^{i\theta})| \ne |g(e^{i\phi})|$. Let A be the set of $(t, e^{i\theta}, e^{i\phi}) \in [-1, 1] \times \mathbb{T}^2$ so that $|f(e^{i\theta}) + tg(e^{i\theta})| = |f(e^{i\phi}) + tg(e^{i\phi})|$. By expanding as a polynomial in t it follows that for almost every $(e^{i\theta}, e^{i\phi})$ the set of t such that $(t, e^{i\theta}, e^{i\phi}) \in A$ has Lebesgue measure zero. Applying Fubini's theorem A has product measure zero and so we may find $t \ne 0$ so that the set of $(e^{i\theta}, e^{i\phi})$ so that either $|f(e^{i\theta})+tg(e^{i\theta})| = |f(e^{i\phi})+tg(e^{i\phi})|$ or $|f(e^{i\theta})-tg(e^{i\theta})| = |f(e^{i\phi})-tg(e^{i\phi})|$ has measure zero. From this it follows that for every $r > 0$ we have $\mu(|f+tg| = r) = \mu(|f-tg| = r) = 0$. Now applying the first part $f \pm tg \in Z_{\mathcal{X}}$ and so $f \in Z_{\mathcal{X}}$. $\square$

Theorem 3.3 has some amusing applications of which we give two.

Corollary 3.5. *Suppose $f \in N^+(\mathbb{D})$ and $\lim_{r\to\infty} r\mu(|f| > r) = 0$. Then*

$$f(0) = \lim_{r\to\infty} \int_{|f|<r} f(e^{i\theta}) \frac{d\theta}{2\pi}.$$

Proof. If $f(0) = 0$ this follows directly from Theorem 3.3 and the remarks at the end of Section 2. If $f(0) = \alpha \neq 0$ then note that

$$\lim_{r\to\infty} \left| \int_{\substack{|f|<r \\ |f-\alpha|\le r}} f \frac{d\theta}{2\pi} \right| + \left| \int_{\substack{|f|\ge r \\ |f-\alpha|<r}} f \frac{d\theta}{2\pi} \right| = 0$$

follows from $\lim_{r\to\infty} r\mu(|f| > r) = 0$. $\square$

Corollary 3.6. *Suppose $\mathcal{X}$ is a geometrically stable symmetric ideal on $\mathbb{T}$ which admits a regular symmetric linear functional φ. Then if $f \in H_{\mathcal{X}}(\mathbb{T})$ we have $\varphi(f) = f(0)$.*

We omit the trivial proof.

4. Conjugate functions

Ceretelli [3] and later (independently) Davis [4] proved that a real function $f \in L_1(\mathbb{T})$ is in $H_{1,0}^{\mathrm{sym}} = Z_{L_1}$ if and only if $f \sim \Re g$ for some $g \in H_1$ with $g(0) = 0$. See also [10] for an alternate proof and [9] for a vector-valued generalization. In [7] there is a general discussion of problems of this type. Let us state our extensions of this result to arbitrary quasi-Banach ideals.

Theorem 4.1. *Let $\mathcal{X}$ be a symmetric quasi-Banach ideal on $\mathbb{T}$ with $\frac{1}{2} < p_{\mathcal{X}} \le q_{\mathcal{X}} < \infty$. If $f \in \mathcal{X}_{\mathbb{R}}$ then in order that there exist $g \in H_{\mathcal{X}}(\mathbb{T})$ with $g(0) = 0$, and $\Re g \sim f$ it is necessary and sufficient that $f \in Z_{\mathcal{X}}$.*

Theorem 4.2. *Let $\mathcal{X}$ be a symmetric quasi-Banach ideal on $\mathbb{R}$ with $q_{\mathcal{X}} < \infty$. If $f \in \mathcal{X}_{\mathbb{R}}$ then in order that there exist $g \in H_{\mathcal{X}}(\mathbb{R})$, with $\Re g \sim f$ it is necessary and sufficient that $f \in Z_{\mathcal{X}}$.*

Before proving this result we note that Davis [4] establishes Theorem 4.1 for the case $\mathcal{X} = L_p$ where $\frac{1}{2} < p < 1$. In this case $Z_{L_p} = L_p$. In fact Aleksandrov proves a stronger theorem [1] that the map $f \to \Re f$ maps $H_{\mathcal{X}}(\mathbb{T})$ *onto* $\Re\mathcal{X}(\mathbb{T})$ when $q_{\mathcal{X}} < 1$. If $1 < p_{\mathcal{X}} < q_{\mathcal{X}} < \infty$ then the Hilbert transform is bounded on $\mathcal{X}$ and the theorems follow easily (cf. Proposition 2.12). Thus the only interesting cases are when $p_{\mathcal{X}} \leq 1 \leq q_{\mathcal{X}}$.

Let $L_{\log}$ denote the (non-locally convex) Orlicz space of all functions $f : \mathbb{R} \to \mathbb{C}$ such that

$$\int \frac{\log_+ |f(x)|}{1+x^2} dx < \infty.$$

This is an F-space when it is equipped with the F-norm

$$f \to \int \frac{\log(1+|f(x)|)}{1+x^2} dx.$$

It is easy to see from (2.17) that $\mathcal{X}$ is continuously embedded into $L_{\log}$. Now, when identified with its boundary values, $N^+(\mathbb{U})$ is a closed subspace of $L_{\log}$ and hence $H_{\mathcal{X}}$ is a closed subspace of $\mathcal{X}$.

Lemma 4.3. *Suppose E is a bounded measurable subset of $\mathbb{R}$ and $N \in \mathbb{N}$. If $f : E \to \mathbb{R}$ is measurable there exists $g : E \to \mathbb{R}$ with $g \sim f$ and*

$$\int_E t^k g(t)\, dt = \frac{1}{\mu(E)} \left(\int_E t^k dt \right) \left(\int_E g(t)\, dt \right), \quad 0 \leq k \leq N.$$

Proof. By repeated applications of Lyapunov's theorem taking $A_1 = E$ we can find Borel sets A_n with $A_n = A_{2n} \cup A_{2n+1}$, $A_{2n} \cap A_{2n+1} = \emptyset$, $\mu(A_{2n}) = \mu(A_{2n+1}) = \frac{1}{2}\mu(A_n)$ and

$$\int_{A_{2n}} t^k\, dt = \int_{A_{2n+1}} t^k\, dt = \frac{1}{2} \int_{A_n} t^k\, dt, \quad 0 \leq k \leq N.$$

The sets $(A_n)_{n=0}^{\infty}$ generate a non-atomic sub$-\sigma$-algebra Σ_0 of measurable subsets of E so that $B \in \Sigma_0$ implies

$$\int_B t^k\, dt = \mu(B) \int_E t^k\, dt, \quad 0 \leq k \leq N.$$

Take $g \sim f$ to be Σ_0-measurable. □

We now turn to the proofs of the theorems. We first consider Theorem 4.2, and then indicate the modifications which establish Theorem 4.1.

Proof. One direction follows immediately from Theorems 3.3 and 3.4. In Theorem 4.1 if $g \in H_{\mathcal{X}}$ and $g(0) = 0$ then Theorem 3.3 implies that $g \in Z_{\mathcal{X}}(\mathbb{T})$. Thus $\Re g \in Z_{\mathcal{X}}$ by Proposition 2.5. A similar argument applies in Theorem 4.2.

We prove first Theorem 4.2. In the proof we use C to denote a constant which depends only on $\mathcal{X}$ but which can vary from line to line. Let us pick $N \in \mathbb{N}$ so that $N + 2 > \frac{1}{p_{\mathcal{X}}}$.

Define f_d as before (see after Proposition 2.12). Now let $J_n = (2^n, 2^{n+1}] \cup [-2^{-n-1}, -2^n)$. Let E_0 be a measurable subset of J_0 so that $\mu(E) = \frac{1}{2}\mu(J_0) = 1$ and

$$\int_E t^k dt = \frac{1}{2}\int_{J_0} t^k\, dt, \quad 0 \le k \le N.$$

Let $F_0 = J_0 \setminus E_0$ and then let $E_n = 2^n E_0,\ F_n = 2^n F_0$ for $n \in \mathbb{Z}$.

By using Lemma 4.3 on each set $E_n \cup F_{n+1}$ (which has measure $(3/2)\mu J_n$) we can find a function f_c so that $f_c\chi_{E_n\cup F_{n+1}} \sim f_d\chi_{J_n}$ for $n \in \mathbb{Z}$ and

$$\int_{E_n\cup F_{n+1}} t^k f_c(t)dt = 2^{-n-1}\int_{E_n\cup F_n} t^k\, dt \int_{J_n} f_c(t)\, dt, \quad 1 \le k \le N,\ n \in \mathbb{Z}.$$

Let us define $K(u, v; f) = -K(v, u; f)$ if $u > v$. Then there exists $h \in \mathcal{X}$ so that

$$|K(u, v; f)| \le uh^*(u) + vh^*(v), \quad u, v > 0.$$

Hence

$$K(u, 1; f) - uh^*(u) \le K(v, 1; f) + vh^*(v), \quad u, v > 0.$$

Now (as in Lemma 2.7) this implies the existence of $\lambda \in \mathbb{R}$ so that

$$|K(u, 1; f) - \lambda| \le uh^*(u), \quad 0 < u < \infty.$$

We may assume h^* is constant on each interval $(2^n, 2^{n+1})$ and that $|f_c(t)| \le h^*(t)$. For $n \in \mathbb{Z}$ we define ϕ_n by

$$\phi_n(t) = \begin{cases} f_c(t) - 2^{-n-1}(K(2^n, 1; f) - \lambda) & t \in E_n \\ -2^{-(n-1)}(K(2^n, 1; f) - \lambda) & t \in F_n \\ 2^{-n-2}(K(2^{n+1}, 1; f) - \lambda) & t \in E_{n+1} \\ f_c(t) + 2^{-n-2}(K(2^{n+1}, 1; f) - \lambda) & t \in F_{n+1} \\ 0 & t \notin J_n \cup J_{n+1}. \end{cases}$$

Then

$$\int \phi_n(t)dt = \int_{E_n\cup F_{n+1}} f_c(t)\, dt + K(2^{n+1}, 1; f) - K(2^n, 1; f) = 0.$$

We thus have

$$\int t^k\phi_n(t)\, dt = 0, \quad 0 \le k \le N. \tag{4.1}$$

Note also that we have an estimate

$$|\phi_n(t)| \le Ch^*(2^n)$$

for a suitable constant C.

Define

$$\psi_n(z) := \frac{i}{\pi}\int_{-\infty}^{\infty}\frac{\phi_n(t)}{z-t}\,dt$$

for $z \in \mathbb{U}$. Then $\psi_n \in H^\infty$ and has boundary values for $\Im z = 0$ such that $\Re\psi_n(x) = \phi_n(x)$ a.e. Note that

$$\Im\psi_n(x) = \frac{1}{\pi}\text{p.v.}\int_{-\infty}^{\infty}\frac{\phi_n(t)}{x-t}dx.$$

We complete the argument by showing that $\sum_{n\in\mathbb{Z}}\psi_n(x)$ converges a.e. to a function g in $H_{\mathcal{X}}$. To show this we show that $\sum_{n\in\mathbb{Z}}|\psi_n(x)|$ converges a.e. to a function in $\mathcal{X}$. It follows then that the series $\sum_{n\in\mathbb{Z}}\psi_n$ converges also in the topology of $L_p + L_q$ provided $p < p_{\mathcal{X}}$ and $q > q_{\mathcal{X}}$ and hence $g \in N^+ \cap \mathcal{X}$. Once this is done it is clear that $\Re g(x) = f_c(x)$ a.e. and so $\Re g \sim f$.

We first observe that if $|x| \le 2^{n-1}$ then we have an easy estimate that

$$|\psi_n(x)| \le Ch^*(2^n).$$

Next suppose $|x| \ge 2^{n+3}$. Then

$$\frac{1}{(x-t)} = \frac{1}{x} + \cdots + \frac{t^N}{x^{N+1}} + \frac{t^{N+1}}{x^{N+1}(x-t)},$$

and so by (4.1),

$$|\psi_n(x)| \le C|x|^{-(N+2)}2^{(N+2)n}h^*(2^n).$$

Thus if $x \in J_n$, and we have

$$|\psi_{n-k}(x)| \le C2^{-k(N+2)}h^*(2^{-k}|x|)$$

for $k \ge 3$ and

$$|\psi_{n+k}(x)| \le Ch^*(2^k|x|)$$

for $k \ge 2$.

Hence if $x \in J_n$,

$$\sum_{k\ge 3}|\psi_{n-k}(x)| \le C\sum_{k\ge 2}2^{-k(N+2)}h^*(2^{-k}|x|) \tag{4.2}$$

and

$$\sum_{k\ge 2}|\psi_{n+k}(x)| \le C\sum_{k=2}^{\infty}h^*(2^k|x|). \tag{4.3}$$

Now by Lemma 2.11 we have that

$$\sum_{k\in\mathbb{Z}}2^{\min(k(N+2),0)}h^*(2^k|x|) < \infty, \quad x \ne 0$$

and

$$\Big\|\sum_{k\in\mathbb{Z}} 2^{\min(k(N+2),0)} h^*(2^k|x|)\Big\|_{\mathcal{X}} \le C\|h\|_{\mathcal{X}}.$$

Consider the cases $-2 \le k \le 1$. Then

$$\left(\int_{J_n} |\psi_{n+k}(x)|^q dx\right)^{\frac{1}{q}} \le C2^{\frac{n+k}{q}} h^*(2^{n+k}).$$

For $r \ge 0$, let A_{nr} be the subset of J_n of all x such that $\sum_{-2}^{1} |\psi_{n+k}(x)| \ge 2^r h^*(2^{n-2})$. Then

$$\mu(A_{nr}) \le C2^{n-rq}.$$

Hence, if $q_{\mathcal{X}} < q' < q$,

$$\Big\|\sum_{n\in\mathbb{Z}}\sum_{k=-2}^{1} h^*(2^{n-2})\chi_{A_{nr}}\Big\|_{\mathcal{X}} \le C2^{-rq/q'}\|h\|_{\mathcal{X}}.$$

This quickly yields the estimate:

$$\Big\|\sum_{n\in\mathbb{Z}}\sum_{k=-2}^{1} |\psi_{n+k}|\chi_{J_n}\Big\|_{\mathcal{X}} \le C\|h\|_{\mathcal{X}}.$$

Combining gives us that $\sum_{n\in\mathbb{Z}} |\psi_n|$ converges a.e. to a function in $\mathcal{X}(\mathbb{R})$. This completes the proof of Theorem 4.2. □

Proof of Theorem 4.1. Only small modifications are necessary for Theorem 4.1. We note that f_d is now supported on a set $[a, 1-a]$ where $0 \le a \le 1$ and h^* is supported on $[0, 1]$. In this case with the additional assumption that $p_{\mathcal{X}} > \frac{1}{2}$ we can take $N = 0$. This allows us to simplify the construction by taking

$$\phi_n(t) = \begin{cases} f_d(t) - 2^{-n-1}K(2^n, 1; f) & t \in J_n \\ 2^{-n-2}K(2^{n+1}, 1; f) & t \in J_{n+1} \\ 0 & t \notin J_n \cup J_{n+1}. \end{cases}$$

The calculations are then very similar. Construct g as above and then define $g_0(0) = 0$ and

$$g_0(re^{i\theta}) = \sum_{n\in\mathbb{Z}} g(\frac{1}{2\pi}(\theta - i\log r) + n).$$

One only really needs to show that $g_0 \in \mathcal{X}(\mathbb{T})$ or equivalently that $\sum_{n\in\mathbb{Z}} g(x+n) \in \mathcal{X}(0, 1]$. Clearly $g\chi_{(n,n+1)} \in \mathcal{X}(n, n+1)$ for all $n \in \mathbb{Z}$. If $|n| \ge 3$ then one gets an estimate

$$\|g\chi_{(n,n+1]}\|_\infty \le C|n|^{-2}$$

by using (4.2) above. It is then easy to see that g_0 solves the problem. □

Let us remark that it should be possible to prove Theorem 4.1 without the restriction $p_X > \frac{1}{2}$ as in the case of Theorem 4.2.

References

[1] A. B. Aleksandrov, Essays on nonlocally convex Hardy classes, in: Complex analysis and spectral theory (Leningrad, 1979/1980), Lecture Notes in Math. 864, Springer-Verlag, Berlin–New York. 1981, 1–89.

[2] M. Cwikel and C. Fefferman, Maximal seminorms on weak L^1, Studia Math. 69 (1980), 149–154.

[3] O. D. Ceretelli, A metric characterization of the set of functions whose conjugate functions are integrable, Bull. Acad. Sci. Georgian S.S.R. 81 (1976), 281–283 (in Russian).

[4] B. Davis, Hardy spaces and rearrangements, Trans. Amer. Math. Soc. 261 (1980), 211–233.

[5] J. Dixmier, Existence de traces non normales, C. R. Acad. Sci. Paris Ser. A-B 262 (1966), A1107–A1108.

[6] K. J. Dykema, T. Figiel, G. Weiss and M. Wodzicki, The commutator structure of operator ideals, preprints 1997, 2001.

[7] A. Gulisashvili, Rearrangement-invariant spaces of functions on LCA groups. J. Funct. Anal. 156 (1998), 384–410.

[8] N. J. Kalton, Banach spaces embedding into L_0, Israel J. Math. 52 (1985), 305–319.

[9] N. J. Kalton, On rearrangements of vector-valued H_1–functions, Ark. Mat. 26 (1988,) 221–229.

[10] N. J. Kalton, Nonlinear commutators in interpolation theory, Mem. Amer. Math. Soc. 85, 1988.

[11] N. J. Kalton, Trace-class operators and commutators, J. Funct. Anal. 86 (1989), 41–74.

[12] N. J. Kalton, Spectral characterization of sums of commutators, I, J. Reine und Angew. Math. 504 (1998), 115–125.

[13] S. Kwapień, Linear functionals invariant under measure preserving transformations. Math. Nachr. 119 (1984), 175–179.

[14] A. Pietsch, Traces and shift invariant functionals, Math. Nachr. 145 (1990), 7–43.

Tadeusz Figiel, Institute of Mathematics of the Polish Academy of Sciences, 81-825 Sopot, Poland

E-mail: `t.figiel@impan.gda.pl`

Nigel Kalton, Department of Mathematics, University of Missouri, Columbia, MS 65211, U.S.A.

E-mail: `nigel@math.missouri.edu`

Axiomatic Sobolev spaces on metric spaces

V. Gol'dshtein and M. Troyanov

Dedicated to Jaak Peetre on his 65th birthday

Abstract. We present an axiomatic approach for the theory of Sobolev spaces on metric spaces and we show that this axiomatic construction covers the main known examples. We also introduce in this general setting the notion of variational p-capacity and study its relation with the geometric properties of the metric space, its connections with pointwise convergence and embedding theorems. The concept of p-parabolic and p-hyperbolic spaces is also introduced and the existence of an extremal function for the different versions of the variational p-capacity under quite general hypothesis is discussed.

2000 Mathematics Subject Classification: Primary 46E35; Secondary 32U20, 31C15

Introduction

Recent years have seen an important activity devoted to geometric analysis on metric spaces. Motivations came from such fields as analysis on singular Riemannian manifolds and rectifiable sets; Carnot–Caratheodory geometries and Hörmander system of vector fields; weighted Sobolev spaces and applications to PDE; graphs and discrete groups, combinatorial Laplacian; analysis on fractal sets; Gromov hyperbolic spaces and their ideal boundaries, etc. We refer to the papers [1], [2], [5], [7] and the references therein for more information in the subject.

A striking fact is that a number of analytical problems transit from one theory to another. For instance the notion of p-capacity of a subset F of a metric space X is more or less defined as the infimum of the Dirichlet p-energy $\mathcal{E}_p(u) = \int_X |\nabla u|^p$ among all functions $u : X \to \mathbb{R}$ which vanish at the boundary (in some sense) of X and such that $u \geq 1$ on F. A precise definition can be given in each special case.

We may then consider a number of classical problems such as

1. Prove the existence and uniqueness of an extremal function for the p-capacity $\operatorname{Cap}_p(F, X)$.

2. Prove that if $\operatorname{Cap}_p(B, X) = 0$ for some ball $B \subset X$, then $\operatorname{Cap}_p(F, X) = 0$ for every bounded subset $F \subset X$.

3. Prove that Cap_p is a Choquet capacity.

4. Prove continuity quasi-everywhere (in the sense of p-capacity) for "weakly" differentiable functions.

5. Give necessary and sufficient geometric or capacitary conditions implying the embedding $W^{1,p}(X) \subset C(X)$.

A natural program is now to obtain precise theorems and proofs which hold in *any* reasonable theory.

An important step was made in 1993 when Piotr Hajłasz introduced a notion of Sobolev space $W^{1,p}(X)$ which makes sense on all metric measure spaces, however Hajłasz's definition does not always coincide with more classic Sobolev spaces. In fact it is important to realize that given a measure metric space (X, d, μ), there is in general not one but several natural notions for the Sobolev space $W^{1,p}(X)$ which may be essentially different. For example, in the plane domain $\Omega := \{z = re^{i\theta} : 0 < r < 1, 0 < \theta < 2\pi\} \subset \mathbb{C}$ (this is the disk minus a radius), we have two natural notions of Sobolev spaces: the first one is the classical Sobolev space $W^{1,p}(\Omega)$ and the second one is the trace space of $W^{1,p}(\mathbb{R}^2)$ on Ω. The corresponding notions of capacities are also different.

To fulfill the proposed program, we develop an axiomatic construction of the Sobolev spaces $W^{1,p}(X)$ associated to a metric measure space X. This construction turns out to be wide enough to cover most examples and yet rich enough so that we can prove significant theorems.

The present paper is but a sketch of the axiomatic theory of Sobolev spaces, more details and proofs are exposed in [3, 4].

1. Axiomatic Sobolev spaces

In the metric measure space (X, d, μ), we select a Boolean ring $\mathcal{K}$ of bounded Borel subsets which generates the Borel σ-algebra (typical examples are the ring of all bounded Borel subsets of X and the ring of all relatively compact Borel subsets if X is locally compact and separable). The ring $\mathcal{K}$ and the measure μ will be assumed to satisfy the following two conditions:

1) every ball $B \subset X$ is measurable and $0 < \mu(B) < \infty$ if B has positive radius;

2) for every $A \in \mathcal{K}$ there exists a finite sequence of open balls $\{B_1, B_2, \dots .B_m\} \subset \mathcal{K}$ such that $A \subset \bigcup_{i=1}^m B_i$ and $\mu(B_i \cap B_{i+1}) > 0$ for $1 \leq i < m$.

We define $L^p_{\mathrm{loc}}(X)$ to be the space of measurable functions which belong to $L^p(K)$ for every subset $K \in \mathcal{K}$. We then assume that to each function $u \in L^p_{\mathrm{loc}}(X)$ is associated (by some way) a set $D[u]$ of functions called the *pseudo-gradients* of u;

intuitively a pseudo-gradient $g \in D[u]$ is a function which exerts some control of the variation of u (for instance in the classical case $X = \mathbb{R}^n$, we have $D[u] := \{g \in L^1_{\mathrm{loc}}(\mathbb{R}^n) : g \geq |\nabla u| \ a.e.\}$). A function u belongs then to $W^{1,p}(X)$ if there exists a pseudo-gradient $g \in D[u]$ such that $g \in L^p(X)$.

This construction yields different types of Sobolev spaces depending on the way the pseudo-gradients are defined.

In the axiomatic approach, we do not specify how the pseudo-gradients are constructed; instead we simply assume that the correspondence $u \to D[u]$ satisfies the following six axioms:

Axiom A1 (Non-triviality). *If $u : X \to \mathbb{R}$ is non-negative and k-Lipschitz, then the function $g = k\,\mathrm{sgn}(u)$ belongs to $D[u]$.*

Axiom A2 (Upper linearity). *If $g_1 \in D[u_1]$, $g_2 \in D[u_2]$ and $g \geq |\alpha| g_1 + |\beta| g_2$ almost everywhere, then $g \in D[\alpha u_1 + \beta u_2]$.*

Axiom A3 (Leibniz rule). *If $u : X \to \mathbb{R}$ is any measurable function and $g \in D[u]$, then for any bounded Lipschitz function $\varphi : X \to \mathbb{R}$ the function $g_1(x) = (\sup|\varphi| g(x) + \mathrm{Lip}(\varphi)|u(x)|)$ belongs to $D[\varphi u]$.*

Axiom A4 (Lattice property). *Let $u := \max\{u_1, u_2\}$ and $v := \min\{u_1, u_2\}$ where $u_1, u_2 \in L^p_{\mathrm{loc}}(X)$. If $g_1 \in D[u_1]$ and $g_2 \in D[u_2]$, then $g := \max\{g_1, g_2\} \in D[v] \cap D[u]$.*

Axiom A5 (Completeness). *Let $\{u_i\}$ and $\{g_i\}$ be two sequences of functions such that $g_i \in D[u_i]$ for all i. Assume that $u_i \to u$ in L^p_{loc} topology and $(g_i - g) \to 0$ in L^p, then $g \in D[u]$.*

The last axiom will be stated later. We now define the notions of Dirichlet energy and Sobolev space based on Axioms 1–5:

Definitions. i.) The *p-Dirichlet energy* of a function u is defined as

$$\mathcal{E}_p(u) = \inf\left\{ \int_X g^p d\mu : g \in D[u] \right\}.$$

ii.) The *p-Dirichlet space* is the space $\mathcal{L}^{1,p}(X)$ of functions $u \in L^p_{\mathrm{loc}}(X)$ with finite p-energy.

iii.) The *Sobolev space* is then defined as

$$W^{1,p}(X) = W^{1,p}(X, d, \mu, D) := \mathcal{L}^{1,p}(X) \cap L^p(X).$$

Theorem 1.1. *$W^{1,p}(X)$ is a Banach space for the norm*

$$\|u\|_{W^{1,p}(X)} = \left(\int_X |u|^p d\mu + \mathcal{E}_p(u) \right)^{1/p}.$$

The proof is given in [3].

The Dirichlet space $\mathcal{L}^{1,p}(X)$ is also a Banach space for the norm given by

$$\|u\|_{\mathcal{L}^{1,p}(X)} := \left(\int_Q |u|^p d\mu + \mathcal{E}_p(u)\right)^{1/p}$$

where $Q \subset X$ is a fixed $\mathcal{K}$-subset of positive measure. This norm depends on the choice of the set Q, but the corresponding topology on $\mathcal{L}^{1,p}(X)$ is independent of that choice.

Our final axiom states that if the energy of a function is small, then this function is close to being a constant.

Axiom A6 (Energy controls variation). *Let $\{u_i\} \subset \mathcal{L}^{1,p}(X)$ be a sequence of functions such that $\mathcal{E}_p(u_i) \to 0$. Then for any metric ball B there exists a sequence of constants $a_i = a_i(B)$ such that $\|u_i - a_i\|_{L^p(B)} \to 0$.*

This axiom turns out to be equivalent to some weak form of the Poincaré inequality, see [3].

The main known constructions of Sobolev spaces on metric spaces fit into this pattern and are thus examples of axiomatic Sobolev spaces.

2. p-capacity and p-parabolicity

We now give some basic results from the theory of axiomatic Sobolev spaces on metric measure spaces. The proofs are given in [3, 4].

We denote by $C_0(X)$ the set of continuous functions $u : X \to \mathbb{R}$ such that $\operatorname{supp}(u)$ is a closed $\mathcal{K}$-set and by $\mathcal{L}_0^{1,p}(X)$ the closure of $C_0(X) \cap \mathcal{L}^{1,p}(X)$ in $\mathcal{L}^{1,p}(X)$.

The *variational p-capacity* of a subset $F \subset X$ is then defined as

$$\operatorname{Cap}_p(F) := \inf\{\mathcal{E}_p(u) : u \in \mathcal{A}_p(F)\},$$

where the set of admissible functions is defined by

$$\mathcal{A}_p(F) := \{u \in \mathcal{L}_0^{1,p}(X) : u \geq 1 \text{ on a neighborhood of } F \text{ and } u \geq 0 \text{ a.e.}\};$$

the variational p-capacity is an outer measure on X.

Definition. The metric measure space X is said to be *p-parabolic* if there exists a $\mathcal{K}$-set $Q \subset X$ such that $\mu(Q) > 0$ and $\operatorname{Cap}_p(Q) = 0$; and *p-hyperbolic* otherwise.

In fact a metric space is p-parabolic if and only if the p-capacity of *all* $\mathcal{K}$-sets is zero. This follows from the next theorem:

Theorem 2.1. *The metric measure space X is* p-parabolic *if and only if* $1 \in \mathcal{L}_0^{1,p}(X)$.

It is well known that a complete Riemannian manifold whose volume growth is polynomial of order at most p is p-parabolic. This facts extends to the case of metric spaces:

Theorem 2.2. *Suppose that the metric measure space X is complete and that $\mathcal{K}$ is the Boolean ring of all bounded Borel subsets of X. If there exists a point $x_0 \in X$ such that*

$$\liminf_{R\to\infty} \frac{\mu(B(x_0, R))}{R^p} = 0,$$

then X is p-parabolic.

If X is σ-compact (i.e. it is the countable union of compact sets), then the p-capacity of an arbitrary Borel set F can be approximated by the p-capacity of its compact subsets:

Theorem 2.3. *Assume that X is σ-compact and that $C(X)$ is dense in $W^{1,p}(X)$. Let $1 < p < \infty$, If $F \subset X$ is a bounded Borel set, then*

$$\mathrm{Cap}_p(F) = \sup\{\mathrm{Cap}_p(K) : K \subset F \textit{ a compact subset}\}.$$

This result follows from Choquet's theory of abstract capacities and the next result:

Theorem 2.4. *Suppose that X is σ-compact and $C(X)$ is dense in $W^{1,p}(X)$. If $1 < p < \infty$, then the variational p-capacity satisfies the following conditions:*

i.) Cap_p *is monotone:* $A \subset B \Rightarrow \mathrm{Cap}_p(A) \le \mathrm{Cap}_p(B)$;

ii.) *If $X \supset K_1 \supset K_2 \supset \cdots$ is a decreasing sequence of compact sets, then*

$$\lim_{i\to\infty} \mathrm{Cap}_p(K_i) = \mathrm{Cap}_p\Big(\bigcap_{i=1}^{\infty} K_i\Big);$$

iii.) *If $A_1 \subset A_2 \subset \cdots \subset X$ is an increasing sequence of non empty sets such that $\bigcup_{i=1}^{\infty} A_i$ is bounded, then*

$$\lim_{i\to\infty} \mathrm{Cap}_p(A_i) = \mathrm{Cap}_p\Big(\bigcup_{i=1}^{\infty} A_i\Big).$$

We finally state a result about the existence and uniqueness of extremal functions for p-capacities. We first need a definition:

Definition A subset F is said to be *p-fat* if it is a Borel subset and there exists a probability measure τ on X which is absolutely continuous with respect to p-capacities (i.e. such that $\tau(S) = 0$ for all subsets $S \subset X$ of local p-capacity zero) and whose support is contained in F.

Theorem 2.5. *Let (X, d) be a σ-compact metric measure space and $F \subset X$ be a p-fat subset ($1 < p < \infty$). Then there exists a unique function $u^* \in \mathcal{L}_0^{1,p}(X)$ such that $u^* = 1$ p-quasi-everywhere on F and $\mathcal{E}_p(u^*) = \mathrm{Cap}_p(F)$. Furthermore $0 \le u^*(x) \le 1$ for all $x \in X$.*

In both previous results, the hypothesis that X is σ-compact can be replaced by the more general condition that X is $\mathcal{K}$-countable (i.e. it is a countable union of $\mathcal{K}$-sets).

3. Examples

Our first example is the classical Sobolev space: let M be a Riemannian manifold and $\mathcal{K}$ be the class of relatively compact Borel subsets of M. We say that a measurable function $g : M \to \mathbb{R}$ is a *classic pseudo-gradient* of a function $u \in L^1_{\text{loc}}(M)$, if and only if $g(x) \geq |\nabla u(x)|$ a.e.

We write $D[u]$ for the set of all classic pseudo-gradients of u, the correspondence $u \to D[u]$ satisfies axioms A1–A6.

Our second example concerns functions on graphs: Let $\Gamma = (V, E)$ be a locally finite connected graph. We define the combinatorial distance between two vertices to be the length of the shortest combinatorial path joining them. The ring $\mathcal{K}$ is the class of all finite subsets of V and the measure μ is the counting measure.

For any function $u : V \to \mathbb{R}$, we define $CD[u]$ to be the set of all functions $g : V \to \mathbb{R}$ such that

$$|u(y) - u(x)| \leq (g(x) + g(y))$$

whenever there is an edge joining x to y.

The correspondence $u \to CD[u]$ satisfies axioms A1–A6.

Our next example is the Sobolev space defined by P. Hajłasz in [5], see also [7]. Let X be an arbitrary measure metric space and $\mathcal{K}$ be the ring of all bounded Borel subsets of X.

A measurable function $g : X \to \mathbb{R}_+$ is said to be a *Hajłasz pseudo-gradient* for a function $u : X \to \mathbb{R}$, if

$$|u(x) - u(y)| \leq d(x, y)(g(x) + g(y))$$

for all $x, y \in X \setminus F$ where $F \subset X$ is some set (called the exceptional set) with $\mu(F) = 0$.

We denote by $HD[u]$ the set of all *Hajłasz pseudo-gradients* of u; the correspondence $u \to HD[u]$ satisfies axioms A1–A6.

Our last example is the notion of upper-gradient, this notion is studied in [1], [6] and [8]. In this section, we assume X to be a rectifiably connected metric space.

We first consider the case of a locally Lipschitz function $u : X \to \mathbb{R}$. A Borel measurable function $g : X \to \mathbb{R}$ is said to be an *upper gradient* for u if for all Lipschitz paths $\gamma : [0, 1] \to X$ we have

$$|u(\gamma(1)) - u(\gamma(0))| \leq \int_0^1 g(\gamma(t))dt.$$

We denote by $UD[u]$ the set of all upper gradients for a locally Lipschitz function u.

We then extend this definition to arbitrary functions $u \in L^p_{\text{loc}}(X)$ as follow: we have $g \in UD[u]$ if and only if there exists two sequences of functions $u_i \to u$ in L^p_{loc} topology and $g_i \to g$ in L^p topology such that u_i is locally Lipschitz and g_i is an upper gradient for u_i.

The correspondence $u \to UD[u]$ satisfies Axioms A1–A5. Axiom A6 may fail to be true, however this axiom is always valid if the space X supports a Poincaré inequality.

4. Polar sets and continuity quasi-everywhere

In this section, we present some results from the non linear potential theory associated to an axiomatic Sobolev space. Details and proofs are given in [4].

It is well known in the classical theory of the Sobolev space $W^{1,p}(\Omega)$, (where $\Omega \subset \mathbb{R}^n$ is an Euclidean domain) that the structure of the set of essential discontinuities of a function $u \in W^{1,p}(\Omega)$ depends on p. For $p > n$ this set is empty and for $p \le n$, the size of this set can be described with the help of capacities. A similar, though more complicate, description exists for axiomatic Sobolev spaces.

We start from definitions of the Sobolev p-capacity and polar sets.

Definition 4.1. The *Sobolev p-capacity* of a pair $F \subset \Omega$ (where $\Omega \subset X$ is open and F is arbitrary) is defined by

$$C_p(F, \Omega) = \inf\{\|u\|^p_{W^{1,p}} : u \in W^{1,p}(\Omega),$$
$$u \ge 1 \text{ on a neighborhood of } F \text{ and } u \ge 0 \; a.e.\}.$$

The Sobolev p-capacity satisfies the same basic properties as the variational p-capacity:

Proposition 4.2. 1.) *The Sobolev p-capacity is an outer measure;*

2.) *for any subset $F \subset X$ we have $C_p(F) = \inf\{C_p(U) : U \supset F$ open $\}$;*

3.) *If $X \supset K_1 \supset K_2 \supset K_3 \supset \cdots$ is a decreasing sequence of compact sets, then*

$$\lim_{i\to\infty} C_p(K_i) = C_p\Big(\bigcap_{i=1}^{\infty} K_i\Big).$$

The next two results show the connection between the set of poles of a Sobolev function and the notion of Sobolev capacity.

Proposition 4.3. *For any $u \in W^{1,p}(X)$, we have $C_p(P_u) = 0$ where*

$$P_u := \{x \in X : \lim_{y\to x} u(y) = \infty\}$$

is the set of poles of u.

Proof. For any $k \geq 1$ the function $u_k(x) := \frac{1}{k}\min(k, u(x))$ is admissible for the Sobolev p-capacity of the P_u. It follows from the axioms that $\|u_k\|_{W^{1,p}(X)} \leq \frac{1}{k}\|u\|_{W^{1,p}(X)}$, we thus have $C_p(P_u) \leq \lim_{k\to\infty} \|u_k\|_{W^{1,p}(X)} = 0$. □

Proposition 4.4. *For any subset $A \subset X$, we have $C_p(A) = 0$ if and only if for any $\varepsilon > 0$ there exists a nonnegative function $u \in W^{1,p}(X)$ such that $\|u\|_{W^{1,p}(X)} \leq \varepsilon$ and $\lim_{y\to x} u(y) = \infty$ for all $x \in A$.*

Proof. Assume $C_p(A) = 0$; by definition of $C_p(A)$, there exists a sequence of nonnegative functions u_n such that $\|u_n\|_{W^{1,p}(X)} \leq 2^{-n}\varepsilon$ and $u_n = 1$ in some neighborhood of A. Then $u = \sum_n u_n$ belongs to $W^{1,p}(X)$ and $\lim_{y\to x} u(y) = \infty$, we also clearly have $\|u\|_{W^{1,p}(X)} \leq \varepsilon$. The converse direction is a consequence of the previous proposition. □

Definition 4.5. a) A set $S \subset X$ is *p-polar* (or *p-null*) if for any pair of open $\mathcal{K}$-sets $\Omega_1 \subset \Omega_2 \neq X$ such that $\operatorname{dist}(\Omega_1, X \setminus \Omega_2) > 0$, we have $\operatorname{Cap}_p(S \cap \Omega_1, \Omega_2) = 0$.

b) A property is said to hold *p-quasi-everywhere* if it holds everywhere except on a p-polar set.

We compare p-polar sets and sets of Sobolev p-capacity zero, we show in particular that in good cases, the p-polar sets and the sets of Sobolev p-capacity zero are the same.

We first formulate a technical lemma with a corollary for Sobolev capacity, the proof of which is based on a familiar cut-off arguments.

Lemma 4.6. *Let $\Omega_1 \subset \Omega_2 \subset X$ be a pair of open sets such that $\Omega_2 \neq X$ and $\delta := \operatorname{dist}(\Omega_1, X \setminus \Omega_2) > 0$. Then for any subset $S \subset \Omega_1$ and every $\varepsilon > 0$, there exists a function $\varphi = \varphi_\varepsilon \in W^{1,p}(X)$ with support in a closed subset of Ω_2, such that $\varphi \geq 1$ in a neighborhood of S and*

$$\|\varphi\|_{W^{1,p}(X)} \leq 2\left(1 + \frac{3}{\delta}\right)\left(C_p(S) + \varepsilon\right)^{1/p}. \tag{4.1}$$

Corollary 4.7. *Let Ω_1, Ω_2 and S be as in the lemma. If $C(X) \cap W^{1,p}(X)$ is dense in $W^{1,p}(X)$ and Ω_2 is a $\mathcal{K}$-set, then*

$$\operatorname{Cap}_p(S, \Omega_2) \leq 2^p \left(1 + \frac{3}{\delta}\right)^p C_p(S).$$

In particular if $C_p(S) = 0$, then S is p-polar.

Proof. By density of $C(X) \cap W^{1,p}(X)$ in $W^{1,p}(X)$, the function φ from the previous lemma belongs to $\mathcal{L}_0^{1,p}(\Omega_2)$. The proof follows then from (4.1) as ε is arbitrarily small. □

A subset S of X is said to be *strongly bounded* if $S \subset \Omega_1 \subset \Omega_2 \subset X$ where Ω_1 and Ω_2 are open $\mathcal{K}$-sets such that $\mu(X \setminus \Omega_2) > 0$ and $\operatorname{dist}(\Omega_1, X \setminus \Omega_2) > 0$. Strongly bounded sets play a role similar to the relatively compact subsets in the classical situation.

We say that a measurable metric space X is *strongly $\mathcal{K}$-coverable* if there exist two countable families of open $\mathcal{K}$-sets $\{U_i\}$ and $\{V_i\}$ such that $X = \bigcup U_i$, $U_i \subset V_i$ for all i, $\operatorname{dist}(U_i, X \setminus V_i) > 0$ and $\mu(V_i \setminus U_i) > 0$.

Proposition 4.8. *Suppose that $C(X) \cap W^{1,p}(X)$ is dense in $W^{1,p}(X)$ and that X is strongly $\mathcal{K}$-coverable. Then*

1.) *If a set $S \subset X$ is p-polar then* $\operatorname{Cap}_p(S, X) = 0$;

2.) *A set $S \subset X$ is p-polar if and only if $C_p(S) = 0$.*

Roughly speaking any function $u \in W^{1,p}(X)$ is continuous outside of a set with arbitrarily small local Sobolev p-capacity. In the classical theory this fact is a generalization of Egorov's theorem from Lebesgue spaces to Sobolev spaces.

We also have Egorov and Lusin type theorems for the Dirichlet space $\mathcal{L}^{1,p}(X)$ with the topology induced by the norm:

$$\|u\|^p_{\mathcal{L}^{1,p}(X,Q)} := \left(\int_Q |u|^p d\mu + \mathcal{E}_p(u)\right)^{1/p},$$

where Q is a fixed $\mathcal{K}$-set such that $\mu(Q) > 0$. This norm is complete and the corresponding Banach space structure is independent of the choice of Q. Any Cauchy sequence in the Dirichlet space $\mathcal{L}^{1,p}(X)$ converges in $W^{1,p}(\Omega)$ for any open $\mathcal{K}$-set $\Omega \subset X$.

Theorem 4.9. *Let $\{u_i\} \subset \mathcal{L}^{1,p}(X) \cap C(X)$ be a Cauchy sequence in $\mathcal{L}^{1,p}(X)$. Then for any open set $\Omega \in \mathcal{K}$ there exists a subsequence $\{u_{i'}\}$ of $\{u_i\}$ and a sequence of open subsets $\Omega \supset U_1 \supset U_2 \supset U_3 \supset \cdots$ such that $\lim_{\nu\to\infty} C_p(U_\nu, \Omega) = 0$ and $\{u_{i'}\}$ converges uniformly in $\Omega \setminus U_\nu$ for all ν. In particular $\{u_{i'}\}$ converges pointwise in the complement of the set of zero Sobolev p-capacity $S := \bigcap_{j=1}^{\infty} U_j$.*

Definition 4.10. A function $v : X \to \mathbb{R}$ is *p-quasi-continuous* if for every point $x \in X$ there exists an open $\mathcal{K}$-set A containing x such that for every $\varepsilon > 0$, we can find a subset $S \subset A$ such that $C_p(S, A) < \varepsilon$ and v is continuous on $A \setminus S$.

Theorem 4.11. *Suppose that X is $\mathcal{K}$-countable. For each $u \in \overline{C(X) \cap \mathcal{L}^{1,p}(X)}$ there is a function $v \in \mathcal{L}^{1,p}(X)$ such that*

1.) *$u = v$ almost everywhere on X and*

2.) *v is p-quasi-continuous.*

This result is proved in [7] for the special case of Hajłasz Sobolev spaces.

The function v is called a *p-quasi-continuous representative* of u. Note in particular that every function $u \in \mathcal{L}_0^{1,p}(X)$ has a p-quasi-continuous representative (since continuous functions are dense in $\mathcal{L}_0^{1,p}(X)$ by definition).

Theorem 4.9 has a version for sequences of quasi-continuous functions in the subspace $\mathcal{L}_0^{1,p}(X)$ of the space $\mathcal{L}^{1,p}(X)$ that is the closure of continuous function which support is a $\mathcal{K}$-set: every Cauchy sequence of quasi-continuous functions in $\mathcal{L}_0^{1,p}(X)$ contains a subsequence which converges uniformly in any set of arbitrary small p-capacity:

Proposition 4.12. *Let $\{u_i\} \subset \mathcal{L}_0^{1,p}(X)$ be a Cauchy sequence of p-quasi-continuous functions. Then for any open set $\Omega \in \mathcal{K}$ and any $\varepsilon > 0$, there exists a subsequence $\{u_{i'}\}$ of $\{u_i\}$ which converges uniformly in $\Omega \setminus F_\varepsilon$, where $F_\varepsilon \subset \Omega$ is a subset such that $C_p(F_\varepsilon, \Omega) \leq 2\varepsilon$.*

If X is $\mathcal{K}$-countable, then we can globalize the previous result:

Corollary 4.13. *Assume that X is $\mathcal{K}$-countable. Let $\{u_i\} \subset \mathcal{L}_0^{1,p}(X)$ be a Cauchy sequence of p-quasi-continuous functions. Then for any $\varepsilon > 0$, there exists a subsequence $\{u_{i'}\}$ of $\{u_i\}$ which converges uniformly in $X \setminus F_\varepsilon$, where $F_\varepsilon \subset X$ is a subset such that $C_p(F_\varepsilon, X) \leq \varepsilon$.*

The proof follows from the previous proposition and countable subadditivity of the Sobolev capacity.

References

[1] J. Cheeger, Differentiability of Lipschitz Functions on Metric Measure Spaces, Geom. Funct. Anal. 9 (195), 428–517.

[2] B. Franchi, P. Hajłasz and P. Koskela, Definitions of Sobolev classes on metric spaces, Ann. Inst. Fourier 49 (1999), 1903-1924.

[3] V. M. Gol'dshtein and M. Troyanov, Axiomatic Theory of Sobolev Spaces, Expo. Math. 19 (2001), 289–336.

[4] V. M. Gol'dshtein and M. Troyanov, Capacities on Metric Spaces, Integral Equations Operator Theory, to appear.

[5] P. Hajłasz, Sobolev spaces on an arbitrary metric space, Potential Anal. 5 (4) (1996), 403–415.

[6] J. Heinonen and P. Koskela, Quasiconformal maps in metric spaces with controlled geometry, Acta Math. 181 (1) (1998), 1–61.

[7] J. Kinnunen and O. Martio, The Sobolev Capacity on Metric Spaces, Ann. Acad. Sci. Fenn. Math. 21 (1996), 367–382.

[8] N. Shanmugalingam, Newtonian Spaces: An Extension of Sobolev Spaces to Metric Measure Spaces, Rev. Mat. Iberoamericana 16 (2) (2000), 243–279.

Vladimir Gol'dshtein, Department of Mathematics, Ben Gurion University of the Negev, P.O. Box 653, 84105 Beer Sheva, Israel
E-mail: vladimir@bgumail.bgu.ac.il

Marc Troyanov, Department de Mathematiques, EPFL, 1015 Lausanne, Suisse
E-mail: marc.troyanov@epfl.ch

Interpolation of subspaces and the unit problem

Sten Kaijser and Peter Sunehag

Dedicated to Jaak Peetre on the occasion of his 65th birthday

Abstract. Let $\bar{X} = (X_0, X_1)$ and $\bar{Y} = (Y_0, Y_1)$ be Banach couples embedded into the same ambient space Ω and also assume that $Y_0 \subset X_0$ and $Y_1 \subset X_1$ are closed subspaces. Then $\bar{Y}$ is called a subcouple to $\bar{X}$. Given an interpolation functor F, what is the relation between $F(\bar{Y})$ and $F(\bar{X})$? The case when $Y_0 = X_0 \cap \operatorname{Ker}\Gamma$ and $Y_1 = X_1$ where $\Gamma \in X_0'$ was recently solved by Ivanov and Kalton in [2]. In another article [4] by Krugljak, Maligranda and Persson a problem concerning Hardy's inequality and interpolation of intersections where $\bar{Y} = (X_0 \cap \operatorname{Ker}\Gamma, X_1 \cap \operatorname{Ker}\Gamma)$ was studied. In there case Γ was not bounded on any of the endpoint spaces.

In this paper we are going to study that case as a generalized subcouple, that is we are going to replace the condition that the couples are embedded into the same ambient space with the condition that $\Delta\bar{Y}$ is closed in $\Delta\bar{X}$. To construct such a couple we only need a linear functional that is bounded on $\Delta\bar{X}$. These constructions are especially natural for Banach algebras and we are here going to restrict our investigation to that case. More general results are going to be presented in a later article. We present two motivating applications, first in Section 2 we give conditions that imply that an interpolation algebra is unital and in Section 4 we present our version of the results by Krugljak, Maligranda and Persson regarding the Hardy operator.

2000 Mathematics Subject Classification: 46M35

1. Interpolation of Banach algebras

We shall say that a Banach couple $\bar{A} = (A_0, A_1)$ is a BCBA (Banach Couple of Banach Algebras) if A_0 and A_1 are Banach algebras and if the multiplication agrees on the intersection. In interpolation theory we usually begin with two Banach spaces and a Hausdorff topological vector space in which the Banach spaces are embedded. That is we start with the triple $(A_0, A_1, \Sigma(\bar{A}))$ together with injections $\sigma_0 : A_0 \to \Sigma(\bar{A})$ and $\sigma_1 : A_1 \to \Sigma(\bar{A})$, but if $\bar{A}$ is a BCBA we know that $\Delta(\bar{A})$ is a BA (Banach algebra) but it is not always true that $\Sigma(\bar{A})$ is a BA. That suggests that we should instead begin with a triple $(\Delta(\bar{A}), A_0, A_1)$ together with injections $\delta_0 : \Delta(\bar{A}) :\to A_0$ and $\delta_1 : \Delta(\bar{A}) :\to A_1$. We can then always embed A_0 and A_1 into the pushout of the diagram $(\Delta(\bar{A}), A_0, A_1, \delta_0, \delta_1)$, and if $\delta_0(\Delta(\bar{A}))$ is dense in A_0 and $\delta_1(\Delta(\bar{A}))$ is dense in A_1 we can embed A_0' and A_1' into $(\Delta(\bar{A}))'$ and then embed A_0 and A_1 into the predualspace of $A_0' \cap A_1'$.

A first question now is for which interpolation methods F must $F(\bar{A})$ be a BA when $\bar{A}$ is BCBA. Two known classes of such methods are the complex method and the real $J(\Theta, 1)$ method. In [1] the $J(\Theta, 1)$ were generalized to the $J(\rho, 1)$ method where ρ is a quasiconcave submultiplicative function. $J(\rho, 1, \bar{A})$ is defined by the norm $\|a\|_\rho = \inf\{\int_0^\infty \rho(1/t)J(t, f(t))\frac{dt}{t} \mid a = \int_0^\infty f(t)\frac{dt}{t}\}$, and it was proved in [1] that $J(\rho, 1, \bar{A})$ is a BA if $\bar{A}$ is a BCBA.

Another question is for which F is $F(\bar{A})$ unital. It is for example not true that $F(\bar{A})$(or even $\Delta(\bar{A})$) has to be unital because A_0 and A_1 are unital and $F(\bar{A})$ can be unital even if not both A_0 and A_1 are unital.

Example 1.1. Let $B_0 = C(T)$, $B_1 = C(2T)$ and $\Delta(\bar{B}) = A_{z_0}(W)$, where $W = \{z \mid 1 \le |z| \le 2\}$, $A(W) = \{f \epsilon C(W) \mid f$ analytic on $\mathrm{Int}(W)\}$ and $A_{z_0}(W) = \{f \epsilon A(W) \mid f(z_0) = 0\}$. Then take $\Delta(\bar{A}) = A(W)$ and $\Gamma \epsilon (\Delta(\bar{A}))'$ is defined by $\Gamma(f) = f(z_0)$. Assume that $z_0 = 2^\gamma e^{i\lambda}$. Then we get that $\bar{B}_{\Theta,1}$ is unital iff $\Theta \neq \gamma$.

2. The unit problem

In this section we will present two theorems dealing with the unit problem. They will both follow directly from theorems concerning interpolation of subspaces of codimension one which will be presented in the next section.

Definition 2.1. Let $\bar{A}$ be a Banach couple and let $\Gamma \epsilon A_0'$. Then define

$$\alpha_0(\Gamma) = \lim_{\tau\to\infty} \inf_{0<\tau t\le 1} \frac{1}{\log\tau} \log\frac{K(\tau t, \Gamma)}{K(t, \Gamma)}$$

$$\alpha_1(\Gamma) = \lim_{\tau\to\infty} \sup_{0<\tau t\le 1} \frac{1}{\log\tau} \log\frac{K(\tau t, \Gamma)}{K(t, \Gamma)}$$

Theorem 2.1. *Let $\bar{B}$ be a BCBA where B_0 is unital but B_1 is not. Define $\bar{A}$ by $A_0 = B_0 \oplus C$, the unitization of B_0, $A_1 = B_1$ and $\Delta(\bar{A}) = \Delta(\bar{B}) \oplus C$. Then define $\Gamma \epsilon A_0'$ by $\Gamma(x, \lambda) = \lambda$ and finally let $0 < \Theta < 1$. Then*

$$\bar{B}_{\Theta,1} \text{ is unital if } \Theta < \alpha_0(\Gamma)$$

and

$$\bar{B}_{\Theta,1} \text{ is nonunital if } \Theta > \alpha_1(\Gamma).$$

Definition 2.2. Let $\bar{A}$ be a BCBA and let $\Gamma \in \Delta(\bar{A})'$ be multiplicative. Then define

$$\beta_0(\Gamma) = \lim_{\tau\to\infty} \frac{\log K(\tau,\Gamma)}{\log\tau}$$

$$\beta_1(\Gamma) = \lim_{\tau\to 0} \frac{\log K(\tau,\Gamma)}{\log\tau}$$

Remark 2.1. These limits exist as proved in [1. Lemma5.1].

Theorem 2.2. *Let $\bar{B} = (B_0, B_1)$ be a BCBA where B_0 and B_1 are unital but $\Delta(\bar{B})$ is nonunital. Define $\bar{A}$ by $A_0 = B_0, A_1 = B_1$ and $\Delta(\bar{A}) = \Delta(\bar{B}) \oplus C$. $\Gamma \epsilon (\Delta(\bar{A}))'$ is defined by $\Gamma(x,\lambda) = \lambda$ and we let $0 < \Theta < 1$. Then $\bar{B}_{\Theta,1}$ is unital if $\Theta \notin [\beta_0(\Gamma), \beta_1(\Gamma)]$*

3. Interpolation of subspaces of codimension one

Let $\bar{X}$ be a regular Banach couple and let $0 \neq \Gamma \epsilon X_0'$. Then define $\bar{Y}$ by $Y_0 = \operatorname{Ker}\Gamma$ and $Y_1 = X_1$. Now if F is an interpolation functor, what is then the relation between $F(\bar{X})$ and $F(\bar{Y})$? For the real (Θ, q)-method the answer was presented in [2], which was a slight improvement of [3].

Theorem 3.1. *Let $0 < \Theta < 1$ and $1 \leq q < \infty$. Then*

$\bar{B}_{\Theta,q} \approx \bar{A}_{\Theta,q}$ (equivalent norms) if $\Theta > \alpha_1$,

$\bar{B}_{\Theta,q}$ is a closed subspace of codimension one in $\bar{A}_{\Theta,q}$ if $\Theta < \alpha_0$

and

$\bar{B}_{\Theta,q}$ is not closed in $\bar{A}_{\Theta,q}$ if $\alpha_0 \leq \Theta \leq \alpha_1$.

Proof of Theorem 2.1. Γ is a bounded, multiplicative linear functional and $\operatorname{Ker}\Gamma = B_1$. Thus if $\bar{B}_{\Theta,1}$ is a closed subspace of codimension one in $\bar{A}_{\theta,1}$ we know that $\bar{B}_{\Theta,1}$ is nonunital and if $\bar{B}_{\Theta,1} \approx \bar{A}_{\Theta,1}$ we know that $\bar{B}_{\Theta,1}$ is unital. The result now follows from Theorem 3.1. □

With our model with triples $(\Delta(\bar{A}), A_0, A_1)$ we get a new situation where Γ is a multiplicative continuous linear functional on $\Delta(\bar{A})$ and $\bar{B} = (\operatorname{Ker}\Gamma, A_0, A_1)$. With the pushout property we get a surjective map $T : \Sigma(\bar{B}) \to \Sigma(\bar{A})$. The question now is what is the relation between $T(F(\bar{B}))$ and $F(\bar{A})$.

Theorem 3.2. *Let $0 < \Theta < 1$ and $1 \leq p < \infty$. Then if $\Theta \notin [\beta_0(\Gamma), \beta_1(\Gamma)]$ we get that*

$$T(\bar{B}_{\Theta,p}) \approx \bar{A}_{\Theta,p}$$

where

$$\|a\|_{T(\bar{B}_{\Theta,p})} = \inf_{Tb=a} \|b\|_{\bar{B}_{\Theta,p}}.$$

Proof. The proof is based on a construction from [2]. Let $w_n = \frac{1}{K(2^{-n},\Gamma)}$ and let $\ell_p(w)$ be the space of all sequences $(\alpha_n)_{n\epsilon Z}$ such that

$$\|(\alpha_n)\| = \Big(\sum_{k=-\infty}^{\infty} |\alpha_k|^p w_n^p\Big)^{1/p} < \infty.$$

Let e_n be the standard basis defined by $e_n = (\delta_{nj})_j$ and define the shift operator S by $S(e_n) = e_{n+1}$. Then since $w_n \le w_{n+1} \le 2w_n$ we can see that S and S^{-1} are both bounded with $\|S\| \le 2$ and $\|S^{-1}\| \le 1$. In [1] it was proved that if Γ is multiplicative, then $K(t,\Gamma)$ is supermultiplicative which implies that $w_{n+k} \le w_n w_k$. Thus $\ell_1(w)$ is a Banach algebra and $\ell_p(w)$ is a module over it. We also get that

$$\frac{1}{w_{-k}} \le \frac{w_{n+k}}{w_n} \le w_k$$

and therefore

$$\inf_n \frac{w_{n+k}}{w_n} = \frac{1}{w_{-k}} \quad \sup_n \frac{w_{n+k}}{w_n} = w_k$$

since $w_0 = 1$. Therefore we can deduce from the spectral radius formula that $2^{-\beta_0} = r(S^{-1})$ and $2^{\beta_1} = r(S)$ so the spectrum of S is contained in $\{z \mid 2^{\beta_0} \le |z| \le 2^{\beta_1}\}$. Thus if $\Theta \notin [\beta_0(\Gamma), \beta_1(\Gamma)]$, we know that $T_\Theta = S - 2^\Theta I$ is an isomorphism. Thus there exists a constant D such that $\|\alpha\| \le D\|T_\Theta \alpha\|\ \forall\ \alpha\epsilon\ell_p(w)$. Now if $a\epsilon \bar{A}_{\Theta,q}$, $\|a\|_{\bar{A}_{\Theta,q}} = 1$ we can find $a_n \epsilon \Delta(\bar{A})$ such that $\sum_{n=-\infty}^{\infty} 2^{n\Theta} a_n = a$ and

$$\Big(\sum_{k=-\infty}^{\infty} J(2^k, a_k)^p\Big)^{1/p} \le 2$$

which implies that

$$\Big(\sum_{k=-\infty}^{\infty} |\Gamma(a_k)|^p w_k^p\Big)^{1/p} \le 2.$$

It now follows that there exists $\alpha\epsilon\ell_p(w)$ with $T_\Theta(\alpha) = (\Gamma(a_n))$ and $\|\alpha\| \le 2D$ and (for every n) $u_n \epsilon \Delta(\bar{A})$ such that $J(2^n, u_n) \le 2|\alpha_n| w_n$ and $\Gamma(u_n) = \alpha_n$. If we now let $v_n = u_{n-1} - 2^\Theta u_n$ we get that

$$\begin{aligned} J(2^k, v_k) &\le \max(\|u_{k-1}\|_0 + 2^\Theta \|u_k\|_0, 2^k(\|u_{k-1}\|_1 + 2^\Theta \|u_k\|_1) \\ &\le 2J(2^{k-1}, u_{k-1}) + 2J(2^k, u_k) \le 4|\alpha_{n-1}| w_{n-1} + 4|\alpha_n| w_n \end{aligned}$$

and hence

$$\Big(\sum_{k=-\infty}^{\infty} J(2^k, v_k)^p \Big)^{1/p} \leq 8\|\alpha\| \leq 16D.$$

We can also see that $\Gamma(v_n) = \Gamma(a_n)$ and $\sum_{n=-\infty}^{\infty} 2^{n\Theta} v_n = 0$. Now let $b \epsilon \Sigma(\bar{B})$ be defined by $b = \sum_{n=-\infty}^{\infty} 2^{n\Theta}(a_n - v_n)$. Then $b \epsilon \bar{B}_{\Theta,p}$, $\|b\|_{\bar{B}_{\Theta,p}} \leq 16D + 2$ and $Tb = a$. Thus $a \epsilon T(\bar{B}_{\Theta,p})$. We have now proved that $\|a\|_{T(\bar{B}_{\Theta,p})} \leq (16D + 2)\|a\|_{\bar{A}_{\theta,p}}$ $\forall\, a \epsilon \bar{A}_{\theta,p}$. Hence $T(\bar{B}_{\Theta,q}) \approx \bar{A}_{\Theta,q}$. □

Proof of Theorem 2.2. $T : \bar{B}_{\Theta,1} \to \bar{A}_{\Theta,1}$ is onto so if we could also prove that it is injective we would be done. First note that the kernel of $T : \Sigma(\bar{B}) \to \Sigma(\bar{A})$ is spanned by the element $u = I_0 - I_1$ where I_0 and I_1 are the units in B_0 and B_1. More precisely every pair $(b, -b)$ such that $\Gamma(b) = 1$ represents u. This implies that $K(t, u) = \inf\{\|b\|_0 + t\|b\|_1 \mid \Gamma(b) = 1\} \geq \inf\{J(t, b) \mid \Gamma(u) = 1\} = \frac{1}{K(\frac{1}{t}, \Gamma)} \geq \max\{t^{\beta_0}, t^{\beta_1}\}$ and it is then clear that $u \notin K(\Theta, 1, \bar{B})$, and therefore T is an isomorphism. □

4. Another application of the $(\Delta(\bar{A}), A_0, A_1)$-approach

In [4] Krugljak, Maligranda and Persson explained the failure of Hardy's inequality by interpolating intersections. But it is also possible to see it as a subspace problem.

Example 4.1. Let $X_0 = L_1(x)$ and $X_1 = L_1(x^{-1})$ where $L_1(w(x))$ is L_1 on $(0, \infty)$ with weight $w(x)$. Define $\Gamma \epsilon \Delta(\bar{X})'$ by $\Gamma(f) = \int_0^\infty f(x)dx$ and $\bar{Y} = (\operatorname{Ker}\Gamma, X_0, X_1)$. Notation is as in Diagram 1.

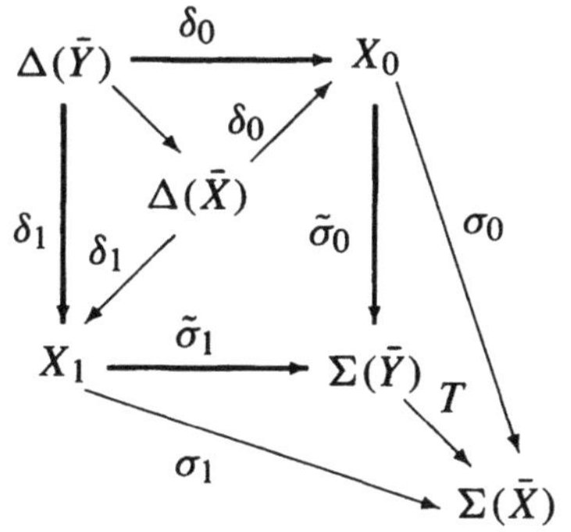

Diagram 1

Ker T is as before spanned by one element $u = \tilde{\sigma}_0(g) - \tilde{\sigma}_1(g)$ where $\int_0^\infty g(x)dx = 1$. We begin by calculating $K(t, u, \bar{Y})$.

$$\begin{aligned} K(t, u, \bar{Y}) &= \inf\{\|g\|_{X_0} + t\|g\|_{X_1} \mid \int_0^\infty g(s)ds = 1\} \\ &= \inf\{\int_0^\infty |g(s)|sds + t\int_0^\infty |g(s)|\frac{ds}{s} \mid \int_0^\infty g(s)ds = 1\} \\ &= \inf\{\int_0^\infty |(g(s)|(s + t/s)ds \mid \int_0^\infty g(s)ds = 1\} \\ &\geq \inf\{2\sqrt{t}\int_0^\infty |g(s)|ds \mid \int_0^\infty g(s)ds = 1\} = 2\sqrt{t}. \end{aligned}$$

Now let $g_\varepsilon = \frac{1}{\varepsilon}\chi_{(\sqrt{t}-\varepsilon,\sqrt{t})}$. Then

$$K(t, u, \bar{Y}) \leq \int_0^\infty |g_\epsilon(s)|(s + t/s)ds \to 2\sqrt{t} \quad as\ \varepsilon \to 0+$$

Thus

$$K(t, u, \bar{Y}) = 2\sqrt{t}$$

and we see that $u \notin \bar{Y}_{\Theta,1}$ so $T : \bar{X}_{\Theta,1} \to \bar{Y}_{\Theta,1}$ is injective $\forall\ \Theta : 0 < \Theta < 1$. We know that for all $f \in \Sigma(\bar{X})$

$$K(t, f, \bar{X}) = \|f_0(t)\|_0 + t\|f_1(t)\|_1$$

where

$$f_0(t) = f\chi_{(0,\sqrt{t})},\ f_1(t) = f\chi_{(\sqrt{t},\infty)}.$$

If $\Theta > 1/2$ and $f \in \bar{X}_{\Theta,1} = L_1(t^{1-2\Theta})$ we get that $\int_0^1 f(s)ds < \infty$. Thus we can define $\tilde{f}_0$ and $\tilde{f}_1$ by

$$\tilde{f}_0(t) = f\chi_{(0,\sqrt{t})} - \frac{2}{\sqrt{t}}(\int_0^1 f(s)ds)\chi_{(\frac{\sqrt{t}}{2},\sqrt{t})}$$

and

$$\tilde{f}_1(t) = f\chi_{(\sqrt{t},\infty)} + \frac{2}{\sqrt{t}}(\int_0^1 f(s)ds)\chi_{(\frac{\sqrt{t}}{2},\sqrt{t})}.$$

Let

$$g = T(\tilde{\sigma}_0(\tilde{f}_0(1)) + \tilde{\sigma}_1(\tilde{f}_1(1))).$$

Then

$$g = T(\tilde{\sigma}_0(\tilde{f}_0(t)) + \tilde{\sigma}_1(\tilde{f}_1(t)))$$

for all t since

$$\Gamma(\tilde{f}_0(t) - \tilde{f}_0(1)) = 0 \,\forall\, t$$

and $Tg = f$ since

$$\sigma_0(\tilde{f}_0) + \sigma_1(\tilde{f}_1) = f.$$

Now

$$\begin{aligned} K(t, g, \bar{Y}) &\leq \|\tilde{f}_0(t)\|_0 + t\|\tilde{f}_1(t)\|_1 \\ &\leq \int_0^{\sqrt{t}} |f(s)| s ds + \frac{2}{\sqrt{t}} |\int_0^{\sqrt{t}} f(s) ds| \frac{\sqrt{t}}{2} \sqrt{t} + t \\ &\qquad \int_1^{\infty} |f(s)| \frac{ds}{s} + t \frac{2}{\sqrt{t}} |\int_0^{\sqrt{t}} f(s) ds| \frac{\sqrt{t}}{2} \frac{2}{\sqrt{t}} \\ &\leq K(t, f, \bar{X}) + 3\sqrt{t} |\int_0^{\sqrt{t}} f(s) ds| \end{aligned}$$

and

$$\begin{aligned} \int_0^{\infty} t^{-\Theta} (\sqrt{t} \int_0^{\sqrt{t}} f(s) ds) \frac{dt}{t} &= \frac{1}{2} \int_0^{\infty} x^{-2\Theta} | \int_0^{x} f(s) ds | dx \\ &\leq \frac{1}{2\Theta - 1} \|f\|_{L_1(x^{1-2\Theta})} \end{aligned}$$

by Hardy's inequality. If $\Theta < 1/2$ and $f \in \bar{X}_{\Theta,1} = L_1(t^{1-2\Theta})$ we get that $\int_1^{\infty} f(s) ds < \infty$. Thus we can define $\tilde{f}_0$ and $\tilde{f}_1$ by

$$\tilde{f}_0(t) = f \chi_{(0,\sqrt{t})} - \frac{2}{\sqrt{t}} (\int_{\sqrt{t}}^{\infty} f(s) ds) \chi_{(\frac{\sqrt{t}}{2}, \sqrt{t})}$$

and

$$\tilde{f}_1(t) = f \chi_{(\sqrt{t}, \infty)} + \frac{2}{\sqrt{t}} (\int_{\sqrt{t}}^{\infty} f(s) ds) \chi_{(\frac{\sqrt{t}}{2}, \sqrt{t})}.$$

Then with

$$g = T(\tilde{\sigma}_0(\tilde{f}_0(1)) + \tilde{\sigma}_1(\tilde{f}_1(1)))$$

we get that

$$\begin{aligned}K(t,g,\bar{Y}) &\le \|\tilde{f}_0(t)\|_0 + t\|\tilde{f}_1(t)\|_1\\ &\le \int_0^{\sqrt{t}} |f(s)|sds + \frac{2}{\sqrt{t}}|\int_{\sqrt{t}}^\infty f(s)ds|\frac{\sqrt{t}}{2}\sqrt{t} + t\\ &\quad \int_{\sqrt{t}}^\infty |f(s)|\frac{ds}{s} + t\frac{2}{\sqrt{t}}|\int_{\sqrt{t}}^\infty f(s)ds|\frac{\sqrt{t}}{2}\frac{2}{\sqrt{t}}\\ &\le K(t,f,\bar{X}) + 3\sqrt{t}|\int_{\sqrt{t}}^\infty f(s)ds|\end{aligned}$$

and

$$\begin{aligned}\int_0^\infty t^{-\Theta}(\sqrt{t}\int_{\sqrt{t}}^\infty f(s)ds)\frac{dt}{t} &= \frac{1}{2}\int_0^\infty x^{-2\Theta}|\int_x^\infty f(s)ds|dx\\ &\le \frac{1}{1-2\Theta}\|f\|_{L_1(x^{1-2\Theta})}\end{aligned}$$

by the other version of Hardy's inequality. Thus $\bar{Y}_{\Theta,1} \approx \bar{X}_{\Theta,1}$ when $\Theta \neq \frac{1}{2}$.

Since $\bar{X}_{1/2,1} = L_1$ we know that Γ is bounded on that space so $\bar{Y}_{1/2,1} \neq \bar{X}_{1/2,1}$. Now define the map $H : \bar{Y} \to \bar{X}$ by letting H on $\Delta(\bar{Y})$ be defined by

$$Hf(x) = \frac{1}{x}\int_0^x f(s)ds.$$

By Hardy's inequality we can then extend H to bounded maps on X_0 and X_1 and by interpolation we also see that H is bounded on $L_1(x^\alpha) = \bar{Y}_{\frac{\alpha+1}{2},1}$ if $\alpha \neq 0$. But we can not say that H is bounded on $L_1 \neq \bar{Y}_{1/2,1}$. In [4] they used the functions $f_n = \chi_{(1,2)} - \chi_{(n,n+1)}$ to prove that it is unbounded.

Remark 4.1. If we define fg by

$$fg(x) = \int_0^\infty f(y)g\left(\frac{x}{y}\right)\frac{dy}{y}$$

we see that $\bar{X}$ is a BCBA and that Γ is multiplicative.

References

[1] S. Kaijser, Interpolation of Banach algebras and open sets, Integral Equations Operator Theory 41 (2001), 189–222.

[2] S. Ivanov, N. Kalton, Interpolation of subspaces and applications to exponential bases, Algebra i Analiz 13 (2) (2001), 93–115.

[3] J. Löfström, Interpolation of subspaces, unpublished but can be found as a ps-file at www.math.chalmers.se/~jorgen.

[4] N. Krugljak, L. Maligranda, L.-E. Persson, The failure of Hardy's inequality and interpolation of intersections, Ark. Mat. 37 (1999), 323–344.

Sten Kaijser, Department of Mathematics, Uppsala University, Box 480, 751 06 Uppsala, Sweden

E-mail: sten@math.uu.se

Peter Sunehag, Department of Mathematics, Uppsala University, Box 480, 751 06 Uppsala, Sweden

E-mail: peter@math.uu.se

On hierarchical structures and reiterated homogenization

Dag Lukkassen and Graeme W. Milton

Dedicated to Professor Jaak Peetre on the occasion of his 65th birthday

Abstract. In this paper we consider the concept of reiterated homogenization, introduced on the physical level by Bruggeman in the 30's and justified mathematically by Bensoussan, Lions and Papanicolaou in 1978. We present and discuss some recent developments of this theory and also give some applications to linear and nonlinear problems.

2000 Mathematics Subject Classification: 35-99, 49-99

1. Introduction

Strongly non-homogeneous structures have fascinated people for a very long time. Archaeological observations in Finland show that fibre-reinforced ceramics were made about 4000 years ago, and that people already at this time had ideas and theories for intelligent combinations of materials and structures. Analysis of the macroscopic properties of composites was investigated by the physicists Maxwell, Rayleigh, and Einstein, among many others. Around 1970 the problem of determining the physical properties of material structures and composites was reformulated in such a way that this field became interesting from a purely mathematical point of view. This formulation initiated a new mathematical discipline called homogenization theory.

Reiterated homogenization was introduced on the physical level by Bruggeman already in the 30's. The mathematical justification of this theory was given by Bensoussan, Lions and Papanicolaou in 1978. In this paper we present and discuss some recent developments of this theory and also give some applications to linear and nonlinear problems.

2. The periodic case

The conductivity problem on a periodic material structure with period equal to $1/h$ can be formulated by the following minimum principle:

$$E_h = \min_u \left(\mathcal{F}_h(u) - \int_\Omega u(x)g(x)\,dx \right), \tag{1}$$

where

$$\mathcal{F}_h(u) = \int_\Omega \left(\lambda(hx)\,|Du(x)|^2 \right) dx.$$

Here, $\lambda(\cdot)$ (the conductivity) is periodic relative to the unit-cube I of $\mathbb{R}^n$ and bounded between two strictly positive constants, Ω is a bounded open subset of $\mathbb{R}^n$ and g is the source-field. The minimization is taken over some suitable (subset of a) Sobolev-space which takes care of the given boundary conditions. It is possible to prove that the energy E_h converges to a "homogenized" energy E_{hom}, as $h \to \infty$, of the form

$$E_{\text{hom}} = \min_u \left(\mathcal{F}_{\text{hom}}(u) - \int_\Omega u(x)g(x)\,dx \right), \tag{2}$$

where $\mathcal{F}_{\text{hom}}$ is of the form

$$\mathcal{F}_{\text{hom}}(u) = \int_\Omega f_{\text{hom}}(Du(x))\,dx$$

and

$$f_{\text{hom}}(\xi) = \min_{W^{1,2}_{\text{per}}(I)} \int_I \lambda(x)\,|\xi + Du(x)|^2\,dx. \tag{3}$$

Here, $W^{1,2}_{\text{per}}(I)$ is the space of I-periodic functions of the Sobolev space $W^{1,2}(I)$. Therefore, since the actual energy E_h (which is interesting for us to determine) is difficult to find when h is large, and since $f_{\text{hom}}(\xi)$ and E_{hom} can be found by many numerical methods, we can use E_{hom} as an approximation of E_h. The convergence of energies is usually seen as a consequence of Γ-convergence of the corresponding Lagrangians (see the definition below).

As an alternative to (1) we can formulate the conductivity problem via the corresponding Euler equation

$$-\operatorname{div} \lambda(hx)Du_h(x) = g,$$

together with the given boundary conditions. It is possible to prove that the solution u_h (in this case the minimizer of (1)) converges weakly in the above mentioned Sobolev space to the solution u (in this case also the minimizer of (2)) of the "homogenized" equation

$$-\operatorname{div} A_{\text{hom}} Du(x) = g.$$

Here, A_{hom} is defined by

$$A_{\mathrm{hom}}\xi = \int_I \lambda(x)\left(Du^{\xi}(x) + \xi\right) dx,$$

where u^{ξ} (in this case also the minimizer of (3)) is the I-periodic solution of

$$\operatorname{div}\left(\lambda(x)(Du^{\xi}(x) + \xi)\right) = 0.$$

The convergence results above were first proved by De Giorgi and Spagnolo in the late 60's. Various kinds of simplifications of this proof were done in the 70's by Murat, Tartar, Bakhvalov, Bensoussan, Lions and Papanicolaou (for more information, see e.g. the book [16]).

3. Reiterated homogenization of integral functionals

Let us consider the class of Lagrangians g such that $g(x, \xi)$ is measurable in x, convex in ξ and satisfying the standard growth condition

$$-c_0 + c_1 |\xi|^p \leq g(x, \xi) \leq c_0 + c_2 |\xi|^p, \tag{4}$$

where $c_1, c_2 > 0$ and $p > 1$. If g_h and g belong to this class we recall that g is the Γ-limit of the sequence g_h, denoted $g = \Gamma - \lim g_h$, if for any bounded open set Ω with Lipschitz boundary the following two conditions hold:

(i) for any $u_h \in W^{1,p}(\Omega)$, $u_h \rightharpoonup u$ weakly in $W^{1,p}(\Omega)$ it holds that

$$\int_\Omega g(x, Du)dx \leq \liminf_{h\to\infty} \int_\Omega g_h(x, Du)\, dx,$$

(ii) for every $u \in W^{1,p}(\Omega)$ there is a sequence u_h such that $u_h \rightharpoonup u$ weakly in $W^{1,p}(\Omega)$ $u_h - u \in W_0^{1,p}(\Omega)$,

$$\int_\Omega g(x, Du)\, dx = \lim_{h\to\infty} \int_\Omega g_h(x, Du)\, dx.$$

Let $f(y, z, \xi)$ be I-periodic and measurable in the first and second variable, respectively. Moreover, assume that f is piecewise continuous in the first variable, i.e. of the form $f(y, z, \xi) = \sum_{i=1}^N \chi_{\Omega_i}(y) f_i(y, z, \xi)$, where f_i satisfies

$$|f_i(y, z, \xi) - f_i(y', z, \xi)| \leq \omega(|y - y'|)(a(z) + f_i(y, z, \xi))$$

for all $y, y', z, \xi \in \mathbb{R}^n$, where ω and a are continuous positive real functions with $\omega(0) = 0$. In the third variable we let f be convex and satisfying the growth condition (4).

Theorem 3.1. *We have the existence of the* Γ*-limit* $f_{\text{hom}} = \Gamma\text{-}\lim f_h$ *where* $f_h(x,\xi) = f(hx, h^2x, \xi)$. *Moreover,*

$$f_{\text{hom}}(\xi) = f^{[2]}(\xi),$$

where $f^{[2]}$ *is found iteratively according to the following scheme:*

$$f^{[2]}(\xi) = \min_{u \in W^{1,p}_{\text{per}}(I)} \int_I f^{[1]}(y, \xi + Du(y))\, dy$$

$$f^{[1]}(x,\xi) = \min_{u \in W^{1,p}_{\text{per}}(I)} \int_I f^{[0]}(x, y, \xi + Du(y))\, dy$$

$$f^{[0]}(x, y, \xi) = f(x, y, \xi).$$

The above theorem is easily extended to the case $f_h(x,\xi) = f(hx, \dots, h^m x, \xi)$ in which

$$f_{\text{hom}}(\xi) = f^{[m]}(\xi),$$

where $f^{[m-j]}$ is found iteratively according to the following scheme:

$$f^{[m-j]}(\mathbf{x}_1, \dots, \mathbf{x}_j, \xi) = \min_{u \in W^{1,p}_{\text{per}}(I)} \int_I f^{[m-j-1]}(\mathbf{x}_1, \dots, \mathbf{x}_j, y, \xi + Du(y))\, dy$$

$$f^{[0]}(\mathbf{x}_1, \dots, \mathbf{x}_m, \xi) = f(\mathbf{x}_1, \dots, \mathbf{x}_m, \xi).$$

In Figures 1 and 2 we have illustrated examples where $f_h(x,\xi) = \lambda(hx, h^2x)\,|\xi|^2$. In Figure 1

$$\lambda(x, y) = 2 - k(x_1)k(y_2)$$

and in Figure 2

$$\lambda(x, y) = 2 - k(x_1)k(x_2)k(y_1)k(y_2),$$

where

$$k(t) = \begin{cases} 0 & t \in [0+n, \frac{1}{3}+n[\\ 1 & t \in [\frac{1}{3}+n, 1+n[\end{cases}, \quad n \text{ is an integer.}$$

The conductivity $\lambda(hx, h^2x)$ takes the value 2 when x is in the black part and the value 1 when x is in the white part, respectively.

Remark 3.2. Theorem 3.1 of Braides and Lukkassen [9, 22] is a generalization of the case of quadratic forms given by Bensoussan, Lions and Papanicolaou in the book [7] where the concept of reiterated homogenization was introduced. Later on Müller, Braides and Defranceschi generalized this result to standard non-convex Lagrangians (see [8]). The corresponding proofs are quite different.

Remark 3.3. In its form Theorem 3.1 (and likewise its non-convex cousin) is a natural generalization of the periodic case, and agrees with the physical intuition that the

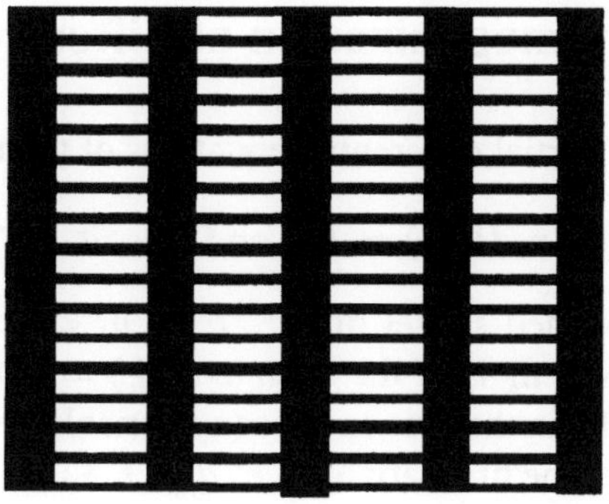

Figure 1: The laminate structure of rank 2. Here, $m = 2$ (h is a fixed number).

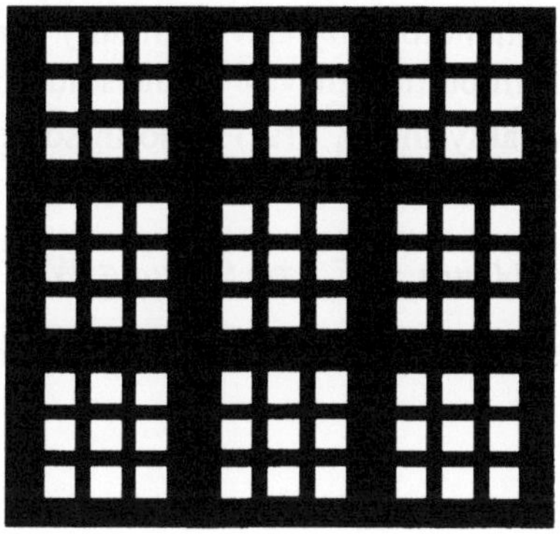

Figure 2: The reiterated cube structure for $m = 2$ (h is a fixed number).

effective properties should be obtained by first homogenizing the medium on the finest microlevel and next on the second one. However, in the case of non-standard Lagrangians, satisfying the growth condition

$$-c_0 + c_1 |\xi|^q \leq f(x, y, \xi) \leq c_0 + c_2 |\xi|^p, \tag{5}$$

an unexpected phenomenon occurs: In contrast to the periodic case the limit Γ–lim f_h may not exist (see [25])! However, by compactness we always have the existence of Γ-converging subsequences. It is possible to show that all such limits are sharply bounded between two Lagrangians whose representation can be found iteratively by a more general scheme than that above. These bounds are strongly dependent on the power q and p of the Sobolev-spaces involved and they are (certainly) equal when $q = p$. Non-trivial examples can be found where these bounds are attained.

4. Reiterated homogenization of differential operators

We consider the class of partial differential equations of the form

$$-\operatorname{div}(a_\varepsilon(x, Du_\varepsilon)) = F \quad \text{on } \Omega,\ u_\varepsilon \in W_0^{1,p}(\Omega), \tag{6}$$

where Ω is an open bounded subset of R^n, $1 < p < \infty$, $1/p + 1/q = 1$ and $F \in W^{-1,q}(\Omega)$. Moreover, a_ε is of the form $a_\varepsilon(x, \xi) = a(x/\varepsilon, x/\varepsilon^2, \xi)$. Here, a is Y-periodic and Z-periodic in the first and second variable, respectively. Moreover, a is piecewise continuous in the first variable, i.e. $a(y, z, \xi) = \sum_{i=1}^{N} \chi_{\Omega_i}(y) a_i(y, z, \xi)$, where a_i satisfies

$$|a_i(y_1, z, \xi) - a_i(y_2, z, \xi)|^q \le \omega(|y_1 - y_2|)\left(1 + |\xi|^p\right),$$

and $\omega : R \to R$ is continuous, increasing, $1/p + 1/q = 1$ and $\omega(0) = 0$. In the third variable a satisfies suitable monotonicity and continuity conditions.

We want to know the asymptotic behavior of the solutions u_ε of (6) as $\varepsilon \to 0$ and prove that u_ε converges weakly in $W_0^{1,p}(\Omega)$ (also in some multiscale sense) to the solution u_0 of the problem

$$-\operatorname{div} b(Du_0) = F \quad \text{on } \Omega,\ u_0 \in W_0^{1,p}(\Omega),$$

whose representation can be obtained from a by reiteration of a "homogenization formula".

Theorem 4.1. *It holds that*

$$u_\varepsilon \to u_0 \quad \textit{weakly in } W_0^{1,p}(\Omega),$$
$$a(\frac{x}{\varepsilon}, \frac{x}{\varepsilon^2}, Du_\varepsilon) \to b(Du_0) \quad \textit{weakly in } L^q(\Omega, R^n),$$

as $\varepsilon \to 0$, where u_0 is the unique solution of the problem

$$\int_\Omega (b(Du_0), D\phi)\, dx = \langle F, \phi\rangle \quad \textit{for every } \phi \in W_0^{1,p}(\Omega),$$

$$u_0 \in W_0^{1,p}(\Omega).$$

Here, the operator $b : R^n \to R^n$ is defined by

$$b(\xi) = \frac{1}{|Y|}\int_Y b_1(y, \xi + Dv^\xi(y))\, dy,$$

where v^ξ is the unique solution of the cell-problem

$$\int_Y (b_1(y, \xi + Dv^\xi(y)), D\phi)\, dy = 0 \quad \textit{for every } \phi \in W_{per}^{1,p}(Y),$$

$$v^\xi \in W_{per}^{1,p}(Y).$$

Moreover, the operator $b_1 : Y \times R^n \to R^n$ *is given by*

$$b_1(y, \xi) = \frac{1}{|Z|} \int_Z a(y, z, \xi + Dv^{\xi, y}(z))\, dz,$$

where $v^{\xi, y}$ *is the unique solution of the cell-problem*

$$\int_Z (a(y, z, \xi + Dv^{\xi, y}(z)), D\phi)\, dz = 0 \quad \textit{for every } \phi \in W^{1,p}_{\text{per}}(Z),$$

$$v^{\xi} \in W^{1,p}_{\text{per}}(Z).$$

Definition 4.2. We say that a sequence u_ε 3-scale converges to

$$u_0(x, y, z) \in L^p(\Omega, Y, Z)$$

if

$$\lim_{\varepsilon \to \infty} \int_\Omega u_\varepsilon(x) \varphi\left(x, \frac{x}{\varepsilon}, \frac{x}{\varepsilon^2}\right) dx = \frac{1}{|Y|\,|Z|} \int_{\Omega \times Y \times Z} u_0(x, y, z) \varphi(x, y, z)\, dx\, dy\, dz$$

for all $\varphi \in \mathcal{D}(\Omega; C^\infty_{\text{per}}(Y, Z))$ (which means functions $\varphi(x, y, z)$ being C^∞ with compact support in x with values of functions which are C^∞ and periodic in y and z).

It is possible to extend Theorem 4.1 to the case $a_\varepsilon(x, \xi) = a(x, x/\varepsilon, x/\varepsilon^2, \xi)$. In the following theorem we let a_ε be of this form.

Theorem 4.3. *Let* u_ε *be a solution of* (6). *Then as* $\varepsilon \to 0$, *it holds that* u_ε *3-scale converges to* $u_0 \in W^{1,p}_0(\Omega)$ *and* Du_ε *3-scale converges to* $Du_0 + D_y u_1 + D_z u_2$, *where* $\{u_0, u_1, u_2\}$ *is a solution of*

$$\frac{1}{|Y|\,|Z|} \int_{\Omega \times Y \times Z} a(x, y, z, Du_0 + D_y u_1 + D_z u_2)(Dv_0 + D_y v_1 + D_z v_2) dx\, dy\, dz =$$

$$= \int_\Omega F v_0\, dx$$

for all $v_0 \in W^{1,p}_0(\Omega)$, $v_1 \in L^p(\Omega; W^{1,p}_{\text{per}}(Y))$, $v_2 \in L^p(\Omega; W^{1,p}_{\text{per}}(Y \times Z))$.

Remark 4.4. The iterated homogenization theorem for monotone operators, Theorem 4.1, was proved by Lions, Lukkassen, Persson and Wall in [20, 21].

Remark 4.5. The concept of multiscale convergence was used by Allaire and Briane [2] to study linear homogenization problems with several scales (which generalizes the concept of 2-scale convergence of Allaire and Nguetseng [1, 37]). The proof of the nonlinear version given in Theorem 4.3 can be found in [21].

Remark 4.6. Theorem 4.1 and Theorem 4.3 can easily be extended to the case when a_ε is on the form $a_\varepsilon(x, \xi) = a(x, x/\varepsilon, \dots, x/\varepsilon^m, \xi)$. Following an idea of Bensoussan and Lions [6] the case of infinitely many scales for linear problems was also studied by Allaire and Briane [2]. It seems to be possible to generalize such results to monotone problems as well.

4.1. Differential effective medium theory

Since reiterated homogenization is quite natural from a physical point of view (see Remark 3.3) it does not come as a surprise that reiterated techniques were used long before the homogenization theory itself was developed. Such techniques led to the so called differential effective medium theory (DEM) introduced in the 30's by Bruggeman [10] for materials with two phases and generalized by Norris to multiphase materials [38]. For two-phase materials the theory is roughly speaking as follows: Let C be the tensor of a "matrix material", let C_1 be the tensor of the "inclusion material" and let c be the volume fraction of the inclusion material. Moreover, suppose that for small c and all C we have

$$C_{\text{eff}} = C + cQ(C) + O(c^2), \tag{7}$$

where Q is a continuous function. Then, the system of ordinary differential equations

$$\frac{dC}{dt} = Q(C(t))$$

gives a representation of the realizable effective properties which can be obtained from an initial material C_0 by iterating (7) (an incremental procedure).

It was rigorously proved by Milton [30, 31] that Bruggeman's differential scheme with spherical inclusions corresponds to a differential microstructure. Later on this result was generalized by Avellaneda [4].

For more detailed information on DEM and its application we refer to Avellaneda [4]. Other interesting variants and applications can be found in the works of Beliaev and Kozlov [5], Jikov and Kozlov [15] and Kozlov [18]. For the case of DEM and reiterated homogenization of linear equations with random coefficients, see Kozlov [17]. DEM can also be used in case of nonlinear problems (see [26]).

5. Further applications

By combining some suitable bounds for nonlinear homogenized functionals with the nonlinear versions of the iterated homogenization above it is possible to give very sharp estimates of the homogenized functional also in cases when it cannot be computed exactly. These estimates can be used to study some two-component reiterated structures with rather surprising macroscopic behavior. In [9] and [22] three types of structures were analyzed: the laminate structure (Figure 1), the iterated cube structure (Figure 2) and a mixed iterated structure (Figure 5), the latter being a mixture of the first two structures and a structure of chess-board type. In particular, in [9] and [22] we pointed out some cases where the macroscopic behaviors of the iterated cube structure and the mixed reiterated structure possess a higher or lower effective energy density than that of the best possible laminate structure of rank n (where n is the dimension

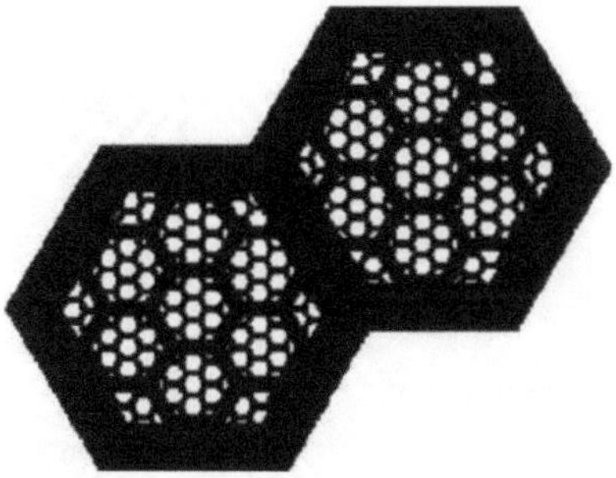

Figure 3: Iterated hexagonal honeycombs (illustrated for $m = 3$).

of the space). The fact that this has proven to be impossible for two-phase linear conductivity problems makes these structures particularly interesting (it should be mentioned that infinite rank laminates do as well as these structures in the non-linear case). Moreover, the results in [9] imply that the effective properties corresponding to the mixed reiterated structure are extremely sensitive to the growth on the local energy density. Other examples of surprising phenomena have been reported for nonlinear iterated checkerboard structures [25].

Reiterated homogenization of linear elliptic operators and effective differential medium theory has proven to be an important tool in the construction of iterated laminates and other iterated structures with optimal effective behavior. The discoveries so far even include globally isotropic structures with negative Poisson's ratio (for more information see Milton [33] and Lakes [19]). Concerning optimal structures and bounds on effective material properties in general we refer to the collection of classical papers [12], where the introduction gives a good selection of references, and particularly the recent book of Milton [34]. Here, we only mention a few examples of optimal iterated structures. Concerning optimal iterated laminates for the conductivity problem see Lurie and Cherkaev [27], Milton [29], Schulgasser [42] and Tartar [45]. On optimal microstructures for the elasticity case in two-dimensions (transversely isotropic case) there exist periodic (i.e. non-iterated) geometries found by Vigdergauz [46, 47, 48, 49, 50] which are optimal for the effective bulk modulus. A very readable treatment of his work can be found in [14]. There is also a new class of optimal iterated structures found recently by Sigmund [43]. For the effective shear modulus, there appear to be no "simple microstructures". All known ones have structure on at least 5 well-separated length scales (iteration levels). They simultaneously attain both the shear and bulk modulus bounds. The result that the Hashin-Shtrikman shear bound is attained is in essence in an article of Roscoe [41]. The only missing part is a rigorous proof that the differential scheme (in the effective differential medium theory) corresponds to some geometry. Such a proof can be found in Avellaneda [4]. Independent of this, the attainability of the shear modulus bound has been shown by Norris [38], Francfort and Murat [13] and Milton [32]. There was also an appendix

Figure 4: Globally isotropic structures with negative Poisson's ratio (of Milton).

in a preprint of Lurie and Cherkaev, written during the same time, that showed realizability. Unfortunately their paper was published without the appendix. Another iterated structure which yields the same effective behavior can be found in [24] (by similar methods one can also verify that this even holds for iterated triangular honeycombs). Moreover, in that paper (see also [22]) an iterated cube structure consisting of m iteration levels was analyzed for the elasticity and the conductivity problem.

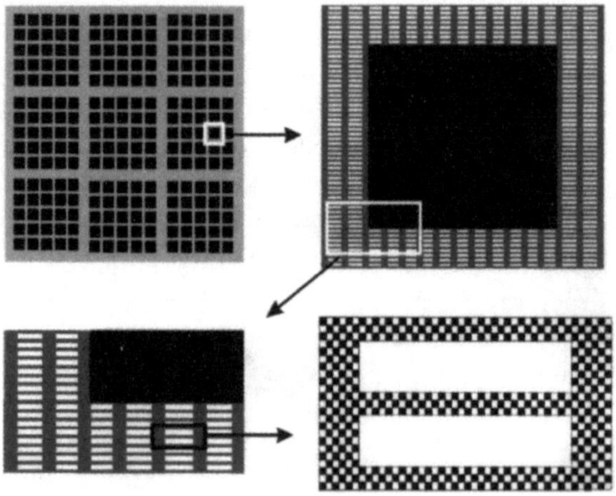

Figure 5: A mixed iterated structure.

In particular it was proven that this structure turns optimal in the class of square symmetric structures when $m \to \infty$ for the elasticity case and also optimal with respect to effective conductivity (recently, we have become aware of the fact that in the conductivity case this was already proven in [17] for a very similar structure. The derivation, though, is different). Similar results were obtained for iterated hexagonal honeycombs in [23] for the conductivity case (see Figure 3). It is also possible to prove optimality for the effective bulk modulus for these structures. For numerical computations on such structures we refer to [11].

For non-well ordered materials the same microstructures attain the bulk modulus bounds. In three dimensions the Walpole bound on the shear modulus can be improved as shown in [28, 35]. In two dimensions it is not known if the Walpole bounds can be improved. Microgeometries found by Sigmund (see above) come close to attaining them. There are also a number of papers on optimal microgeometries for multiphase materials (see [34] and the references given there). Besides these structures there is also another family of microstructures introduced in [34] in the name of *partial differential microstructures.*

6. A final comment

The homogenized integrand $f_{\text{hom}}(\xi)$ can be seen as some kind of average of the local function $f_h(\cdot, \xi)$. In the laminate case we even have a direct link to the well known arithmetic and harmonic mean. A similar example is the checkerboard structure for which $f_{\text{hom}}(\xi) = g(\lambda)\,|\xi|^2$, where $g(\lambda)$ denotes the geometric mean. Power means and compositions of such means often serve as upper and lower bounds for $f_{\text{hom}}(\xi)$ (see e.g. [22], [26], [36] and [44]). Hence, there is an obvious link between reiterated homogenization and iteration of means. The latter subject has fascinated many (see e.g. Jaak Peetre [3], [39], [40] and the references given there).

References

[1] G. Allaire. Homogenization and two-scale convergence, SIAM J. Math. Anal. 23 (1992), 1482–1518.

[2] G. Allaire and M. Briane, Multiscale convergence and reiterated homogenization, Proc. Roy. Soc. Edinburgh Ser. A 126 (1996), 297–342.

[3] J. Arazy, T. Claesson and J. Peetre, Means and their iterations, Proceedings of the Nineteenth Nordic Congress of Mathematics, Reykjavik (1984), 191–212.

[4] M. Avellaneda, Iterated homogenization, differential effective medium theory and applications, Commun. Pure Appl. Math. 40 (1987), 527–554.

[5] A. Beliaev and S.M. Kozlov, Hierarchical structures and estimates for homogenized coefficients, Russian J. Math. Phys. 1 (1993), 5–18.

[6] A. Bensoussan and J. L. Lions, Homogenization with an infinite number of periodic arguments, unpublished notes.

[7] A. Bensoussan, J. L. Lions and G. C. Papanicolaou, Asymptotic Analysis for Periodic Structures, North Holland, Amsterdam, 1978.

[8] A. Braides and A. Defranceschi, Homogenization of Multiple Integrals, Oxford University Press, Oxford, 1998.

[9] A. Braides and D. Lukkassen, Reiterated homogenization of integral functionals, Math. Mod. Meth. Appl. Sci, 10 (2000), 47–71.

[10] D. A. G. Bruggeman, Berechnung verschiedener physikalisher konstanten von heterogenen substanzen, Ann. Phys. 24 (1935), 634.

[11] J. Bystrøm, J. Helsing and A. Meidell, Some computational aspects of iterated structures, Composites Part B 32 (2001), 485–490.

[12] A. Cherkaev and R. Kohn (eds), Topics in mathematical modelling of composite materials, Birkhäuser, Bosten, 1997.

[13] G. A. Francfort and F. Murat, Homogenization and optimal bounds in linear elasticity, Arch. Rat. Mech. Anal. 94 (1986), 161–177.

[14] Y. Grabovsky and R. V. Kohn, Microstructures minimizing the energy of a two phase elastic composite in two space dimensions. II: The Vigdergauz microstructure, J. Mech. Phys. Solids 43 (1995), 949–972.

[15] V. V. Jikov and S. M. Kozlov, Multiscaled homogenization, in: Homogenization, Serguei Kozlov memorial volume, Ser. Adv. Math. Appl. Sci. 50, World Scientific, 1999.

[16] V. V. Jikov, S. M. Kozlov and O. A. Oleinik, Homogenization of differential operators and integral functionals, Springer-Verlag, Berlin, 1994.

[17] S. M. Kozlov, A central limit theorem for multiscaled permeability, Transaction of the Oberwolfach Porous Medium Conference, Birkhäuser, Basel, 1993.

[18] S. M. Kozlov, Multiscaled approach in homogenization, Proceedings of the second workshop on Composite media and homogenization theory, 217-227, World Scientific, 1995.

[19] R. Lakes, Materials with structural hierarchy, Nature 361 (1993), 511–515.

[20] J.-L. Lions, D. Lukkassen, L.-E. Persson and P. Wall, Reiterated homogenization of monotone operators, C. R. Acad. Sci. Paris Ser. I Math. 330 (2000), 675–680.

[21] J.-L. Lions, D. Lukkassen, L.-E. Persson and P. Wall, Reiterated homogenization of nonlinear monotone operators, Chinese Ann. Math. Ser. B 22 (2001), 1–14.

[22] D. Lukkassen, Formulae and bounds connected to optimal design and homogenization of partial differential operators and integral functionals, Ph.D. thesis, Dept. of Math., Tromsö University, Norway, 1996.

[23] D. Lukkassen, Bounds and homogenization of optimal reiterated honeycombs, in: Computer aided optimum design of structures V (S. Hernándes, C. A. Breddia, eds.), Computational Mechanical Publications, Southampton, 1997, 267–276 .

[24] D. Lukkassen, A new reiterated structure with optimal macroscopic behaviour, SIAM J. Appl. Math. 59, 5 (1999), 1825–1842.

[25] D. Lukkassen, Reiterated homogenization of non-standard Lagrangians. C. R. Acad. Sci. Paris Ser. I Math. 332 (2001), 999–1004.

[26] D. Lukkassen, J. Peetre and L. E. Persson, On some iterated means from homogenization theory, submitted.

[27] K.A. Lurie and A. V. Cherkaev Exact estimates of conductivity of composites formed by two isotropically conducting media taken in prescribed proportion. Proc. Roy. Soc. Edinburgh A 99 (1984), 71–87.

[28] G. W. Milton, Bounds on the electromagnetic, elastic, and other properties of two-component composites, Phys. Rev. Lett. 46 (1981), 542–545.

[29] G. W. Milton, Bounds on the complex permittivity of a two-component composite material, J. Appl. Phys. 52 (1981), 5286–5293.

[30] G. W. Milton, Some exotic models in statistical physics. I. The coherent potential approximation is a realizable effective medium scheme, II. Anomalous first-order transitions, Ph.D. thesis, Inst-Cornell, 1985.

[31] G. W. Milton, The coherent potential approximation is a realizable effective medium scheme, J. Commun. Math. Phys. 99 (1985), 4, 463–500.

[32] G. W. Milton, Modeling the properties of composites by laminates. in: Homogenization and Effective Moduli of Materials and Media (J. L. Ericksen, D. Kinderlehrer, R. Kohn, and J.-L. Lions, eds.), IMA Vol. Math. Appl. 1, Springer-Verlag, 1986, 150–174.

[33] G. W. Milton, Composite materials with Poissons ratio close to -1, J. Mech. Phys. Solids 49 (1992), 1105–37.

[34] G. W. Milton, The theory of composites, Cambridge University Press, Cambridge, 2001.

[35] G. W. Milton and N. Phan-Thien, New bounds on effective elastic moduli of two-component materials, Proc. Roy. Soc. London A 380 (1982), 305–331.

[36] S. Mortola and S. Steffe, A two-dimensional homogenization problem (Italian), Atti Accad. Naz. Lincei Cl. Sci. Fis. Mat. Natur. Rend. Lincei (8) Mat. Appl. 78 (3) (1985), 77–82.

[37] G. Nguetseng. A general convergence result for a functional related to the theory of homogenization. SIAM J. Math. Anal. 20 (1989), 608–623.

[38] A. N. Norris, A differential scheme for effective moduli of composites, Mechanics of Materials, 4 (1985), 1–16.

[39] J. Peetre, Generalizing the arithmetic-geometric mean-a hapless computer experiment, Internat. J. Math. Sci. 12 (1989), 235–246.

[40] J. Peetre, Some observations on algorithms of the Gauss-Borchardt type, Proc. Edinburgh Math. Soc. 34 (1991), 415–431.

[41] R. Roscoe, Isotropic composites with elastic or viscoelastic phases: General bounds for the moduli and solutions for special geometries, J. Rheol-Acta 12 (1973), 404–411.

[42] K. Schulgasser, Bounds on the conductivity of statistically isotropic polycrystals, J. Phys. C 10 (1977), 407–417.

[43] O. Sigmund, A new class of extremal composites, J. Mech. Phys. Solids 48 (2000), 397–428.

[44] L. Tartar, Estimation de coefficients homogeneises, Lecture Notes in Math. 704, Springer-Verlag, Berlin, 1979, 364–373.

[45] L. Tartar, Estimation fines de coefficients homogénéisés, in: Ennio De Giorgi's Colloquium (P. Kree, ed.), Pitman Research Notes in Math., Pitman, London, 1985.

[46] S. B. Vigdergauz, Effective elastic parameters of a plate with a regular system of equal-strength holes, Mech. Solids 21 (1986), 165–169.

[47] S. B. Vigdergauz, Two-dimensional grained composites of extreme rigidity, J. Appl. Mech. 61 (1994), 390–394.

[48] S. B. Vigdergauz, Rhombic lattice of equi-stress inclusions in an elastic plate, Quart. J. Mech. Appl. Math, 49 (1996), 565–580.

[49] S. B. Vigdergauz, Energy-minimizing inclusions in a planar elastic structure with macroisotropy, J. Struct. Optim. 17 (1999), 104–112.

[50] S. B. Vigdergauz, Complete elasticity solution to the stress problem in a planar grained structure, Math. Mech. Solids 4 (1999), 407–439.

Dag Lukkassen, Narvik University College, P.O. Box 385, 8505 Narvik, Norway
E-mail: Dag.Lukkassen@hin.no

Graeme W. Milton, Department of Mathematics, The University of Utah, Salt Lake City, UT 84112-0090, U.S.A.
E-mail: milton@math.utah.edu

Boundedness and compactness criteria for the one-dimensional Schrödinger operator

Vladimir G. Maz'ya and Igor E. Verbitsky *

Dedicated to Jaak Peetre on the occasion of his 65th birthday

Abstract. We establish necessary and sufficient conditions for the boundedness of the one-dimensional Schrödinger operator $\mathcal{L} = -d^2/dx^2 + Q$ from the Sobolev space $\overset{\circ}{L}{}^1_p$ to L_p^{-1}, $1 < p < \infty$, on a half-line or a finite line segment, for an arbitrary complex-valued potential Q. Analogous results for Sobolev spaces W_p^1 on the real line, as well as the corresponding compactness criteria, and related two weight inequalities are discussed.

2000 Mathematics Subject Classification: 26D10, 34L40, 35J10, 47E05.

1. Introduction

We are concerned with the problem of characterizing "indefinite weights" $Q(x)$ on $\Omega \subset \mathbb{R}^n$ such that the weighted inequality

$$\left| \int_\Omega |u(x)|^2 \, Q(x) \, dx \right| \le \text{const} \int_\Omega |\nabla u(x)|^2 \, dx, \tag{1.1}$$

holds for all $u \in C_0^\infty(\Omega)$. Here Q may be a real- or complex-valued function, or even a distribution, without any a priori regularity assumptions on Q.

This inequality, together with its inhomogeneous analogue,

$$\left| \int_\Omega |u(x)|^2 \, Q(x) \, dx \right| \le \text{const} \left(\int_\Omega |\nabla u(x)|^2 \, dx + \int_\Omega |u(x)|^2 \, dx \right), \tag{1.2}$$

*Supported in part by NSF Grant DMS-0070623.

plays a major role in spectral theory of the Schrödinger operator $\mathcal{L} = -\Delta + Q$. (See [1–3, 6, 7, 9, 11, 12].) In particular, (1.1) is equivalent to the boundedness of $\mathcal{L} : \overset{\circ}{L}{}^1_2(\Omega) \to L_2^{-1}(\Omega)$, where $\overset{\circ}{L}{}^1_2(\Omega)$ is the completion of $C_0^\infty(\Omega)$ with respect to the Dirichlet norm $||\nabla u||_{L_2(\Omega)}$, and $L_2^{-1}(\Omega)$ is the dual of $\overset{\circ}{L}{}^1_2(\Omega)$.

The usual approach, in case Q is real-valued, is to represent it in the form $Q = Q_+ - Q_-$, where Q_+ and Q_- are respectively the positive and negative parts of Q, and then treat them separately. However, this procedure ignores a possible cancellation between Q_+ and Q_-, and diminishes the class of admissible potentials Q.

For general Q, this problem was solved recently by the authors in [9] where necessary and sufficient conditions were found which ensure that (1.1) holds in the case $\Omega = \mathbb{R}^n$, $n \geq 3$; analogous characterizations were obtained for (1.2). Similar results for a wide class of bounded domains $\Omega \subset \mathbb{R}^n$, including those with Lipschitz boundaries, were established as well.

In this paper, we treat analogous problems for the Schrödinger operator $\mathcal{L} = -d^2/dx^2 + Q$ on the positive real axis $\mathbb{R}_+$, the real line $\mathbb{R}$, and a finite interval. In particular, we characterize all admissible potentials Q such that

$$\mathcal{L} : \overset{\circ}{L}{}^1_2 \to L_2^{-1} \tag{1.3}$$

is a bounded operator. Moreover, we actually develop a more general theory which includes the corresponding L^p inequalities and weighted norms.

Obviously, one encounters fewer technical difficulties in the one-dimensional case, because there is no need to employ advanced methods of potential theory and weighted norm inequalities. However, it is instructive to consider it in detail since many features of the multidimensional characterizations obtained in [9] are already present in this setting. Both the statements of the main results and the proofs become more transparent and elementary. Nevertheless, they seem to be new, at least in the necessity part, even for experts in spectral theory of differential operators.

We would like to mention that our investigation started with a search for boundedness and compactness criteria for the Schrödinger operator on a half-line, which were established during a visit of the first author to the University of Missouri-Columbia in March, 1997.

We first state our main results for $\Omega = \mathbb{R}^n$, $n \geq 3$. Generally, the left-hand side of (1.1) is understood as the absolute value of the quadratic form $\langle Qu, u\rangle$ generated by the corresponding multiplication operator (rigorous definitions are given in [9]).

Theorem I. *The inequality* (1.1) *holds if and only if Q is the divergence of a vector-field $\vec{\Gamma} : \mathbb{R}^n \to \mathbb{C}^n$ $(n \geq 3)$ such that*

$$\int_{\mathbb{R}^n} |u(x)|^2 \, |\vec{\Gamma}(x)|^2 \, dx \leq \text{const} \int_{\mathbb{R}^n} |\nabla u(x)|^2 \, dx, \tag{1.4}$$

for all $u \in C_0^\infty(\mathbb{R}^n)$. *The vector-field* $\vec{\Gamma} \in \mathbf{L}_2^{\mathrm{loc}}(\mathbb{R}^n)$ *can be chosen in the form* $\vec{\Gamma} = \nabla \Delta^{-1} Q$.

Thus, the problem has been reduced to a trace inequality with nonnegative weight $|\vec{\Gamma}(x)|^2$ which has been very well studied by now. (See, e.g., [3, 8, 9, 13], and the literature cited there.) A number of explicit characterizations, including simpler sufficient and necessary conditions which follow immediately from Theorem I are thoroughly discussed in [9].

We now state an analogue of Theorem I for a half-line. As above, Q can be a real- or complex-valued locally integrable function, or a distribution. We use the standard notation $\mathcal{D} = C_0^\infty(\mathbb{R}_+)$ for C^∞ functions with compact support in $\mathbb{R}_+ = (0, +\infty)$, and $\mathcal{D}' = \mathcal{D}'(\mathbb{R}_+)$ for the corresponding class of distributions.

Theorem II. *Let* $Q \in \mathcal{D}'(\mathbb{R}_+)$. *The following statements are equivalent.*

(i) *The Schrödinger operator* $\mathcal{L} = -\frac{d^2}{dx^2} + Q$ *is a bounded operator from* $\overset{\circ}{L}{}_2^1(\mathbb{R}_+)$ *to* $L_2^{-1}(\mathbb{R}_+)$.

(ii) *The inequality*

$$|\langle Qu, u\rangle|^2 \le \text{const} \int_{\mathbb{R}_+} |u'(x)|^2 \, dx \tag{1.5}$$

holds for all $u \in C_0^\infty(\mathbb{R}_+)$.

(iii) *There exists* $\Gamma \in L_2^{\mathrm{loc}}(\mathbb{R}_+)$ *so that* $Q = \Gamma'$ *in* $\mathcal{D}'(\mathbb{R}_+)$, *and the inequality*

$$\int_{\mathbb{R}_+} |u(x)|^2 \, |\Gamma(x)|^2 \, dx \le \text{const} \int_{\mathbb{R}_+} |u'(x)|^2 \, dx \tag{1.6}$$

holds for all $u \in C_0^\infty(\mathbb{R}_+)$.

The preceding inequality holds if and only if

$$\sup_{a>0} a \int_a^\infty |\Gamma(x)|^2 \, dx < \infty. \tag{1.7}$$

Furthermore, the multiplication operator Q *mapping* $\overset{\circ}{L}{}_2^1(\mathbb{R}_+)$ *to* $L_2^{-1}(\mathbb{R}_+)$ *is compact if and only if*

$$\lim_{a\to 0+} a \int_a^\infty |\Gamma(x)|^2 \, dx = 0, \quad \text{and} \quad \lim_{a\to +\infty} a \int_a^\infty |\Gamma(x)|^2 \, dx = 0. \tag{1.8}$$

Remark 1. A distribution Γ such that $Q = \Gamma'$ is defined up to a constant. It follows from (1.7) that actually Γ is determined uniquely. If $Q \in L_1^{\mathrm{loc}}(\mathbb{R}_+)$, and $\lim_{b\to+\infty} \int_a^b Q(t)\,dt$ exists for every $a > 0$, then

$$\Gamma(x) = -\int_x^{+\infty} Q(t)\,dt.$$

We will use this notation sometimes even if Q is a distribution.

Remark 2. For *nonnegative* Q, (1.7) is easily seen to be equivalent to the classical Hille condition (see [4, 5]):

$$\sup_{a>0} a \int_a^\infty Q(x)\,dx < \infty. \tag{1.9}$$

A similar statement is true for the compactness criterion (1.8).

Remark 3. For general Q, the following simple condition obviously implies (1.7):

$$\sup_{a>0} a \left| \int_a^\infty Q(x)\,dx \right| < \infty.$$

However, it is not necessary for any one of the equivalent statements of Theorem II to hold. See Example 2 at the end of this section.

We observe that our boundedness and compactness results can be carried over to Sobolev spaces $\overset{\circ}{L}{}^1_p(\mathbb{R}_+)$ where $1 < p < \infty$. Let $\langle Q\cdot, \cdot\rangle$ be a bilinear form generated by the corresponding multiplication operator, with Q being a complex-valued function or distribution on $\mathbb{R}_+$ as above.

Theorem III. *Let $Q \in \mathcal{D}'(\mathbb{R}_+)$. Let $p > 1$, $\frac{1}{p} + \frac{1}{p'} = 1$, and $p^* = \max(p, p')$. Then the following statements are equivalent.*

(i) *The Schrödinger operator $\mathcal{L} = -\frac{d^2}{dx^2} + Q$ is a bounded operator from $\overset{\circ}{L}{}^1_p(\mathbb{R}_+)$ to $L_p^{-1}(\mathbb{R}_+)$.*

(ii) *The inequality*

$$|\langle Qu, v\rangle| \le \text{const}\, ||u'||_{L_p(\mathbb{R}_+)}\, ||v'||_{L_{p'}(\mathbb{R}_+)}, \tag{1.10}$$

holds for all $u, v \in C_0^\infty(\mathbb{R}_+)$.

(iii) *There exists $\Gamma \in L^{\text{loc}}_{p^*}(\mathbb{R}_+)$ so that $\Gamma' = Q$ in $\mathcal{D}'(\mathbb{R}_+)$, and both the inequalities*

$$\int_{\mathbb{R}_+} |u(x)|^p\, |\Gamma(x)|^p\, dx \le \text{const}\, ||u'||^p_{L_p(\mathbb{R}_+)}, \tag{1.11}$$

and

$$\int_{\mathbb{R}_+} |v(x)|^{p'}\, |\Gamma(x)|^{p'}\, dx \le \text{const}\, ||v'||^{p'}_{L_{p'}(\mathbb{R}_+)}, \tag{1.12}$$

are valid with the constants which do not depend on $u, v \in C_0^\infty(\mathbb{R}_+)$.

Inequalities (1.11) *and* (1.12) *are equivalent to the condition*

$$\sup_{a>0} a^{p^*-1} \int_a^\infty |\Gamma(x)|^{p^*}\, dx < \infty. \tag{1.13}$$

Furthermore, the multiplication operator $Q : \overset{\circ}{L}{}^1_p(\mathbb{R}_+) \to L_p^{-1}(\mathbb{R}_+)$ *is compact if and only if*

$$\lim_{a\to 0+} a \int_a^\infty |\Gamma(x)|^{p^*}\, dx = 0, \quad \text{and} \quad \lim_{a\to +\infty} a \int_a^\infty |\Gamma(x)|^{p^*}\, dx = 0. \tag{1.14}$$

Remark 4. For any $p \in (1, \infty)$, a simple condition

$$\sup_{a>0} a\, |\Gamma(a)| < \infty \tag{1.15}$$

is sufficient, but generally not necessary for (1.13) to hold. However, for *nonnegative* Q, condition (1.13) is equivalent to (1.15), and hence is actually independent of $p \in (1, \infty)$.

Note that in the multidimensional case the problem becomes nontrivial even for nonnegative measures Q on $\mathbb{R}^n$ (see [8]).

The following examples demonstrate the difference between sharp results which follow from Theorem III, and the usual approach where Q_+ and Q_- are treated separately.

Example 1. Let

$$Q(x) = \frac{\sin x}{x^{1+\epsilon}}, \qquad \epsilon > 0.$$

Then

$$-\Gamma(x) = \int_x^{+\infty} \frac{\sin t}{t^{1+\epsilon}} dt = \frac{\cos x}{x^{1+\epsilon}} + O\Big(\frac{1}{x^{2+\epsilon}}\Big) \quad \text{as } x \to +\infty.$$

As $x \to 0+$, clearly, $\Gamma(x) = O(1)$ for $\epsilon < 1$, $\Gamma(x) = O(\log x)$ for $\epsilon = 1$, and $\Gamma(x) = O(x^{1-\epsilon})$ for $\epsilon > 1$. From this it is easy to see that (1.13) is valid if and only if $0 \le \epsilon \le 2$, and hence by Theorem III, $\mathcal{L} : \overset{\circ}{L}{}^1_p(\mathbb{R}_+) \to L_p^{-1}(\mathbb{R}_+)$ is bounded for $1 < p < \infty$. Moreover, the multiplication operator $Q : \overset{\circ}{L}{}^1_p(\mathbb{R}_+) \to L_p^{-1}(\mathbb{R}_+)$ is compact if and only if $0 < \epsilon < 2$.

Note that the same Theorem III applied separately to Q_+ and Q_- gives a satisfactory result only for $1 \le \epsilon \le 2$.

In the next example, Q is a (complex-valued) measure on $\mathbb{R}_+$, and the condition imposed on Q depends explicitly on p.

Example 2. Let

$$Q = \sum_{j=1}^{\infty} c_j\, (\delta_j - \delta_{j+1}),$$

where δ_a is a unit point mass at $x = a$, and c_j are complex numbers. Then clearly

$$\Gamma(x) = \sum_{j=1}^{\infty} c_j \, \chi_{(j,j+1)}(x).$$

It follows that (1.13) holds if and only if

$$\sup_{n \geq 1} n^{p^*-1} \sum_{j=n}^{\infty} |c_j|^{p^*} < \infty.$$

In particular, for $1 < r \leq 2$, let $c_j = j^{-1/r}$ if $j = 2^m$, and $c_j = 0$ otherwise. Then $\mathcal{L} : \overset{\circ}{L}{}^1_p(\mathbb{R}_+) \to L^{-1}_p(\mathbb{R}_+)$ if and only if $r \leq p \leq r/(r-1)$. Note that in this example condition (1.15) fails for all $r > 1$.

In the next section we prove Theorem III, and also discuss related results for a finite interval and the real line, as well as some two weight generalizations.

2. Boundedness and compactness criteria

We start with the operator $\mathcal{L} = -\frac{d^2}{dx^2} + Q$ on a half-line. Define by $\overset{\circ}{L}{}^1_2(\mathbb{R}_+)$ the completion of $C_0^\infty(\mathbb{R}_+)$ with respect to the Dirichlet norm $||u'||_{L_2(\mathbb{R}_+)}$. Equivalently, $\overset{\circ}{L}{}^1_2(\mathbb{R}_+)$ can be defined as the space of absolutely continuous functions $u : \mathbb{R}_+ \to \mathbb{C}$ such that $u(0) = 0$, and

$$||u||_{\overset{\circ}{L}{}^1_2(\mathbb{R}_+)} = \left[\int_{\mathbb{R}_+} (x^{-2} \, |u(x)|^2 + |u'(x)|^2) \, dx \right]^{\frac{1}{2}} < \infty. \tag{2.1}$$

By $L^{-1}_2(\mathbb{R}_+)$ we denote the dual space $[\overset{\circ}{L}{}^1_2(\mathbb{R}_+)]^*$.

As was already pointed out in the Introduction, the boundedness of the operator $\mathcal{L} : \overset{\circ}{L}{}^1_2(\mathbb{R}_+) \to L^{-1}_2(\mathbb{R}_+)$ is equivalent to the quadratic form inequality

$$\left| \int_{\mathbb{R}_+} |u(x)|^2 \, Q(x) \, dx \right| \leq \text{const} \int_{\mathbb{R}_+} |u'(x)|^2 \, dx, \quad u \in C_0^\infty(\mathbb{R}_+), \tag{2.2}$$

and hence by the polarization identity, to the boundedness of the corresponding sesquilinear form:

$$\left| \int_{\mathbb{R}_+} u(x) \, \overline{v(x)} \; Q(x) \, dx \right| \leq \text{const} \, ||u'||_{L_2(\mathbb{R}_+)} \, ||v'||_{L_2(\mathbb{R}_+)}, \tag{2.3}$$

where the constant is independent of $u, v \in C_0^\infty(\mathbb{R}_+)$.

More generally, denote by $\overset{\circ}{L}{}^1_p(\mathbb{R}_+)$, $1 < p < \infty$, the completion of $C_0^\infty(\mathbb{R}_+)$ with respect to the norm $||u'||_{L_p(\mathbb{R}_+)}$, and let $L_p^{-1}(\mathbb{R}_+) = [\overset{\circ}{L}{}^1_{p'}(\mathbb{R}_+)]^*$, $1/p + 1/p' = 1$. The operator $\mathcal{L} : \overset{\circ}{L}{}^1_p(\mathbb{R}_+) \to L_p^{-1}(\mathbb{R}_+)$ is bounded if and only if

$$\left| \int_{\mathbb{R}_+} u(x)\, \overline{v(x)}\; Q(x)\, dx \right| \le \text{const}\, ||u'||_{L_p(\mathbb{R}_+)}\, ||v'||_{L_{p'}(\mathbb{R}_+)}, \tag{2.4}$$

for all $u, v \in C_0^\infty(\mathbb{R}_+)$.

Let us assume first that Q is a locally integrable real- or complex-valued function such that

$$\lim_{b\to+\infty} \int_a^b Q(x)\, dx = \int_a^\infty Q(x)\, dx \tag{2.5}$$

exists for every $a > 0$.

Notice that a similar argument works with minor changes when the integration in (2.4) is performed against a locally finite complex-valued measure dQ in place of $Q(x)dx$.

Theorem 2.1. *Under the above assumptions on Q, let*

$$\Gamma(x) = -\int_x^\infty Q(t)\, dt, \quad x > 0.$$

Let $1 < p < \infty$, and $p^* = \max(p, p')$. Then (2.4) is valid if and only if

$$\sup_{a>0} a^{p^*-1} \int_a^\infty |\Gamma(x)|^{p^*}\, dx < \infty. \tag{2.6}$$

It is not difficult to see that (2.6) is equivalent to a pair of conditions,

$$\sup_{a>0} a^{p-1} \int_a^\infty |\Gamma(x)|^{p}\, dx < \infty, \quad \sup_{a>0} a^{p'-1} \int_a^\infty |\Gamma(x)|^{p'}\, dx < \infty. \tag{2.7}$$

Proof. For $u, v \in C_0^\infty(\mathbb{R}_+)$, let

$$\langle Qu, v\rangle = \int_0^\infty Q(x)\, u(x)\, \overline{v(x)}\, dx.$$

We can extend $\langle Qu, v\rangle$ by continuity to the case where

$$u(x) = \int_0^x f(t)\, dt, \quad v(x) = \int_0^x g(\tau)\, d\tau,$$

for $f, g \in C_0^\infty(\mathbb{R}_+)$, by setting:

$$\langle Qu, v\rangle = \lim_{a\to+\infty} \int_0^a Q(x)\, u(x)\, \overline{v(x)}\, dx.$$

To show that the limit on the right-hand side exists, assume that both f and g are supported in $(\delta,\ b) \subset \mathbb{R}_+$. Then clearly,

$$\lim_{a\to+\infty} \int_0^a Q(x)\, u(x)\, \overline{v(x)}\, dx = \int_\delta^b Q(x) \left(\int_\delta^x f(t)\, dt \int_\delta^x \overline{g(\tau)}\, d\tau \right) dx$$

$$+ \int_b^\infty Q(x)\, dx \int_\delta^b f(t)\, dt \int_\delta^b \overline{g(\tau)}\, d\tau.$$

Observe that we have to be careful here: in what follows one cannot estimate the two terms on the right-hand side of the preceding equation separately because this would lead to the restriction:

$$\sup_{b>0} b \left| \int_b^\infty Q(x)\, dx \right| < \infty,$$

which is not necessary for the boundedness of the bilinear form. (See Remark 2 in the Introduction.)

Using Fubini's theorem, we obtain:

$$\begin{aligned}\langle Qu, v\rangle &= \int_\delta^b \int_\delta^b f(t)\, \overline{g(\tau)} \int_{\max(t,\tau)}^b Q(x)\, dx\ dt\, d\tau \\ &\quad + \int_b^\infty Q(x)\, dx \int_\delta^b f(t)\, dt \int_\delta^b \overline{g(\tau)}\, d\tau \\ &= \int_\delta^b \int_\delta^b f(t)\, \overline{g(\tau)} \int_{\max\{t,\tau\}}^\infty Q(x)\, dx\, dt\, d\tau.\end{aligned}$$

Thus, (2.4) is equivalent to the inequality

$$\left| \int_0^\infty \int_0^\infty f(t)\, \overline{g(\tau)}\, \Gamma(\max\{t, \tau\})\, dt\, d\tau \right| \le \text{const}\, \|f\|_{L_p(\mathbb{R}_+)}\, \|g\|_{L_{p'}(\mathbb{R}_+)}, \tag{2.8}$$

for compactly supported f, g.

Using the reverse Hölder inequality, the preceding estimate can be rewritten in the equivalent form:

$$\int_0^\infty \left| \int_0^\infty \Gamma(\max\{t, \tau\})\, f(t)\, dt \right|^p d\tau \le c\, \|f\|^p_{L_p(\mathbb{R}_+)}. \tag{2.9}$$

Clearly,

$$\int_0^\infty \Gamma(\max\{t, \tau\})\, f(t)\, dt = \Gamma(\tau) \int_0^\tau f(t)\, dt + \int_\tau^\infty f(t)\, \Gamma(t)\, dt. \tag{2.10}$$

Suppose now that (2.6), or equivalently, both inequalities in (2.7) hold. Then the estimate involving the first term in (2.10) is established by means of the weighted Hardy inequality:

$$\int_0^\infty \left| \int_0^\tau f(t)\,dt \right|^p |\Gamma(\tau)|^p\,d\tau \le C\,\|f\|^p_{L_p(\mathbb{R}_+)}, \tag{2.11}$$

which holds if and only if the first part of condition (2.7) is valid (see, e.g., [7]).

The second term in (2.10) is estimated by using a similar weighted Hardy inequality:

$$\int_0^\infty \left| \int_\tau^\infty f(t)\,\Gamma(t)\,dt \right|^p d\tau \le C\,\|f\|^p_{L_p(\mathbb{R}_+)},$$

which is known [7] to be equivalent to the second part of condition (2.7). This proves the "if" part of the theorem.

To prove the "only if" part, it suffices to assume that $f(x)$ in (2.9) is supported on an interval $[\delta, a]$, $a > 0$, and restrict the domain of integration in τ on the left-hand side of (2.9) to $\tau \in (a, +\infty)$. Taking into account that the second term in (2.10) vanishes, we get:

$$\int_a^\infty \left| \int_\delta^a \Gamma(\max\{t,\tau\}) f(t)\,dt \right|^p d\tau = \left| \int_\delta^a f(t)\,dt \right|^p \int_a^\infty |\Gamma(\tau)|^p d\tau \le C\|f\|^p_{L_p(\mathbb{R}_+)}.$$

Applying the reverse Hölder inequality again, we obtain the first part of (2.7):

$$a^{p-1} \int_a^\infty |\Gamma(\tau)|^p\,d\tau \le C.$$

Since (2.9) is symmetric, a dual estimate in $L^{p'}$-norm yields the second part of (2.7):

$$a^{p'-1} \int_a^\infty |\Gamma(\tau)|^{p'}\,d\tau \le C.$$

Hence (2.6) holds. □

Remark 1. The corresponding compactness criterion for the multiplication operator $Q : \overset{\circ}{L}{}^1_p(\mathbb{R}_+) \to L_p^{-1}(\mathbb{R}_+)$ is given by the following conditions:

$$\lim_{a\to 0+} a^{p^*-1} \int_a^\infty |\Gamma(x)|^{p^*}\,dx = 0, \qquad \lim_{a\to +\infty} a^{p^*-1} \int_a^\infty |\Gamma(x)|^{p^*}\,dx = 0. \tag{2.12}$$

The proof is analogous to the argument in [9], Sec. 3.

The general case of Theorem III stated in the Introduction, where Q is a distribution, involves some additional technical problems considered in the proof of the following statement.

Theorem 2.2. *Let $Q \in \mathcal{D}'(\mathbb{R}_+)$, and let $1 < p < \infty$. Then the inequality*

$$|\langle Qu, v\rangle| \le \text{const}\, ||u'||_{L_p(\mathbb{R}_+)}\, ||v'||_{L_{p'}(\mathbb{R}_+)} \tag{2.13}$$

holds for all $u, v \in C_0^\infty(\mathbb{R}_+)$ if and only if there exists $\Gamma \in L_{p^*}(\mathbb{R}_+)$ so that $Q = \Gamma'$ in the distributional sense, and (2.6) *holds.*

Proof. We have to show first that there exists $\Gamma \in L_1^{\text{loc}}(\mathbb{R}_+)$ so that $Q = \Gamma'$ in the sense of distributions provided (2.13) holds. Notice that (2.13) ensures that $\langle Q, w\rangle$ is well defined for every $w = uv$, where

$$u \in \overset{\circ}{L}{}_p^1(\mathbb{R}_+) \cap C^\infty(\mathbb{R}_+), \quad v \in \overset{\circ}{L}{}_{p'}^1(\mathbb{R}_+) \cap C^\infty(\mathbb{R}_+).$$

This can be seen from the following approximation argument (cf. [9], Lemma 2.3). Let $\psi_n \in C_0^\infty(\mathbb{R}_+)$ be a sequence of functions such that $0 \le \psi_n(x) \le 1$, $|\psi_n'(x)| \le C/x$, and $\operatorname{supp} \psi_n \subset [1/n, n]$, $n = 1, 2, \ldots$. Then $u_n = \psi_n u$ and $v_n = \psi_n v$ lie in $C_0^\infty(\mathbb{R}_+)$, and

$$\lim_{n\to\infty} ||u - u_n||_{\overset{\circ}{L}{}_p^1(\mathbb{R}_+)} = \lim_{n\to\infty} ||v - v_n||_{\overset{\circ}{L}{}_{p'}^1(\mathbb{R}_+)} = 0.$$

Hence

$$\langle Q, w\rangle = \lim_{n\to\infty} \langle Q, u_n v_n\rangle = \lim_{n\to\infty} \langle Q, w\psi_n^2\rangle$$

exists, and is independent of the choice of ψ_n, and of the factorization $w = uv$.

Next we show that $\langle Q, w\rangle$ is well defined for $w = \int_0^x f(t)\,dt$ where $f \in C_0^\infty(\mathbb{R}_+)$. To this end suppose that $\operatorname{supp} f \subset [a, b]$, where $0 < a < b < +\infty$. Let $\eta \in C^\infty(\mathbb{R}_+)$ be an increasing cut-off function such that $0 \le \eta(x) \le 1$; $\eta(x) = 0$ for $0 \le x \le a/2$; $\eta(x) = 1$ for $x \ge 2b$; and $\eta(a) > 0$. Then $w(x) = \int_0^x f(t)\,dt$ can be represented in the form $w = uv$, where $u = \eta$ and $v = \chi_{(a,+\infty)} w\, \eta^{-1}$. Obviously, $\eta \in \overset{\circ}{L}{}_p^1(\mathbb{R}_+)$, and $||\eta||_{\overset{\circ}{L}{}_p^1(\mathbb{R}_+)} \le C_{a,b}$, where the constant $C_{a,b}$ depends on a and b. Also, since $\eta'(x) = 0$ for $x > 2b$, it follows

$$||v||_{\overset{\circ}{L}{}_{p'}^1(\mathbb{R}_+)} = ||\eta(x)^{-1}\chi_{(a,+\infty)}(x)\int_a^x f(t)\,dt||_{\overset{\circ}{L}{}_{p'}^1(\mathbb{R}_+)} \le C_{a,b}\, ||f||_{L_{p'}(\mathbb{R}_+)}.$$

Thus, $u, v \in C^\infty(\mathbb{R}_+)$ satisfy the conditions stated above, and hence Γ can be defined by

$$\langle \Gamma, f\rangle = -\langle Q, uv\rangle = -\langle Q, \int_0^x f(t)dt\rangle,$$

for all $f \in C_0^\infty(\mathbb{R}_+)$. Moreover,

$$|\langle \Gamma, f\rangle| \le C_{a,b}\, ||f||_{L_{p'}(\mathbb{R}_+)},$$

which yields $\Gamma \in L_p(a, b)$. Interchanging the roles of u and v, one can verify that $\Gamma \in L_{p'}^{\mathrm{loc}}(\mathbb{R}_+)$ as well. Clearly,

$$\langle \Gamma', u\rangle = -\int_0^\infty \Gamma(x)u'(x)\,dx = \langle Q, u\rangle$$

for all $u \in C_0^\infty(\mathbb{R}_+)$, i.e., $\Gamma' = Q$ in the distributional sense.

The rest of the proof is completed in the same way as in Theorem 2.1. Indeed, let

$$u(x) = \int_0^x f(t)\,dt, \quad v(x) = \int_0^x g(\tau)\,d\tau,$$

for $f, g \in C_0^\infty(\mathbb{R}_+)$. Then

$$\begin{aligned}\langle Q, uv\rangle &= -\langle \Gamma, u'v\rangle - \langle \Gamma, uv'\rangle \\ &= -\int_0^\infty \Gamma(x)f(x)\int_0^x g(\tau)d\tau dx - \int_0^\infty \Gamma(x)g(x)\int_0^x f(t)dtdx \\ &= -\int_0^\infty g(\tau)\int_\tau^\infty \Gamma(x)f(x)\,dx - \int_0^\infty \Gamma(x)g(x)\int_0^x f(t)dt.\end{aligned}$$

Choosing f and g as in the proof of Theorem 2.1 and using a pair of the weighted Hardy inequalities, we complete the proof of Theorem 2.2. □

Theorem 2.1 is easily carried over to the two weight setting.

Theorem 2.3. *Let $W_i \geq 0$, $i = 1, 2$, be locally integrable weight functions on $\mathbb{R}_+$ such that*

$$\int_0^a W_1(x)^{1-p'}\,dx < +\infty, \quad \int_0^a W_2(x)^{1-p}\,dx < +\infty, \tag{2.14}$$

for every $a > 0$, and

$$\int_0^{+\infty} W_1(x)^{1-p'}\,dx = \int_0^{+\infty} W_2(x)^{1-p}\,dx = +\infty. \tag{2.15}$$

Then the two weight bilinear inequality

$$|\langle Qu, v\rangle| \leq \text{const}\left(\int_0^\infty |u'(x)|^p\, W_1(x)\,dx\right)^{1/p} \times \left(\int_0^\infty |v'(x)|^{p'}\, W_2(x)\,dx\right)^{1/p'} \tag{2.16}$$

holds for all $u, v \in C_0^\infty(\mathbb{R}_+)$ if and only if $Q = \Gamma'$ in the distributional sense, where $\Gamma \in L_1^{\mathrm{loc}}(\mathbb{R}_+)$, and the following pair of conditions hold:

$$\sup_{a>0}\left(\int_0^a W_1(x)^{1-p'}\,dx\right)^{p-1}\int_a^\infty |\Gamma(x)|^p\, W_2(x)^{1-p}dx < \infty, \tag{2.17}$$

and

$$\sup_{a>0} \left(\int_0^a W_2(x)^{1-p} \right)^{p'-1} \int_a^\infty |\Gamma(x)|^{p'} W_1(x)^{1-p'} dx < \infty. \tag{2.18}$$

It is not difficult to obtain similar theorems under other possible restrictions on the weights W_i in place of (2.14) and (2.15), as in standard two weight Hardy inequalities (see [10]).

For functions defined on a finite interval (a, b), Theorem 2.1 can be recast in a similar way:

Theorem 2.4. *The inequality*

$$|\langle Qu, v\rangle| \le \text{const}||u'||_{L^p(a,b)}||v'||_{L^{p'}(a,b)} \tag{2.19}$$

holds for all $u, v \in C^\infty(a, b)$ such that $u(a) = 0$, $v(a) = 0$ if and only if Q can be represented in the form $Q = \Gamma'$, where

$$\sup_{x>a} x^{p-1} \int_x^b |\Gamma(t)|^{p^*} dt < \infty. \tag{2.20}$$

The corresponding compactness criterion holds with the preceding condition complemented by

$$\lim_{x\to a+} x^{p^*-1} \int_x^b |\Gamma(t)|^{p^*} dt = 0.$$

It is not difficult to obtain similar criteria for spaces of functions with zero boundary values at both endpoints.

We now state an analogue of Theorem 2.1 on the real line $\mathbb{R}$ for the inhomogeneous Sobolev space $W_p^1(\mathbb{R})$ with norm

$$||u||_{W_p^1(\mathbb{R})} = ||u||_{L_p(\mathbb{R})} + ||u'||_{L_p(\mathbb{R})}.$$

We assume here that Q is a distribution in the Schwartz class $\mathcal{S}'(\mathbb{R})$. Define $G_\pm \in \mathcal{S}'(\mathbb{R})$ by

$$G_+(t) = e^t \int_t^{+\infty} e^{-s} Q(s)\, ds, \quad G_-(t) = e^{-t} \int_{-\infty}^t e^s Q(s)\, ds, \tag{2.21}$$

where the preceding expressions are understood in the distributional sense.

Theorem 2.5. *Let $1 < p < \infty$, and $p^* = \max(p, p')$. Then the Schrödinger operator $\mathcal{L} : W_p^1(\mathbb{R}) \to W_p^{-1}(\mathbb{R})$ is bounded, or equivalently, the inequality*

$$|\langle Qu, v\rangle| \le \text{const}\, ||u||_{W_p^1(\mathbb{R})}\, ||v||_{W_{p'}^1(\mathbb{R})}, \tag{2.22}$$

holds for all $u, v \in C_0^\infty(\mathbb{R})$, *if and only if* Q *can be represented in the form* $Q = \Gamma' + \Gamma_0$, *where* Γ *and* Γ_0 *satisfy the following conditions:*

$$\sup_{x\in\mathbb{R}} \int_{x-a}^{x+a} |\Gamma(t)|^{p^*} dt < \infty, \quad \sup_{x\in\mathbb{R}} \int_{x-a}^{x+a} |\Gamma_0(t)|^{p^*} dt < \infty, \tag{2.23}$$

for every fixed $a > 0$.

Equivalently, $G_\pm$ *defined by* (2.21) *are functions in* $L_{p^*}^{\text{loc}}(\mathbb{R})$, *and, for a fixed* $a > 0$,

$$\sup_{x\in\mathbb{R}} \int_{x-a}^{x+a} |G_\pm(t)|^{p^*} dt < \infty. \tag{2.24}$$

Furthermore, the multiplication operator $Q : W_p^1(\mathbb{R}) \to W_{p}^{-1}(\mathbb{R})$ *is compact if and only if*

$$\lim_{x\to\pm\infty} \int_{x-a}^{x+a} |G_\pm(t)|^{p^*} dt = 0.$$

Remark 2. It can be shown that (2.24) is equivalent to the following estimate:

$$\sup_{x,y\in\mathbb{R}} \int_{x-a}^{x+a} \int_{y-a}^{y+a} \left| \int_t^s Q(\tau)d\tau \right|^{p^*} dt\, ds \le c_1 + c_2|x-y|^{p^*}, \tag{2.25}$$

for a fixed $a > 0$, where c_1, c_2 are nonnegative constants which may depend on a. Here $\int_t^s Q(\tau)d\tau$ is understood in the distributional sense.

Note that the simpler conditions:

$$\sup_{x\in\mathbb{R}} \int_{x-a}^{x+a} \left| \int_{t-a}^{t+a} Q(\tau)d\tau \right|^{p^*} dt < \infty, \quad \text{or} \quad \sup_{x\in\mathbb{R}} \int_{x-a}^{x+a} \left| \int_x^t Q(\tau)d\tau \right|^{p^*} dt < \infty,$$

are either not sufficient, or not necessary for the statements of Theorem 2.5 to hold.

The proofs of Theorem 2.5 and other related results using the approach outlined above will appear elsewhere.

References

[1] Birman, M. S., and Solomyak, M. Z., Schrödinger operator. Estimates for number of bound states as function-theoretical problem, in: Spectral Theory of Operators (S. G. Gindikin, ed.), Amer. Math. Soc. Transl. Ser. 2, 150, Amer. Math. Soc., Providence, RI, 1992, 1–54.

[2] Davies, E. B., L^p spectral theory of higher order elliptic differential operators, Bull. London Math. Soc. 29 (1997), 513–546.

[3] Fefferman, C., The uncertainty principle, Bull. Amer. Math. Soc. 9 (1983), 129–206.

[4] Hartman, P., Ordinary Differential Equations, Birkhäuser, Boston, 1982.

[5] Hille, E., Non-oscillation theorems, Trans. Amer. Math. Soc. 64 (1948), 234–252.

[6] Maz'ya, V. G., On the theory of the n-dimensional Schrödinger operator, Izv. Akad. Nauk SSSR Ser. Mat. 28 (1964), 1145–1172.

[7] —, Sobolev Spaces, Springer-Verlag, Berlin–Heidelberg–New York, 1985.

[8] Maz'ya, V. G., and Verbitsky, I. E., Capacitary estimates for fractional integrals, with applications to partial differential equations and Sobolev multipliers. Ark. Mat. 33 (1995), 81–115.

[9] —, The Schrödinger operator on the energy space: boundedness and compactness criteria, Acta Math., to appear.

[10] Opic, B., and Kufner, A., Hardy-type Inequalities, Pitman Research Notes in Math. 219, Longman Scientific and Technical, Harlow, 1990.

[11] Reed, M., and Simon, B., Methods of Modern Mathematical Physics. II: Fourier Analysis, Self-Adjointness, Academic Press, New York–London, 1975.

[12] Simon, B., Schrödinger semigroups, Bull. Amer. Math. Soc. 7 (1982), 447–526.

[13] Verbitsky, I. E., Nonlinear potentials and trace inequalities, in: The Maz'ya Anniversary Collection (J. Rossmann, P. Takáč, and G. Wildenhain, eds.), Oper. Theory Adv. App. 110, Birkhäuser, Basel–Boston–Berlin, 1999, 323–343.

Vladimir G. Maz'ya, Department of Mathematics, Linköping University, 58183 Linköping, Sweden

E-mail: `vlmaz@mai.liu.se`

Igor E. Verbitsky, Department of Mathematics, 202 Mathematical Sciences Building, University of Missouri-Columbia, Columbia, MO 65211, U.S.A.

E-mail: `igor@math.missouri.edu`

On Gaussian-summing identity maps between Lorentz sequence spaces

Carsten Michels

To Professor Jaak Peetre on the occasion of his 65th birthday

Abstract. Using formulae for the interpolation of spaces of operators, we show that for $1 < p < 2$ and $1 \leq q < 2$ the identity map $\mathrm{id} : \ell_{p,q} \hookrightarrow \ell_{\tilde{p},\tilde{q}}$ is Gaussian-summing, where $1/\tilde{p} = 1/p - 1/2$ and $1/\tilde{q} = 1/q - 1/2$. This extends the corresponding result for ℓ_p-spaces due to Linde and Pietsch.

2000 Mathematics Subject Classification: 47B10, 46B70, 60B99

1. The main result

A linear and continuous operator $T : X \to Y$ between Banach spaces X and Y (we will write $T \in \mathcal{L}(X, Y)$ in the following) is called Gaussian-summing if there exists a constant $c > 0$ such that for all $x_1, \dots, x_n \in X$

$$\Big(\int_\Omega \| \sum_{i=1}^n g_i T x_i \|_Y^2 dP \Big)^{1/2} \leq c \cdot \sup_{x' \in B_{X'}} \Big(\sum_{i=1}^n |x'(x_i)|^2 \Big)^{1/2},$$

where (g_i) are independent standard Gaussian variables on some probability space (Ω, Σ, P); here, $B_{X'}$ denotes the unit ball of the dual space X' of X. We write $\pi_\gamma(T)$ for the smallest constant c satisfying the inequality above; in this way we obtain the maximal Banach operator ideal (Π_γ, π_γ) which was originally introduced by Linde and Pietsch (see [15]). Substituting in the above definition the Gaussian variables by the Rademacher functions on [0, 1], this gives the same operator ideal (with an equivalent ideal norm), see e. g. [11, 12.12]. Hence, a Gaussian-summing operator transfers weakly 2-summable sequences into almost unconditionally summable ones. Linde and Pietsch in [15] proved a connection to probability theory: A Gaussian-summing operator $T : X \to Y$ transforms Gaussian cylindrical measures on X into Radon measures on Y'' (endowed with the weak*-topology). For operators $T : \ell_2 \to Y$ the Gaussian-summing norm is also known as the ℓ-norm (see e. g. [11, 12.15] and

[24, Chapter 12]) which turned out to be of great importance within the geometry of Banach spaces.

Amongst other examples, Linde and Pietsch in [15] proved the following result, using a factorization theorem by Maurey: For $1 \leq p < 2$ the identity map $\mathrm{id} : \ell_p \hookrightarrow \ell_{\tilde{p}}$ is Gaussian-summing, where $1/\tilde{p} = 1/p - 1/2$.

In this note we use the above result together with interpolation techniques to prove the following analogue for Lorentz sequence spaces:

Theorem 1.1. *For $1 < p < 2$ and $1 \leq q < 2$ the identity map $\mathrm{id} : \ell_{p,q} \hookrightarrow \ell_{\tilde{p},\tilde{q}}$ is Gaussian-summing, where $1/\tilde{p} = 1/p - 1/2$ and $1/\tilde{q} = 1/q - 1/2$.*

Recall that for $1 < p < \infty$, $1 \leq q \leq \infty$ a zero-sequence (x_n) is contained in $\ell_{p,q}$ if and only if

$$\varphi_{p,q}((x_n^*)) := \Big(\sum_{n=1}^{\infty} |x_n^*|^q n^{q/p-1} \Big)^{1/q} < \infty, \quad q < \infty,$$

and

$$\varphi_{p,\infty}((x_n^*)) := \sup_n |x_n^*| \, n^{1/p} < \infty, \quad q = \infty,$$

where (x_n^*) denotes the decreasing rearrangement of the sequence (x_n). For $q \leq p$ the quasi-Banach space $(\ell_{p,q}, \varphi_{p,q})$ is normed, and more general, for all p, q there exists a canonical norm on $\ell_{p,q}$ equivalent to $\varphi_{p,q}$ (see e. g. [13, 1.c.11]). Henceforth, we will consider $\ell_{p,q}$ as a Banach space.

2. Real interpolation of spaces of operators

We refer to [1] for all information on interpolation of Banach spaces. For a Banach couple (A_0, A_1) and $0 < \theta < 1$, $1 \leq q \leq \infty$ the associated real interpolation space is denoted by $(A_0, A_1)_{\theta,q}$ (for the definition see e. g. [1, Chapter 3]).

For two symmetric Banach sequence spaces E and F (for this notion see e. g. [23]) the symmetric Banach sequence space containing all $\lambda \in \ell_\infty$ such that the multiplier $M_\lambda : E \to F$, $\mu \mapsto \lambda\mu$ is defined (and continuous) is denoted by $M(E, F)$, equipped with the norm $\|\lambda\|_{M(E,F)} := \|M_\lambda : E \to F\|$. A lattice E is said to be 2-concave if there exists $c > 0$ such that for all choices of finitely many $x_1, \ldots, x_n \in E$ the inequality $\big(\sum_{i=1}^n \|x_i\|_E^2\big)^{1/2} \leq c \cdot \| \big(\sum_{i=1}^n |x_i|^2\big)^{1/2} \|_E$ holds. In this case, we denote by $\mathbf{M}_{(2)}(E)$ the smallest constant c satisfying the above inequality. This is equivalent to E being of cotype 2 (see e. g. [16, 1.f.16]).

The following lemma is a counterpart to results of Pisier, Kouba and Defant–Michels on the complex interpolation of spaces of operators (see [14], [22] and [9]). Earlier results for real interpolation methods by Peetre [21] have recently been extended by Cobos and Signes [2]. Essentially, the proof of the lemma is contained in

[6] and [5] within a more general theory of interpolation of spaces of operators, but for the sake of completeness we state it here.

Lemma 2.1. *Let* $1 \leq p_0, p_1, q \leq 2$ *such that* $p_0 \neq p_1$. *Then for all* $0 < \theta < 1$

$$\sup_{m,n} \| \mathcal{L}(\ell_2^m, \ell_{p_\theta,q}^n) \hookrightarrow (\mathcal{L}(\ell_2^m, \ell_{p_0}^n), \mathcal{L}(\ell_2^m, \ell_{p_1}^n))_{\theta,\tilde{q}} \| < \infty,$$

where $1/p_\theta = (1-\theta)/p_0 + \theta/p_1$ *and* $1/\tilde{q} = 1/q - 1/2$.

Proof. As usual, for $1 \leq p \leq \infty$ its conjugated number p' is determined by $1/p + 1/p' = 1$. For $2 \leq r \leq \infty$ define $\tilde{r}$ by $1/\tilde{r} = 1/r - 1/2$. We consider powers of sequence spaces: For a Banach sequence space E and $0 < p < \infty$ let E^p be the quasi-Banach sequence space of all $x = (x_n)$ for which $|x|^{1/p} \in E$, endowed with the quasi-norm $\|x\|_{E^p} := \||x|^{1/p}\|_E^p$. The quasi-norm $\|\cdot\|_{E^p}$ is a norm provided that $p \leq 1$, and for $p > 1$ it is equivalent to a norm provided that E is p-convex. An easy calculation shows $\ell_{r,s}^p = \ell_{r/p,s/p}$ (provided that $r > p$ and $s \geq p$) and $M(\ell_2, E) = (((E^\times)^2)^\times)^{1/2}$ for any 2-concave Banach sequence space E (see [7, (3.2)]; here, $X^\times$ denotes the Köthe dual of a Banach lattice X), hence, for $1 < r < 2$, $1 \leq s \leq 2$ or $1 \leq r = s \leq 2$,

$$M(\ell_2, \ell_{r,s}) = (((\ell_{r,s}^\times)^2)^\times)^{1/2} = (\ell_{r'/2,s'/2}^\times)^{1/2} = \ell_{2(r'/2)',2(s'/2)'} = \ell_{\tilde{r},\tilde{s}}.$$

Obviously $1 < p_\theta < 2$, hence $\ell_{p_\theta,q}$ is 2-concave (see e. g. [4]). Together with the formula for the real interpolation of ℓ_p-spaces (see e. g. [1, 5.2.1]; note that there the condition on q can be dropped) this gives

$$M(\ell_2, \ell_{p_\theta,q}) = \ell_{\tilde{p}_\theta,\tilde{q}} = (\ell_{\tilde{p}_0}, \ell_{\tilde{p}_1})_{\theta,\tilde{q}} = (M(\ell_2, \ell_{p_0}), M(\ell_2, \ell_{p_1}))_{\theta,\tilde{q}}. \tag{1}$$

Now let $T \in \mathcal{L}(\ell_2^m, \ell_{p_\theta,q}^n)$. By a variant of the Maurey–Rosenthal Factorization Theorem (see [4, 4.2] and also [18]) there exist an operator $R \in \mathcal{L}(\ell_2^m, \ell_2^n)$ and $\lambda \in \mathbb{R}^n$ such that $T = M_\lambda \circ R$ with $\|R\| \cdot \|\lambda\|_{M(\ell_2^n, \ell_{p_\theta,q}^n)} \leq \sqrt{2} \cdot \mathbf{M}_{(2)}(\ell_{p_\theta,q}) \cdot \|T\|$. Obviously the map Φ defined by $\Phi(\mu) := M_\mu \circ R$, $\mu \in \mathbb{R}^n$ maps the couple $(M(\ell_2^n, \ell_{p_0}^n), M(\ell_2^n, \ell_{p_1}^n))$ into the couple $(\mathcal{L}(\ell_2^m, \ell_{p_0}^n), \mathcal{L}(\ell_2^m, \ell_{p_1}^n))$ such that both restrictions have norm less or equal $\|R\|$. Hence by the interpolation property and (1) the map $\Phi : M(\ell_2^n, \ell_{p_\theta,q}^n) \to (\mathcal{L}(\ell_2^m, \ell_{p_0}^n), \mathcal{L}(\ell_2^m, \ell_{p_1}^n))_{\theta,\tilde{q}}$ has norm less or equal $C \cdot \|R\|$, where $C > 0$ is some constant involved in (1). Thus we obtain

$$\begin{aligned}\|T\|_{(\mathcal{L}(\ell_2^m,\ell_{p_0}^n),\mathcal{L}(\ell_2^m,\ell_{p_1}^n))_{\theta,\tilde{q}}} &= \|M_\lambda \circ R\|_{(\mathcal{L}(\ell_2^m,\ell_{p_0}^n),\mathcal{L}(\ell_2^m,\ell_{p_1}^n))_{\theta,\tilde{q}}} \\ &\leq C \cdot \|R\| \cdot \|\lambda\|_{M(\ell_2^n,\ell_{p_\theta,q}^n)} \\ &\leq \sqrt{2} \cdot C \cdot \mathbf{M}_{(2)}(\ell_{p_\theta,q}) \cdot \|T\|_{\mathcal{L}(\ell_2^m,\ell_{p_\theta,q}^n)},\end{aligned}$$

which gives the claim. □

3. The proof of the main result

For a Banach space X we denote by $\ell_2^{m,weak}(X)$ the vector space X^m equipped with the norm $\|(x_1,\ldots,x_m)\| := \sup_{x'\in B_{X'}} \left(\sum_{i=1}^m |x'(x_i)|^2\right)^{1/2}$. By the isometric identification $\mathcal{L}(\ell_2^m, X) = \ell_2^{m,weak}(X)$, $S \mapsto (Se_i)_{i=1}^m$ an operator $T \in \mathcal{L}(X,Y)$ is Gaussian-summing if and only if the maps $\Phi_m : \mathcal{L}(\ell_2^m, X) \to L_2(Y)$, $S \mapsto \sum_{i=1}^m g_i TSe_i$ are uniformly bounded, and in this case $\pi_\gamma(T) = \sup_m \|\Phi_m\|$. Now consider for any fixed r such that $p < r < 2$ the maps

$$\Phi_{m,n} : (\mathcal{L}(\ell_2^m, \ell_1^n), \mathcal{L}(\ell_2^m, \ell_r^n)) \to (L_{\tilde{q}}(\ell_2^n), L_{\tilde{q}}(\ell_{\tilde{r}}^n))$$

defined by $\Phi_{m,n}(S) := \sum_{i=1}^m g_i Se_i$. By the results of Linde and Pietsch in [15] (with the proof belonging to Maurey) the identity maps $\mathrm{id} : \ell_1 \hookrightarrow \ell_2$ and $\mathrm{id} : \ell_r \hookrightarrow \ell_{\tilde{r}}$ are Gaussian-summing. The Kahane-type inequality for Gaussian variables due to Hoffmann-Jørgensen [12] states that there exists a constant $C_{\tilde{q}} > 0$ such that

$$\left(\int_\Omega \Big\|\sum_{i=1}^n g_i x_i\Big\|_X^{\tilde{q}} dP\right)^{1/\tilde{q}} \le C_{\tilde{q}} \cdot \left(\int_\Omega \Big\|\sum_{i=1}^n g_i x_i\Big\|_X^2 dP\right)^{1/2}$$

for any Banach space X and every finite choice of $x_1,\ldots,x_n \in X$. We therefore conclude that

$$c_1 := \sup_{m,n} \|\Phi_{m,n} : \mathcal{L}(\ell_2^m, \ell_1^n) \to L_{\tilde{q}}(\ell_2^n)\| < \infty$$

and

$$c_2 := \sup_{m,n} \|\Phi_{m,n} : \mathcal{L}(\ell_2^m, \ell_r^n) \to L_{\tilde{q}}(\ell_{\tilde{r}}^n)\| < \infty.$$

By the fact that the functor $(\cdot,\cdot)_{\theta,\tilde{q}}$ is of exponent θ (see e. g. [1, Chapter 3]) we obtain for any $0 < \theta < 1$

$$\sup_{m,n} \|\Phi_{m,n} : (\mathcal{L}(\ell_2^m, \ell_1^n), \mathcal{L}(\ell_2^m, \ell_r^n))_{\theta,\tilde{q}} \to (L_{\tilde{q}}(\ell_2^n), L_{\tilde{q}}(\ell_{\tilde{r}}^n))_{\theta,\tilde{q}}\| \le c_1^{1-\theta} c_2^\theta.$$

Now fix $0 < \theta < 1$ such that $1/p = (1-\theta)/1 + \theta/r$. The Lions–Peetre formula for the real interpolation of vector-valued L_p's (see [17] and also [3]) gives

$$(L_{\tilde{q}}(\ell_2^n), L_{\tilde{q}}(\ell_{\tilde{r}}^n))_{\theta,\tilde{q}} = L_{\tilde{q}}((\ell_2^n, \ell_{\tilde{r}}^n)_{\theta,\tilde{q}}) = L_{\tilde{q}}(\ell_{\tilde{p},\tilde{q}}^n),$$

where the constant arising in the equivalence of the norms does not depend on the dimension n. Together with the preceding lemma and the triviality $L_{\tilde{q}}(X) \subset L_2(X)$ for any Banach space X and any finite measure space it follows that

$$\sup_n \pi_\gamma(\mathrm{id} : \ell_{p,q}^n \hookrightarrow \ell_{\tilde{p},\tilde{q}}^n) = \sup_{m,n} \|\Phi_{m,n} : \mathcal{L}(\ell_2^m, \ell_{p,q}^n) \to L_2(\ell_{\tilde{p},\tilde{q}}^n)\| \le C \cdot c_1^{1-\theta} c_2^\theta$$

for some $C > 0$. The maximality of the operator ideal Π_γ now gives the claim. □

Note that the index $\tilde{p}$ in our theorem is best possible (this follows by factorization from the ℓ_p-case where it is known by [15] that $\tilde{p}$ is best possible). However, we do not know whether this is also true for $\tilde{q}$. We believe that, under certain assumptions on the symmetric Banach sequence space E (2-concave, properly contained in ℓ_2), the identity map id : $E \hookrightarrow M(\ell_2, E)$ is Gaussian-summing and that the sequence space $M(\ell_2, E)$ is best possible in some sense.

For some more recent applications of interpolation formulae for spaces of operators within the theory of summing operators we refer to [5], [6], [7], [8], [10], [19] and [20].

References

[1] J. Bergh and J. Löfström, Interpolation spaces, Grundlehren Math. Wiss. 223, Springer-Verlag, Berlin–Heidelberg–New York, 1978.

[2] F. Cobos and T. Signes, On a result of Peetre about interpolation of operator spaces, Publ. Mat. 44 (2000), 457–481.

[3] M. Cwikel, On $(L^{p_0}(A_0), L^{p_1}(A_1))_{\theta,q}$, Proc. Amer. Math. Soc., 44 (1974), 286–292.

[4] A. Defant, Variants of the Maurey–Rosenthal theorem for quasi Köthe function spaces, Positivity 5 (2001), 153–175.

[5] A. Defant, M. Mastyło, and C. Michels, Eigenvalue estimates of operators between symmetric sequence spaces, preprint.

[6] A. Defant, M. Mastyło, and C. Michels, Summing inclusion maps between symmetric sequence spaces, Trans. Amer. Math. Soc., to appear.

[7] A. Defant, M. Mastyło, and C. Michels, Summing inclusion maps between symmetric sequence spaces, a survey, Recent progress in Functional Analysis (Valencia, 2000), North-Holland Math. Stud. 189, North-Holland, Amsterdam, 2001, 43–60.

[8] A. Defant and C. Michels, Bennett–Carl inequalities for symmetric Banach sequence spaces and unitary ideals, `math.FA/9904176`.

[9] A. Defant and C. Michels, A complex interpolation formula for tensor products of vector-valued Banach function spaces, Arch. Math. 74 (2000), 441–451.

[10] A. Defant and C. Michels. Complex interpolation of spaces of operators on ℓ_1, Bull. Pol. Acad. Sci. Math. 48 (2000), 303–318.

[11] J. Diestel, H. Jarchow, and A. Tonge, Absolutely Summing Operators, Cambridge University Press, Cambridge, 1995.

[12] J. Hoffmann-Jørgensen, Sums of independent Banach space valued random variables, Studia Math. 52 (1974), 159–186.

[13] Her. König, Eigenvalue distributions of compact operators, Birkhäuser, 1986.

[14] O. Kouba, On the interpolation of injective or projective tensor products of Banach spaces, J. Funct. Anal. 96 (1991), 38–61.

[15] W. Linde and A. Pietsch, Mappings of Gaussian cylindrical measures in Banach spaces, Theory Prob. Appl. 19 (1974), 445–460.

[16] J. Lindenstrauss and L. Tzafriri, Classical Banach Spaces II: Function Spaces, Ergeb. Math. Grenzgeb. 97, Springer-Verlag, Berlin–Heidelberg–New York, 1979.

[17] J. L. Lions and J. Peetre, Sur une classe d'espaces d'interpolation, Inst. Hautes Etudes Sci. Publ. Math. 19 (1964), 5–68, .

[18] F. Lust-Picard and G. Pisier, Non commutative Khintchine and Paley inequalities, Ark. Mat. 29 (1991), 241–260.

[19] C. Michels, $\Lambda(p)$-sets and the limit order of operator ideals, Math. Nachr., to appear.

[20] C. Michels, Complex interpolation of tensor products and applications to summing norms, Carl von Ossietzky Universität Oldenburg, Thesis, 1999.

[21] J. Peetre, Zur Interpolation von Operatorenräumen, Arch. Math. 21 (1970), 601–608.

[22] G. Pisier, A remark on $\Pi_2(\ell_p, \ell_p)$, Math. Nachr. 148 (1990), 243–245.

[23] B. Simon, Trace ideals and their applications, London Math. Soc. Lecture Notes Ser. 35, Cambridge University Press, Cambridge, 1979.

[24] N. Tomczak-Jaegermann, Banach–Mazur Distances and Finite-Dimensional Operator Ideals, Pitman Monogr. Surveys Pure Appl. Math. 38. Longman Scientific & Technical, Harlow, 1989.

Carsten Michels, Fachbereich 6, Carl von Ossietzky University of Oldenburg, P.O. Box 2503, 26111 Oldenburg, Germany.

E-mail: `michels@mathematik.uni-oldenburg.de`

E-mail: `balin@amsta.leeds.ac.uk`.

Current address: School of Mathematics, University of Leeds, Leeds LS2 9JT, United Kingdom.

On generalized fractional integrals on the weak Orlicz spaces, BMO_ϕ, the Morrey spaces and the Campanato spaces

Eiichi Nakai

Dedicated to Professor Jaak Peetre on his 65th birthday

Abstract. It is known that the fractional integral I_α $(0 < \alpha < n)$ is bounded from $L^p(\mathbb{R}^n)$ to $L^q(\mathbb{R}^n)$ when $p > 1$ and $n/p - \alpha = n/q > 0$, from $L^p_{\text{weak}}(\mathbb{R}^n)$ to $\mathrm{BMO}(\mathbb{R}^n)$ when $p > 1$ and $n/p - \alpha = 0$, from $L^p_{\text{weak}}(\mathbb{R}^n)$ to $\mathrm{Lip}_\beta(\mathbb{R}^n)$ when $p > 1$ and $-1 < n/p - \alpha = -\beta < 0$, from $\mathrm{BMO}(\mathbb{R}^n)$ to $\mathrm{Lip}_\alpha(\mathbb{R}^n)$ when $0 < \alpha < 1$, and from $\mathrm{Lip}_\beta(\mathbb{R}^n)$ to $\mathrm{Lip}_\gamma(\mathbb{R}^n)$ when $0 < \alpha + \beta = \gamma < 1$. In [4] the author introduced generalized fractional integrals and extended the above boundedness to the Orlicz spaces and BMO_ϕ. In this paper, we investigate the boundedness of generalized fractional integrals on the Morrey and Campanato spaces with $p = 1$. As corollaries, we have the boundedness from the weak Orlicz spaces to BMO_ϕ and from BMO_ϕ to BMO_ψ. For this purpose, we also give the relation between the Morrey space and the weak Orlicz space.

2000 Mathematics Subject Classification: 26A33, 46E30, 42B35, 46E15.

1. Introduction

The fractional integral I_α $(0 < \alpha < n)$ is defined by

$$I_\alpha f(x) = \int_{\mathbb{R}^n} \frac{f(y)}{|x-y|^{n-\alpha}}\, dy.$$

It is known that I_α is bounded from $L^p(\mathbb{R}^n)$ to $L^q(\mathbb{R}^n)$ when $p > 1$ and $n/p - \alpha = n/q > 0$ as the Hardy–Littlewood–Sobolev theorem. The fractional integral was studied by many authors (see, for example, Rubin [10] or Chapter 5 in Stein [11]). The Hardy–Littlewood–Sobolev theorem is an important result in the fractional integral theory and the potential theory.

In [4] the author introduced generalized fractional integrals and extended the Hardy–Littlewood–Sobolev theorem to the Orlicz spaces. It was showed, for example, that a generalized fractional integral is bounded from $\exp L^p$ to $\exp L^q$ (see

Theorem 3.1 and Example 3.1). In [5] the author also extended to spaces of homogeneous type and gave several examples.

Let $B(a,r)$ be the ball $\{x \in \mathbb{R}^n : |x-a| < r\}$ with center a and of radius $r > 0$, and $B_0 = B(0,1)$ with center the origin and of radius 1. The modified fractional integral $\tilde{I}_\alpha$ $(0 < \alpha < n+1)$ is defined by

$$\tilde{I}_\alpha f(x) = \int_{\mathbb{R}^n} f(y) \left(\frac{1}{|x-y|^{n-\alpha}} - \frac{1-\chi_{B_0}(y)}{|y|^{n-\alpha}} \right) dy,$$

where χ_{B_0} is the characteristic function of B_0. It is known that the modified fractional integral $\tilde{I}_\alpha$ is bounded from $L^p_{\text{weak}}(\mathbb{R}^n)$ to $\text{BMO}(\mathbb{R}^n)$ when $p > 1$ and $n/p - \alpha = 0$, from $L^p_{\text{weak}}(\mathbb{R}^n)$ to $\text{Lip}_\beta(\mathbb{R}^n)$ when $p > 1$ and $-1 < n/p - \alpha = -\beta < 0$, from $\text{BMO}(\mathbb{R}^n)$ to $\text{Lip}_\alpha(\mathbb{R}^n)$ when $0 < \alpha < 1$, and from $\text{Lip}_\beta(\mathbb{R}^n)$ to $\text{Lip}_\gamma(\mathbb{R}^n)$ when $0 < \alpha + \beta = \gamma < 1$. In [4] the author extended these boundedness to the Orlicz spaces and BMO_ϕ.

In this paper, we investigate the boundedness of generalized fractional integrals on the Morrey and Campanato spaces with $p = 1$. As corollaries, we have the boundedness from $L^\Phi_{\text{weak}}(\mathbb{R}^n)$ to $\text{BMO}_\phi(\mathbb{R}^n)$ and from $\text{BMO}_\phi(\mathbb{R}^n)$ to $\text{BMO}_\psi(\mathbb{R}^n)$. For this purpose, we also give the relation between the weak Orlicz space and the Morrey space. If $\phi(r) \equiv 1$, then $\text{BMO}_\phi(\mathbb{R}^n) = \text{BMO}(\mathbb{R}^n)$. If $\phi(r) = r^\alpha$ $(0 < \alpha \leq 1)$, then $\text{BMO}_\phi(\mathbb{R}^n) = \text{Lip}_\alpha(\mathbb{R}^n)$. Therefore our results are generalization of the known results.

2. Notation and definitions

For a function $\rho : (0, +\infty) \to (0, +\infty)$, let

$$I_\rho f(x) = \int_{\mathbb{R}^n} f(y) \frac{\rho(|x-y|)}{|x-y|^n} dy.$$

We consider the following conditions on ρ:

$$\int_0^1 \frac{\rho(t)}{t} dt < +\infty, \tag{2.1}$$

$$\frac{1}{A_1} \leq \frac{\rho(s)}{\rho(r)} \leq A_1 \quad \text{for } \frac{1}{2} \leq \frac{s}{r} \leq 2, \tag{2.2}$$

$$\frac{\rho(r)}{r^n} \leq A_2 \frac{\rho(s)}{s^n} \quad \text{for } s \leq r, \tag{2.3}$$

where $A_1, A_2 > 0$ are independent of $r, s > 0$. If $\rho(r) = r^\alpha$, $0 < \alpha < n$, then I_ρ is the fractional integral or the Riesz potential denoted by I_α.

We define the modified version of I_ρ as follows:

$$\tilde{I}_\rho f(x) = \int_{\mathbb{R}^n} f(y) \left(\frac{\rho(|x-y|)}{|x-y|^n} - \frac{\rho(|y|)(1-\chi_{B_0}(y))}{|y|^n} \right) dy.$$

We consider the following conditions on ρ: (2.1), (2.2) and

$$\frac{\rho(r)}{r^{n+1}} \le A_2' \frac{\rho(s)}{s^{n+1}} \quad \text{for } s \le r, \tag{2.4}$$

$$\int_r^{+\infty} \frac{\rho(t)}{t^2}\,dt \le A_2'' \frac{\rho(r)}{r}, \tag{2.5}$$

$$\left|\frac{\rho(r)}{r^n} - \frac{\rho(s)}{s^n}\right| \le A_3 |r-s| \frac{\rho(r)}{r^{n+1}} \quad \text{for } \frac{1}{2} \le \frac{s}{r} \le 2, \tag{2.6}$$

where $A_2', A_2'', A_3 > 0$ are independent of $r, s > 0$. If $\rho(r) r^{\alpha}$ is increasing for some $\alpha \ge 0$ and $\rho(r)/r^{\beta}$ is decreasing for some $\beta \ge 0$, then ρ satisfies (2.2) and (2.6). If $\rho(r) = r^{\alpha}$, $0 < \alpha < n+1$, then $\tilde{I}_\rho = \tilde{I}_\alpha$. If $\tilde{I}_\rho f$ and $I_\rho f$ are well defined, then $\tilde{I}_\rho f - I_\rho f$ is a constant.

For functions $\theta, \kappa : (0, +\infty) \to (0, +\infty)$, we denote $\theta(r) \sim \kappa(r)$ if there exists a constant $C > 0$ such that

$$C^{-1}\theta(r) \le \kappa(r) \le C\theta(r) \quad \text{for } r > 0.$$

A function $\theta : (0, +\infty) \to (0, +\infty)$ is said to be almost increasing (almost decreasing) if there exists a constant $C > 0$ such that

$$\theta(r) \le C\theta(s) \ (\theta(r) \ge C\theta(s)) \quad \text{for } r \le s.$$

A function $\theta : (0, +\infty) \to (0, +\infty)$ is said to satisfy the doubling condition if there exists a constant $C > 0$ such that

$$C^{-1} \le \frac{\theta(r)}{\theta(s)} \le C \quad \text{for } \frac{1}{2} \le \frac{r}{s} \le 2.$$

A function $\Phi : [0, +\infty] \to [0, +\infty]$ is called a Young function if Φ is convex, $\lim_{r \to +0} \Phi(r) = \Phi(0) = 0$ and $\lim_{r \to +\infty} \Phi(r) = \Phi(+\infty) = +\infty$. Any Young function is increasing. For a Young function Φ, the complementary function is defined by

$$\widetilde{\Phi}(r) = \sup\{rs - \Phi(s) : s \ge 0\}, \quad r \ge 0.$$

For example, if $\Phi(r) = r^p/p$, $1 < p < \infty$, then $\widetilde{\Phi}(r) = r^{p'}/p'$, $1/p + 1/p' = 1$. If $\Phi(r) = r$, then $\widetilde{\Phi}(r) = 0$ $(0 \le r \le 1)$, $= +\infty$ $(r > 1)$.

For a Young function Φ, let

$$L^{\Phi}(\mathbb{R}^n) = \left\{f \in L^1_{\text{loc}}(\mathbb{R}^n) : \int_{\mathbb{R}^n} \Phi(\epsilon|f(x)|)\,dx < +\infty \text{ for some } \epsilon > 0\right\},$$

$$\|f\|_{\Phi} = \inf\left\{\lambda > 0 : \int_{\mathbb{R}^n} \Phi\left(\frac{|f(x)|}{\lambda}\right) dx \le 1\right\},$$

$$L^{\Phi}_{\text{weak}}(\mathbb{R}^n) = \left\{f \in L^1_{\text{loc}}(\mathbb{R}^n) : \sup_{r>0} \Phi(r)\, m(r, \epsilon f) < +\infty \text{ for some } \epsilon > 0\right\},$$

$$\|f\|_{\Phi,\text{weak}} = \inf\left\{\lambda > 0 : \sup_{r>0} \Phi(r)\, m\left(r, \frac{f}{\lambda}\right) \le 1\right\},$$

where $m(r, f) = |\{x \in \mathbb{R}^n : |f(x)| > r\}|$.

Then

$$L^{\Phi}(\mathbb{R}^n) \subset L^{\Phi}_{\text{weak}}(\mathbb{R}^n) \quad \text{and} \quad \|f\|_{\Phi,\text{weak}} \le \|f\|_{\Phi}.$$

We have Hölder's inequality for Orlicz spaces:

$$\int_{\mathbb{R}^n} |f(x)g(x)|\,dx \le 2\|f\|_{\Phi}\|g\|_{\tilde{\Phi}}. \tag{2.7}$$

If a Young function Φ satisfies

$$0 < \Phi(r) < +\infty \quad \text{for } 0 < r < +\infty, \tag{2.8}$$

then Φ is continuous and bijective from $[0, +\infty)$ to itself. The inverse function Φ^{-1} is increasing, continuous and concave, so it satisfies the doubling condition.

Let χ_B be the characteristic function of $B(a, r)$. Then

$$\|\chi_B\|_{\tilde{\Phi}} \sim \Phi^{-1}(1/r^n)r^n. \tag{2.9}$$

A function Φ is said to satisfy the ∇_2-condition, denoted $\Phi \in \nabla_2$, if

$$\Phi(r) \le \frac{1}{2k}\Phi(kr), \quad r \ge 0,$$

for some $k > 1$.

If Φ is a Young function with (2.8) and

$$\Phi(r) = \begin{cases} 1/\exp(1/r^p) & \text{for small } r, \\ \exp(r^p) & \text{for large } r, \end{cases} \qquad p > 0,$$

then we denote the Orlicz space L^{Φ} and the weak Orlicz space L^{Φ}_{weak} by $\exp L^p$ and $\exp L^p_{\text{weak}}$, respectively. In this case, Φ satisfies the ∇_2-condition.

Let $Mf(x)$ be the maximal function, i.e.

$$Mf(x) = \sup_{B \ni x} \frac{1}{|B|}\int_B |f(y)|\,dy,$$

where the supremum is taken over all balls B containing x.

We assume that Φ satisfies (2.8). Then M is bounded from $L^{\Phi}(\mathbb{R}^n)$ to $L^{\Phi}_{\text{weak}}(\mathbb{R}^n)$ and

$$\|Mf\|_{\Phi,\text{weak}} \le C_0 \|f\|_{\Phi}. \tag{2.10}$$

If $\Phi \in \nabla_2$, then M is bounded on $L^{\Phi}(\mathbb{R}^n)$ and

$$\|Mf\|_{\Phi} \le C_0 \|f\|_{\Phi}. \tag{2.11}$$

For $1 \le p < \infty$ and a function $\phi : (0, +\infty) \to (0, +\infty)$, let

$$\|f\|_{L_{p,\phi}} = \sup_{B=B(a,r)} \frac{1}{\phi(r)} \left(\frac{1}{|B|} \int_B |f(x)|^p \, dx \right)^{1/p},$$
$$L_{p,\phi}(\mathbb{R}^n) = \left\{ f \in L^p_{\text{loc}}(\mathbb{R}^n) : \|f\|_{L_{p,\phi}} < +\infty \right\}.$$

We assume that ϕ satisfies the doubling condition and that $\phi(r)r^{n/p}$ is almost increasing. If $\phi(r) = r^{(\lambda-n)/p}$ $(0 \le \lambda \le n)$, then $L_{p,\phi}(\mathbb{R}^n) = L^{p,\lambda}(\mathbb{R}^n)$ which is the classical Morrey space. If $\lambda = 0$, then $L^{p,\lambda}(\mathbb{R}^n) = L^p(\mathbb{R}^n)$. If $\lambda = n$, then $L^{p,\lambda}(\mathbb{R}^n) = L^{\infty}(\mathbb{R}^n)$.

For $1 \le p < \infty$ and a function $\phi : (0, +\infty) \to (0, +\infty)$, let

$$\|f\|_{\mathcal{L}_{p,\phi}} = \sup_{B=B(a,r)} \frac{1}{\phi(r)} \left(\frac{1}{|B|} \int_B |f(x) - f_B|^p \, dx \right)^{1/p},$$
$$\mathcal{L}_{p,\phi}(\mathbb{R}^n) = \left\{ f \in L^p_{\text{loc}}(\mathbb{R}^n) : \|f\|_{\mathcal{L}_{p,\phi}} < +\infty \right\},$$
$$\text{where } f_B = \frac{1}{|B|} \int_B f(x) \, dx.$$

We assume that ϕ satisfies the doubling condition and that $\phi(r)r^{n/p}$ is almost increasing. If $\phi(r) = r^{(\lambda-n)/p}$ $(0 \le \lambda \le n+1)$, then $\mathcal{L}_{p,\phi}(\mathbb{R}^n) = \mathcal{L}^{p,\lambda}(\mathbb{R}^n)$ which is the classical Campanato space.

If ϕ is almost increasing, then $\mathcal{L}_{p,\phi}(\mathbb{R}^n) = \mathcal{L}_{1,\phi}(\mathbb{R}^n)$ for all $p > 1$. We denote $\mathcal{L}_{1,\phi}(\mathbb{R}^n)$ by $\text{BMO}_{\phi}(\mathbb{R}^n)$. If $\phi \equiv 1$, then $\text{BMO}_{\phi}(\mathbb{R}^n) = \text{BMO}(\mathbb{R}^n)$. If $\phi(r) = r^{\alpha}$, $0 < \alpha \le 1$, then it is known that $\text{BMO}_{\phi}(\mathbb{R}^n) = \text{Lip}_{\alpha}(\mathbb{R}^n)$.

The letter C shall always denote a constant, not necessarily the same one.

3. Main results

In [4] the author extended the Hardy–Littlewood–Sobolev theorem to the Orlicz spaces as follows:

Theorem 3.1 ([4]). *Let ρ satisfy* (2.1)–(2.3). *Let Φ and Ψ be Young functions with* (2.8). *Assume that there exist constants A, A', $A'' > 0$ such that, for all $r > 0$,*

$$\int_r^{+\infty} \widetilde{\Phi}\left(\frac{\rho(t)}{A \int_0^r (\rho(s)/s)\, ds\ \Phi^{-1}(1/r^n) t^n}\right) t^{n-1}\, dt \le A', \tag{3.1}$$

$$\int_0^r \frac{\rho(t)}{t}\, dt\ \Phi^{-1}\left(\frac{1}{r^n}\right) \le A''\ \Psi^{-1}\left(\frac{1}{r^n}\right), \tag{3.2}$$

where $\widetilde{\Phi}$ is the complementary function with respect to Φ. Then, for any $C_0 > 0$, there exists a constant $C_1 > 0$ such that, for $f \in L^\Phi(\mathbb{R}^n)$,

$$\Psi\left(\frac{|I_\rho f(x)|}{C_1 \|f\|_\Phi}\right) \le \Phi\left(\frac{Mf(x)}{C_0 \|f\|_\Phi}\right). \tag{3.3}$$

Therefore I_ρ is bounded from $L^\Phi(\mathbb{R}^n)$ to $L^\Psi_{\text{weak}}(\mathbb{R}^n)$. Moreover, if $\Phi \in \nabla_2$, then I_ρ is bounded from $L^\Phi(\mathbb{R}^n)$ to $L^\Psi(\mathbb{R}^n)$.

O'Neil [7] showed the boundedness for convolution operators on the Orlicz spaces. Cianchi [1] gave a necessary and sufficient condition on Φ and Ψ so that the fractional integral I_α is bounded from L^Φ to L^Ψ.

Example 3.1 ([5]). Let ρ satisfy the doubling condition and

$$\rho(r) = \begin{cases} 1/(\log(1/r))^{\alpha+1} & \text{for small } r, \\ (\log r)^{\alpha-1} & \text{for large } r, \end{cases} \qquad \alpha > 0. \tag{3.4}$$

Then

$$\int_0^r \frac{\rho(t)}{t}\, dt \sim \begin{cases} 1/(\log(1/r))^{\alpha} & \text{for small } r, \\ (\log r)^{\alpha} & \text{for large } r. \end{cases}$$

If $0 < p < 1/\alpha$, $1/q = 1/p - \alpha$, then the generalized fractional integral I_ρ is bounded from $\exp L^p(\mathbb{R}^n)$ to $\exp L^q(\mathbb{R}^n)$.

For other applications of Theorem 3.1, see [5].

Now, we investigate the boundedness on Morrey and Campanato spaces.

Theorem 3.2. *Let ρ satisfy* (2.1)–(2.3). *Let ϕ and ψ satisfy the doubling condition, and let $\phi(r)r^n$ and $\psi(r)r^n$ be almost increasing. Assume that there exists a constant $A > 0$ such that, for all $r > 0$,*

$$\int_0^r \frac{\rho(t)}{t}\, dt\ \phi(r) + \int_r^{+\infty} \frac{\rho(t)\phi(t)}{t}\, dt \le A\psi(r). \tag{3.5}$$

Then I_ρ is bounded from $L_{1,\phi}(\mathbb{R}^n)$ to $L_{1,\psi}(\mathbb{R}^n)$.

Theorem 3.3. *Let ρ satisfy* (2.1), (2.2), (2.4) *and* (2.6). *Let ϕ and ψ satisfy the doubling condition, and let $\phi(r)r^n$ and $\psi(r)r^n$ be almost increasing. Assume that* *there exists a constant* $A > 0$ *such that, for all* $r > 0$,

$$\int_0^r \frac{\rho(t)}{t}\,dt\;\phi(r) + r\int_r^{+\infty} \frac{\rho(t)\phi(t)}{t^2}\,dt \le A\psi(r). \tag{3.6}$$

Then $\tilde{I}_\rho$ *is bounded from* $L_{1,\phi}(\mathbb{R}^n)$ *to* $\mathcal{L}_{1,\psi}(\mathbb{R}^n)$.

We have the following relation between a weak Orlicz space and a Morrey space:

Theorem 3.4. *Let* Φ *be a Young function with* (2.8) *and* $\phi(r) = \Phi^{-1}(1/r^n)$. *Then*

$$L^{\Phi}(\mathbb{R}^n) \subset L^{1,\phi}(\mathbb{R}^n), \quad \textit{and} \quad \|f\|_{L^{1,\phi}} \le C\|f\|_{\Phi}. \tag{3.7}$$

Moreover, if $\Phi \in \nabla_2$, *then*

$$L^{\Phi}_{\text{weak}}(\mathbb{R}^n) \subset L^{1,\phi}(\mathbb{R}^n), \quad \textit{and} \quad \|f\|_{L^{1,\phi}} \le C\|f\|_{\Phi,\text{weak}}. \tag{3.8}$$

Combining Theorems 3.3 and 3.4, we have the following:

Corollary 3.5. *Let* ρ *satisfy* (2.1), (2.2), (2.4) *and* (2.6). *Let* Φ *be a Young function with* (2.8) *and let* ϕ *satisfy the doubling condition and be almost increasing. Assume that there exists a constant* $A > 0$ *such that, for all* $r > 0$,

$$\int_0^r \frac{\rho(t)}{t}\,dt\;\Phi^{-1}\left(\frac{1}{r^n}\right) + r\int_r^{+\infty} \frac{\rho(t)\Phi^{-1}(1/t^n)}{t^2}\,dt \le A\phi(r). \tag{3.9}$$

Then $\tilde{I}_\rho$ *is bounded from* $L^{\Phi}(\mathbb{R}^n)$ *to* $\mathrm{BMO}_\phi(\mathbb{R}^n)$. *Moreover, if* $\Phi \in \nabla_2$, *then* $\tilde{I}_\rho$ *is bounded from* $L^{\Phi}_{\text{weak}}(\mathbb{R}^n)$ *to* $\mathrm{BMO}_\phi(\mathbb{R}^n)$.

Example 3.2. Let ρ be as in (3.4). Let ϕ_β satisfy the doubling condition and

$$\phi_\beta(r) = \begin{cases} 1/(\log(1/r))^\beta & \text{for small } r, \\ (\log r)^\beta & \text{for large } r. \end{cases} \tag{3.10}$$

If $p > 0$, $0 \le \beta = \alpha - 1/p < 1$, then $\tilde{I}_\rho$ is bounded from $\exp L^p_{\text{weak}}(\mathbb{R}^n)$ to $\mathrm{BMO}_{\phi_\beta}(\mathbb{R}^n)$.

The next result is for the Campanato spaces:

Theorem 3.6. *Let* ρ *satisfy* (2.1), (2.2), (2.5) *and* (2.6). *Let* ϕ *and* ψ *satisfy the doubling condition, and let* $\phi(r)r^n$ *and* $\psi(r)r^n$ *be almost increasing. Assume that there exists a constant* $A > 0$ *such that, for all* $r > 0$,

$$\int_0^r \frac{\rho(t)}{t}\,dt\;\phi(r) + r\int_r^{+\infty} \frac{\rho(t)\phi(t)}{t^2}\,dt \le A\psi(r). \tag{3.11}$$

Then $\tilde{I}_\rho$ *is bounded from* $\mathcal{L}_{1,\phi}(\mathbb{R}^n)$ *to* $\mathcal{L}_{1,\psi}(\mathbb{R}^n)$.

Remark 3.1. From Lemma 4.3 it follows that $\tilde{I}_\rho 1$ is a constant. Hence $\tilde{I}_\rho$ is well defined as an operator from $\mathcal{L}_{1,\phi}(\mathbb{R}^n)$ to $\mathcal{L}_{1,\psi}(\mathbb{R}^n)$.

The boundedness of the fractional integral I_α on a Campanato space is known (Peetre [8, Theorem 5.4]).

Corollary 3.7 ([4]). *Let ρ satisfy* (2.1), (2.2), (2.5) *and* (2.6). *Let ϕ and ψ satisfy the doubling condition and be almost increasing. Assume that there exists a constant $A > 0$ such that, for all $r > 0$,*

$$\int_0^r \frac{\rho(t)}{t}\, dt\, \phi(r) + r \int_r^{+\infty} \frac{\rho(t)\phi(t)}{t^2}\, dt \le A\psi(r). \tag{3.12}$$

Then $\tilde{I}_\rho$ is bounded from $\mathrm{BMO}_\phi(\mathbb{R}^n)$ *to* $\mathrm{BMO}_\psi(\mathbb{R}^n)$.

Example 3.3. Let ρ be as in (3.4). Let ϕ_β and ϕ_γ be as in (3.10). If $\beta \ge 0$ and $\gamma = \alpha + \beta$, then $\tilde{I}_\rho$ is bounded from $\mathrm{BMO}_{\phi_\beta}(\mathbb{R}^n)$ to $\mathrm{BMO}_{\phi_\gamma}(\mathbb{R}^n)$.

4. Proofs

First, we state a lemma to prove Theorems 3.2 and 3.3.

Lemma 4.1. *Let ρ, ϕ and ψ be functions from $(0, +\infty)$ to itself with the doubling condition. Let $k = 0$ or $k = 1$. If there exists a constant $C > 0$ such that, for all $r > 0$,*

$$r^k \int_r^{+\infty} \frac{\rho(t)\phi(t)}{t^{1+k}}\, dt \le C\psi(r),$$

then there exists a constant $C' > 0$ such that, for all $f \in L_{1,\phi}(\mathbb{R}^n)$, $a \in \mathbb{R}^n$ and $r > 0$,

$$r^k \int_{B(a,r)^C} \frac{\rho(|a-y|)}{|a-y|^{n+k}} |f(y)|\, dy \le C'\psi(r) \|f\|_{L_{1,\phi}}.$$

Proof. We have

$$r^k \int_{B(a,r)^C} \frac{\rho(|a-y|)}{|a-y|^{n+k}} |f(y)|\, dy = r^k \sum_{j=1}^{+\infty} \int_{2^{j-1}r \le |a-y| \le 2^j r} \frac{\rho(|a-y|)}{|a-y|^{n+k}} |f(y)|\, dy$$

$$\le Cr^k \sum_{j=1}^{+\infty} \frac{\rho(2^j r)}{(2^j r)^{n+k}} \int_{B(a,2^j r)} |f(y)|\, dy \le Cr^k \sum_{j=1}^{+\infty} \frac{\rho(2^j r)\phi(2^j r)}{(2^j r)^k} \|f\|_{L_{1,\phi}}$$

$$\sim r^k \int_r^{+\infty} \frac{\rho(t)\phi(t)}{t^{1+k}} \|f\|_{L_{1,\phi}} \le C'\psi(r) \|f\|_{L_{1,\phi}}.$$

□

Proof of Theorem 3.2. For any ball $B = B(a, r)$, let

$$E_B{}^1(x) = \int_B f(y)\frac{\rho(|x-y|)}{|x-y|^n}\,dy, \quad E_B{}^2(x) = \int_{B^C} f(y)\frac{\rho(|x-y|)}{|x-y|^n}\,dy.$$

By (3.5) we have

$$\begin{aligned}
&\int_B\left(\int_B |f(y)|\frac{\rho(|x-y|)}{|x-y|^n}\,dx\right)dy \\
&\quad\le \int_B |f(y)|\left(\int_{B(y,2r)}\frac{\rho(|x-y|)}{|x-y|^n}\,dx\right)dy \le \int_B |f(y)|\,dy\int_0^{2r}\frac{\rho(t)}{t}\,dt \\
&\quad\le C\|f\|_{L_{1,\phi}}r^n\phi(r)\int_0^r\frac{\rho(t)}{t}\,dt \le C\|f\|_{L_{1,\phi}}r^n\psi(r).
\end{aligned}$$

From Fubini's theorem it follows that $E_B{}^1$ is well defined and that

$$\int_B |E_B{}^1(x)|\,dx \le C\psi(r)r^n\|f\|_{L_{1,\phi}}. \tag{4.1}$$

From Lemma 4.1 with $k = 0$ it follows that $E_B{}^2$ is well defined and

$$|E_B{}^2(x)| \le C\psi(r)\|f\|_{L_{1,\phi}}. \tag{4.2}$$

Therefore we have the desired result. □

Proof of Theorem 3.3. For any ball $B = B(a, r)$, let $\tilde{B} = B(a, 2r)$ and

$$\begin{aligned}
E_B(x) &= \int_{\mathbb{R}^n} f(y)\left(\frac{\rho(|x-y|)}{|x-y|^n} - \frac{\rho(|a-y|)(1-\chi_{\tilde{B}}(y))}{|a-y|^n}\right)dy, \\
C_B &= \int_{\mathbb{R}^n} f(y)\left(\frac{\rho(|a-y|)(1-\chi_{\tilde{B}}(y))}{|a-y|^n} - \frac{\rho(|y|)(1-\chi_{B_0}(y))}{|y|^n}\right)dy, \\
E_B{}^1(x) &= \int_{\tilde{B}} f(y)\frac{\rho(|x-y|)}{|x-y|^n}\,dy, \\
E_B{}^2(x) &= \int_{\tilde{B}^C} f(y)\left(\frac{\rho(|x-y|)}{|x-y|^n} - \frac{\rho(|a-y|)}{|a-y|^n}\right)dy.
\end{aligned}$$

Then

$$\tilde{I}_\rho f(x) - C_B = E_B(x) = E_B{}^1(x) + E_B{}^2(x) \quad \text{for } x \in B.$$

By (2.6) we have

$$\begin{aligned}
&\left|\frac{\rho(|a-y|)(1-\chi_{\tilde{B}}(y))}{|a-y|^n} - \frac{\rho(|y|)(1-\chi_{B_0}(y))}{|y|^n}\right| \\
&\qquad\qquad \le \begin{cases} C, & |a-y| \le \max(2|a|, 2r) \\ C|a|\dfrac{\rho(|a-y|)}{|a-y|^{n+1}}, & |a-y| \ge \max(2|a|, 2r). \end{cases}
\end{aligned}$$

From Lemma 4.1 it follows that C_B is well defined. By (3.6) we have

$$\int_{\tilde{B}}\left(\int_B |f(y)|\frac{\rho(|x-y|)}{|x-y|^n}\,dx\right)dy$$
$$\le \int_{\tilde{B}}|f(y)|\left(\int_{B(y,3r)}\frac{\rho(|x-y|)}{|x-y|^n}\,dx\right)dy \le \int_{\tilde{B}}|f(y)|\,dy\int_0^{3r}\frac{\rho(t)}{t}\,dt$$
$$\le C\|f\|_{L_{1,\phi}}r^n\phi(r)\int_0^r\frac{\rho(t)}{t}\,dt \le C\|f\|_{L_{1,\phi}}r^n\psi(r).$$

From Fubini's theorem it follows that ${E_B}^1$ is well defined and that

$$\int_B |{E_B}^1(x)|\,dx \le C\psi(r)r^n\|f\|_{L_{1,\phi}}. \tag{4.3}$$

From (2.6) and Lemma 4.1 with $k=1$ it follows that ${E_B}^2$ is well defined and

$$|{E_B}^2(x)| \le C\psi(r)\|f\|_{L_{1,\phi}}. \tag{4.4}$$

By (4.3) and (4.4) we have

$$\frac{1}{|B|}\int_B\left|\tilde{I}_\rho f(x)-C_B\right|\,dx \le C\psi(r)\|f\|_{L_{1,\phi}},$$

$$\|\tilde{I}_\rho f\|_{\mathcal{L}_{1,\psi}} \le C\|f\|_{L_{1,\phi}}.$$ □

Lemma 4.2. *Let Φ be a Young function. If $\Phi\in\nabla_2$, then there exists a constant $C>0$ such that*

$$\int_r^\infty\frac{dt}{\Phi(t)} \le C\frac{r}{\Phi(r)}, \quad r>0.$$

Proof. By definition, there exists a constant $k>1$ such that

$$\frac{1}{\Phi(k^j r)} \le \frac{1}{(2k)^j}\frac{1}{\Phi(r)}.$$

Then

$$\int_{k^j r}^{k^{j+1}r}\frac{dt}{\Phi(t)} \le \frac{k^{j+1}r-k^j r}{\Phi(k^j r)} \le \frac{k-1}{2^j}\frac{r}{\Phi(r)}.$$

Hence

$$\int_r^\infty\frac{dt}{\Phi(t)} \le \sum_{j=0}^\infty\frac{k-1}{2^j}\frac{r}{\Phi(r)} = 2(k-1)\frac{r}{\Phi(r)}.$$ □

Proof of Theorem 3.4. Let χ_B be the characteristic function of $B = B(a,r)$. By (2.7) and (2.9), we have

$$\int_B |f(x)|\,dx \le 2\|\chi_B\|_{\widetilde{\Phi}}\|f\|_\Phi \sim \phi(r)r^n\|f\|_\Phi,$$

and so (3.7).

Let $\Phi \in \nabla_2$. For $B = B(a,r)$ and for $\lambda > \|f\|_{\Phi,\mathrm{weak}}$, let

$$\eta = \lambda\Phi^{-1}\left(\frac{1}{r^n}\right) = \lambda\phi(r),$$

and

$$f = f^\eta + f_\eta, \quad f^\eta(x) = \begin{cases} f(x) & |f(x)| \ge \eta, \\ 0 & |f(x)| < \eta. \end{cases}$$

By Lemma 4.2 we have

$$\begin{aligned}\int_B |f^\eta(x)|\,dx &\le \int_0^\infty m(t, f^\eta)\,dt = \int_0^\eta m(\eta, f)\,dt + \int_\eta^\infty m(t, f)\,dt \\ &\le \frac{\eta}{\Phi(\eta/\lambda)} + \int_\eta^\infty \frac{dt}{\Phi(t/\lambda)} \le C\frac{\eta}{\Phi(\eta/\lambda)} = C\lambda\phi(r)r^n.\end{aligned}$$

Let

$$\Phi_1(t) = \frac{\epsilon}{n+\epsilon}r^\epsilon\Phi(t)^{(n+\epsilon)/n}, \quad \epsilon > 0.$$

Then Φ_1 is a Young function and

$$\int_B |f_\eta(x)|\,dx \le 2\|f_\eta\|_{\Phi_1}\|\chi_B\|_{\widetilde{\Phi}_1}.$$

Since

$${\Phi_1}^{-1}\left(\frac{1}{r^n}\right) = \Phi^{-1}\left(\left(\frac{(n+\epsilon)}{\epsilon}\right)^{n/(n+\epsilon)}\frac{1}{r^n}\right) \sim \Phi^{-1}\left(\frac{1}{r^n}\right) = \phi(r),$$

we have $\|\chi_B\|_{\widetilde{\Phi}_1} \sim {\Phi_1}^{-1}(1/r^n)r^n \sim \phi(r)r^n$. Since

$$d\Phi_1(t) = (\epsilon/n)r^\epsilon(\Phi(t))^{\epsilon/n}\,d\Phi(t),$$

we have

$$\begin{aligned}\int_{\mathbb{R}^n} \Phi_1\left(\frac{|f_\eta(x)|}{\lambda}\right)dx &= \int_0^{\eta/\lambda} m\left(t, \frac{f}{\lambda}\right)d\Phi_1(t) \le \int_0^{\eta/\lambda}\frac{1}{\Phi(t)}\,d\Phi_1(t) \\ &= \int_0^{\eta/\lambda}\frac{\epsilon}{n}r^\epsilon(\Phi(t))^{-1+\epsilon/n}\,d\Phi(t) = \int_0^{\Phi(\eta/\lambda)}\frac{\epsilon}{n}r^\epsilon\tau^{-1+\epsilon/n}\,d\tau = 1.\end{aligned}$$

Hence $\|f_\eta\|_{\Phi_1} \leq \lambda$. Therefore we have

$$\int_B |f_\eta(x)|\,dx \leq C\lambda\phi(r)r^n,$$

and so (3.8) follows. □

From the next lemma it follows that $\tilde{I}_\rho$ is well-defined as an operator from $\mathcal{L}_{1,\phi}(\mathbb{R}^n)$ to $\mathcal{L}_{1,\psi}(\mathbb{R}^n)$.

Lemma 4.3 ([4]). *If ρ satisfies* (2.1), (2.2), (2.5) *and* (2.6), *then*

$$\frac{\rho(|x_1-y|)}{|x_1-y|^n} - \frac{\rho(|x_2-y|)}{|x_2-y|^n} \tag{4.5}$$

is integrable on $\mathbb{R}^n$ as a function of y and its value is equal to 0 *for every choice of x_1 and x_2.*

The assumption in Theorem 3.6 is weaker than [4, Theorem 3.4]. Replacing $\|f\|_{\mathrm{BMO}_\phi}$ by $\|f\|_{\mathcal{L}_{1,\phi}}$ in [4, the proof of Lemma 3.4], and using (3.11), we have the following:

Lemma 4.4. *Under the assumption in Theorem* 3.6, *there exists a constant $C > 0$ such that, for all $a \in \mathbb{R}^n$ and $r > 0$,*

$$\int_{B(a,r)^C} \frac{\rho(|a-y|)}{|a-y|^{n+1}} \left|f(y) - f_{B(a,r)}\right| dy \leq C\frac{\psi(r)}{r}\|f\|_{\mathcal{L}_{1,\phi}}.$$

Proof of Theorem 3.6. Replacing $\|f\|_{\mathrm{BMO}_\phi}$ by $\|f\|_{\mathcal{L}_{1,\phi}}$ in [4, the proof of Theorem 3.4], and using Lemmas 4.3 and 4.4, we have the desired result. □

Acknowledgement. The author would like to thank Professor Kôzô Yabuta for his suggestions.

This research was partially supported by the Grant-in-Aid for Scientific Research (C), No. 11640165, 1999–2001, the Ministry of Education, Science, Sports and Culture, Japan.

References

[1] A. Cianchi, Strong and weak type inequalities for some classical operators in Orlicz spaces, J. London Math. Soc. 60 (1999), 187–202.

[2] I. Genebashvili, A. Gogatishvili, V. Kokilashvili and M. Krbec, Weight theory for integral transforms on spaces of homogeneous type, Longman, Harlow, 1998.

[3] V. Kokilashvili and M. Krbec, Weighted inequalities in Lorentz and Orlicz spaces, World Scientific, Singapore–New Jersey–London–Hong Kong, 1991.

[4] E. Nakai, On generalized fractional integrals, Proceedings of the International Conference on Mathematical Analysis and its Applications 2000, Kaohsiung, Taiwan, Taiwanese J. Math. 5 (2001), 587–602.

[5] E. Nakai, On generalized fractional integrals in the Orlicz spaces on spaces of homogeneous type, Sci. Math. 54 (2001), 473–487.

[6] E. Nakai and H. Sumitomo, On generalized Riesz potentials and spaces of some smooth functions, Sci. Math. 54 (2001), 463–472.

[7] R. O'Neil, Fractional integration in Orlicz spaces. I, Trans. Amer. Math. Soc. 115 (1965), 300–328.

[8] J. Peetre, On the theory of $\mathcal{L}_{p,\lambda}$ spaces, J. Funct. Anal. 4 (1969), 71–87.

[9] M. M. Rao and Z. D. Ren, Theory of Orlicz Spaces, Marcel Dekker, Inc., New York–Basel–Hong Kong, 1991.

[10] B. Rubin, Fractional integrals and potentials, Addison Wesley Longman Limited, Essex, 1996.

[11] E. M. Stein, Singular integrals and differentiability Properties of functions, Princeton University Press, Princeton, NJ, 1970.

[12] E. M. Stein, Harmonic Analysis, real-variable methods, orthogonality, and oscillatory integrals, Princeton University Press, Princeton, NJ, 1993.

[13] A. Torchinsky, Interpolation of operations and Orlicz classes, Studia Math. 59 (1976), 177–207.

Eiichi Nakai, Department of Mathematics, Osaka Kyoiku University, Kashiwara, Osaka 582-8582, Japan,

E-mail: `enakai@cc.osaka-kyoiku.ac.jp`

Optimal Sobolev embeddings – old and new

Luboš Pick *

Dedicated to Jaak Peetre on the occasion of his 65th birthday

Abstract. We study Sobolev-type embeddings of the first degree on domains of finite measure in $\mathbb{R}^n$. We develop an elementary unified proof of a sharp Sobolev embedding which can be used to prove most of the known results on embeddings of Sobolev spaces built upon spaces of various types, namely Orlicz, Lorentz and Lorentz–Zygmund spaces. We show the optimality of function spaces involved. As a by-product we obtain in the limiting case of embedding a new function class of independent interest. We point out some of its basic properties and relations to known function spaces.

2000 Mathematics Subject Classification: Primary 46E30; Secondary 26D10, 47G10

1. Introduction

Let $\Omega \subset \mathbb{R}^n$, $n \geq 2$, be a bounded open set. Without loss of generality we shall throughout assume that $|\Omega| = 1$. Let $W_0^{1,p}(\Omega)$, $1 \leq p \leq \infty$, be the closure of $C_0^1(\Omega)$ in the norm

$$\|u\|_{W^{1,p}(\Omega)} = \|u\|_{L^p(\Omega)} + \|Du\|_{L^p(\Omega)},$$

where D denotes the distributional gradient of u. The classical Sobolev embedding theorem states that, if $1 \leq p < n$, then

$$W_0^{1,p}(\Omega) \hookrightarrow L^{p^*}(\Omega) \quad \text{where } p^* = \frac{np}{n-p}. \tag{1.1}$$

In the limiting situation when $p = n$, we have

$$W_0^{1,n}(\Omega) \hookrightarrow L^q(\Omega) \quad \text{for every } 1 \leq q < \infty, \tag{1.2}$$

and, in the super-limiting case when $p > n$,

$$W_0^{1,p}(\Omega) \hookrightarrow L^\infty(\Omega). \tag{1.3}$$

*This research was partly supported by the grant no. 201/01/0333 of the Grant Agency of the Czech Republic and by the grant no. MSM 113200007 of the Czech Ministry of Education

All the embeddings (1.1)–(1.3) can be sharpened as far as the target space is concerned, although they cannot be sharpened within the context of L^p spaces. In the limiting case when $p = n$, we have

$$W_0^{1,n}(\Omega) \hookrightarrow \exp L^{\frac{n}{n-1}}(\Omega), \tag{1.4}$$

where $\exp L^{\frac{n}{n-1}}(\Omega)$ is the exponential-type Orlicz space endowed with the norm

$$\|u\|_{\exp L^{\frac{n}{n-1}}(\Omega)} = \inf\left\{\lambda > 0;\ \int_\Omega \exp\left(\left(\frac{|u(x)|}{\lambda}\right)^{\frac{n}{n-1}}\right) dx \le 1\right\}.$$

The embedding (1.4) is usually associated to Trudinger [25], similar results had been obtained earlier by Pokhozhaev [21] and Yudovich [26].

In general, let A be a *Young function*, that is, a convex strictly increasing function A on $[0, \infty)$ such that $A(0) = 0$. We define the Orlicz space $L_A(\Omega)$ as the set of all measurable functions u on Ω for which there exists a positive λ such that

$$\int_\Omega A\left(\frac{|u(x)|}{\lambda}\right) dx < \infty.$$

The Orlicz space is furnished with the Luxemburg norm

$$\|u\|_{L_A(\Omega)} = \inf\left\{\lambda > 0;\ \int_\Omega A\left(\frac{|u(x)|}{\lambda}\right) dx \le 1\right\}.$$

Since every L^p–space is Orlicz space, both (1.1) and (1.4) can be considered as embeddings of Sobolev spaces into Orlicz spaces. It is thus reasonable to ask whether the target spaces in these embeddings are optimal in the context of Orlicz spaces. The answer is in both cases known to be positive: there is no Orlicz space $L_A(\Omega)$ strictly smaller than $L^{p^*}(\Omega)$ for which we would have

$$W_0^{1,p}(\Omega) \hookrightarrow L_A(\Omega), \tag{1.5}$$

and, similarly, there is no Orlicz space $L_A(\Omega)$ strictly smaller than $\exp L^{\frac{n}{n-1}}(\Omega)$ such that

$$W_0^{1,n}(\Omega) \hookrightarrow L_A(\Omega). \tag{1.6}$$

Put another way, if (1.5) or (1.6) holds, then necessarily $L^{p^*}(\Omega) \hookrightarrow L_A(\Omega)$ or $\exp L^{\frac{n}{n-1}}(\Omega) \hookrightarrow L_A(\Omega)$, respectively. This is a result of Hempel, Morris and Trudinger [12] in the case of (1.2). A general construction (covering in particular both the above cases) of the optimal Orlicz target space was given by Andrea Cianchi [6].

It is now natural to ask about the optimality of the domain, too. This however requires some explanation. We can think of $W_0^{1,p}(\Omega)$ as the first-order Sobolev space built upon the Lebesgue space $L^p(\Omega)$. In this sense, $L^p(\Omega)$ is a *domain space* of the Sobolev embedding. We would like to know whether or not this space is optimal, that is, whether or not it can be replaced by a bigger space. While it is quite clear that both $L^p(\Omega)$ in (1.1) and $L^n(\Omega)$ in (1.4) are optimal domain *Lebesgue* spaces, it is an interesting question whether the same is true also in the broader context of *Orlicz* spaces.

This problem was solved in a joint work with Ron Kerman. The solution is rather surprising; it turns out that the answer is positive in the case of (1.1), but it is negative in the case of (1.4). Let us be more precise.

Theorem 1.1 (Kerman, Pick 1999). (i) *Let* $p \in [1, n)$. *Then,* $L^{p^*}(\Omega)$ *is the largest Orlicz space which renders* (1.1) *true.*

(ii) *The space* $L^n(\Omega)$ *is* not *the largest Orlicz space which renders* (1.2) *true.*

The negative result in (ii) raises the question what is the optimal (largest) Orlicz domain space $L_A(\Omega)$ such that

$$W_0^1 L_A(\Omega) \hookrightarrow \exp L^{\frac{n}{n-1}}(\Omega). \tag{1.7}$$

Our next result shows that there does not exist such a space.

Theorem 1.2 (Kerman, Pick 1999). *Let* A *be a Young function such that* (1.7) *holds. Then there is another Young function* B *such that* $L_A(\Omega) \subsetneqq L_B(\Omega)$ *and*

$$W_0^1 L_B(\Omega) \hookrightarrow \exp L^{\frac{n}{n-1}}(\Omega).$$

To prove Theorem 1.1 (i) we combine the results from [9], where an optimal rearrangement-invariant domain space is found (see below) with the idea that there is always only one Orlicz space corresponding to a given fundamental function. Theorem 1.2 is proved by the construction of the Young function B; Theorem 1.1 (ii) then clearly follows. Details can be found for example in [19] or [20], see also [7].

The situation described in Theorem 1.2 can be interpreted as that there exists an *open set* (with no endpoint) of Orlicz domain spaces satisfying the limiting Sobolev embedding.

This motivates us to broaden still the context of function spaces in which we look for optimal results. The appropriate next step is to consider rearrangement-invariant Banach function spaces.

Denote by $\mathfrak{M}(\Omega)$ the class of real-valued measurable functions on Ω and by $\mathfrak{M}^+(\Omega)$ the class of nonnegative functions in $\mathfrak{M}(\Omega)$. Given $f \in \mathfrak{M}(\Omega)$, its *non-increasing rearrangement* is defined by

$$f^*(t) = \inf\{\lambda > 0;\ |\{x \in \Omega;\ |f(x)| > \lambda\}| \le t\}, \quad 0 < t < |\Omega|.$$

Definition 1.3. A *rearrangement-invariant norm* ϱ on $\mathfrak{M}^+(0,1)$ is defined by the following seven axioms:

(A$_1$) $\varrho(f) \ge 0$ with $\varrho(f) = 0$ if and only if $f = 0$ a.e.;

(A$_2$) $\varrho(af) = a\varrho(f)$ for all $a \ge 0$;

(A$_3$) $\varrho(f+g) \le (\varrho(f) + \varrho(g))$ for all $f, g \in \mathfrak{M}^+(0,1)$;

(A$_4$) $f_n \nearrow f$ implies $\varrho(f_n) \nearrow \varrho(f)$;

(A$_5$) $\varrho(\chi_{(0,1)}) < \infty$;

(A$_6$) there exists $C > 0$ such that

$$\int_0^1 f(x)\,dx \le C\varrho(f) \quad \text{for all } f \in \mathfrak{M}_+(0,1);$$

(A$_7$) $\varrho(f) = \varrho(f^*)$.

As an example let us mention the familiar *Lorentz* (quasi-)norms, given for every $p, q \in (0, \infty]$ by

$$\|u\|_{L^{p,q}(\Omega)} = \varrho_{p,q}(u^*) = \|t^{\frac{1}{p}-\frac{1}{q}} u^*(t)\|_{L^q(0,1)}, \quad u \in \mathfrak{M}^+(\Omega),$$

and their generalization, *Lorentz–Zygmund* (quasi-)norms, given for every $p, q \in (0, \infty]$ and $\alpha \in \mathbb{R}$ by

$$\|u\|_{L^{p,q;\alpha}(\Omega)} = \varrho_{p,q;\alpha}(u^*) = \|t^{\frac{1}{p}-\frac{1}{q}} \left(\log(\tfrac{e}{t})\right)^{\alpha} u^*(t)\|_{L^q(0,1)}, \quad u \in \mathfrak{M}^+(\Omega).$$

In the new context of r.i. spaces, the target spaces in (1.1) and (1.4) are no longer optimal. As for the sublimiting embedding (1.1), we have ([17], [18])

Theorem 1.4 (O'Neil 1963, Peetre 1966). *Let* $1 \le p < n$, *then*

$$W_0^{1,p}(\Omega) \hookrightarrow L^{p^*,p}(\Omega). \tag{1.8}$$

For the limiting embedding (1.4), we have ([16], [11], [5])

Theorem 1.5 (Maz'ya 1964, Hansson 1979, Brézis–Wainger 1980).

$$W_0^{1,n}(\Omega) \hookrightarrow L^{\infty,n;-1}(\Omega). \tag{1.9}$$

It is not difficult to verify that (1.8) and (1.9) give sharper results than (1.1) and (1.4), respectively. This follows immediately from the well-known embeddings $L^{p^*,p}(\Omega) \hookrightarrow L^{p^*}(\Omega)$ and $L^{\infty,n;-1}(\Omega) \hookrightarrow \exp L^{\frac{n}{n-1}}(\Omega)$ (cf. for instance [3]).

Now, it was shown by Michael Cwikel and Evgenniy Pustylnik [8] that the target space in (1.9) is optimal (even in a broader sense). In [9], a general method, based on reduction of Sobolev embeddings to Hardy-type inequalities, is developed which allows us to conclude our analysis of optimality in the following way.

Theorem 1.6 (Edmunds, Kerman, Pick 2000). (i) *Let* $1 \le p < n$. *Then both the domain space* $L^p(\Omega)$ *and the target space* $L^{p^*,p}(\Omega)$ *are optimal for the Sobolev embedding* (1.8).

(ii) *The target space* $L^{\infty,n;-1}(\Omega)$ *is an optimal r.i. space in the Sobolev embedding* (1.9), *but the domain space* $L^n(\Omega)$ *is* not optimal *as it can be replaced by a strictly larger space* $X(\Omega)$ *endowed with the norm*

$$\|u\|_{X(\Omega)} = \varrho_{\infty,n;-1}\left(\int_t^1 s^{\frac{1}{n}-1} u^*(s)\,ds\right).$$

2. Elementary proof of Sobolev embeddings (1.8) and (1.9)

Now we know that the embeddings (1.8) and (1.9) are sharp as far as their target spaces are concerned. In a recent joint work with Jan Malý [14] we compiled an elementary and unified proof of both of these embeddings, which we shall sketch in this section. The proof is based on a general idea that strong-type estimates can be obtained from their weak-type analogues if an appropriate strategy is adopted. Such idea is not new. The fact that strong-type Sobolev estimates can be derived from weak ones has been observed by Federer [10] when deriving the Sobolev–Gagliardo–Nirenberg inequality ($p = 1$) from the isoperimetric inequality. The co-area integral argument can be simplified by the truncation trick which has been invented by Maz'ya [15] in connection with capacitary estimates. Its application to getting strong-type Sobolev embeddings from their weak forms is pursued for example by Bakry, Coulhon, Ledoux and Saloff-Coste in [1]. The point of departure in [1] is the well-known equivalence of the Sobolev embedding to Nash and Moser inequalities, which is avoided here. Weak estimates are usually being proved by means of Hardy–Littlewood maximal inequalities based on covering techniques. Our proof avoids the use of covering tricks. Another elementary proof based on a simple interpolation can be found in the book [23] by Stein.

In the limiting situation our approach gives a slight improvement of the target space. Of course, as we already know, this is not possible in the context of r.i. spaces.

Definition 2.1. Let $0 < p \leq \infty$. We define $W_p(\Omega)$ as the family of all measurable functions on Ω for which

$$\|u\|_{W_p(\Omega)} := \begin{cases} \left(\int_0^{|\Omega|} \frac{(u^*(\frac{t}{2})-u^*(t))^p}{t}\,dt\right)^{\frac{1}{p}} < \infty & \text{when } p < \infty; \\ \sup\limits_{0<t<1} \left(u^*(\tfrac{t}{2}) - u^*(t)\right) & \text{when } p = \infty. \end{cases}$$

Theorem 2.2 (Malý, Pick 2000). *If* $|\Omega| < \infty$, *then*

$$W_0^{1,n}(\Omega) \hookrightarrow W_n(\Omega). \tag{2.1}$$

It is not difficult to verify that $W_n(\Omega) \subsetneqq L^{\infty,n;-1}(\Omega)$, hence Theorem 2.2 improves Theorem 1.5. Let us note that this is not a contradiction with Theorem 1.6 (ii), because $W_n(\Omega)$ is not an r.i. space. (Of course it is obviously a rearrangement-invariant structure but it is not a linear set; see Theorem 3.1 below.)

Our first step is the following lemma.

Lemma 2.3. *For every* $u \in W_0^{1,1}(\Omega)$ *and* $\lambda > 0$, *the weak Sobolev–Gagliardo–Nirenberg inequality*

$$\lambda\Big(|\{|u| \geq \lambda\}|\Big)^{\frac{1}{n'}} \leq C \int_\Omega |\nabla u|\,dx \tag{2.2}$$

holds.

Next, we show that the weak Sobolev–Gagliardo–Nirenberg inequality contains already all the information needed to prove the desired sharp Sobolev embeddings. Our next step is a lemma based on the Maz'ya's truncation trick.

Lemma 2.4. *Let* $|\Omega| < \infty$ *and* $1 \le p \le n$. *Let* $u \in W_0^{1,p}(\Omega)$ *and denote*

$$t_k = 2^{1-k}|\Omega| \quad \text{and} \quad a_k = u^*(t_k), \quad k \in \mathbb{N}. \tag{2.3}$$

Then

$$\sum_{k=1}^{\infty} t_k^{\frac{p}{p^*}} \left(a_{k+1} - a_k\right)^p \le C \int_\Omega |\nabla u|^p \, dx \tag{2.4}$$

with C *depending only on* p *and* n.

It is worth noticing that (2.4) is a universal estimate covering both sublimiting and limiting cases.

Proof of Theorem 1.5. Let t_k and a_k have the same meaning as in (2.3). Given $\varepsilon > 0$, the convexity of t^p yields ([13, Lemma 1.1])

$$a_{k+1}^p \le (1 + \frac{1}{\varepsilon})^{p-1}\left(a_{k+1} - a_k\right)^p + (1+\varepsilon)^{p-1} a_k^p.$$

Hence (taking into account that $a_1 = u^*(t_1) = 0$)

$$\begin{aligned} 2^{\frac{p}{p^*}} \sum_{k=1}^{\infty} t_{k+1}^{\frac{p}{p^*}} a_k^p &= \sum_{k=1}^{\infty} t_{k+1}^{\frac{p}{p^*}} a_{k+1}^p \\ &\le (1+\varepsilon)^{p-1} \sum_{k=1}^{\infty} t_{k+1}^{\frac{p}{p^*}} a_k^p + \left(1 + \frac{1}{\varepsilon}\right)^{p-1} \sum_{k=1}^{\infty} t_{k+1}^{\frac{p}{p^*}} \left(a_{k+1} - a_k\right)^p. \end{aligned}$$

Choosing $\varepsilon > 0$ so small that $(1+\varepsilon)^{p-1} < 2^{\frac{p}{p^*}}$, we obtain

$$\sum_{k=1}^{\infty} t_{k+1}^{\frac{p}{p^*}} \left(u^*(t_k)\right)^p \le C \int_\Omega |\nabla u|^p \, dx,$$

which is a discrete version of (2.1). □

Proof of Theorem 2.2. Note that, for $p = n$, (2.4) reads as

$$\sum_{k=1}^{\infty} \left(a_{k+1} - a_k\right)^n \le C \int_\Omega |\nabla u|^n \, dx, \tag{2.5}$$

which is just a discrete version of (2.1). □

3. A new function space

Finally, let us mention some basic properties of the new function space $W_p(\Omega)$.

Theorem 3.1. (i) $\|\chi_E\|_{W_p(\Omega)} = (\log 2)^{\frac{1}{p}}$ *for every measurable* $E \subset \Omega$;

(ii) $L^\infty(\Omega) = W_1(\Omega)$;

(iii) *for* $p \in [1, \infty)$, *each integer-valued* $u \in W_p(\Omega)$ *is bounded;*

(iv) *for* $p \in (1, \infty)$, $W_p(\Omega)$ *is not a linear set;*

(v) *for* $p \in (1, \infty)$, $W_p(\Omega) \subsetneqq BW_p(\Omega)$;

(vi) $W_p(\Omega) \subsetneqq W_q(\Omega)$ *for every* $0 < p < q \leq \infty$.

Proof. To show (i), (ii) and (v) is an easy exercise.

(iii) Suppose that u is an integer-valued unbounded function on Ω. Then there are $\alpha_1 > \alpha_2 > \cdots > 0$ such that $u^*(\alpha_j-) - u^*(\alpha_j+) \geq 1$, $j \in \mathbb{N}$. For each $j = 1, 2, \ldots$, we have

$$u^*(\tfrac{t}{2}) - u^*(t) \geq 1, \quad \alpha_j \leq t < 2\alpha_j,$$

and thus

$$\int_{\alpha_j}^{2\alpha_j} \frac{\left(u^*(\frac{t}{2}) - u^*(t)\right)^p}{t}\, dt \geq \int_{\alpha_j}^{2\alpha_j} \frac{dt}{t} = \log 2.$$

From the system $\{(\alpha_j, 2\alpha_j)\}$ of intervals we may obviously select an infinite disjoint subsystem, and thus

$$\int_0^{|\Omega|} \frac{\left(u^*(\frac{t}{2}) - u^*(t)\right)^p}{t}\, dt = \infty,$$

whence $u \notin W_p(\Omega)$.

(iv) It follows from (ii) that there is a nonnegative function $u \in W_p(\Omega)$ such that u^* is unbounded. Let $[u]$ be the integer part of u and set $w := [u] + 1$, $v := w - u$. Then $|v| \leq 1$, thus $v \in W_p(\Omega)$ by (ii). On the other hand, $w = u + v$ is integer valued but unbounded, hence $w \notin W_p(\Omega)$ by (iii).

(vi) Fix $t \in (0, 1)$, then

$$u^*(\tfrac{t}{2}) - u^*(t) \leq u^*(\tfrac{s}{2}) - u^*(2s) \quad \text{for every } s \in (\tfrac{t}{2}, t),$$

hence

$$\left(u^*(\tfrac{t}{2}) - u^*(t)\right)^p \leq \int_{t/2}^{t} \left(u^*(\tfrac{s}{2}) - u^*(2s)\right)^p \frac{ds}{s},$$

which yields

$$u^*(\tfrac{t}{2}) - u^*(t) \leq C\|u\|_{W_p(\Omega)}$$

by the triangle inequality. On taking supremum over t we get

$$\|u\|_{W_\infty(\Omega)} \leq \|u\|_{W_p(\Omega)}. \tag{3.1}$$

Now, let $q > p$, then

$$\begin{aligned}\|u\|_{W_q(\Omega)} &= \left(\int_0^1 \left[u^*(\tfrac{t}{2}) - u^*(t)\right]^{q-p} \left[u^*(\tfrac{t}{2}) - u^*(t)\right]^p \frac{dt}{t}\right)^{1/q} \\ &\leq \|u\|_{W_\infty(\Omega)}^{1-\frac{p}{q}} \|u\|_{W_p(\Omega)}^{\frac{p}{q}} \leq \|u\|_{W_p(\Omega)} \quad \text{by (3.1).}\end{aligned}$$

To see that the inclusion in (vi) is proper, note that, for $\alpha \in \mathbb{R}$, the function u_α, such that $u_\alpha^* = (\log(e/t))^\alpha$ for small t, belongs to $W_p(\Omega)$ if and only if $\alpha < 1 - \frac{1}{p}$. □

References

[1] S. Bakry, T. Coulhon, M. Ledoux and L. Saloff-Coste, Sobolev inequalities in disguise, Indiana Univ. Math. J. 44 (1995), 1033–1074.

[2] C. Bennett, R. De Vore and R. Sharpley, Weak–L^∞ and BMO, Ann. of Math. 113 (1981), 601–611.

[3] C. Bennett and K. Rudnick, On Lorentz–Zygmund spaces, Dissertationes Math. (Rozprawy Mat.) 175 (1980), 1–72.

[4] C. Bennett and R. Sharpley, Interpolation of Operators, Pure Appl. Math. 129, Academic Press, Boston, 1988.

[5] H. Brézis and S. Wainger, A note on limiting cases of Sobolev embeddings and convolution inequalities, Comm. Partial Differential Equations 5 (1980), 773–789.

[6] A. Cianchi, A sharp embedding theorem for Orlicz–Sobolev spaces, Indiana Univ. Math. J. 45 (1996), 39–65.

[7] A. Cianchi and L. Pick, Sobolev embeddings into BMO, VMO and L_∞, Ark. Mat. 36 (1998), 317–340.

[8] M. Cwikel and E. Pustylnik, Sobolev type embeddings in the limiting case, J. Fourier Anal. Appl. 4 (1998), 433–446.

[9] D. E. Edmunds, R. Kerman and L. Pick, Optimal Sobolev embeddings involving rearrangement-invariant quasinorms, J. Funct. Anal. 170 (2000), 307–355.

[10] H. Federer, Geometric Measure Theory, Springer-Verlag, Berlin, 1969 (Second edition 1996).

[11] K. Hansson, Imbedding theorems of Sobolev type in potential theory, Math. Scand. 45 (1979), 77–102.

[12] J. A. Hempel, G. R. Morris and N. S. Trudinger, On the sharpness of a limiting case of the Sobolev imbedding theorem, Bull. Australian Math. Soc. 3 (1970), 369–373.

[13] J. Malý and W. P. Ziemer, Fine regularity of solutions of elliptic partial differential equations, Math. Surveys and Monogr. 51, Amer. Math. Soc., Providence, RI, 1997.

[14] J. Malý and L. Pick, An elementary proof of sharp Sobolev embeddings, Proc. Amer. Math. Soc., to appear.

[15] V. G. Maz'ya, A theorem on the multidimensional Schrödinger operator (Russian), Izv. Akad. Nauk. 28 (1964), 1145–1172.

[16] V. G. Maz'ya, Sobolev Spaces, Springer-Verlag, Berlin, 1975.

[17] R. O'Neil, Convolution operators and $L_{(p,q)}$ spaces, Duke Math. J. 30 (1963), 129–142.

[18] J. Peetre, Espaces d'interpolation et théorème de Soboleff, Ann. Inst. Fourier 16 (1966), 279–317.

[19] L. Pick, Optimal Sobolev embeddings, in: Nonlinear Analysis, Function Spaces and Applications 6 (M. Cianchi and A. Kufner, eds.), Proceedings of the 6th International Spring School held in Prague, Czech Republic, May–June 1998, Olympia Press, Prague, 1999, 156–199.

[20] L. Pick, Supremum operators and optimal Sobolev inequalities, in: Function Spaces, Differential Operators and Nonlinear Analysis (V. Mustonen and J. Rákosník, eds.), Proceedings of the Spring School held in Syöte Centre, Pudasjärvi (Northern Finland), June 1999, Mathematical Institute AS CR, Prague, 2000, 207–219.

[21] S. I. Pokhozhaev, On eigenfunctions of the equation $\Delta u + \lambda f(u) = 0$, Dokl. Akad. Nauk SSR 165 (1965), 36–39.

[22] S. L. Sobolev, Applications of Functional Analysis in Mathematical Physics, Transl. Math. Monogr. 7, Amer. Math. Soc., Providence, RI, 1963.

[23] E. M. Stein, Singular Integrals and Differentiability Properties of Functions, Princeton University Press, Princeton, NJ,.

[24] L. Tartar, Imbedding theorems of Sobolev spaces into Lorentz spaces, Boll. Un. Mat. Ital. 8 1–B (1998), 479–500.

[25] N. S. Trudinger, On imbeddings into Orlicz spaces and some applications, J. Math. Mech. 17 (1967), 473–483.

[26] V. I. Yudovich, Some estimates connected with integral operators and with solutions of elliptic equations, Soviet Math. Doklady 2 (1961), 746–749.

Luboš Pick, Department of Mathematical Analysis, Charles University, Sokolovská 83, 186 75 Praha 8, Czech Republic

E-mail: pick@karlin.mff.cuni.cz

Moduli of smoothness and the interrelation of some classes of functions

Sergey Yu. Tikhonov *

Dedicated to Jaak Peetre

Abstract. In this article the functions are considered which have generalized derivative in Weyl's sense. The lower and upper estimates for the modules of smoothness $\omega_\beta(f,t)_p$ of these derivatives are expressed in terms of the modules of smoothness of the function itself. Our results provide embedding theorems of generalized Besov class and generalized Weyl class.

2000 Mathematics Subject Classification: 46E30, 42A99, 46E35, 42A08.

1. Introduction

Let L_p $(1 < p < \infty)$ be the space of 2π-periodic, measurable functions $f(x)$ such that $\|f\|_p = \left(\int_0^{2\pi} |f(x)|^p\,dx\right)^{\frac{1}{p}} < \infty$; $\omega_\beta(f,t)_p$ is the modulus of smoothness of order β $(\beta > 0)$ of function $f \in L_p$, that is, $\omega_\beta(f,t)_p = \sup_{|h|\le t} \|\Delta_h^\beta\|_p$, where

$$\Delta_h^\beta = \sum_{\nu=0}^{\infty} (-1)^\nu \frac{\beta(\beta-1)\dots(\beta-\nu+1)}{\nu!} f(x+(\beta-\nu)h).$$

For the sake of brevity, the trigonometric series

$$\frac{a_0}{2} + \sum_{n=1}^{\infty} (a_n \cos nx + b_n \sin nx)$$

*Partially supported by the Russian Foundation for Fundamental Research (grant no. 00-01-00042) and the Leading Scientific Schools (grant no. 00-15-96143).

will be written in a form $\sum_{k=0}^{\infty} A_k(x)$. We shall also use the notation $\Delta_0 = A_0(x)$, $\Delta_m = \sum_{k=2^{m-1}}^{2^m-1} A_k(x)$, $k = 1, 2, \ldots$; and $\theta = \min(2, p)$, $\tau = \max(2, p)$, where $1 < p < \infty$.

If $F(\eta) \geq 0$ and $G(\eta) \geq 0$ for all η, then the notation $F(\eta) \ll G(\eta)$ will mean that there exists a positive constant c not depending on η and such that $F(\eta) \leq c\,G(\eta)$ for all η. If $F(\eta) \ll G(\eta)$ and $G(\eta) \ll F(\eta)$ hold simultaneously, then we shall write $F(\eta) \asymp G(\eta)$.

Given a numerical sequence $\{\lambda_n\}$, let $\sigma(f, \lambda_n)$ be the transformed Fourier series, that is, such that, if $f(x) \sim \sum_{n=0}^{\infty} A_n(x)$, then $\sigma(f, \lambda_n) := \sum_{n=0}^{\infty} \lambda_n A_n(x)$.

We note that for $\omega_\beta(f^{(r)}, \delta)_p$ $(r > 0)$ the following estimates can be obtained (see [1]):

$$\left\{\int_0^\delta t^{-r\tau-1}\omega_{r+\beta}^\tau(f,t)_p dt\right\}^{\frac{1}{\tau}} \ll \omega_\beta(f^{(r)}, \delta)_p \ll \left\{\int_0^\delta t^{-r\theta-1}\omega_{r+\beta}^\theta(f,t)_p dt\right\}^{\frac{1}{\theta}}.$$

If we consider the estimates of $\omega_\beta(f^{(r-\varepsilon)}, \delta)_p$ and $\omega_\beta(f^{(r+\varepsilon)}, \delta)_p$ under the condition of the fixed $\omega_{r+\beta}(f, t)_p$ then the inequalities change (see [2],[3]):

$$1)\ \left\{\int_0^\delta t^{-(r-\varepsilon)\tau-1}\omega_{r+\beta}^\tau(f,t)_p dt + \delta^{\beta\tau}\int_\delta^1 t^{-(r-\varepsilon+\beta)\tau-1}\omega_{r+\beta}^\tau(f,t)_p dt\right\}^{\frac{1}{\tau}}$$
$$\ll \omega_\beta(f^{(r-\varepsilon)}, \delta)_p$$
$$\ll \left\{\int_0^\delta t^{-(r-\varepsilon)\theta-1}\omega_{r+\beta}^\theta(f,t)_p dt + \delta^{\beta\theta}\int_\delta^1 t^{-(r-\varepsilon+\beta)\theta-1}\omega_{r+\beta}^\theta(f,t)_p dt\right\}^{\frac{1}{\theta}}$$
$$(0 < \varepsilon < r),$$

$$2)\ \left\{\int_0^\delta t^{-(r+\varepsilon)\tau-1}\omega_{r+\beta}^\tau(f,t)_p dt\right\}^{\frac{1}{\tau}} \ll \delta^{\beta-\varepsilon}\left\{\int_\delta^1 t^{-(\beta-\varepsilon)\theta-1}\omega_\beta^\theta(f^{(r+\varepsilon)},t)_p dt\right\}^{\frac{1}{\theta}};$$

$$\delta^{\beta-\varepsilon}\left\{\int_\delta^1 t^{-(\beta-\varepsilon)\tau-1}\omega_\beta^\tau(f^{(r+\varepsilon)},t)_p dt\right\}^{\frac{1}{\tau}} \ll \left\{\int_0^\delta t^{-(r+\varepsilon)\theta-1}\omega_{r+\beta}^\theta(f,t)_p dt\right\}^{\frac{1}{\theta}}$$
$$(0 < \varepsilon < \beta).$$

In these cases the sequences $\{\lambda_n\}$ transforming the Fourier series of $f(x)$ are $\lambda_n^{(1)} = n^{r-\varepsilon}$ and $\lambda_n^{(2)} = n^{r+\varepsilon}$.

We shall obtain the estimates of the modulus of a function which has Fourier series $\sum_{n=1}^{\infty} \lambda_n A_n(x)$, where either $\lambda_n^{(1)} = \frac{n^r}{(\ln_N(nd_N))^A}$, or $\lambda_n^{(2)} = n^r \cdot (\ln_N(nd_N))^A$ respectively. Here $\ln_1 u = \ln u$, $\ln_i u = \ln(\ln_{i-1} u)$ $(i = 2, 3, \ldots, N)$ and the constants d_i such that $\ln_i d_i \geq 1$.

We shall use the notation $P_0(t) = 1$, $P_N(t) = \prod_{i=1}^N \ln_i \frac{d_i}{t}$ for $\forall N \in \mathbb{N}$ and $\forall t \in (0, 2\pi]$. Let us introduce some classes of functions.

- By $W(p, \lambda_n)$ denote the generalized Weyl class, that is, the class of functions $f(x) \in L_p$ with Fourier series $\sum_{n=0}^{\infty} A_n(x)$, for which the series $\sigma(f, \lambda_n)$ is the Fourier series of some function $\varphi(x) \in L_p$.

- Let $B^{(1)}(p, r, A, \beta, s)\, [B^{(2)}(p, r, A, \beta, s)]$ be the generalized Besov class of functions $f(x) \in L_p$ such that $I^{(1)}(p, r, A, \beta, s) < \infty\, [I^{(2)}(p, r, A, \beta, s) < \infty]$, where

$$I^{(1)}(p, r, A, \beta, s) = \left(\int_0^1 \frac{t^{-rs-1}}{P_N(t)\,(\ln_N \frac{d_N}{t})^{As}} \omega_\beta^s(f, t)_p dt \right)^{\frac{1}{s}}$$

$$\left[I^2(p, r, A, \beta, s) = \left(\int_0^1 t^{-rs-1} (\ln_N \frac{d_N}{t})^{As} \omega_\beta^s(f, t)_p dt \right)^{\frac{1}{s}} \right]$$

and $N \in \mathbb{N}$, s, β, r and $A \in (0, +\infty)$.

- M_γ denotes the class of functions, for which the Fourier series is $\sum_{n=1}^{\infty} a_n \cos nx$, where $a_n n^{-\gamma} \downarrow 0$ $(n \to \infty)$ for some given number $\gamma \geq 0$.

- Λ denotes the class of functions, for which the Fourier series is $\sum_{n=1}^{\infty} a_n \cos 2^n x$.

Note, that if $\beta < r$, then $B^{(1)}(p, r, A, \beta, s)\, [B^{(2)}(p, r, A, \beta, s)]$ consist of the functions which are equivalent constants.

The author is grateful to professor M. K. Potapov for constant attention to this work.

2. Fourier series transformed by means of $\lambda_n^{(1)} = \frac{n^r}{(\ln_N(nd_N))^A}$

Theorem 2.1. *If $s \leq \theta$, then $B^{(1)}(p, r, A, r, s) \subset W(p, \lambda^{(1)})$, in other words, if $f(x) \in B^{(1)}(p, r, A, r, s)$, then $\sigma(f, \lambda_n^{(1)})$ is the Fourier series of some function $\varphi(x) \in L_p$. Also*

$$\|\varphi\|_p \ll I^{(1)}(p, r, A, r, s)$$

and $\forall\, \beta > 0$

$$\omega_\beta^s(\varphi, \delta)_p \ll \int_0^\delta \frac{t^{-rs-1}}{(\ln_N \frac{d_N}{t})^{As}} \omega_{r+\beta}^s(f, t)_p dt + \delta^{\beta s} \int_\delta^1 \frac{t^{-(r+\beta)s-1}}{P_N(t)\,(\ln_N \frac{d_N}{t})^{As}} \omega_{r+\beta}^s(f, t)_p dt.$$

Theorem 2.2. *If $s \geq \tau$, then $W(p,\lambda^{(1)}) \subset B^{(1)}(p,r,A,r,s)$, besides,*

$$I^{(1)}(p,r,A,r,s) \ll \|\varphi\|_p$$

and $\forall\, \beta > 0$

$$\int\limits_0^\delta \frac{t^{-rs-1}}{(\ln_N \frac{d_N}{t})^{As}} \omega^s_{r+\beta}(f,t)_p dt + \delta^{\beta s} \int_\delta^1 \frac{t^{-(r+\beta)s-1}}{P_N(t)\,(\ln_N \frac{d_N}{t})^{As}} \omega^s_{r+\beta}(f,t)_p dt \ll \omega^s_\beta(\varphi,\delta)_p.$$

Theorem 2.3. *If $f(x) \in M_\gamma$, then*

a) *the conclusion of Theorem 2.1 will hold if we substitute the condition $0 < s \leq \theta$ by $0 < s \leq p$;*

b) *the conclusion of Theorem 2.2 will hold if we substitute the condition $\tau \leq s < +\infty$ by $p \leq s < +\infty$.*

Theorem 2.4. *If $f(x) \in \Lambda$, then*

a) *the conclusion of Theorem 2.1 will hold if we substitute the condition $0 < s \leq \theta$ by $0 < s \leq 2$;*

b) *the conclusion of Theorem 2.2 will hold if we substitute the condition $\tau \leq s < +\infty$ by $2 \leq s < +\infty$.*

3. Fourier series transformed by means of $\lambda_n^{(2)} = n^r \cdot (\ln_N(nd_N))^A$

Theorem 3.1. *If $s \leq \theta$ and at some $\beta > 0$ $f(x) \in B^{(2)}(p,r,A,r+\beta,s)$, then $\sigma(f,\lambda_n^{(2)})$ is the Fourier series of some function $\varphi(x) \in L_p$. That is, $B^{(2)}(p,r,A,r+\beta,s) \subset W(p,\lambda^{(2)})$. Also*

$$\|\varphi\|_p \ll I^{(2)}(p,r,A,r+\beta,s)$$

and

$$\left\{\delta^{\beta\tau}(\ln_N \frac{d_N}{\delta})^{A\tau} \int\limits_\delta^1 \frac{t^{-\beta\tau-1}}{P_N(t)(\ln_N \frac{d_N}{t})^{A\tau}} \omega^\tau_\beta(\varphi,t)_p dt + \omega^\tau_\beta(\varphi,\delta)_p\right\}^{\frac{1}{\tau}}$$

$$\ll \left\{\int\limits_0^\delta t^{-rs-1}(\ln_N \frac{d_N}{t})^{As} \omega^s_{r+\beta}(f,t)_p dt\right\}^{\frac{1}{s}}.$$

Theorem 3.2. *If $s \geq \tau$, then $W(p, \lambda^{(2)}) \subset B^{(2)}(p, r, A, r+\beta, s)$, besides,*

$$I^{(2)}(p, r, A, r+\beta, s) \ll \|\varphi\|_p$$

and $\forall \beta > 0$

$$\left\{\int_0^{\delta} t^{-rs-1}(\ln_N \frac{d_N}{t})^{As}\omega_{r+\beta}^s(f,t)_p dt\right\}^{\frac{1}{s}}$$

$$\ll \left\{\delta^{\beta\theta}(\ln_N \frac{d_N}{\delta})^{A\theta}\int_{\delta}^{1}\frac{t^{-\beta\theta-1}}{P_N(t)(\ln_N \frac{d_N}{t})^{A\theta}}\omega_{\beta}^{\theta}(\varphi,t)_p dt + \omega_{\beta}^{\theta}(\varphi,\delta)_p\right\}^{\frac{1}{\theta}}.$$

Theorem 3.3. *If $f(x) \in M_\gamma$, then*

a) *the conclusion of Theorem 3.1 will hold if we substitute the conditions $0 < s \leq \theta$ and $\tau = \max(2, p)$ by $0 < s \leq p$ and $\tau = p$.*

b) *the conclusion of Theorem 3.2 will hold if we substitute the conditions $\tau \leq s < +\infty$ and $\theta = \min(2, p)$ by $p \leq s < +\infty$ and $\theta = p$.*

Theorem 3.4. *If $f(x) \in \Lambda$, then*

a) *the conclusion of Theorem 3.1 will hold if we substitute the conditions $0 < s \leq \theta$ and $\tau = \max(2, p)$ by $0 < s \leq 2$ and $\tau = 2$;*

b) *the conclusion of Theorem 3.2 will hold if we substitute the conditions $\tau \leq s < +\infty$ and $\theta = \min(2, p)$ by $2 \leq s < +\infty$ and $\theta = 2$.*

4. Auxiliary results

Lemma 1 (see [4]). *Assume that $f(x) \in L_p$, $1 < p < \infty$, possesses the Fourier series $\sum_{n=1}^{\infty} A_n(x)$. Then*

$$\|f\|_p \asymp \left(\int_0^{2\pi}\left(\sum_{n=0}^{\infty}\Delta_n^2\right)^{\frac{p}{2}}dx\right)^{\frac{1}{p}}.$$

Lemma 2 (see [5]). *Assume that $f(x) \in L_p$, $1 < p < \infty$, possesses the Fourier series $\sum_{n=0}^{\infty} A_n(x)$, and let the sequence $\{\lambda_n\}$ be such that*

$$|\lambda_n| \leq M,\ n \in \mathbb{N};\quad \sum_{n=2^\nu}^{2^{\nu+1}-1}|\lambda_n - \lambda_{n+1}| \leq M,\ \nu = 0, 1, 2, \ldots$$

Then the series $\sum_{n=1}^{\infty}\lambda_n A_n(x)$ is the Fourier series of some function $\varphi(x) \in L_p$ and $\|\varphi\|_p \leq c\,\|f\|_p$, where $c = c(p, \{\lambda_n\})$.

Lemma 3 (see [1]). *Assume that $f(x) \in L_p$, $1 < p < \infty$, possesses the Fourier series $\sum_{n=1}^{\infty} A_n(x)$. Then*

(a)
$$\omega_\beta(f, \tfrac{1}{m})_p \asymp m^{-\beta}\Big\|\sum_{n=1}^{m} n^\beta A_n(x)\Big\|_p + \Big\|\sum_{n=m+1}^{\infty} A_n(x)\Big\|_p ;$$

(b) *If $f(x) \in M_\gamma$, then*

$$\omega_\beta(f, \frac{1}{m})_p \asymp m^{-\beta}\Big(\sum_{n=1}^{m} a_n^p n^{\beta p+p-2}\Big)^{\frac{1}{p}} + \Big(\sum_{n=m+1}^{\infty} a_n^p n^{p-2}\Big)^{\frac{1}{p}};$$

(c) *If $f(x) \in \Lambda$, then*

$$\omega_\beta(f, \frac{1}{2^m})_p \asymp 2^{-m\beta}\Big(\sum_{n=1}^{m} a_n^2 2^{2\beta n}\Big)^{\frac{1}{2}} + \Big(\sum_{n=m+1}^{\infty} a_n^2\Big)^{\frac{1}{2}}.$$

5. Proofs

We will prove only Theorem 2.2 in detail.

Proof of Theorem 2.2. Using method of work [6], it is not difficult to show that $W(p, \lambda^{(1)}) \subset B^{(1)}(p, r, A, r, s)$ and the estimate $I^{(1)}(p, r, A, r, s) \ll \|\varphi\|_p$. Let us show that $I_1 + I_2 \ll \omega_\beta^s(\varphi, \delta)_p$, where

$$I_1+I_2 := \int_0^\delta \frac{t^{-rs-1}}{(\ln_N \frac{d_N}{t})^{As}}\omega_{r+\beta}^s(f,t)_p dt + \delta^{\beta s}\int_\delta^1 \frac{t^{-(r+\beta)s-1}}{P_N(t)\,(\ln_N \frac{d_N}{t})^{As}}\omega_{r+\beta}^s(f,t)_p dt.$$

Let n be a natural number such that $\frac{1}{2^{n+1}} \le \delta < \frac{1}{2^n}$.

$$I_1 + I_2 \ll \sum_{\nu=n}^{\infty} \frac{2^{\nu rs}}{(\ln_N (d_N 2^\nu))^{As}}\omega_{r+\beta}^s(f, \frac{1}{2^\nu})_p$$
$$+ 2^{-n\beta s}\sum_{\nu=1}^{n} \frac{2^{\nu(r+\beta)s}}{P_N(\frac{1}{2^\nu})\,(\ln_N (d_N 2^\nu))^{As}}\omega_{r+\beta}^s(f, \frac{1}{2^\nu})_p.$$

According to Lemma 3(a),

$$I_1 \ll \sum_{\nu=n}^{\infty} \frac{2^{\nu r s}}{(\ln_N (d_N 2^{\nu}))^{As}} \Big\| \sum_{\xi=2^{\nu}+1}^{\infty} A_{\xi}(x) \Big\|_p^s + \sum_{\nu=n}^{\infty} \frac{2^{-\nu\beta s}}{(\ln_N (d_N 2^{\nu}))^{As}} \Big\| \sum_{\xi=1}^{2^{\nu}} A_{\xi}(x) \xi^{r+\beta} \Big\|_p^s =: I_3 + I_4.$$

Making use of Lemma 1 and the generalized Minkowski inequality first for sum, then for integrals and sums ($\frac{s}{2} \geq 1$, $\frac{s}{p} \geq 1$) gives

$$I_3 \ll \sum_{\nu=n}^{\infty} \frac{2^{\nu r s}}{(\ln_N (d_N 2^{\nu}))^{As}} \Big(\int_0^{2\pi} \Big\langle \sum_{m=\nu}^{\infty} \Delta_m^2 \Big\rangle^{\frac{p}{2}} dx \Big)^{\frac{s}{p}}$$

$$\ll \Big\{ \int_0^{2\pi} \Big(\sum_{\nu=n+1}^{\infty} \frac{2^{\nu r s}}{(\ln_N (d_N 2^{\nu}))^{As}} \Big\langle \sum_{m=\nu+1}^{\infty} \Delta_m^2 \Big\rangle^{\frac{s}{2}} \Big)^{\frac{p}{s}} dx \Big\}^{\frac{s}{p}}$$

$$\ll \Big\{ \int_0^{2\pi} \Big(\sum_{m=n+1}^{\infty} \Delta_m^2 \Big\langle \sum_{\nu=n}^{m-1} \frac{2^{\nu r s}}{(\ln_N (d_N 2^{\nu}))^{As}} \Big\rangle^{\frac{2}{s}} \Big)^{\frac{p}{2}} dx \Big\}^{\frac{s}{p}}.$$

Using $\Big\langle \sum_{\nu=n}^{m-1} \frac{2^{\nu r s}}{(\ln_N (d_N 2^{\nu}))^{As}} \Big\rangle^{\frac{2}{s}} \ll \frac{2^{2mr}}{(\ln_N (d_N 2^{m}))^{2A}}$, we get

$$I_3 \ll \Big\{ \int_0^{2\pi} \Big(\sum_{m=n+1}^{\infty} \Delta_m^2 \frac{2^{2mr}}{(\ln_N (d_N 2^{m}))^{2A}} \Big)^{\frac{p}{2}} dx \Big\}^{\frac{s}{p}}.$$

Let us estimate I_4.

$$I_4 \ll \sum_{\nu=n}^{\infty} \frac{2^{-\nu\beta s}}{(\ln_N (d_N 2^{\nu}))^{As}} \Big\| \sum_{\xi=1}^{2^{n}} A_{\xi}(x) \xi^{r+\beta} \Big\|_p^s + \sum_{\nu=n}^{\infty} \frac{2^{-\nu\beta s}}{(\ln_N (d_N 2^{\nu}))^{As}} \Big\| \sum_{\xi=2^{n}+1}^{2^{\nu}} A_{\xi}(x) \xi^{r+\beta} \Big\|_p^s =: I_5 + I_6.$$

An argument, analogous to the one that was used in the proof of the upper estimate for I_3, provides the estimate for I_5:

$$I_5 \ll \sum_{\nu=n}^{\infty} \frac{2^{-\nu\beta s}}{(\ln_N (d_N 2^{\nu}))^{As}} \Big(\int_0^{2\pi} \Big\langle \sum_{m=1}^{n} \Delta_m^2 2^{2m(\beta+r)} \Big\rangle^{\frac{p}{2}} dx \Big)^{\frac{s}{p}}$$

$$\ll \Big\{ \int_0^{2\pi} \Big(\sum_{\nu=n}^{\infty} \frac{2^{-\nu\beta s}}{(\ln_N (d_N 2^{\nu}))^{As}} \Big\langle \sum_{m=1}^{n} \Delta_m^2 2^{2m(\beta+r)} \Big\rangle^{\frac{s}{2}} \Big)^{\frac{p}{s}} dx \Big\}^{\frac{s}{p}}.$$

Since

$$\sum_{\nu=n}^{\infty} \frac{2^{-\nu\beta s}}{(\ln_N (d_N 2^\nu))^{As}} \ll \frac{2^{-n\beta s}}{(\ln_N (d_N 2^n))^{As}},$$

it follows that

$$I_5 \ll 2^{-n\beta s} \left\{ \int_0^{2\pi} \Big(\sum_{m=1}^{n} \Delta_m^2 \frac{2^{2m(r+\beta)}}{(\ln_N (d_N 2^m))^{2A}} \Big)^{\frac{p}{2}} dx \right\}^{\frac{s}{p}}.$$

We shall estimate I_6 as follows

$$I_6 \ll \sum_{\nu=n}^{\infty} \frac{2^{-\nu\beta s}}{(\ln_N (d_N 2^\nu))^{As}} \Big(\int_0^{2\pi} \Big\langle \sum_{m=n}^{\nu} \Delta_m^2 2^{2m(\beta+r)} \Big\rangle^{\frac{p}{2}} dx \Big)^{\frac{s}{p}}$$

$$\ll \left\{ \int_0^{2\pi} \Big(\sum_{\nu=n}^{\infty} \frac{2^{-\nu\beta s}}{(\ln_N (d_N 2^\nu))^{As}} \Big\langle \sum_{m=n}^{\nu} \Delta_m^2 2^{2m(\beta+r)} \Big\rangle^{\frac{s}{2}} \Big)^{\frac{p}{s}} dx \right\}^{\frac{s}{p}}$$

$$\ll \left\{ \int_0^{2\pi} \Big(\sum_{m=n}^{\infty} \Delta_m^2 2^{2m(\beta+r)} \Big\langle \sum_{\nu=m}^{\infty} \frac{2^{-\nu\beta s}}{(\ln_N (d_N 2^\nu))^{As}} \Big\rangle^{\frac{2}{s}} \Big)^{\frac{p}{2}} dx \right\}^{\frac{s}{p}}$$

$$\ll \left\{ \int_0^{2\pi} \Big(\sum_{m=n}^{\infty} \Delta_m^2 \frac{2^{2mr}}{(\ln_N (d_N 2^m))^{2A}} \Big)^{\frac{p}{2}} dx \right\}^{\frac{s}{p}}.$$

Now, we shall estimate I_2. Repeating the same argument which resulted above in the estimate for I_1 proves the fulfilment of the following chain of inequalities:

$$I_2 \ll 2^{-n\beta s} \sum_{\nu=1}^{n} \frac{2^{\nu(r+\beta)s}}{P_N(\frac{1}{2^\nu}) (\ln_N (d_N 2^\nu))^{As}} \Big\| \sum_{\xi=2^\nu+1}^{\infty} A_\xi(x) \Big\|_p^s$$

$$+ 2^{-n\beta s} \sum_{\nu=n+1}^{\infty} \frac{1}{P_N(\frac{1}{2^\nu}) (\ln_N (d_N 2^\nu))^{As}} \Big\| \sum_{\xi=1}^{2^\nu} A_\xi(x) \xi^{r+\beta} \Big\|_p^s =: I_7 + I_8.$$

Clearly,

$$I_7 \ll 2^{-n\beta s} \sum_{\nu=1}^{n} \frac{2^{\nu(r+\beta)s}}{P_N(\frac{1}{2^\nu}) (\ln_N (d_N 2^\nu))^{As}} \Big\| \sum_{\xi=2^\nu+1}^{2^{n+1}-1} A_\xi(x) \Big\|_p^s$$

$$+ 2^{-n\beta s} \sum_{\nu=1}^{n} \frac{2^{\nu(r+\beta)s}}{P_N(\frac{1}{2^\nu}) (\ln_N (d_N 2^\nu))^{As}} \Big\| \sum_{\xi=2^n+1}^{\infty} A_\xi(x) \Big\|_p^s =: I_9 + I_{10}.$$

Making use of Lemma 1 and the generalized Minkowski inequality twice gives

$$I_9 \ll 2^{-n\beta s} \sum_{\nu=1}^{n+1} \frac{2^{\nu(r+\beta)s}}{P_N(\frac{1}{2^\nu})\,(\ln_N\ (d_N 2^\nu))^{As}} \Big(\int_0^{2\pi} \Big\langle \sum_{m=\nu}^{n+1} \Delta_m^2 \Big\rangle^{\frac{p}{2}} dx \Big)^{\frac{s}{p}}$$

$$\ll 2^{-n\beta s} \Big\{ \int_0^{2\pi} \Big(\sum_{\nu=1}^{n+1} \frac{2^{\nu(r+\beta)s}}{P_N(\frac{1}{2^\nu})\,(\ln_N\ (d_N 2^\nu))^{As}} \Big\langle \sum_{m=\nu}^{n+1} \Delta_m^2 \Big\rangle^{\frac{s}{2}} \Big)^{\frac{p}{s}} dx \Big\}^{\frac{s}{p}}$$

$$\ll 2^{-n\beta s} \Big\{ \int_0^{2\pi} \Big(\sum_{m=1}^{n+1} \Delta_m^2 \Big\langle \sum_{\nu=1}^{m} \frac{2^{\nu(r+\beta)s}}{P_N(\frac{1}{2^\nu})\,(\ln_N\ (d_N 2^\nu))^{As}} \Big\rangle^{\frac{2}{s}} \Big)^{\frac{p}{2}} dx \Big\}^{\frac{s}{p}}.$$

Since

$$\Big\langle \sum_{\nu=1}^{m} \frac{2^{\nu(r+\beta)s}}{P_N(\frac{1}{2^\nu})\,(\ln_N\ (d_N 2^\nu))^{As}} \Big\rangle^{\frac{2}{s}} \ll \frac{2^{2m(r+\beta)}}{(\ln_N\ (d_N 2^m))^{2A}} \qquad (*)$$

holds, we have

$$I_9 \ll 2^{-n\beta s} \Big\{ \int_0^{2\pi} \Big(\sum_{m=1}^{n} \Delta_m^2 \frac{2^{2m(r+\beta)}}{(\ln_N\ (d_N 2^m))^{2A}} \Big)^{\frac{p}{2}} dx \Big\}^{\frac{s}{p}}.$$

Further, we shall estimate I_{10} and I_8. Using $(*)$ and Lemma 1, we get

$$I_{10} \ll \frac{2^{nrs}}{(\ln_N\ (d_N 2^n))^{2A}} \Big\| \sum_{\xi=2^n+1}^{\infty} A_\xi(x) \Big\|_p^s$$

$$\ll \frac{2^{nrs}}{(\ln_N\ (d_N 2^n))^{2A}} \Big(\int_0^{2\pi} \Big\langle \sum_{m=n}^{\infty} \Delta_m^2 \Big\rangle^{\frac{p}{2}} dx \Big)^{\frac{s}{p}}.$$

Thus,

$$I_{10} \ll \Big\{ \int_0^{2\pi} \Big(\sum_{m=n+1}^{\infty} \Delta_m^2 \frac{2^{2mr}}{(\ln_N\ (d_N 2^m))^{2A}} \Big)^{\frac{p}{2}} dx \Big\}^{\frac{s}{p}}.$$

We have

$$I_8 \ll 2^{-n\beta s}\sum_{\nu=1}^{n+1}\frac{1}{P_N(\frac{1}{2^\nu})\,(\ln_N\ (d_N 2^\nu))^{As}}\Big(\int_0^{2\pi}\Big\{\sum_{m=1}^{\nu}\Delta_m^2 2^{2m(\beta+r)}\Big\}^{\frac{p}{2}}dx\Big)^{\frac{s}{p}}$$

$$\ll 2^{-n\beta s}\Big\{\int_0^{2\pi}\Big(\sum_{\nu=1}^{n+1}\frac{1}{P_N(\frac{1}{2^\nu})\,(\ln_N\ (d_N 2^\nu))^{As}}\Big(\sum_{m=1}^{\nu}\Delta_m^2 2^{2m(\beta+r)}\Big)^{\frac{s}{2}}\Big)^{\frac{p}{s}}dx\Big\}^{\frac{s}{p}}$$

$$\ll 2^{-n\beta s}\Big\{\int_0^{2\pi}\Big(\sum_{m=1}^{n+1}\Delta_m^2 2^{2m(\beta+r)}\Big(\sum_{\nu=m}^{n+1}\frac{1}{P_N(\frac{1}{2^\nu})\,(\ln_N\ (d_N 2^\nu))^{As}}\Big)^{\frac{2}{s}}\Big)^{\frac{p}{2}}dx\Big\}^{\frac{s}{p}}$$

$$\ll 2^{-n\beta s}\Big\{\int_0^{2\pi}\Big(\sum_{m=1}^{n+1}\Delta_m^2\frac{2^{2m(r+\beta)}}{(\ln_N\ (d_N 2^m))^{2A}}\Big)^{\frac{p}{2}}dx\Big\}^{\frac{s}{p}}.$$

Thus we have

$$I_1 + I_2 \ll I_3 + I_5 + I_6 + I_8 + I_9 + I_{10}$$

$$\ll 2^{-n\beta s}\Big\{\int_0^{2\pi}\Big(\sum_{m=1}^{n+1}\Delta_m^2\frac{2^{2m(r+\beta)}}{(\ln_N\ (d_N 2^m))^{2A}}\Big)^{\frac{p}{2}}dx\Big\}^{\frac{s}{p}}$$

$$+\Big\{\int_0^{2\pi}\Big(\sum_{m=n+1}^{\infty}\Delta_m^2\frac{2^{2mr}}{(\ln_N\ (d_N 2^m))^{2A}}\Big)^{\frac{p}{2}}dx\Big\}^{\frac{s}{p}}.$$

From the definition of $W(p,\lambda^{(1)})$ it follows that we have $\varphi(x)\in L_p$ and $\varphi(x)\sim \sum_{\nu=1}^{\infty}\lambda_n A_n(x)$ exists.

Hence the sequence $\lambda_\nu^* = \lambda_2^m/\lambda_\nu$ $(2^{m-1}\le \nu < 2^m)$ and the function $\varphi(x)$ satisfy the conditions of Lemma 2.

Using Lemmas 2, 1 and 3(a), we get

$$I_1 + I_2 \ll 2^{-n\beta s}\Big\|\sum_{m=1}^{2^n} A_m(x)\frac{m^{r+\beta}}{(\ln_N\ (d_N m))^A}\Big\|_p^s$$

$$+\Big\|\sum_{m=2^n+1}^{\infty} A_m(x)\frac{m^r}{(\ln_N\ (d_N m))^A}\Big\|_p^s$$

$$\ll \omega_\beta^s(\varphi,\delta)_p.$$

This completes the proof of Theorem 2.2.

Other theorems can be proved in the same way as this one but we have to apply Lemma 3(b) (in place of Lemma 3(a)) in the proofs of Theorems 2.3, 3.3 and Lemma 3(c) in the proofs of Theorems 2.4 and 3.4. In the proofs of Theorems 2.1, 3.1 and 3.2 we have to apply Lemma 3(a).

References

[1] Simonov, B. V., About the properties of the transformed Fourier series, Deposited in VINITI, No 3031-81. Moscow 22.06.1981.

[2] Potapov, M. K., Lakovic, B., About the embedding and coincidence of Besov-Nikol'skii class and Weyl-Nikol'skii class of functions, Vestnik Moskov. Univ. Ser. I Mat. Mekh. 4 (1992), 44–52.

[3] Potapov, M. K., Lakovic, B., Simonov, B. V., On the estimates of moduli of smoothness of functions with a fractional derivatives, Math. Montisnigri, VII (1996), 41–52.

[4] Littlewood, J., Paley, R., Theorems on Fourier series and power series, J. London Math. Soc. 6 (1931), 230–233.

[5] Marcinkiewcz, J., Sur une nouvelle condition pour la convergence presque partout des series de Fourier, Ann. Scuola Norm. Sup. Pisa 8 (1939), 239–240.

[6] Lakovic, B., Potapov, M. K., On the question of correlation of some classes of functions, Mat. Vesnik, 3 (16) (31) (1979), 295–312.

Sergey Yu. Tikhonov, Chair of Theory of Functions and Functional Analysis, Department of Mechanics and Mathematics, Moscow State University, Vorob'yovy gory, Moscow 119899, Russia

E-mail: tikhonov@mccme.ru

Towards a Gausslet analysis: Gaussian representations of functions

Hans Triebel

Dedicated to Professor Jaak Peetre on his sixty-fifth birthday

Abstract. We survey in Section 2 some types of analysis: Gabor, wavelet, quarkonial, Gausslet. Section 3 deals with Gaussian representations of functions belonging to some function spaces of Sobolev–Besov type on Euclidean n-space.

2000 Mathematics Subject Classification: 46E35, 42C40 [MSC 2000]

1. Introduction

This paper is the outgrowth of some relevant parts of the two books [Tri97], [Tri01], and of [Tri98]. In [Tri98] we complemented quarkonial decompositions in the spaces $B^s_{pq}(\mathbb{R}^n)$ and $F^s_{pq}(\mathbb{R}^n)$ as considered in [Tri97] and [Tri01] by Gaussian decompositions based on the building blocks

$$x^\beta e^{-\frac{|x|^2}{2}}, \quad \text{where } x^\beta = \prod_{j=1}^n x_j^{\beta_j}, \ \beta \in \mathbb{N}_0^n, \ x = (x_1, \ldots, x_n) \in \mathbb{R}^n. \tag{1}$$

As it stands, there is a problem with the convergence of some series. One aim of this paper is to correct this weak point. But we use this opportunity to incorporate our results in the realm of different types of representations of functions now available in the literature and characterised by such key words as Gabor analysis (or time-frequency analysis), wavelet analysis and quarkonial analysis. They have all in common that one starts with one or a few functions g which are multiplied typically

$$\text{by } e^{ikx} \text{ where } k \in \mathbb{Z}^n \quad \text{or} \quad \text{by } x^\beta \text{ where } \beta \in \mathbb{N}_0^n, \tag{2}$$

and which are afterwards subject to translations, dilations or both translations and dilations typically of the form

$$x \mapsto x + m \text{ where } m \in \mathbb{Z}^n \quad \text{and} \quad x \mapsto 2^j x \text{ where } j \in \mathbb{N}_0, \tag{3}$$

with $x \in \mathbb{R}^n$. It is the question to find appropriate functions g where one is starting from and to find out what can be expected. In this general scheme fit the Gabor analysis (time-frequency analysis), wavelet analysis, quarkonial analysis and, what we wish to outline here, Gausslet analysis. In Section 2 we give a brief description of a few relevant points, always restricted to our intentions. We are not experienced in Gabor analysis or in wavelet analysis and therefore we mainly rely on the recent well-written book by K. Gröchenig, [Gro01], and the references given there. Note however that we rephrase some standard definitions and assertions from Gabor analysis and wavelet analysis in order to emphasize the intended unified scheme we have in mind. In this spirit we set in Section 2 the stage for Section 3, dealing there with Gaussian representations of functions belonging to some function spaces $B^s_{pq}(\mathbb{R}^n)$ and $F^s_{pq}(\mathbb{R}^n)$. We define there what we call now Gausslets, in modification of [Tri98]. Afterwards we prove in detail a representation theorem in some spaces $B^s_{pq}(\mathbb{R}^n)$ and $F^s_{pq}(\mathbb{R}^n)$ in terms of Gausslets. Looking only at the specific formulation of the theorem given, one can simplify the offered proof. But this will not be done, because at the end of this paper we have a discussion, based on the proof given, about further possibilities, parallel to what is available now in Gabor analysis and wavelet analysis. At this moment there are no further results in this new direction beyond those stated explicitly in this paper. But the examination of the proofs provides the feeling that there might be a Gausslet analysis similar to Gabor analysis and wavelet analysis (as far as representations of functions belonging to distinguished function spaces are concerned) with the quarkonial analysis and the Gaussian representations from the theorem as special cases.

Acknowledgement. I would like to thank Hans G. Feichtinger (Vienna, Austria) and Karlheinz Gröchenig (Vienna, Austria; Storrs, USA). They read the first draft of this paper and made many valuable suggestions which have been incorporated in the text.

2. Types of analysis

2.1. Gabor analysis

In connection with functions and distributions on Euclidean n-space we always use standard notation without further explanations. In case of doubt one may consult [Tri97], [Tri01]. Let Q be the cube in $\mathbb{R}^n$, centred at the origin, with side-length 1 and sides parallel to the axes of coordinates. Let χ be the characteristic function of Q. Then $f \in L_2(\mathbb{R}^n)$ can be represented by

$$f(x) = \sum_{m\in\mathbb{Z}^n} \sum_{k\in\mathbb{Z}^n} c_{mk}\, e^{2\pi ikx} \chi(x-m), \quad x \in \mathbb{R}^n, \tag{4}$$

where we can use the coefficients

$$c_{mk} = \int_{\mathbb{R}^n} e^{-2\pi iky} \chi(y-m)\, f(y)\, dy, \quad m \in \mathbb{Z}^n,\ k \in \mathbb{Z}^n. \tag{5}$$

Then

$$\|f\,|L_2(\mathbb{R}^n)\| = \Big(\sum_{m,k} |c_{mk}|^2\Big)^{\frac{1}{2}}. \tag{6}$$

In Gabor analysis one studies the question under which circumstances one can replace $\chi(x)$ by an L_∞ function $g(x)$ (*window*) with a counterpart of (4) (unconditional convergence in $L_2(\mathbb{R}^n)$), of (5) (a linear procedure with respect to f to calculate *frame-coefficients* c_{mk}) and of (6), now as an equivalence, (*frame*) instead of equality. More precisely, let

$$g \in L_\infty(\mathbb{R}^n) \cap L_2(\mathbb{R}^n), \quad \alpha > 0,\ \beta > 0. \tag{7}$$

If any $f \in L_2(\mathbb{R}^n)$ can be represented by

$$f(x) = \sum_{m\in\mathbb{Z}^n} \sum_{k\in\mathbb{Z}^n} c_{mk}\, e^{2\pi i\beta kx} g(x - \alpha m), \quad x \in \mathbb{R}^n, \tag{8}$$

(unconditional convergence in $L_2(\mathbb{R}^n)$) with a linear procedure, (scalar product in $L_2(\mathbb{R}^n)$),

$$f \mapsto c_{mk}(f) = (f, e_{mk}) \in \mathbb{C}, \quad m \in \mathbb{Z}^n,\ k \in \mathbb{Z}^n, \tag{9}$$

(*frame coefficients*) and

$$\|f\,|L_2(\mathbb{R}^n)\| \sim \Big(\sum_{m\in\mathbb{Z}^n} \sum_{k\in\mathbb{Z}^n} |c_{mk}|^2\Big)^{\frac{1}{2}}, \tag{10}$$

(*frame*), then g, α, β in (7) is called a *Gabor frame* in $L_2(\mathbb{R}^n)$ and denoted by $\mathcal{G}(g, \alpha, \beta)$. Here $\sim$ means that there are two positive constants c_1 and c_2 such that for all $f \in L_2(\mathbb{R}^n)$,

$$c_1\,\|f\,|L_2(\mathbb{R}^n)\| \le \Big(\sum_{m,k} |c_{mk}(f)|^2\Big)^{\frac{1}{2}} \le c_2\,\|f\,|L_2(\mathbb{R}^n)\|. \tag{11}$$

In addition it is always assumed that the *synthesis operator D*,

$$(Dc)(x) = \sum_{m\in\mathbb{Z}^n} \sum_{k\in\mathbb{Z}^n} c_{mk}\, e^{2\pi i\beta kx}\, g(x - \alpha m), \quad x \in \mathbb{R}^n, \tag{12}$$

where $c = \{c_{mk} \ :\ m \in \mathbb{Z}^n, k \in \mathbb{Z}^n\}$, is a bounded map from ℓ_2 into $L_2(\mathbb{R}^n)$. Altogether this is equivalent to the usual way to say what is meant by Gabor analysis. We refer to [Gro01], in particular Sections 7.3 and 5.2. The generating functions $e_{mk}(x)$ in (9) have the same structure as the corresponding functions in (8). The above reformulation is convenient for us and adapted to our later intentions. Otherwise we refer, in addition to [Gro01], to [FeS98], and in particular to the papers [BHW98] and [FeZ98] in this book. There one finds a detailed discussion of diverse aspects of Gabor frames and synthesis operators.

Hence the first natural question is to ask for conditions on g such that the synthesis operator D in (12) is a bounded map from ℓ_2 into $L_2(\mathbb{R}^n)$. This is the case if $g \in W(\mathbb{R}^n)$, the Wiener algebra, that means

$$\|g\,|W(\mathbb{R}^n)\| = \sum_{m\in\mathbb{Z}^n} \operatorname*{ess\,sup}_{x\in Q} |g(x-m)| < \infty. \tag{13}$$

To justify this well-known claim we may assume $\alpha = \beta = 1$. Let $Q_l = l + Q$ be the cube centred at $l \in \mathbb{Z}^n$ with side-length 1 (and sides parallel to the axes). Then

$$\begin{aligned}\|Dc\,|L_2(Q_l)\| &\le \sum_{m\in\mathbb{Z}^n} \Big\| \sum_{k\in\mathbb{Z}^n} c_{mk}\, e^{2\pi ikx}\, g(x-m)\,|L_2(Q_l)\Big\| \\ &\le \sum_{m\in\mathbb{Z}^n} \|g\,|L_\infty(Q_{l-m})\| \Big(\sum_{k\in\mathbb{Z}^n} |c_{mk}|^2\Big)^{\frac12} \\ &= \sum_{m\in\mathbb{Z}^n} \|g\,|L_\infty(Q_m)\| \Big(\sum_{k\in\mathbb{Z}^n} |c_{l-m,k}|^2\Big)^{\frac12} \\ &\le \|g\,|W(\mathbb{R}^n)\|^{\frac12} \Big(\sum_{m,k} \|g|L_\infty(Q_m)\|\, |c_{l-m,k}|^2\Big)^{\frac12}.\end{aligned} \tag{14}$$

Now it follows that

$$\|Dc\,|L_2(\mathbb{R}^n)\|^2 = \sum_l \|Dc\,|L_2(Q_l)\|^2 \le \|g|W(\mathbb{R}^n)\|^2\, \|c|\ell_2\|^2. \tag{15}$$

Hence, (12) *makes sense and D is a bounded map from ℓ_2 into $L_2(\mathbb{R}^n)$ if $g \in W(\mathbb{R}^n)$.* As for conditions under which $\mathcal{G}(g,\alpha,\beta)$ is a Gabor frame we have so far the above example $g = \chi$. We formulate the more substantial criterion by D. F. Walnut, [Wal92], which may be found in [Gro01], Theorem 6.5.1, p. 121:

Let $g \in W(\mathbb{R}^n)$ and let for some $\alpha > 0$ and $a > 0$,

$$\sum_{m\in\mathbb{Z}^n} |g(x-\alpha m)|^2 \ge a, \quad x \in \mathbb{R}^n,\ a.e. \tag{16}$$

Then there is an $\beta_0 = \beta_0(\alpha) > 0$, such that $\mathcal{G}(g,\alpha,\beta)$ is a Gabor frame for all $0 < \beta \le \beta_0$.

This applies especially to related compactly supported functions g. An other distinguished example is the (adapted) Gauss function

$$g_G(x) = e^{-|x|^2\pi}, \quad x \in \mathbb{R}^n, \tag{17}$$

from which the whole theory started from. In other words, one asks for representations of all $f \in L_2(\mathbb{R}^n)$ by

$$f(x) = \sum_{m\in\mathbb{Z}^n} \sum_{k\in\mathbb{Z}^n} c_{mk}\, e^{2\pi i\beta kx}\, e^{-|x-\alpha m|^2\pi}, \quad x \in \mathbb{R}^n, \tag{18}$$

with (9), (10). By the above criterion, $\mathcal{G}(g_G, \alpha, \beta)$ is a Gabor frame if $\alpha\beta > 0$ is small. If $n = 1$ then one has the following definitive improvement:

$\mathcal{G}(g_G, \alpha, \beta)$ *is a Gabor frame if, and only if,* $\alpha\beta < 1$.

We refer to [Gro01], Theorem 7.5.3, p. 140, and the literature mentioned there. The case $n = 1$ and $\alpha = \beta = 1$, hence,

$$f(x) = \sum_{m\in\mathbb{Z}} \sum_{k\in\mathbb{Z}} c_{mk}\, e^{2\pi i k x}\, e^{-\pi(x-m)^2}, \quad x \in \mathbb{R}, \tag{19}$$

is the original representation suggested by J. v. Neumann, [Neu32], and D. Gabor, [Gab46]. It came out later that any $f \in L_2(\mathbb{R})$ can be represented in this way, as Gabor conjectured, but is not stable since it is not a frame, in particular (11) does not hold. A discussion of these points may be found in [FeZ98] and in [Gro01], 7.5. In particular it was shown in [Jan81] that for some $f \in S(\mathbb{R})$ the series in (19) converges only in $S'(\mathbb{R})$.

Gabor analysis preferably takes place in $L_2(\mathbb{R}^n)$. An attempt to extend this theory to $L_p(\mathbb{R}^n)$ or, more generally, to $B^s_{pq}(\mathbb{R}^n)$ and $F^s_{pq}(\mathbb{R}^n)$, collides with the dyadic structure of these spaces according to the Littlewood–Paley theory. To make clear what is meant we look at the Fourier transform of (8) and obtain that

$$\widehat{f}(\xi) = \sum_{m\in\mathbb{Z}^n} \sum_{k\in\mathbb{Z}^n} c'_{mk}\, e^{-i\alpha m\xi}\, \widehat{g}\,(\xi - 2\pi\beta k). \tag{20}$$

Hence also on the Fourier side one has translations and modulations as structural elements. On the other hand the structure of the spaces $B^s_{pq}(\mathbb{R}^n)$ and $F^s_{pq}(\mathbb{R}^n)$ is dominated on the Fourier side by dyadic dilations $\xi \mapsto 2^j\xi$, where $j \in \mathbb{N}_0$. This observation, combined with (20), suggests the escape route: One replaces on the Fourier side dyadic annuli, which are characteristic for the spaces $B^s_{pq}(\mathbb{R}^n)$ and $F^s_{pq}(\mathbb{R}^n)$, by congruent cubes. Then one gets the so-called *modulation spaces* of type $M^s_{pq}(\mathbb{R}^n)$, where s might be a function or a real number and $0 < p \leq \infty$, $0 < q \leq \infty$. We do not go into detail. The theory of these spaces goes back to H. G. Feichtinger in the early eighties in the larger context of abelian groups. We refer to [Fei83], [Fei89], [FeG88]. The present state of the art may be found in [Gro01], Sections 11–14, concentrating on the Gabor analysis in these spaces and including striking applications to pseudo-differential operators. There one finds also the relevant references.

2.2. Wavelet analysis

Wavelet analysis is a very fashionable subject nowadays. After a slow start this theory gained speed in the last two decades. There are numerous books and many papers dealing with all aspects, theory and a wide range of applications. From our point of view (the theory of function spaces) we refer in particular to [Dau90], [Mey92], [Dau92], [HeW96], [Woj97], [Mal99]. We give a rough description of some aspects of relevance for our later intentions adapted to our purposes. Let $g(x)$ be a continuous function in $\mathbb{R}^n$. Then the (discrete) wavelet system is given by

$$2^{jt}\, g(2^j x - m) \quad \text{with } j \in \mathbb{Z} \text{ and } m \in \mathbb{Z}^n, \tag{21}$$

where $t \in \mathbb{R}$ is a normalising factor, for example $t = \frac{n}{2}$ in case of $L_2(\mathbb{R}^n)$. We say that g generates a *wavelet frame* in $L_2(\mathbb{R}^n)$ if any $f \in L_2(\mathbb{R}^n)$ can be represented as

$$f(x) = \sum_{j\in\mathbb{Z}} \sum_{m\in\mathbb{Z}^n} c_{jm}\, 2^{j\frac{n}{2}}\, g(2^j x - m), \quad x \in \mathbb{R}^n, \tag{22}$$

(unconditional convergence in $L_2(\mathbb{R}^n)$) with the counterpart of (9), (10), this means, a linear procedure, (scalar product in $L_2(\mathbb{R}^n)$),

$$f \mapsto c_{jm}(f) = (f, e_{jm}) \in \mathbb{C}, \quad m \in \mathbb{Z}^n,\ j \in \mathbb{Z}, \tag{23}$$

(*frame coefficients*) and

$$\|f\,|L_2(\mathbb{R}^n)\| \sim \Big(\sum_{j\in\mathbb{Z}} \sum_{m\in\mathbb{Z}^n} |c_{jm}|^2\Big)^{\frac{1}{2}}, \tag{24}$$

(*frame*). Again we modified what is usually called a wavelet frame, but the outcome is the same. In particular, the generating functions $e_{jm}(x)$ in (23) have the same wavelet structure as in (21).

Comparing the Gabor representation (8) with the wavelet representation (22) one has in both cases translations. But frequency or oscillation is expressed rather differently. As a result that wavelet transforms are better adapted to the type of function spaces we have in mind. But first of all one has to ask whether there are frame-generating functions g with the desired properties. This has been studied in detail in the books mentioned above. There are necessary conditions and (different) sufficient conditions. A typical ingredient of these conditions is given by

$$|\widehat{g}(\xi)| \le \frac{c\,|\xi|^r}{1+|\xi|^s}, \quad \xi \in \mathbb{R}^n, \text{ for some } r > 0 \text{ and } s > r + n. \tag{25}$$

This formulation may be found in [Gro01], Theorem 10.5, p. 212, where also precise assertions and references are given. One has in particular

$$\int_{\mathbb{R}^n} g(x)\, dx = 0. \tag{26}$$

Let $n = 1$. The question of whether

$$\big\{2^{\frac{j}{2}}\, g(2^j x - m) : j \in \mathbb{Z},\ m \in \mathbb{Z}\big\} \tag{27}$$

is an orthonormal basis in $L_2(\mathbb{R})$ goes far beyond the problem of frame-generating procedures. It is one of the central topics of the books and papers mentioned at the beginning of this subsection. It results in the so-called multi-resolution analysis and the restrictions for g are even more severe. There is an extension from the one-dimensional case $L_2(\mathbb{R})$ to the n-dimensional case $L_2(\mathbb{R}^n)$.

In contrast to Gabor representations, wavelet representations can be extended from $L_2(\mathbb{R}^n)$ to other function spaces of interest in our later context such as $L_p(\mathbb{R}^n)$, $1 < p < \infty$, Hardy spaces, Hölder–Zygmund spaces, Besov spaces, BMO, (fractional) Sobolev spaces. This applies both the bases and to frames. We refer to [Mey92] and [Woj97]. Some assertions may also be found in [Dau92], Section 9. A corresponding theory for the spaces $B^s_{pq}(\mathbb{R}^n)$ and $F^s_{pq}(\mathbb{R}^n)$ has been developed in [FJW91], Section 7, and in [RuS96], 2.3.3, pp. 62–66, and recently in [KyP99]. One may also consult [Gro91], where related problems are treated from the unifying point of view of coorbit spaces actually showing that in any such context there is always an appropriate family of spaces surrounding the Hilbert space just as the above-mentioned spaces are surrounding L_2 in the wavelet context.

2.3. Quarkonial analysis

Temporarily we suppose that the reader knows what is meant by $B^s_{pq}(\mathbb{R}^n)$ and $F^s_{pq}(\mathbb{R}^n)$. One may also consult 3.2 below. First we collect a few notation which will be useful now and in the later considerations. We follow [Tri01], Section 2. We use standard symbols. In case of doubt one should consult [Tri01], Section 2. Let Q_{jm} (in case when extra clarity is desirable written as $Q_{j,m}$) be a cube in $\mathbb{R}^n$ with sides parallel to the axes of coordinates, centred at $2^{-j}m$, with side length 2^{-j}, where $m \in \mathbb{Z}^n$ and $j \in \mathbb{N}_0$. If Q is a cube in $\mathbb{R}^n$ and $r > 0$ then rQ is the cube in $\mathbb{R}^n$ concentric with Q and with side length r times the side length of Q. We denote by $\chi^{(p)}_{jm}$ the p-normalised characteristic function of the cube Q_{jm}, which means

$$\chi^{(p)}_{jm}(x) = 2^{\frac{jn}{p}} \text{ if } x \in Q_{jm} \quad \text{and} \quad \chi^{(p)}_{jm}(x) = 0 \text{ if } x \notin Q_{jm}, \tag{28}$$

where $j \in \mathbb{N}_0$, $m \in \mathbb{Z}^n$, and $0 < p \le \infty$. Let $0 < p \le \infty$, $0 < q \le \infty$ and

$$\lambda = \left\{\lambda_{jm} \in \mathbb{C} : j \in \mathbb{N}_0,\ m \in \mathbb{Z}^n\right\}. \tag{29}$$

We introduce the sequence spaces

$$b_{pq} = \left\{\lambda : \|\lambda \,|b_{pq}\| = \Big(\sum_{j=0}^{\infty}\Big(\sum_{m\in\mathbb{Z}^n} |\lambda_{jm}|^p\Big)^{\frac{q}{p}}\Big)^{\frac{1}{q}} < \infty\right\} \tag{30}$$

and

$$f_{pq} = \left\{\lambda : \|\lambda \,| f_{pq}\| = \Big\|\Big(\sum_{j=0}^{\infty}\sum_{m\in\mathbb{Z}^n} \Big|\lambda_{jm}\, \chi^{(p)}_{jm}(\cdot)\Big|^q\Big)^{\frac{1}{q}} \,|\, L_p(\mathbb{R}^n)\Big\| < \infty\right\} \tag{31}$$

with the usual modifications if $p = \infty$ and/or $q = \infty$. As for more details and a few properties we refer to [Tri01], 2.2, 2.3. To understand what follows one may always think about $p = q$; then $b_{pp} = f_{pp} = \ell_p$, appropriately interpreted. Let ψ be a non-negative C^∞ function in $\mathbb{R}^n$ with

$$\operatorname{supp}\psi \subset \left\{y \in \mathbb{R}^n : |y| < 2^r\right\} \tag{32}$$

for some $r > 0$ and

$$\sum_{m\in\mathbb{Z}^n} \psi(x-m) = 1 \quad \text{if } x \in \mathbb{R}^n \tag{33}$$

(resolution of unity). Let

$$s \in \mathbb{R},\ 0 < p \le \infty,\ \beta \in \mathbb{N}_0^n,\ j \in \mathbb{N}_0,\ m \in \mathbb{Z}^n, \tag{34}$$

and $\psi^\beta(x) = x^\beta\, \psi(x)$ with $x^\beta = x_1^{\beta_1} \dots x_n^{\beta_n}$. Then

$$(\beta qu)_{jm}(x) = 2^{-j(s-\frac{n}{p})}\, \psi^\beta(2^j x - m), \quad x \in \mathbb{R}^n, \tag{35}$$

is called an (s,p)-β-quark related to Q_{jm}. We refer to [Tri01], 2.4, 2.5. Let $0 < p \le \infty$, $0 < q \le \infty$, and $a_+ = \max(a,0)$ if $a \in \mathbb{R}$. Then we put

$$\sigma_p = n\left(\frac{1}{p} - 1\right)_+ \quad \text{and} \quad \sigma_{pq} = n\left(\frac{1}{\min(p,q)} - 1\right)_+. \tag{36}$$

Let $\varrho > 0$ and

$$\lambda = \left\{\lambda^\beta : \beta \in \mathbb{N}_0^n\right\} \quad \text{with } \lambda^\beta = \left\{\lambda^\beta_{jm} \in \mathbb{C} : j \in \mathbb{N}_0,\ m \in \mathbb{Z}^n\right\}. \tag{37}$$

We put

$$\|\lambda\, |b_{pq}\|_\varrho = \sup_{\beta\in\mathbb{N}_0^n} 2^{\varrho|\beta|}\, \|\lambda^\beta\, |b_{pq}\| \tag{38}$$

and

$$\|\lambda\, |f_{pq}\|_\varrho = \sup_{\beta\in\mathbb{N}_0^n} 2^{\varrho|\beta|}\, \|\lambda^\beta\, |f_{pq}\|\,. \tag{39}$$

One of the main aims of [Tri01], Section 2, is the proof of the following quarkonial representations of some spaces $B^s_{pq}(\mathbb{R}^n)$ and $F^s_{pq}(\mathbb{R}^n)$:

2.3 (i). *Let $\varrho > r$, where r is given by* (32) *and let*

$$0 < p \le \infty, \quad 0 < q \le \infty, \quad s > \sigma_p. \tag{40}$$

Then $B^s_{pq}(\mathbb{R}^n)$ is the collection of all $g \in S'(\mathbb{R}^n)$ which can be represented as

$$f = \sum_{\beta\in\mathbb{N}_0^n} \sum_{j=0}^{\infty} \sum_{m\in\mathbb{Z}^n} \lambda^\beta_{jm}\, (\beta qu)_{jm}(x), \quad x \in \mathbb{R}^n, \tag{41}$$

where $(\beta qu)_{jm}$ are (s,p)-β-quarks according to (35) *and*

$$\|\lambda\, |b_{pq}\|_\varrho < \infty. \tag{42}$$

Furthermore,

$$\|f \mid B^s_{pq}(\mathbb{R}^n)\| \sim \inf \|\lambda \mid b_{pq}\|_\varrho \tag{43}$$

(equivalent quasi-norms), where the infimum is taken over all admitted representations (41), (42).

2.3 (ii). *Let $\varrho > r$, where r is given by* (32) *and let*

$$0 < p < \infty, \quad 0 < q \leq \infty, \; s > \sigma_{pq}. \tag{44}$$

Then $F^s_{pq}(\mathbb{R}^n)$ is the collection of all $f \in S'(\mathbb{R}^n)$ which can be represented as in (41) *where $(\beta qu)_{jm}$ are (s, p)-β-quarks according to* (35) *and*

$$\|\lambda \mid f_{pq}\|_\varrho < \infty. \tag{45}$$

Furthermore,

$$\|f \mid F^s_{pq}(\mathbb{R}^n)\| \sim \inf \|\lambda \mid f_{pq}\|_\varrho \tag{46}$$

(equivalent quasi-norms), where the infimum is taken over all admitted representations (41), (45).

2.3.1. Comments. We discuss the convergence of the right-hand side of (41) in $S'(\mathbb{R}^n)$ under the restriction (42) or (45). If $\beta \in \mathbb{N}^n_0$ is fixed and if $p < \infty$ and $q < \infty$ then the inner sums over $j \in \mathbb{N}_0$ and $m \in \mathbb{Z}^n$ converge in $B^s_{pq}(\mathbb{R}^n)$ or $F^s_{pq}(\mathbb{R}^n)$, respectively, to some function f^β. If $p = \infty$ or $q = \infty$ then one has at least convergence in $S'(\mathbb{R}^n)$ (even in $B^{s-\varepsilon}_{pq}(\mathbb{R}^n)$ for any $\varepsilon > 0$). This follows from the usual atomic decomposition as it may be found in [Tri97], Theorem 13.8, p. 75. Afterwards one gets immediately that

$$f = \sum_{\beta \in \mathbb{N}^n_0} f^\beta, \quad \text{convergence being in } S'(\mathbb{R}^n),$$

[if $p < \infty$, $q < \infty$, then one has covergence even in $B^s_{pq}(\mathbb{R}^n)$ or $F^s_{pq}(\mathbb{R}^n)$, respectively]. But one can say more. Under the above circumstances the right-hand side of (41) converges absolutely and, hence, unconditionally, in some spaces $L_{\overline{p}}(\mathbb{R}^n)$ with $1 \leq \overline{p} \leq \infty$, and hence unconditionally in $S'(\mathbb{R}^n)$. (If $p \geq 1$, then one may choose $\overline{p} = p$; if $p < 1$ then one can take $\overline{p} = 1$). In particular, we may shorten (41) by

$$f = \sum_{\beta, j, m} \lambda^\beta_{jm}\, (\beta qu)_{jm}(x), \quad x \in \mathbb{R}^n, \tag{47}$$

without any ambiguity. We discussed this point in detail in [Tri01], 1.4, 2.6, 2.7, 2.11. In general, however there is no claim that (41) and (47) converge unconditionally with respect to all three indices β, j, m, in, say, $B^s_{pq}(\mathbb{R}^n)$ with $p < \infty$, $q < \infty$.

2.3.2. In analogy to what has been said in connection with Gabor analysis and wavelet analysis there are again frame coefficients, this means linear procedures

$$f \mapsto \lambda^\beta_{jm}(f) \in \mathbb{C}, \quad \beta \in \mathbb{N}^n_0, \; j \in \mathbb{N}_0, \; m \in \mathbb{Z}^n, \tag{48}$$

and related frames. More precisely:

Under the above respective hypotheses in 2.3 (i) *and* 2.3 (ii), *there are universal frame coefficients*

$$\lambda^{\beta}_{jm}(f) = 2^{j(s-\frac{n}{p})}\, 2^{-\varrho|\beta|} \left(f, \Psi^{\beta,\varrho}_{jm}\right), \quad f \in S'(\mathbb{R}^n), \tag{49}$$

with respect to the dual pairing in $S(\mathbb{R}^n)$ *and* $S'(\mathbb{R}^n)$ *where*

$$\Psi^{\beta,\varrho}_{jm} \in S(\mathbb{R}^n),\ \beta \in \mathbb{N}^n_0,\ j \in \mathbb{N}_0,\ m \in \mathbb{Z}^n, \tag{50}$$

such that (frame)

$$\|f \,|B^s_{pq}(\mathbb{R}^n)\| \sim \|\lambda(f)\,|b_{pq}\|_{\varrho}, \quad \|f \,|F^s_{pq}(\mathbb{R}^n)\| \sim \|\lambda(f)\,|f_{pq}\|_{\varrho}, \tag{51}$$

respectively.

Again we refer to [Tri01], in particular 2.12.

2.3.3. To avoid a misunderstanding we add a comment about the miraculous role of ϱ. If $\varrho > r$ is fixed then we have (43), (46), or the frame versions in (51). But the equivalence constants depend on ϱ. In [Tri01], 2.11, we did some calculations with the outcome that these coefficients depend exponentially on ϱ, this means as $2^{c\varrho}$ for some $c > 0$.

2.3.4. The theory can be extended to $B^s_{pq}(\mathbb{R}^n)$ with $s \le \sigma_p$ and $F^s_{pq}(\mathbb{R}^n)$ with $s \le \sigma_{pq}$. The outcome looks a little bit more complicated (one needs quarks satisfying some moment conditions). This is more or less a technical matter. We refer to [Tri01], Section 3. In this paper we always restrict ourselves to $s > \sigma_p$ and $s > \sigma_{pq}$, respectively.

2.4. Discussion; Gausslet analysis

2.4.1. Discussion. First we say a few words about the origin of and the motivations resulting in the quarkonial analysis outlined in 2.3. Atomic decompositions for $B^s_{pq}(\mathbb{R}^n)$ and $F^s_{pq}(\mathbb{R}^n)$ with the full range of the admitted parameters s, p, q go back to [FrJ85], [FrJ90], [FJW91]. We refer also to [Tri97], Section 13, where one finds new proofs. In contrast to the diverse frames considered so far, [Gabor in (8) and (18); wavelet in (22); quarkonial in (41)], atoms are not constructive building blocks; they are defined in qualitative terms. But this is not very suitable if one wishes to use decomposition techniques in connection with entropy numbers of compact embeddings between function spaces. Let Ω be a bounded domain in $\mathbb{R}^n$ (no smoothness assumptions are needed) and let $B^s_p(\Omega)$ be the restriction of $B^s_{pp}(\mathbb{R}^n)$ to Ω, where $s \in \mathbb{R}$ and $0 < p \le \infty$. Then the embedding

$$\mathrm{id}: \ B^{s_1}_p(\Omega) \mapsto B^{s_2}_p(\Omega) \quad \text{with } 0 < p \le \infty,\ \infty > s_1 > s_2 > -\infty, \tag{52}$$

is compact and one has for its entropy numbers e_k ,

$$e_k \sim k^{-\frac{s_1-s_2}{n}}, \quad k \in \mathbb{N}. \tag{53}$$

This is the proto-type of more general assertions dealing with all compact embeddings between $B^s_{pq}(\Omega)$ and $F^s_{pq}(\Omega)$ spaces. We proved this assertion in [ET96], 3.3, avoiding explicit decompositions, by direct reductions to the original Fourier-analytical definitions of the spaces involved. If one replaces Ω by, say, compact fractal sets Γ in $\mathbb{R}^n$, or by some manifolds, then the situation is different. The techniques developed in [ET96] (and the forerunners mentioned there) do not work any longer to get assertions of type (53). One needs rigid decompositions of function spaces (frames in the above understanding) which allow to reduce problems of type (53) to sequence spaces, preferably of type b_{pq} in (30) and their weighted generalisations. This cannot be done by atoms. On the other hand, frames provide constructive building blocks. However by our remarks at the end of 2.1 it is clear that Gabor frames do not fit in our scheme. Wavelet frames as outlined in 2.2, are better adapted. However dealing with function spaces near and on fractals Γ, or on manifolds, or asking for intrinsic characterisation of spaces of type $B^s_{pq}(\Omega)$ or $F^s_{pq}(\Omega)$ in smooth and non-smooth bounded domains Ω in $\mathbb{R}^n$, one needs permanently local diffeomorphic distortions, cutting functions in pieces, gluing together, and all that based on adapted irregular dyadic lattices. It is quite clear that wavelets, generating frames or bases, subject to more or less severe conditions, for example of type (25) or (26), cannot survive this torture. One needs simple robust localised but flexible building blocks such as adapted resolutions of unity. This became clear in the middle of the nineties and resulted in the theory of subatomic decompositions as presented in [Tri97], Section 14. Originally invented as a tool it developed its own live in the years after. This theory maybe found in [Tri01], including new applications, for example to semi-linear equations.

2.4.2. Gausslet analysis. We just explained the (or our) necessity to develop quarkonial decompositions when dealing not only with $\mathbb{R}^n$ but also with fractals, manifolds, (irregular) domains etc. But in this paper we are back to $\mathbb{R}^n$. Accepting that quarkonial analysis might well be the beginning of a self-contained theory parallel to Gabor analysis and wavelet analysis as outlined in 2.1 and 2.2 one can ask the same questions as there, taking in account what had been said in 2.3. Let again $g(x)$ be a bounded continuous function in $\mathbb{R}^n$. Gabor analysis and wavelet analysis are characterised by the search for conditions for g such that (8) and (22) are frames as explained there. Relying on the constructions in 2.3 we assume in addition

$$\sum_{m \in \mathbb{Z}^n} g(x-m) = 1 \quad \text{if } x \in \mathbb{R}^n \tag{54}$$

(resolution of unity). Again let $g^\beta(x) = x^\beta\, g(x)$ and let $s,\ p,\ \beta,\ j,\ m$ be as in (34). Recall $x^\beta = x_1^{\beta_1} \dots x_n^{\beta_n}$. Then we put

$$(\beta g)_{jm}(x) = 2^{-j(s-\frac{n}{p})}\, g^\beta(2^j x - m), \quad x \in \mathbb{R}^n. \tag{55}$$

In analogy to (41) one may ask for (frame) representations

$$f = \sum_{\beta \in \mathbb{N}_0^n} \sum_{j=0}^{\infty} \sum_{m \in \mathbb{Z}^n} \lambda_{jm}^{\beta} \, (\beta g)_{jm}(x), \quad x \in \mathbb{R}^n, \tag{56}$$

of functions belonging to $B^s_{pq}(\mathbb{R}^n)$ or $F^s_{pq}(\mathbb{R}^n)$. If $g = \psi$ with (32), (33), then one gets the quarkonial analysis as outlined in 2.3, based on the sequence spaces b_{pq}, f_{pq}, and (42), (45). Instead of tame functions of type e^{ikx} in the Gabor analysis (8) one has now to cope with the much more aggressive monomials x^{β}. This must be reflected by appropriately chosen sequence spaces. We do not deal in general here with such questions. In analogy to the original set-up in Gabor analysis we concentrate on the modified Gauss function

$$G(x) = e^{-\frac{|x|^2}{2}} H^{-1}(x), \quad x \in \mathbb{R}^n, \tag{57}$$

with

$$H(x) = \sum_{m \in \mathbb{Z}^n} e^{-\frac{|x-m|^2}{2}}, \quad x \in \mathbb{R}^n. \tag{58}$$

Then we have the counterpart of (54). The corresponding functions according to (55) are called *Gausslets* (in modification of [Tri98]). Section 3 deals with representations of type (56) and corresponding sequence spaces. The subsequent discussion gives a few hints on possible further generalisations.

This might be called *Gausslet analysis* in the same way as Gabor analysis stands not only for (18) but also for (8). The constructions given in [KyP99] of more general representations in $B^s_{pq}(\mathbb{R}^n)$ and $F^s_{pq}(\mathbb{R}^n)$, based on special functions, include also the Gauss function $e^{-\frac{|x|^2}{2}}$ as a building block.

3. Gaussian representations

3.1. Some preparations

We collect a few estimates which will be useful in the later considerations.

3.1.1. The function $\varkappa$. We need an improvement of (2.62) in [Tri01]. Let $\varkappa \in D(\mathbb{R}^n)$.

Then

$$\begin{aligned} x^\beta D^\alpha \varkappa^\vee(x) &= x^\beta i^{|\alpha|} (2\pi)^{-\frac{n}{2}} \int_{\mathbb{R}^n} e^{ix\xi} \xi^\alpha \varkappa(\xi)\, d\xi \\ &= i^{|\alpha|-|\beta|} (2\pi)^{-\frac{n}{2}} \int_{\mathbb{R}^n} D_\xi^\beta \left(e^{ix\xi}\right) \xi^\alpha \varkappa(\xi)\, d\xi \\ &= i^{|\alpha|+|\beta|} (2\pi)^{-\frac{n}{2}} \int_{\mathbb{R}^n} e^{ix\xi} D_\xi^\beta \left[\xi^\alpha \varkappa(\xi)\right] d\xi. \end{aligned} \tag{59}$$

We need an estimate for some β, say, with $|\beta| \le K$, and all $\alpha \in \mathbb{N}_0^n$. Then, for some $c \ge 0$,

$$\begin{aligned} \left|D_\xi^\beta \left[\xi^\alpha \varkappa(\xi)\right]\right| &\le c_K \sum_{\gamma+\eta=\beta} \left|D^\gamma \xi^\alpha\right| \left|D^\eta \varkappa(\xi)\right| \\ &\le c'_K \, (1+|\alpha|)^K \, 2^{c|\alpha|} \quad \text{if } \xi \in \operatorname{supp} \varkappa. \end{aligned} \tag{60}$$

Inserting (60) in (59) and again using that $\varkappa$ has a compact support it follows that

$$|x^\beta D^\alpha \varkappa^\vee(x)| \le c_K \, 2^{C|\alpha|}, \quad x \in \mathbb{R}^n, \tag{61}$$

for some $C > 0$. Hence, for given $b > 0$, we have

$$\left|D^\alpha \varkappa^\vee(x)\right| \le c_b \, 2^{C|\alpha|} \left(1+|x|^2\right)^{-b}, \quad x \in \mathbb{R}^n, \ \alpha \in \mathbb{N}_0^n, \tag{62}$$

where c_b is independent of x and α. This is the improvement of (2.62) in [Tri01] we are looking for.

3.1.2. The function *H*. Let

$$H(x) = \sum_{m \in \mathbb{Z}^n} e^{-\frac{|x-m|^2}{2}} = (2\pi)^{\frac{n}{2}} \sum_{m \in \mathbb{Z}^n} e^{-2\pi^2 |m|^2} e^{2\pi i m x}. \tag{63}$$

The first equality can be taken as a definition. The second one follows from Poisson's summation formula. We will not need this assertion explicitly (so far). But it sheds light on what is going on. We give a short proof. Let Q be a cube centred at the origin with side-length 1 and sides parallel to the axes of coordinates. Obviously, $H(x)$ is periodic. We expand $H(x)$ in Q in its Fourier series. The Fourier coefficient c_m with respect to $e^{2\pi i m x}$ is given by

$$\begin{aligned} c_m &= \int_Q e^{-2\pi i m x} \sum_{k \in \mathbb{Z}^n} e^{-\frac{|x-k|^2}{2}} \, dx \\ &= \int_{\mathbb{R}^n} e^{-2\pi i m x} e^{-\frac{|x|^2}{2}} \, dx = (2\pi)^{\frac{n}{2}} e^{-2\pi^2 |m|^2}, \end{aligned} \tag{64}$$

where we used that $e^{-\frac{|x|^2}{2}}$ coincides with its Fourier transform. This proves (63).

3.1.3. Modified Gauss functions. Let

$$G^\beta(x) = \frac{x^\beta}{\sqrt{\beta!}} e^{-\frac{|x|^2}{2}} H^{-1}(x), \quad \beta \in \mathbb{N}_0^n,\ x \in \mathbb{R}^n, \tag{65}$$

where $H(x)$ is given by (63). To get a feeling we derive some elementary estimates. The maximum of the one-dimensional function $G_\beta(t) = t^\beta e^{-\frac{t^2}{2}}$ can be calculated by

$$0 = G'_\beta(t_0) = t_0^{\beta-1}(\beta - t_0^2) e^{-\frac{t_0^2}{2}}, \quad \text{hence } t_0 = \sqrt{\beta}, \tag{66}$$

and

$$\max_{t\in\mathbb{R}} G_\beta(t) = \beta^{\frac{\beta}{2}} e^{-\frac{\beta}{2}}, \quad \beta \in \mathbb{N}. \tag{67}$$

Recall Stirling's formula

$$\Gamma(t) = e^{-t} t^{t-\frac{1}{2}} \sqrt{2\pi}\, e^{\frac{\Theta(t)}{t}} \quad \text{with } 0 < \Theta(t) < \frac{1}{12}, \tag{68}$$

where $t > 0$. This may be found in [ET96], p. 98, with a reference to [WhW52], 12.33. In particular,

$$k! = \Gamma(k+1) \sim e^{-k} k^{k+\frac{1}{2}}, \quad k \in \mathbb{N}. \tag{69}$$

Hence, by (67), (65),

$$\max_{x\in\mathbb{R}^n} G^\beta(x) \sim \prod_{j=1}^n (1+\beta_j)^{-\frac{1}{4}}, \quad \beta \in \mathbb{N}_0^n, \tag{70}$$

where the equivalence constants are independent of β. This is sharper than needed. On the other hand, this estimate must be extended to $D^\gamma G^\beta(x)$ for some derivatives $\gamma \in \mathbb{N}_0^n$ with, say, $|\gamma| \le K$. By the above considerations and (65) it follows that for given $\varepsilon > 0$ there is a constant c (depending on ε and K) such that

$$\left| D^\gamma G^\beta(x) \right| \le c\, 2^{\varepsilon|\beta|}, \quad x \in \mathbb{R}^n,\ |\gamma| \le K,\ \beta \in \mathbb{N}_0^n. \tag{71}$$

3.2. Function spaces

The theory of the spaces $B^s_{pq}(\mathbb{R}^n)$ and $F^s_{pq}(\mathbb{R}^n)$ has been developed in detail in [Tri83], [Tri92], [Tri97]. To fix our notation we repeat the Fourier-analytical definition of these spaces. Recall that we normalise the Fourier transform in $\mathbb{R}^n$ by

$$\widehat{\varphi}(\xi) = (F\varphi)(\xi) = (2\pi)^{-\frac{n}{2}} \int_{\mathbb{R}^n} e^{-ix\xi} \varphi(x)\, dx, \quad \xi \in \mathbb{R}^n. \tag{72}$$

Then the inverse Fourier transform $\varphi^\vee(\xi)$ or $(F^{-1}\varphi)(\xi)$ is given by (72) with i in place of $-i$. As usual, F and F^{-1} are defined first on $S(\mathbb{R}^n)$ and then extended to

$S'(\mathbb{R}^n)$. Let $\varphi \in S(\mathbb{R}^n)$ with

$$\varphi(x) = 1 \text{ if } |x| \le 1 \quad \text{and} \quad \varphi(x) = 0 \quad \text{if } |x| \ge \frac{3}{2}. \tag{73}$$

We put $\varphi_0 = \varphi$, $\varphi_1(x) = \varphi(\frac{x}{2}) - \varphi(x)$, and

$$\varphi_k(x) = \varphi_1\big(2^{-k+1}x\big), \quad x \in \mathbb{R}^n,\ k \in \mathbb{N}. \tag{74}$$

Then, since

$$1 = \sum_{k=0}^{\infty} \varphi_k(x) \quad \text{for all } x \in \mathbb{R}^n, \tag{75}$$

the φ_k form a dyadic resolution of unity in $\mathbb{R}^n$. Recall that $(\varphi_k \widehat{f})^\vee$ is an entire analytic function on $\mathbb{R}^n$ for any $f \in S'(\mathbb{R}^n)$. In particular, $(\varphi_k \widehat{f})^\vee(x)$ makes sense pointwise. Let

$$s \in \mathbb{R}, \quad 0 < p \le \infty,\ 0 < q \le \infty, \tag{76}$$

(with $p < \infty$ in case of F-spaces). Then $B^s_{pq}(\mathbb{R}^n)$ is the collection of all $f \in S'(\mathbb{R}^n)$ with

$$\|f \,|B^s_{pq}(\mathbb{R}^n)\| = \Big(\sum_{j=0}^{\infty} 2^{jsq} \Big\|(\varphi_j \widehat{f})^\vee \,|L_p(\mathbb{R}^n)\Big\|^q\Big)^{\frac{1}{q}} < \infty \tag{77}$$

and $F^s_{pq}(\mathbb{R}^n)$ is the collection of all $f \in S'(\mathbb{R}^n)$ with

$$\|f \,|F^s_{pq}(\mathbb{R}^n)\| = \Big\|\Big(\sum_{j=0}^{\infty} 2^{jsq} |\,(\varphi_j \widehat{f})(\cdot)|^q\Big)^{\frac{1}{q}} \,|L_p(\mathbb{R}^n)\Big\| < \infty \tag{78}$$

(usual modification if $q = \infty$). We introduced σ_p and σ_{pq} in (36). Recall

$$B^s_{pq}(\mathbb{R}^n) \subset L_1^{\mathrm{loc}}(\mathbb{R}^n) \quad \text{and} \quad F^s_{pq}(\mathbb{R}^n) \subset L_1^{\mathrm{loc}}(\mathbb{R}^n) \quad \text{if } s > \sigma_p. \tag{79}$$

Furthermore no moment conditions for atoms, quarks, and Gausslets are needed

$$\text{if } s > \sigma_p \text{ in case of } B^s_{pq}(\mathbb{R}^n) \quad \text{and} \quad \text{if } s > \sigma_{pq} \text{ in case of } F^s_{pq}(\mathbb{R}^n). \tag{80}$$

We deal here only with these cases. The extension of what follows to $s \in \mathbb{R}$ is a technical matter (lifting) which may be found in [Tri97] and [Tri01] and which will not be repeated here.

3.3. Gausslets

We combine the structure of the (s, p)-β-quarks in (35) with the estimates concerning the modified Gauss function given in 3.1.3. Let, as in (34),

$$s \in \mathbb{R}, \quad 0 < p \le \infty, \quad \beta \in \mathbb{N}_0^n, \quad j \in \mathbb{N}_0, \quad m \in \mathbb{Z}^n, \tag{81}$$

and let $G^\beta(x)$ be the modified Gauss functions introduced in (65). Then

$$G^\beta_{jm}(x) = 2^{-j(s-\frac{n}{p})}\, G^\beta(2^j x - m) \tag{82}$$

$$= 2^{-j(s-\frac{n}{p})}\, \frac{(2^j x - m)^\beta}{\sqrt{\beta!}}\, e^{-\frac{|2^j x - m|^2}{2}}\, H^{-1}(2^j x), \quad x \in \mathbb{R}^n,$$

are called *Gausslets*, or, more precisely (s, p)-β-*Gausslets* related to the cubes Q_{jm}, introduced in 2.3. We used that $H(x)$ in (63), and hence also $H^{-1}(x)$, is periodic with respect to $\mathbb{Z}^n$. We are looking for the counterpart of the quarkonial representations in $B^s_{pq}(\mathbb{R}^n)$ and $F^s_{pq}(\mathbb{R}^n)$ as described in 2.3(i) and 2.3(ii), respectively. As discussed in 2.3.1 the writing in (41) can be shortened by (47) since the series involved converge absolutely and, hence, unconditionally in some spaces $L_{\overline{p}}(\mathbb{R}^n)$ with $1 \le \overline{p} < \infty$. In particular it converges unconditionally in $S'(\mathbb{R}^n)$. This will also be the case here and it follows as a by-product of the estimates of the proof of the theorem below. This justifies to use the counterpart of the short version (47). Otherwise the sequence spaces b_{pq} and f_{pq} have the same meaning as in 2.3, including the abbreviations (38), (39).

3.4. Theorem

(i) *Let* $\varrho > 0$ *and let*

$$0 < p \le \infty, \quad 0 < q \le \infty, \quad s > \sigma_p. \tag{83}$$

Then $B^s_{pq}(\mathbb{R}^n)$ *is the collection of all* $f \in S'(\mathbb{R}^n)$ *which can be represented as*

$$f = \sum_{\beta, j, m} \lambda^\beta_{jm}\, G^\beta_{jm}(x), \quad x \in \mathbb{R}^n, \tag{84}$$

unconditional convergence in $S'(\mathbb{R}^n)$, *where* $G^\beta_{jm}(x)$ *are* (s, p)-β-*Gausslets according to* (82), *and*

$$\|\lambda \,|b_{pq}\|_\varrho < \infty\,. \tag{85}$$

Furthermore,

$$\|f \,|B^s_{pq}(\mathbb{R}^n)\| \sim \inf \|\lambda \,|b_{pq}\|_\varrho \tag{86}$$

(equivalent quasi-norms), where the infimum is taken over all admitted representations (84), (85).

(ii) *Let* $\varrho > 0$ *and let*

$$0 < p < \infty, \quad 0 < q \le \infty, \quad s > \sigma_{pq}\,. \tag{87}$$

Then $F^s_{pq}(\mathbb{R}^n)$ *is the collection of all* $f \in S'(\mathbb{R}^n)$ *which can be represented as* (84), *unconditional convergence in* $S'(\mathbb{R}^n)$, *where* $G^\beta_{jm}(x)$ *are* (s, p)-β-*Gausslets according to* (82), *and*

$$\|\lambda \,|f_{pq}\|_{\varrho} < \infty. \tag{88}$$

Furthermore,

$$\|f \,|F^s_{pq}(\mathbb{R}^n)\| \sim \inf \|\lambda \,|f_{pq}\|_{\varrho} \tag{89}$$

(*equivalent quasi-norms*), *where the infimum is taken over all admitted representations* (84), (88).

Proof. Step 1. First we prove that the right-hand side of (84) with (85) converges unconditionally in $S'(\mathbb{R}^n)$ to some $f \in B^s_{pq}(\mathbb{R}^n)$. Furthermore there is a number $c > 0$ such that for all admitted representations

$$\|f \,|B^s_{pq}(\mathbb{R}^n)\| \le c\, \|\lambda \,|b_{pq}\|_{\varrho}. \tag{90}$$

According to [FJW91], Section 5, the Gausslets $G^\beta_{jm}(x)$, given by (82), are molecules, and (84) can be interpreted as molecular representations. But we do not rely on this observation and give a detailed proof reducing (84) to atomic representations. We fix $\varrho > 0$ and choose $0 < \varepsilon < \varrho$. In modification of 3.1.3 we need some estimates of $G^\beta_{jm}(x)$ given by (82). Let

$$l = 1, \dots, n; \quad j \in \mathbb{N}_0 \quad \text{and} \quad \beta = (\beta_1, \dots, \beta_n) \in \mathbb{N}^n_0.$$

Then, using the Taylor series for e^y,

$$\begin{aligned} &2^{\varepsilon\beta_l} \left(\frac{\left[2^{-\varepsilon}\,(2^j x_l - m_l)\right]^{2\beta_l}}{\beta_l!} \right)^{\frac{1}{2}} e^{-\frac{(2^j x_l - m_l)^2}{2}} \\ &\qquad \le 2^{\varepsilon\beta_l}\, e^{2^{-2\varepsilon-1}(2^j x_l - m_l)^2}\, e^{-\frac{(2^j x_l - m_l)^2}{2}} \le 2^{\varepsilon\beta_l}\, e^{-a(2^j x_l - m_l)^2} \end{aligned} \tag{91}$$

for some $a = a_\varepsilon = \frac{1}{2}(1 - 2^{-2\varepsilon}) > 0$. Now we fix $k \in \mathbb{Z}^n$, and also $j \in \mathbb{N}_0$ and $\beta \in \mathbb{N}^n_0$. To convert (84) in an atomic decomposition we use the resolution of unity given by (32), (33). Then it follows by (84), (82), and (91) that

$$\begin{aligned} &\psi(2^j x - k) \Big| \sum_{m \in \mathbb{Z}^n} \lambda^\beta_{j,m+k}\, G^\beta_{j,m+k}(x) \Big| \\ &\le c\, 2^{-j(s-\frac{n}{p})}\, \psi(2^j x - k) \sum_{m \in \mathbb{Z}^n} \big|\lambda^\beta_{j,m+k}\big|\, \frac{(2^j x - m - k)^\beta}{\sqrt{\beta!}}\, e^{-\frac{|2^j x - m - k|^2}{2}} \\ &\le c\, 2^{\varepsilon|\beta|}\, 2^{-j(s-\frac{n}{p})}\, \psi(2^j x - k) \sum_{m \in \mathbb{Z}^n} \big|\lambda^\beta_{j,m+k}\big|\, e^{-a|2^j x - m - k|^2} \\ &\le c'\, 2^{\varepsilon|\beta|}\, 2^{-j(s-\frac{n}{p})}\, \psi(2^j x - k) \sum_{m \in \mathbb{Z}^n} \big|\lambda^\beta_{j,m+k}\big|\, e^{-a|m|^2}. \end{aligned} \tag{92}$$

Again for fixed $\beta \in \mathbb{N}_0^n$ we put

$$\lambda_j^\beta = \{\lambda_{jm}^\beta : m \in \mathbb{Z}^n\}, \tag{93}$$

$$\Lambda_{jk}^\beta = \sum_{m\in\mathbb{Z}^n} |\lambda_{j,m+k}^\beta|\, e^{-a|m|^2}\, 2^{\varepsilon|\beta|}, \tag{94}$$

and

$$\begin{aligned} G_{jk}^\beta(\lambda_j^\beta, x) &= (\Lambda_{jk}^\beta)^{-1} \sum_{m\in\mathbb{Z}^n} \lambda_{j,m+k}^\beta\, G_{j,m+k}^\beta(x) \\ &= 2^{-j(s-\frac{n}{p})}\, H^{-1}(2^j x)\, (\Lambda_{jk}^\beta)^{-1} \sum_{m\in\mathbb{Z}^n} \lambda_{j,m+k}^\beta \frac{(2^j x - m - k)^\beta}{\sqrt{\beta!}}\, e^{-\frac{|2^j x - m - k|^2}{2}}. \end{aligned} \tag{95}$$

Then we have

$$\begin{aligned} f &= \sum_{\beta,j,m} \lambda_{jm}^\beta\, G_{jm}^\beta(x) \\ &= \sum_{\beta\in\mathbb{N}_0^n} \sum_{j=0}^\infty \Big[\sum_{k\in\mathbb{Z}^n} \psi(2^j x - k) \sum_{m\in\mathbb{Z}^n} \lambda_{j,m+k}^\beta\, G_{j,m+k}^\beta(x) \Big] \\ &= \sum_{\beta\in\mathbb{N}_0^n} \sum_{j=0}^\infty \sum_{k\in\mathbb{Z}^n} \Lambda_{jk}^\beta\, a_{jk}^\beta(x) = \sum_{\beta\in\mathbb{N}_0^n} f^\beta \end{aligned} \tag{96}$$

with

$$a_{jk}^\beta(x) = \psi(2^j x - k)\, G_{jk}^\beta(\lambda_j^\beta, x), \quad x \in \mathbb{R}^n. \tag{97}$$

We claim that a_{jk}^β are essentially (s, p)-atoms with respect to the cubes Q_{jk}. Of course, they are correctly located. By (92)–(95) we have

$$|a_{jk}^\beta(x)| \le c\, 2^{-j(s-\frac{n}{p})}, \quad x \in \mathbb{R}^n, \tag{98}$$

where c is independent of β, j, k. We need a corresponding estimate for the derivatives of $a_{jk}^\beta(x)$ up to order $K = 1 + [s]$. By (95) it follows that

$$|D^\gamma a_{jk}^\beta(x)| \le c\,(1 + |\beta|)^K\, 2^{j|\gamma|}\, 2^{-j(s-\frac{n}{p})} \le c_\delta\, 2^{\delta|\beta|}\, 2^{j|\gamma| - j(s-\frac{n}{p})} \tag{99}$$

for some $\delta > 0$ and some $c_\delta > 0$. We may assume that $\delta > 0$ is included in the above $\varepsilon > 0$. Then we have by the usual atomic decomposition for $B_{pq}^s(\mathbb{R}^n)$ as it may be found, for example, in [Tri97], Section 13, $f^\beta \in B_{pq}^s(\mathbb{R}^n)$, and

$$\| f^\beta \,|B_{pq}^s(\mathbb{R}^n)\| \le c\, \|\Lambda^\beta \,|b_{pq}\|, \tag{100}$$

where we used (30) with Λ^β in place of λ, in analogy to (37). By (94) and standard arguments it follows that

$$\| f^\beta \, | B^s_{pq}(\mathbb{R}^n) \| \le c \, 2^{\varepsilon|\beta|} \, \| \lambda^\beta \, | b_{pq} \| \le c \, \| \lambda \, | b_{pq} \|_\varrho \, 2^{(\varepsilon-\varrho)|\beta|}. \tag{101}$$

Since $\varepsilon < \varrho$ we get $f \in B^s_{pq}(\mathbb{R}^n)$ and

$$\| f \, | B^s_{pq}(\mathbb{R}^n) \| \le c \, \| \lambda \, | b_{pq} \|_\varrho . \tag{102}$$

In case of the spaces $F^s_{pq}(\mathbb{R}^n)$ we must replace the sequence spaces b_{pq} in (38) by the more complicated sequence spaces f_{pq} in (39). This causes some technical difficulties. We dealt with this complication in detail in [Tri01], in Step 2 of the proof of Theorem 2.9, in particular in connection with the formula (2.71). Then we get the counterpart of (102),

$$\| f \, | F^s_{pq}(\mathbb{R}^n) \| \le c \, \| \lambda \, | f_{pq} \|_\varrho . \tag{103}$$

In all cases under consideration we have absolute and (hence) unconditional convergence of the series involved in $L_{\overline{p}}(\mathbb{R}^n)$ with $\overline{p} = \max(1, p)$, and in particular unconditional convergene in $S'(\mathbb{R}^n)$. We refer for details to [Tri01], 2.7. This justifies the writing in (84).

Step 2. We prove the converse assertion: Let $f \in B^s_{pq}(\mathbb{R}^n)$ with (83) or $f \in F^s_{pq}(\mathbb{R}^n)$ with (87). Then we wish to prove that f can be represented by (84) with

$$\| \lambda \, | b_{pq} \|_\varrho \le c \, \| f \, | B^s_{pq}(\mathbb{R}^n) \|, \quad f \in B^s_{pq}(\mathbb{R}^n), \tag{104}$$

where c is independent of f. Similarly for $F^s_{pq}(\mathbb{R}^n)$ with the converse of (103). The corresponding quarkonial decomposition maybe found in 2.3 above. A detailed proof of this quarkonial decomposition has been given in [Tri01], Theorem 2.9. We modify Step 2 of this proof. As said above we are interested to give a proof which does not only cover the above specific situation but which indicates also possible generalisations. Let $f \in B^s_{pq}(\mathbb{R}^n)$ or $f \in F^s_{pq}(\mathbb{R}^n)$ with (83) or (87), respectively. We have (77) or (78), where the φ_k are defined by (73)–(75). Hence,

$$\widehat{f}(\xi) = \sum_{k=0}^{\infty} \varphi_k(\xi) \, \widehat{f}(\xi), \quad \xi \in \mathbb{R}^n, \tag{105}$$

(convergence in $S'(\mathbb{R}^n)$). Let Q_k be a cube in $\mathbb{R}^n$ centred at the origin and with side-length, say, $2\pi 2^k$. In particular, $\operatorname{supp} \varphi_k \subset Q_k$. We interpret $\varphi_k \widehat{f}$ as a periodic distribution and expand it in Q_k by

$$\left(\varphi_k \widehat{f} \right)(\xi) = \sum_{m \in \mathbb{Z}^n} b_{km} \exp(-i 2^{-k} m \xi), \quad \xi \in Q_k, \tag{106}$$

with

$$b_{km} = c\,2^{-kn} \int_{Q_k} \exp\big(i2^{-k}m\xi\big)\,\big(\varphi_k \widehat{f}\big)(\xi)\,d\xi$$
$$= c'\,2^{-kn}\,\big(\varphi_k\widehat{f}\big)^{\vee}\big(2^{-k}m\big). \tag{107}$$

Let

$$\Lambda = \big\{\Lambda_{km} : k \in \mathbb{N}_0,\ m \in \mathbb{Z}^n\big\} \tag{108}$$

with

$$\Lambda_{km} = 2^{k\left(s-\frac{n}{p}\right)}\,\big(\varphi_k\widehat{f}\big)^{\vee}\big(2^{-k}m\big). \tag{109}$$

We may assume that

$$\|f\,|B^s_{pq}(\mathbb{R}^n)\| \sim \|\Lambda\,|b_{pq}\| \quad \text{and} \quad \|f\,|F^s_{pq}(\mathbb{R}^n)\| \sim \|\Lambda\,|f_{pq}\|, \tag{110}$$

respectively (equivalent quasi-norms). Details and references may be found in [Tri01], 2.10. Let $\varkappa \in S(\mathbb{R}^n)$,

$$\varkappa_k(\xi) = \varkappa\big(2^{-k}\xi\big), \quad \varkappa_k(\xi) = 1 \quad \text{if } \xi \in \operatorname{supp}\varphi_k,\ \operatorname{supp}\varkappa_k \subset Q_k, \tag{111}$$

where $k \in \mathbb{N}_0$. We multiply (106) with $\varkappa_k$ and extend it by zero from Q_k to $\mathbb{R}^n$. Then we have

$$\big(\varphi_k\widehat{f}\big)^{\vee}(x) = \sum_{m\in\mathbb{Z}^n} b_{km}\,\varkappa_k^{\vee}\big(x - 2^{-k}m\big)$$
$$= 2^{kn} \sum_{m\in\mathbb{Z}^n} b_{km}\varkappa^{\vee}\big(2^k x - m\big) \tag{112}$$
$$= c \sum_{m\in\mathbb{Z}^n} \Lambda_{km}\,2^{-k\left(s-\frac{n}{p}\right)}\,\varkappa^{\vee}\big(2^k x - m\big), \quad x \in \mathbb{R}^n,$$

where we used (107), (109). We expand the entire analytic function $\varkappa^{\vee}(2^k x - m)$ in (112) at the point $2^{-k-K}l$, where $l \in \mathbb{Z}^n$ and $K \in \mathbb{N}$ is fixed. Let

$$G(x) = e^{-\frac{|x|^2}{2}}\,H^{-1}(x) = G^0(x) \tag{113}$$

according to (65). We have by (65),

$$G\big(2^{k+K}x - l\big)\,\varkappa^{\vee}\big(2^k x - m\big) \tag{114}$$
$$= \sum_{\beta\in\mathbb{N}_0^n} \frac{2^{k|\beta|}}{\beta!}\,(D^\beta\varkappa^{\vee})(2^{-K}l - m)(x - 2^{-k-K}l)^\beta\,G(2^{k+K}x - l)$$
$$= \sum_{\beta\in\mathbb{N}_0^n} \frac{(D^\beta\varkappa^{\vee})(2^{-K}l - m)}{\sqrt{\beta!}}\,G^\beta\big(2^{k+K}x - l\big)\,2^{-K|\beta|}.$$

By (63), (113),

$$1 = \sum_{l_0} \sum_{l \in \mathbb{Z}^n} G\big(2^{k+K}x - 2^K l - l_0\big) = \sum_{l \in \mathbb{Z}^n} G\big(2^{k+K}x - 2^K l\big) + \cdots \tag{115}$$

is a resolution of unity, where the finite sum with respect to l_0 is taken over all lattice points $l_0 = (l_{0,1}, \ldots, l_{0,n}) \in \mathbb{Z}^n$ with $0 \le l_{0,j} < 2^K$ where $j = 1, \ldots, n$. As indicated in (115) we concentrate on the term with $l_0 = 0$, where $+\cdots$ stands for the remaining terms which can be treated in the same way. Then we obtain by (112),

$$(\varphi_k \widehat{f})^\vee(x) = c\, 2^{-k\left(s-\frac{n}{p}\right)} \sum_{m \in \mathbb{Z}^n} \Lambda_{km} \sum_{l \in \mathbb{Z}^n} G\big(2^{k+K}x - 2^K l\big)\, \varkappa^\vee\big(2^k x - m\big) + \cdots \tag{116}$$

and hence by (114),

$$\begin{aligned}
&\big(\varphi_k \widehat{f}\big)^\vee(x) \\
&= c\, 2^{-k\left(s-\frac{n}{p}\right)} \sum_{\beta \in \mathbb{N}_0^n} \sum_{l \in \mathbb{Z}^n} G^\beta\big(2^{k+K}x - 2^K l\big) \sum_{m \in \mathbb{Z}^n} \frac{D^\beta \varkappa^\vee(l-m)}{\sqrt{\beta!}}\, \Lambda_{km}\, 2^{-K|\beta|} \\
&\quad + \cdots \\
&= c' \sum_{\beta \in \mathbb{N}_0^n} \sum_{l \in \mathbb{Z}^n} \lambda^\beta_{k+K,2^K l}\, G^\beta_{k+K,2^K l}(x)\, 2^{K\left(s-\frac{n}{p}\right)} + \cdots,
\end{aligned} \tag{117}$$

where $G^\beta_{k+K,2^K l}(x)$ are the (s,p)-β-Gausslets according to (82) and

$$\lambda^\beta_{k+K,2^K l} = 2^{-K|\beta|} \sum_{m \in \mathbb{Z}^n} \frac{D^\beta \varkappa^\vee(l-m)}{\sqrt{\beta!}}\, \Lambda_{km}. \tag{118}$$

Using (62) we get

$$\big|\lambda^\beta_{k+K,2^K l}\big| \le c_b\, 2^{(C-K)|\beta|}\, \frac{1}{\sqrt{\beta!}} \sum_{m \in \mathbb{Z}^n} \frac{|\Lambda_{km}|}{1 + |l-m|^b}. \tag{119}$$

Recall that K and b can be chosen arbitrarily large. Similar estimates hold for $\lambda^\beta_{k+K,2^K l + l_0}$ as described above. Using the notation introduced in (37), (30), it follows by (110) that

$$\|\lambda^\beta \,|b_{pq}\| \le c\, 2^{-\varrho|\beta|}\, \frac{1}{\sqrt{\beta!}}\, \|\Lambda \,|b_{pq}\| \le c\, \frac{2^{-\varrho|\beta|}}{\sqrt{\beta!}}\, \|f \,|B^s_{pq}(\mathbb{R}^n)\|. \tag{120}$$

This proves, in particular, (104) for any given $\varrho > 0$. In case of the spaces $F^s_{pq}(\mathbb{R}^n)$ there are again some technical complications. We again refer to Step 2 of the proof of Theorem 2.9 in [Tri01], especially to formula (2.71). Then we get a similar assertion for the spaces $F^s_{pq}(\mathbb{R}^n)$. The proof is complete.

3.5. Discussion

3.5.1. Frames. The procedure how to get the coefficients λ^{β}_{jm} in Step 2 of the above proof is constructive and linear with respect to f. As in (49), (50), one gets universal *frame coefficients*

$$\lambda^{\beta}_{jm}(f) = 2^{j(s-\frac{n}{p})}\,\frac{2^{-\varrho|\beta|}}{\sqrt{\beta!}}\,\big(f\,,\,\widetilde{\Psi}^{\beta,\varrho}_{jm}\big), \quad f \in S'(\mathbb{R}^n), \tag{121}$$

with respect to the dual pairing in $S(\mathbb{R}^n)$ and $S'(\mathbb{R}^n)$, where

$$\widetilde{\Psi}^{\beta,\varrho}_{jm} \in S(\mathbb{R}^n), \quad \beta \in \mathbb{N}^n_0\,, \; j \in \mathbb{N}_0, \; m \in \mathbb{Z}^n, \tag{122}$$

such that (*frame*)

$$\|f\,|B^s_{pq}(\mathbb{R}^n)\| \sim \|\lambda(f)\,|b_{pq}\|_{\varrho}, \quad \|f\,|F^s_{pq}(\mathbb{R}^n)\| \sim \|\lambda(f)\,|f_{pq}\|_{\varrho}, \tag{123}$$

respectively. As for the convergence in (84) we have the same situation as discussed in 2.3.1. Here we used the word *frame* in a wider sense. Commonly, this notation is restricted to the case of Hilbert spaces. Its extension to Banach spaces, then called *Banach frames*, goes back to [Gro91].

3.5.2. Improvement. We proved more than stated. Let $\varrho > 0$. By Step 2 of the proof of the theorem it is clear that one can replace (85), (88), by

$$\sup_{\beta\in\mathbb{N}^n_0} 2^{\varrho|\beta|}\sqrt{\beta!}\;\|\lambda^{\beta}\,|b_{pq}\|\,, \quad \sup_{\beta\in\mathbb{N}^n_0} 2^{\varrho|\beta|}\sqrt{\beta!}\;\|\lambda^{\beta}\,|f_{pq}\|\,, \tag{124}$$

respectively. Hence, the decay of $\|\lambda^{\beta}\,|b_{pq}\|$ and of $\|\lambda^{\beta}\,|f_{pq}\|$ with respect to $|\beta|$ can be assumed to be even more rapid than in (85), (88), at the expense of the equivalence constants. A reformulation of the theorem with the sequence spaces from (124) in place of $\|\lambda\,|b_{pq}\|_{\varrho}$ and $\|\lambda\,|f_{pq}\|_{\varrho}$ in (86) and (89), respectively, is not only possible, but might be even more natural.

3.5.3. Quarkonial analysis. We described in 2.3 what is meant by quarkonial analysis in $\mathbb{R}^n$. We formulated the outcome in 2.3(i) and 2.3(ii) with a reference to [Tri01], Theorem 2.9. Checking now the proof of the above theorem it comes out that

$$\|\lambda\,|b_{pq}\|_{\varrho} \text{ in (42)} \quad \text{and} \quad \|\lambda\,|f_{pq}\|_{\varrho} \text{ in (45)}$$

can be even strengthened by

$$\sup_{\beta\in\mathbb{N}^n_0} 2^{\varrho|\beta|}\,\beta!\;\|\lambda^{\beta}\,|b_{pq}\| \quad \text{and} \quad \sup_{\beta\in\mathbb{N}^n_0} 2^{\varrho|\beta|}\,\beta!\;\|\lambda^{\beta}\,|f_{pq}\|\,, \tag{125}$$

respectively. Here ϱ can be any positive number. The main point is to replace formula (2.62) in [Tri01] by (62) above. Again a reformulation of 2.3(i) and 2.3(ii) or of Theorem 2.9 in [Tri01] with the sequence spaces from (125) in place of $\|\lambda\,|b_{pq}\|_{\varrho}$ and $\|\lambda\,|f_{pq}\|_{\varrho}$, respectively, might be even more natural.

3.6. Towards a Gausslet analysis

We return to 2.4.2. The question arises of whether quarks in 2.3(i), 2.3(ii) and Gausslets in the above Theorem 3.4 can be replaced by more general functions (windows) g parallel to the Gabor analysis and wavelet analysis as outlined in 2.1 and 2.2. What we have in mind is described in 2.4.2, especially in (54)–(56). For this purpose one can examine the proof of the above theorem. In Step 1 one needs normalising factors in order to get atoms. This is, with respect to β, the factor $(\beta!)^{-\frac{1}{2}}$ in (95). Secondly one must be sure that one can prove (101) as a consequence of (100) and (94). Here one does not need a decay of type $e^{-a|m|^2}$ in (94). Something like $(1+|m|^2)^{-b}$ with a sufficiently large $b > 0$ is sufficient. Glancing at Step 2 of the proof of the above theorem it comes out that one always has the factors $2^{-\varrho|\beta|}$ with arbitrarily large $\varrho > 0$ at hand. The crucial point here is (114). One has to use the factor $(\beta!)^{-1}$ to compensate the needed normalising factors for the building blocks involved; in our case $(\beta!)^{-\frac{1}{2}}$, in connection with $G^\beta(2^{k+K}x - l)$. Replacing the above Gausslets by more general building blocks g with (54)–(56), then in the respective counterpart of (114) at most $(\beta!)^{-1}$ is available to normalise the resulting atoms, not to speak about $2^{-\varrho|\beta|}$. Looking at the considerations in 3.1.3 then the decay $e^{-\frac{|x|^2}{2}}$ can be replaced by a decay of type $e^{-|x|}$ if $|x| > 1$, but not less. Hence an exponential decay of $g(x)$ fits in the arguments of the both steps of the proof of the theorem. One can follow the proof starting with a more general function $g(x)$ in place of $G(x)$ in (113) and one can check which conditions for $g(x)$ are needed. Then one gets a Gausslet analysis according to (55), (56). We do not go into detail. In connection with quarkonial decompositions we explained in 2.4.1 our own way. It was not so much $\mathbb{R}^n$, but smooth and non-smooth domains, manifolds, and fractals which triggered off the search for β-quarks. In the present paper we are exclusively in $\mathbb{R}^n$. Gabor analysis and wavelet analysis have not only their intrinsic, self-contained theories, but also many applications. On the other hand, for Gaussian representations according to the above Theorem 3.4 and for the just outlined more general Gausslet analysis it is not so clear, so far, whether there are good applications.

References

[BHW98] J. J. Benedetto, C. Heil, D. F. Walnut, Gabor systems and the Balian-Low theorem, in; [FeS98], 85–122.

[Dau90] I. Daubechies, The wavelet transform, time-frequency localization and signal analysis, IEEE Trans. Inform. Theory 36(5) (1990), 961–1005.

[Dau92] I. Daubechies, Ten Lectures on Wavelets, CBMS-NSF Regional Conf. Ser. in Appl. Math. 61, SIAM, Philadelphia, PA, 1992.

[ET96] D. E. Edmunds, H. Triebel, Function Spaces, Entropy Numbers, Differential Operators, Cambridge Tracts in Math. 120, Cambridge University Press, Cambridge, 1996.

[FeG88] H. G. Feichtinger, K. Gröchenig, A unified approach to atomic decompositions via integrable group representations, in: Function spaces and applications (M. Cwikel, J. Peetre, Y. Sagher, H. Wallin, eds.), Lecture Notes in Math. 1302, Springer-Verlag, Berlin, 1988, 52–73.

[Fei83] H. G. Feichtinger, Modulation spaces on locally compact abelian groups, Technical report, Univ. Vienna, 1983.

[Fei89] H. G. Feichtinger, Atomic characterizations of modulation spaces through Gabor-type representations, Rocky Mountain J. Math. 19 (1989), 113–126.

[FeS98] H. G. Feichtinger, T. Strohmer (eds.), Gabor Analysis and Algorithms, Theory and Applications, Birkhäuser, Boston, 1998.

[FeZ98] H. G. Feichtinger, G. Zimmermann, A Banach space of test functions for Gabor analysis, in: [FeS98], 123–170.

[FJW91] M. Frazier, B. Jawerth, G. Weiss, Littlewood–Paley theory and the study of function spaces, CBMS-AMS Regional Conf. Ser. 79, SIAM, Philadelphia, PA, 1991.

[FrJ85] M. Frazier, B. Jawerth, Decomposition of Besov spaces, Indiana Univ. Math. J. 34 (1985), 777–799.

[FrJ90] M. Frazier, B. Jawerth, A discrete transform and decompositions of distribution spaces, J. Funct. Analysis 93 (1990), 34–170.

[Gab46] D. Gabor, Theory of communication, J. Inst. Electr. Engrg. (London) 93 III (1946), 429–457.

[Gro91] K. Gröchenig, Describing functions: atomic decompositions versus frames, Monatsh. Math. 112 (1991), 1–42.

[Gro01] K. Gröchenig, Foundations of Time-Frequency Analysis, Birkhäuser, Boston, 2001.

[HeW96] E. Hernández, G. Weiss, A First Course on Wavelets, CRC Press, Boca Raton, 1996.

[Jan81] A. J. E. M. Janssen, Gabor representation of generalized functions, J. Math. Anal. Appl. 83 (1981), 377–394.

[KyP99] G. Kyriazis, P. Petrushev, New bases for Triebel–Lizorkin and Besov spaces, IMI Research Reports 6 (1999), Columbia, Dep. Math. Univ. South Carolina.

[Mal99] S. Mallat, A Wavelet Tour of Signal Processing, Academic Press, San Diego, 1999.

[Mey92] Y. Meyer, Wavelets and Operators, Cambridge Stud. Adv. Math. 37, Cambridge University Press, Cambridge, 1992; French original, 1990.

[Neu32] J. v. Neumann, Mathematische Grundlagen der Quantenmechanik, Springer-Verlag, Berlin, 1932.

[RuS96] T. Runst, W. Sickel, Sobolev Spaces of Fractional Order, Nemytskij Operators, and Nonlinear Partial Differential Equations, de Gruyter Ser. Nonlinear Anal. Appl. 3, Walter de Gruyter, Berlin, 1996.

[Tri83] H. Triebel, Theory of Function Spaces, Monogr. Math. 78, Birkhäuser, Basel, 1983.

[Tri92] H. Triebel, Theory of Function Spaces II, Monogr. Math. 84, Birkhäuser, Basel, 1992.

[Tri97] H. Triebel, Fractals and Spectra, Monogr. Math. 91, Birkhäuser, Basel, 1997.

[Tri98] H. Triebel, Gaussian decompositions in function spaces, Result. Math. 34 (1998), 174–184.

[Tri01] H. Triebel, The Structure of Functions, Monogr. Math. 97, Birkhäuser, Basel, 2001.

[Wal92] D. F. Walnut, Continuity properties of the Gabor frame operator, J. Math. Anal. Appl. 165(2) (1992), 479–504.

[WhW52] E. T. Whittaker, G. N. Watson, Modern Analysis, Cambridge University Press, Cambridge, 1952.

[Woj97] P. Wojtaszczyk, A Mathematical Introduction to Wavelets, Cambridge University Press, Cambridge, 1997.

Hans Triebel, Mathematisches Institut, Universität Jena, 07740 Jena, Germany

E-mail: `triebel@minet.uni-jena.de`

List of talks

(Invited speakers are marked with an asterisk)

Aksoy, Asuman, *On s-numbers and Bernstein pairs.* (Co-author: G. Lewicki)

Ameur, Yacin, *Interpolation of Hilbert spaces.*

*Arazy, Jonathan, *Invariant symbolic calculi and eigenvalues of link operators on symmetric domains.*

Barza, Ilie, *Elements of measure and integration theory on nonorientable surfaces.*

Barza, Sorina, *A new two-dimensional decreasing rearrangement and the corresponding function spaces.*

Bondesson, Lennart, *Multiplicative Properties of Generalized Gamma Convolutions.*

Bricchi, Michele, *Besov spaces defined on (d, Ψ)-sets.*

*Brudnyi, Yuri, *Approximation spaces – an overview.*

Cianchi, Andrea, *Rearrangement estimates and second order Sobolev inequalities.*

Cobos, Fernando, *Interpolation of strictly singular operators.* (Co-authors: A. Manzano, A. Martinez, P. Matos)

Connett, William, *Measure algebras with multivariate polynomial characters.*

*Cotlar, Mischa, *Some complements to the method of almost orthogonal sums and non-commutative versions of the Helson–Szegö theorem.*

Coulhon, Thierry, *Sobolev algebras on Lie groups and Riemannian manifolds.*

Cwikel, Michael, *A unified theory of commutator estimates for a broad class of old and new interpolation methods.*

Dachkovski, Serguei, *Smoothness of solutions of some semi-linear equations.*

Engliš, Miroslav, *Reproducing kernels and mean value properties.*

Feichtinger, Hans G., *Interpolation Theory and Banach Frames.*

*Foias, Ciprian, *On the surviving time of the regular solutions of the 3D Navier–Stokes equations.*

Fontes, Magnus, *Elliptic–Parabolic Equations with Non-Standard Growth.*

Gogatishvili, Amiran, *Duality principles and reduction theories.* (Co-author: Luboš Pick)

Gol'dshtein, Vladimir, *Axiomatic theory of Sobolev spaces on metric spaces.* (Co-author: Marc Troyanov)

Gustafsson, Björn, *On a Hilbert space related to quadrature domains and hyponormal operators.* (Co-author: Mihai Putinar)

Hedenmalm, Håkan, *Dirichlet series and functional analysis.*

Iwaniec, Tadeusz, *Nonlinear commutators and interpolation.*

*Janson, Svante, *Identities for Schatten norms of Hankel operators.*

Johnsen, Jon, *Parametrices of semi-linear boundary problems in Sobolev spaces over* L_p.

Kaijser, Sten, *Interpolation of Banach algebras.*

*Kalton, Nigel, *Interpolation of subspaces and exponential bases.*

Karlovich, Alexei, *On the interpolation constant for sublinear operators in Orlicz spaces.* (Co-author: Lech Maligranda)

Kato, Mikio, *Some geometric properties of absolute norms on* $\mathbb{C}^2$. (Co-authors: Kichi-Suke Saito and Yasuji Takahashi)

Kerman, Ron, *Optimal Sobolev and Besov Imbedding Norms.*

*Kislyakov, Sergey, *Interpolation, extension of operators, and Grothendieck's theorem.*

Komatsu, Hikosaburo, *Generalized functions and operational calculus discussed by Fourier and Heaviside.*

Krbec, Miroslav, *On extrapolation and decomposition in function spaces.*

Krugljak, Natan, *On* K*-divisibility for couples and triples.*

Kryakin, Yuri, K*-functional,* q*-moduli of continuity in* $H^p(D)$, $p > 0$, *and an inequality of Hardy and Littlewood.*

Kühn, Thomas, *On compact embeddings of Besov spaces into spaces of Lipschitz type.*

Kurasov, Pavel, *Few-body operators with singular perturbations in the scale of Hilbert spaces.*

*Lindqvist, Peter, *The derivative of the maximal function.*

Lukkassen, Dag, *On reiterated homogenization.*

Maligranda, Lech, *The failure of the Hardy inequality and interpolation of intersections.*

Mastyło, Mieczyslaw, *Interpolation spaces containing* ℓ_1.

*Maz'ya, Vladimir, *The Schrödinger operator on the energy space: Boundedness and compactness criteria.* (Co-author: I. Verbitsky)

Michels, Carsten, *The identity map* $E \hookrightarrow \ell_2$ *is* $(E, 1)$*-summing whenever* E *is 2-concave.* (Co-authors: A. Defant and M. Mastyło)

Nakai, Eiichi, *On generalized fractional integrals on the Orlicz spaces and* BMO_ϕ.

Opic, Bohumír, *Boundedness of fractional maximal operators between classical and weak type Lorentz spaces.*

*Ovchinnikov, Vladimir, *On parameters of the real method spaces.*

Overgaard, Niels Christian, *On the asymptotic expansion of the pushforward map given by a real homogeneous polynomial.*

*Peller, Vladimir, *Hankel operators in noncommutative analysis.*

Persson, Lars-Erik, *Some remarkable inequalities connected to* L^p*-spaces and fractional order Hardy inequalities.*

Pesenson, Isaac, *A sampling theorem on manifolds.*

Pick, Luboš, *Optimal Sobolev embeddings – old and new.*

Pustylnik, Evgeniy, *Generalized Marcinkiewicz spaces.*

Rantzer, Anders, *On stability and convergence of nonlinear systems.*

*Rochberg, Richard, *Carleson measure for Dirichlet space.*

Roginskaya, Maria, *On the growth of harmonic and* $\mathcal{M}$*-harmonic functions near* $\mathbb{S}^n$.

*Semenov, Evgueni, *The space* $\exp L_2$ *and its extremal properties.*

Shaposhnikova, Tatyana, *Pointwise interpolation inequalities for derivatives and potentials.*

Shimorin, Serguei, *Biharmonic Hadamard maximum principle and weighted Bergman spaces.* (Co-authors: Håkan Hedenmalm and Stefan Jakobsson)

Shvartsman, Pavel, *Interpolation of weighted Lipschitz spaces and extension of functions.*

Sickel, Winfried, *Pointwise multipliers of Besov spaces of smoothness zero and spaces of continuous functions.* (Co-author: Herbert Koch)

Signes, Teresa, *Some interpolation formulae for operator spaces.*

Sigstam, Kibret, *A pseudo-metric structure on interpolation functors.* (Co-author: Sten Kaijser)

Silvestrov, Sergei, *Topological dynamics, deformation quantization, and crossed product* C^**-algebras.*

Stepanov, Vladimir, *On two-sided estimates of the singular and approximation numbers of certain Volterra integral operators.*

Sunehag, Peter, *Interpolation of subspaces of codimension one.* (Co-author: Sten Kaijser)

Tikhonov, Sergey, *Moduli of smoothness and the interrelation of some classes of functions.*

Tolsa, Xavier, *Littlewood–Paley theory and the* $T(1)$ *theorem for non-doubling measures.*

*Triebel, Hans, *Spectral theory for the fractal Laplacian.*

*Wallin, Hans, *Wavelets and distortion invariant properties of signals and images.* (Co-authors: Amiran Ambroladze and Vassil Bugadze)

Wingren, Peter, *Zygmund and Lipschitz space on subsets.*

Zhang, Genkai, *Homogeneous multiplication operators on bounded symmetric domains.* (Co-author: Jonathan Arazy)

*Zobin, Nahum, *Extension of smooth functions and a Fourier analysis on Fock spaces.*

List of participants

Abrahamsson, Mikael, Centre for Mathematical Sciences, Lund University, Box 118, 221 00 Lund, Sweden, mikab@maths.lth.se

Aksoy, Asuman, Department Of Mathematics, Claremont McKenna College, 850 Columbia Ave., Claremont, CA 91711, U.S.A., asuman.aksoy@claremontmckenna.edu

Aleman, Alexandru, Centre for Mathematical Sciences, Lund University, Box 118, 221 00 Lund, Sweden, aleman@maths.lth.se

Ambroladze, Amiran, Centre för Mathematical Sciences, Lund University, Box 118, 221 00 Lund, Sweden, amiran.ambroladze@math.lth.se

Ameur, Yacin, Department of Mathematics, Uppsala University, Box 480, 751 06 Uppsala. Sweden, yacin@math.uu.se

Andersson, Mats Erik, Department of Mathematics, Stockholm University, 106 91 Stockholm, Sweden, matsa@matematik.su.se

Arazy, Jonathan, Department of Mathematics, University of Haifa, Haifa 31905, Israel, jarazy@math.haifa.ac.il

Arcozzi, Nicola, Department of Mathematics, University of Bologna, Piazza di Porta S. Donato, 540127 Bologna, Italy, arcozzi@dm.unibo.it

Barza, Ilie, Division for Engineering Sciences, Physics and Mathematics, Karlstad University, 651 88 Karlstad, Sweden, Ilie.Barza@kau.se

Barza, Sorina, Division for Engineering Sciences, Physics and Mathematics, Karlstad University, 651 88 Karlstad, Sweden, sorina.barza@kau.se

Berggren, Stein, Department of Mathematics, Narvik University College, P.O.Box 385, 8505 Narvik, Norway, sab@hin.no

Bergh, Jöran, Department of Mathematics, Chalmers University of Technology/ Gothenburg University, 412 96 Göteborg, Sweden, bergh@math.chalmers.se

Bernhardsson, Bo, Department of Automatic Control, Lund University, Box 118, 221 00 Lund, Sweden, bob@control.lth.se

Björn, Jana, Centre for Mathematical Sciences, Lund University, Box 118, 22100 Lund, Sweden, jabjo@maths.lth.se

Boman, Jan, Department of Mathematics, Stockholm University, 106 91 Stockholm, Sweden, jabo@matematik.su.se

Bondesson, Lennart, Departments of Mathematical Statistics, Umeå University, 90187 Umeå, Sweden, lennart.bondesson@math.uu.se

Bricchi, Michele, Department of Mathematics, Friedrich-Schiller-Universität, Ernst-Abbe Platz 1-4, 07743 Jena, Germany, mbricchi@minet.uni-jena.de

Brudnyi, Yuri, Department of Mathematics, Technion - Israel Institute of Technology, Haifa 32000, Israel, ybrudnyi@math.technion.ac.il

Carlsson, Linus P, Departments of Mathematics, Umeå University, 90187 Umeå, Sweden, linus@math.umu.se

Carro, Maria, Departament de Matemàtica Aplicada i Anàlisi, Universidad de Barcelona, Gran via 585, Barcelona 08071, Spain, carro@cerber.mat.ub.es

Cianchi, Andrea, Dipartimento di matematica e applicazioni per l' architettura, Universitá di Firenze, Piazza Ghiberti 27, 50122, Firenze, Italy, cianchi@cesit1.unifi.it

Cobos, Fernando, Departamento de Analisis Matematico, Facultad de Matematicas, Universidad Complutense de Madrid, 28040 Madrid, Spain, cobos@eucmax.sim.ucm.es

Connett, William, Department of Mathematics, University of Missouri, St. Louis, St. Louis, MO 63121, U.S.A., connett@math.umsl.edu

Cordeiro, Jose M., Departamento de Matematica Aplicada, Universidad de Vigo, E.T.S.I. Industriales, 36200 Vigo, Spain, cordeiro@uvigo.es

Cotlar, Mischa, Central University of Venezuela, UCV, Residencias Paris ap.82, Av. El Mirador, La Campiía, Caracas 1050, Venezuela, mcotlar@euler.ciens.ucv.ve

Coulhon, Thierry, Département de Mathématiques, Université de Cergy-Pontoise, 2, rue Adolphe Chauvin, 95302 Pontoise, France, Thierry.Coulhon@math.u-cergy.fr

Cwikel, Michael, Department of Mathematics, Technion - Israel Institute of Technology, Haifa 32000, Israel, mcwikel@math.technion.ac.il

Dachkovski, Serguei, Mathematisches Institut, Friedrich-Schiller-Universität, Ernst-Abbe Platz 1-4, 07743 Jena, Germany, dsn@minet.uni-jena.de

Dahlner, Anders, Centre for Mathematical Sciences, Lund University, Box 118, 221 00 Lund, Sweden, dahlner@maths.lth.se

Dencker, Nils, Centre for Mathematical Sciences, Lund University, Box 118, 221 00 Lund, Sweden, dencker@maths.lth.se

Engliš, Miroslav, Institute of Mathematics, Czech Academy of Sciences, Zitna 25, 11567 Prague 1, Czech Republic, englis@math.cas.cz

Engström, Jonas, Department of Mathematics, Luleå University of Technology, 97187 Luleå, Sweden, jonase@sm.luth.se

Feichtinger, Hans G., Department of Mathematics, University of Vienna, Strudlhofsgasse 4, 1090 Vienna, Austria, fei@tyche.mat.univie.ac.at

Fernandez-Cabrera, Luz Maria, Seccion Departamental de Matematica Aplicada, Escuela Universitaria de Estadistica, Ciudad Universitaria, Universidad Complutense de Madrid, 28040 Madrid, Spain, luzfer@mat.ucm.es

Fernandez-Martinez, Pedro, Facultad de Informática, Campus de Espinardo, Universidad de Murcia, 30071 Espinardo, Murcia, Spain, pedrofdz@fcu.um.es

Finol, Carlos, School of Mathematics, Faculty of Sciences, Central University of Venezuela, UCV, Apartado 48059, Caracas 1040-A, Venezuela, cfinol@euler.ciens.ucv.ve

Foias, Ciprian, Department of Mathematics, Indiana University, Bloomington, IN 47405, U.S.A., foias@indiana.edu

Fontes, Magnus, Centre for Mathematical Sciences, Lund University, Box 118, 221 00 Lund, Sweden, fontes@maths.lth.se

Fujii, Nobuhiko, Department of Mathematics, Tokai University, 3-20-1 Orido, Shimizu, Shizuoka 424-8610, Japan, nfujii@scc.u-tokai.ac.jp

Gogatishvili, Amiran, Institute of Mathematics, Czech Academy of Sciences, Zitna 25, 11567 Prague 1, Czech Republic, gogatish@math.cas.cz

Gol'dshtein, Vladimir, Department of Mathematics, Ben Gurion University of the Negev, P.O.Box 653, Beer Sheva, 84105, Israel, vladimir@bgumail.bgu.ac.il

Grip, Niklas, Department of Mathematics, Luleå University of Technology, 971 87, Luleå, Sweden, grip@sm.luth.se

Gustafsson, Björn, Department of Mathematics, Royal Institute of Technology, KTH, 100 44, Stockholm, Sweden, gbjorn@math.kth.se

Gustavsson, Jan, Centre for Mathematical Sciences, Lund University, Box 118, 221 00 Lund, Sweden, jang@maths.lth.se

Gårding, Lars, Centre for Mathematical Sciences, Lund University, Box 118, 221 00 Lund, Sweden, larsg@maths.lth.se

Hare, Kathryn, Department of Pure Mathematics, University of Waterloo, Waterloo, Ontario N2L 3G1, Canada, kehare@uwaterloo.ca

Harjulehto, Petteri, Department of Mathematics, University of Helsinki, P.O.Box 4, 00014 Helsinki, Finland, petteri.harjulehto@helsinki.fi

Hedenmalm, Håkan, Centre for Mathematical Sciences, Lund University, Box 118, 221 00 Lund, Sweden, haakan@maths.lth.se

Holmstedt, Tord, Centre for Mathematical Sciences, Lund University, Box 118, 221 00 Lund, Sweden, tord@maths.lth.se

Holst, Anders, Centre for Mathematical Sciences, Lund University, Box 118, 221 00 Lund, Sweden, anders.holst@math.lu.se

Hörmander, Lars, Centre for Mathematical Sciences, Lund University, Box 118, 221 00 Lund, Sweden, Lars.Hormander@math.lu.se

Iwaniec, Tadeusz, Department of Mathematics, Syracuse University, Syracuse NY 13244, U.S.A., tiwaniec@mailbox.syr.edu

Janson, Svante, Departments of Mathematics, Uppsala University, Box 480, 751 06 Uppsala, Sweden, svante.janson@math.uu.se

Jawerth, Björn, Summus Inc, 434 Fayetteville Street Mall, Rayleigh, NC 27601, U.S.A., bj@summus.com

Johansson, Maria, Department of Mathematics, Luleå University of Technology, 971 87 Luleå, Sweden, Maria.Johansson@sm.luth.se

Johnsen, Jon, Department of Mathematics, Aalborg University, Fredrik Bajers Vej 7E, 9220 Aalborg, Denmark, jjohnsen@math.auc.dk

Kaijser, Sten, Department of Mathematics, Uppsala University, Box 480, 751 06 Uppsala, Sweden, sten@math.uu.se

Kalton, Nigel, Department of Mathematics, University of Missouri, 2116 Oakcliff Dr, Columbia, MO 65203, U.S.A., nigel@math.missouri.edu

Karlovych, Oleksiy, Departamento de Matematica, Instituto Superior Tecnico, Av. Rovisco Pais, 1049-001 Lisboa, Portugal, akarlov@math.ist.utl.pt

Kato, Mikio, Department of Mathematics, Kyushu Institute of Technology, Kitakyushu 804-8550, Japan, katom@tobata.isc.kyutech.ac.jp

Kerman, Ron, Department of Mathematics, Brock University, 500 Glenridge Avenue, St. Catharines, Ontario L2S 3A1, Canada, rkerman@spartan.ac.brocku.ca

Kislyakov, Sergey, V.A. Steklov Mathematical Institute, Fontanka 27, St. Petersburg 191 011, Russia, skis@pdmi.ras.ru

Komatsu, Hikosaburo, Department of Mathematics, Science University of Tokyo, Wakamiya-cho 26, Shinjuku-ku, Tokyo, 162-0827, Japan, komatsu@ma.kagu.sut.ac.jp

Koskela, Pekka, Department of Mathematics, University of Jyväskylä, P.O.Box 35, 40351 Jyväskylä, Finland, pkoskela@math.jyu.fi

Koskenoja, Mika, Department of Mathematics, University of Helsinki, P.O.Box 4, 00014 Helsinki, Finland, mika.koskenoja@helsinki.fi

Krbec, Miroslav, Institute of Mathematics, Czech Academy of Sciences, Zitna 25, 115 67 Prague 1, Czech Republic, krbecm@matsrv.math.cas.cz

Krugljak, Natan, Department of Mathematics, Yaroslavl State University, ul. Kalinina 31, apt.152, Yaroslavl 150035, Russia, natan@univ.uniyar.ac.ru

Kryakin, Yuri, Odessa University, Malinovski 35/2, ap.83, 65074, Odessa, Ukraine, kryakin@paco.net

Kufner, Alois, Institute of Mathematics, Czech Academy of Sciences, Zitna 25, 11567 Praha 1, Czech Republic, kufner@math.cas.cz

Kühn, Thomas, Mathematisches Institut, Universität Leipzig, Augustusplatz 10/11, 04109 Leipzig, Germany, kuehn@mathematik.uni-leipzig.de

Kurasov, Pavel, Department of Mathematics, University of Stockholm, 106 91 Stockholm, Sweden, pak@matematik.su.se

Lewis, John, Department of Mathematics, University of Kentucky, Lexington, KY 40506, U.S.A., john@ms.uky.edu

Lindqvist, Peter, Department of Mathematical Sciences, Norwegian University of Science and Technology, 7491 Trondheim, Norway, lqvist@math.ntnu.no

Lions, Jacques-Louis, College de France, 11 place Marcelin Berthelot, 75231 Paris Cedex 05, France

Lukkassen, Dag, Department of Mathematics, Narvik University College, P.O.Box 385, 8505 Narvik, Norway, dl@hin.no

Maligranda, Lech, Department of Mathematics, Luleå University of Technology, 971 87 Luleå, Sweden, lech@sm.luth.se

Martin, Joaquim, Departament de Matemàtica Aplicada i Anàlisi, Universidad de Barcelona, Gran Via de les Corts Catalanes, 585, 08071 Barcelona, Spain, jmartin@mat.ub.es

Martinez, Anton, Departamento de Matematica Aplicada, Universidad de Vigo, E.T.S. Ingenieros Industriales, 36200-Vigo, Spain, antonmar@uvigo.es

Mastyło, Mieczyslaw, Faculty of Mathematics and Computer Science, Adam Mickiewicz University, Matejki 48/49, 60-769 Poznan, Poland, mastylo@amu.edu.pl

Maz'ya, Vladimir, Department of Mathematics, Linköping University, 58183 Linköping, Sweden, vlmaz@mai.liu.se

Meidell, Annette, Department of Mathematics, Narvik University College, P.O.Box 385, 8505 Narvik, Norway, am@hin.no

Meurman, Arne, Centre for Mathematical Sciences, Lund University, Box 118, 22100 Lund, Sweden, arnem@maths.lth.se

Michels, Carsten, Fachbereich Mathematik, Carl von Ossietzky University of Oldenburg, Postfach 2503, 26111 Oldenburg, Germany, michels@mathematik.uni-oldenburg.de; Current address: School of Mathematics, University of Leeds, Leeds LS2 9JT, United Kingdom

Nakai, Eiichi, Department of Mathematics, Osaka Kyoiku University, Kashiwara, Osaka 582-8582, Japan, enakai@cc.osaka-kyoiku.ac.jp

Natiello, Mario, Centre for Mathematical Sciences, Lund University, Box 118, 221 00 Lund, Sweden, Mario.Natiello@math.lth.se

Nilsson, Nils, Centre for Mathematical Sciences, Lund University, Box 118, 221 00 Lund, Sweden, nn@maths.lth.se

Nilsson, Per G, Belle Systems A/S, Universitetsparken 7, 4000 Roskilde, Denmark, pnilsson@bellesystems.com

Olofsson, Anders, Department of Mathematics, Stockholm University, 106 91 Stockholm, Sweden, anderso@matematik.su.se

Opic, Bohumír, Institute of Mathematics, Czech Academy of Sciences, Zitna 25, 115 67 Praha 1, Czech Republic, opic@math.cas.cz

Ovchinnikov, Vladimir, Department of Mathematics, Voronezh State University, Plehanovskaia, 1-1, Voronezh, 394006, Russia, vio@func.vsu.ru

Overgaard, Niels Christian, Centre for Mathematical Sciences, Lund University, Box 118, 221 00 Lund, Sweden, nco@maths.lth.se

Peller, Vladimir, Mathematics Department, Kansas State University, 138 Cardwell Hall, Manhattan, KS 66506-2602, U.S.A., peller@maths.ksu.edu

Pere, Mikko, Department of Mathematics, University of Helsinki, P.O.Box 4, 00014 Helsinki, Finland, mikko.pere@helsinki.fi

Persson, Lars-Erik, Department of Mathematics, Luleå University of Technology, 971 87 Luleå, Sweden, larserik@sm.luth.se

Pesenson, Isaac, Department of Mathematics, Temple University, Philadelphia, PA 19122, U.S.A., pesenson@math.temple.edu

Petersson, Joakim, Centre for Mathematical Sciences, Lund University, Box 118, 221 00 Lund, Sweden, joakimp@maths.lth

Pettersson, Pelle, Centre for Mathematical Sciences, Lund University, Box 118, 221 00 Lund, Sweden, pelle@maths.lth.se

Pick, Luboš, Department of Mathematical Analysis, Charles University in Prague, Sokolovska 83, 186 75 Praha 8, Czech Republic, pick@karlin.mff.cuni.cz

Pustylnik, Evgeniy, Department of Mathematics, Technion - Israel Institute of Technology, Haifa 32000, Israel, evg@techunix.technion.ac.il

Rantzer, Anders, Department of Automatic Control, Lund University, Box 118, 221 00 Lund, Sweden, rantzer@control.lth.se

Rochberg, Richard, Department of Mathematics, Washington University in St Louis, Campus Box 1146, 1 Brookings Drive, St.Louis, MO 63130, U.S.A., rr@math.wustl.edu

Roginskaya, Maria, Department of Mathematics, Chalmers University of Technology/Gothenburg University, 41296 Göteborg, Sweden, maria@math.chalmers.se

Semenov, Evgueni, Department of Mathematics, Voronezh State University, 394042 Voronezh, Leninskii prospekt, 152-8, Russia, root@func.vsu.ru

Shaposhnikova, Tatyana, Department of Mathematics, Linköping University, 58183 Linköping, Sweden, tasha@mai.liu.se

Shimorin, Serguei, Centre for Mathematical Sciences, Lund University, Box 118, 221 00 Lund, Sweden, shimorin@maths.lth.se

Shvartsman, Pavel, Department of Mathematics, Technion - Israel Institute of Technology, 32000 Haifa, Israel, pshv@tx.technion.ac.il

Sickel, Winfried, Matematisches Institut, Friedrich-Schiller-Universität, Ernst-Abbe Platz 1-4, 07743 Jena, Germany, sickel@minet.uni-jena.de

Signes, Teresa, Departamento de Matemática Aplicada, C.E.S. - Centro Universitario Felipe II, Capitan 39, 28300 Aranjuez, Spain, Teresa_Signes@mat.ucm.es

Sigstam, Kibret, Department of Mathematics, Uppsala University, 751 06 Uppsala, Sweden, kibret@math.uu.se

Silvestrov, Sergei, Centre for Mathematical Sciences, Lund University, Box 118, 221 00 Lund, Sweden, ssilvest@maths.lth.se

Simula, Leon, Department of Mathematics, Narvik University College, P.O.Box 385, 8505 Narvik, Norway, leon.simula@hin.no

Sjölin, Per, Department of Mathematics, Royal Institute of Technology, KTH, 10044 Stockholm, Sweden, pers@math.kth.se

Sjöstrand, Sigrid, Centre for Mathematical Sciences, Lund University, Box 118, 221 00 Lund, Sweden, sigrid@maths.lth.se

Sobukawa, Takuya, Department of Mathematics, Okayama University, Tsushima-naka 3-1-1, Okayama, 700-8530, Japan, sobu@cc.okayama-u.ac.jp

Spanne, Sven, Centre for Mathematical Sciences, Lund University, Box 118, 221 00 Lund, Sweden, sven@maths.lth.se

Sparr, Annika, Centre for Mathematical Sciences, Lund University, Box 118, 221 00 Lund, Sweden, annika@maths.lth.se

Sparr, Gunnar, Centre for Mathematical Sciences, Lund University, Box 118, 221 00 Lund, Sweden, gunnar@maths.lth.se

Stenlund, Mikael, Department of Mathematics, Luleå University of Technology, 971 87 Luleå, Sweden, stenlund@sm.luth.se

Stepanov, Vladimir, Computer Centre, Russian Academy of Sciences, Far Eastern Branch, Shelest 118-205, Khabarovsk 680042, Russia, stepanov@as.fe.ru

Stokolos, Alexander, Department of Mathematics, University of Missouri - Kansas City, 206 HH, 5120 Rockhill, Kansas City, MO 64110-2499, U.S.A., stokolos@excite.com

Sunehag, Peter, Department of Mathematics, Uppsala University, Box 480, 751 06 Uppsala, Sweden, peter@math.uu.se

Tikhonov, Sergey, GZ-MSU, Moscow State University, Vorob'jovy Gory, Moscow 117234, Russia, ser_tikhonov@mail.ru

Tolsa, Xavier, Department of Mathematics, Chalmers University of Technology/ Gothenburg University, 412 96 Göteborg, Sweden, xavier@math.chalmers.se

Torres, Rodolfo H., Department of Mathematics, University of Kansas, Lawrence, KS 66045, U.S.A., torres@math.ukans.edu

Triebel, Hans, Mathematisches Institut, Universität Jena, 07740 Jena, Germany, triebel@minet.uni-jena.de

Vretare, Lars, Centre for Mathematical Sciences, Lund University, Box 118, 221 00 Lund, Sweden, lv@maths.lth.se

Wallin, Hans, Department of Mathematics, Umeå University, 90187 Umeå, Sweden, Hans.Wallin@math.umu.se

Wingren, Peter, Department of Mathematics, Umeå University, 90187 Umeå, Sweden, Peter.Wingren@math.umu.se

Zhang, Genkai, Department of Mathematics, Chalmers University of Technology/ Gothenburg University, 412 96 Göteborg, Sweden, genkai@math.chalmers.se

Zobin, Nahum, Department of Mathematics, College of William and Mary, Williamsburg, VA 23187, U.S.A., zobin@math.wm.edu

Bei Fragen zur Produktsicherheit wenden Sie sich bitte an:
If you have any questions regarding product safety,
please contact:

Walter de Gruyter GmbH
Genthiner Straße 13
10785 Berlin
productsafety@degruyterbrill.com